Recent Advances in Civil Engineering

Chief Patron

Prof. A. Venugopal Reddy

Vice-Chancellor, J.N.T. University Hyderabad

Patron

Dr. A. Govardhan

Rector, J.N.T. University Hyderabad

Chairman

Dr. N. Yadaiah

Registrar, J.N.T. University Hyderabad

Co-Chairman

Dr. E. Saibaba Reddy

Principal, JNTUH-CEH

Convener

Dr. K. Manjula Vani

Professor of Civil Engineering-JNTUH-CEH.

Organized by

Department of Civil Engineering,
JNTUH-CEH, Kukatpally,
Hyderabad-500 085.

Disclaimer

The Publisher and Organisers do not claim any responsibility for the accuracy of the data, statements made, and opinions expressed by authors. The authors are solely responsible for the contents published in the paper.

Published by

BSP BS Publications

A unit of **BSP Books Pvt. Ltd.**

4-4-309/316, Giriraj Lane, Sultan Bazar,
Hyderabad - 500 095
Phone : 040 - 23445600, 23445688
e-mail : info@bspbooks.net
www.bspbooks.net

ISBN: 978-93-89354-21-8

Preface

A National Conference in Civil Engineering gives us an opportunity to exchange ideas amongst the peer groups. Such interaction not only leads to a feeling of fraternity amongst the Civil engineers, but also may open the avenue for inter-institute collaboration in research, pedagogy and consultancy.

The National Conference on **"Recent Advances in Civil Engineering-NCRACE 2019"** is dedicated to golden jubilee 50th anniversary of JNTUH-CEH. Over the period of 50 years, the sector of building and the fields closely related to it – landscape, environment and land management have strengthened their positions and have become internationally acknowledged and appreciated.

National conference on **"Recent Advances in Civil Engineering-NCRACE 2019"** covers 5 themes

- **Structural Engineering**
- **Remote Sensing and GIS Applications**
- **Environmental Engineering**
- **Disaster Management**
- **Transportation Engineering**

More than 150 delegates and 45 technical papers are being presented in this technical sessions.

I wish and expect that the participants will find this conference useful and give their total participation to make it a grand success.

It is with this great pleasure: I extend a warm welcome to all the delegates , speakers and participants to NCRACE-2019.

K.Manjulavani

-Editor

JAWAHARLAL NEHRU TECHNOLOGICAL UNIVERSITY HYDERABAD

COLLEGE OF ENGINEERING HYDERABAD
(AUTONOMOUS)
Kukatpalli, Hyderabad – 500 085, Telangana State, India.

Dr. E.SAIBABA REDDY,
B.Tech.,M.E (Hons.Roorkee), Ph.D. (Nottingham,UK),
Post Doc. (Halifax, canada),Post Doc. (Birmingham,UK)
CE,FIE,FIGS,MIWRS,MISRTMT,MISFET,FIAH,MINCA,MABS
PROFESSOR OF CIVIL ENGG &
PRINCIPAL

<u>Message</u>

I am glad to announce a national conference in civil engineering. The convener and faculty of Department of Civil Engineering, JNTUH-CEH are to be appreciated for organizing the one day   1st National conference **"Recent Advances in Civil Engineering-NCRACE 2019 "**on 05th July 2019.

Civil engineering is arguably the oldest engineering discipline. It is everything we see that's been built around us. Though the history dates back three thousand years in constructions, the changes need to be reviewed from time to time. The present conference covers all core civil engineering fields through different themes .I hope that this National conference will light up the knowledge to the upcoming researchers.

I congratulate the convener of this conference and wish them all success.

DR.E.SAIBABA REDDY

Web   : www.jntuhceh.ac.in   Phone: Off: +91–40–23057787

E Mail: principal.ceh @ jntuh.ac.in   Fax: +91–40–23057787

Off: 23158439,32408664

Mob: +91-9866053569
Email: sravana.jntu@gmail.com

DEPARTMENT OF CIVIL ENGINEERING
JAWAHARLAL NEHRU TECHNOLOGICAL UNIVERSITY
HYDERABAD

Dr.P.Sravana
M.Tech, Ph.D
Prof. of Civil Engineering & Head

<u>Message</u>

It is a pleasure to note that Department of Civil Engineering, JNTUH-CEH is organizing its 1st National conference on **"Recent Advances in Civil Engineering-NCRACE 2019"** on 05th July 2019.

Civil Engineering is a professional engineering discipline that deals with the design, construction and maintenance of physically built works like roads, bridges, canals, dams and buildings. As we are in 21st century need for study of infrastructure is increased wisely environmental issues to be drop down.

I hope the National conference will bring out such issues we are facing and hope to find productive measures in Environmental issues.

I congratulate the organizers of this conference for all their hard work and wish the National Conference would be a grand success.

Dr.P.Sravana

श्री माता वैष्णो देवी विश्वविद्यालय

SHRI MATA VAISHNO DEVI UNIVERSITY

(Approved Under 12(b) & 2(f) of UGC Act 1956)

SCHOOL OF CIVIL ENGINEERING

SUB POST OFFICE, KATRA-182320, JAMMU & KASHMIR, INDIA

Dr. Vaibhav Rajiv Sapkal
(Ph. D. in Civil Engineering from IIT Roorkee)
Assistant Professor
School of Civil Engineering

Phone (Office) 01991-285524 (Ext. – 2836)
M – 9960735164
Fax – 0191-2430067
Email – vaibhav.sapkal@smvdu.ac.in

Message

It is universally accepted that the role of Civil Engineer in development of any nation is very important. '**Civil Engineer C**reates with **I**nnovation a **V**irgin **I**dea for **L**eaving human being'.

Civil engineering education is also very essential for catering the need of country for excellent manpower for empowerment.

JNTUH – CEH and Shri Mata Vaishno Devi University (SMVDU), Katra, Jammu (J&K) under the twinning activity of TEQIP Phase III is pleased to organize NCRACE -2019 in beautiful campus of this historical city Hyderabad.

On the behalf of both this organizations I extend warm welcome to all delegates who have arrived from the various corners of the country to this well known reputed institute JNTUH – CEH, Hyderabad.

This conference is for the various deliberations in recent advances in Civil Engineering. During the conference many innovative ideas, outcome of various researches, academically administered projects, various experiences with different thoughts, concrete outcome of various case studies etc. of recent development of Civil Engineering will be shared by the delegates for future strategies both for input in research and direction for development of branch. The main goal of organizing this conference is to share and enhance the knowledge of each and every individual in this sustainable world. The conference aims to bridge the researchers working in academia and other professionals through research presentations and keynote addresses in current technological trends.

It is needless to say that the Civil Engineers of the country are contributing in various minor, major and mega projects for development of infrastructures in the country which is structurally stable, environmentally sustainable as well as economical viable with energy saving concept.

Water recourses, water supply, rain water harvesting, treatment and recycling of water and wastewater are on the top most priority of the researcher which will be taken care by the delegates by the recent advanced techniques to fulfill the need of human being as well as water conservation. Solid waste utilization, disposal and treatment are another challenge inviting the attention of Civil Engineer for clean and pollution free environment with energy generation.

Proper plantation of trees on the earth where 'soil' development is needed by new innovative techniques is the major thrust area of research and technology transfer for the participants.

On behalf of SMVDU and JNTUH-CEH I welcome all of you and wish you a very comfortable stay as well as fruitful discussion during the deliberations for concrete outcome of this conference.

(Dr. Vaibhav Sapkal)
Co - Convener

(ix)

Acknowledgements

I would like to express their gratitude to all the people that have helped us during these months for the organization of the conference. The 1st National conference on "Recent Advances in Civil Engineering-NCRACE 2019 has been made possible with the support of many technical experts, individuals and organizations both in man power and finance. This support is gratefully acknowledged.

I owe a deep sense of gratitude to *Prof. A. Venugopal Reddy*, Vice-Chancellor, Jawaharlal Nehru Technological University Hyderabad and Chief patron of the conference for his constant encouragement valuable guidance in organizing the conference in most efficient way.

I am very thankful to *Dr. A. Govardhan*, Rector, Jawaharlal Nehru Technological University Hyderabad for his precious support as Patron of Conference.

My sincere and special thanks to *Dr. N Yadaiah*, Registrar, Jawaharlal Nehru Technological University Hyderabad as the Chairman of the conference for his cordial, time to time permissions and support.

I am deeply indebted to *Dr. E. Saibaba Reddy* Principal, JNTUH-CEH and Co-Chairman of this conference for having taken every responsibility for completing this task through various stages.

I would like to extend my grateful thanks to Head of the Department *Dr. P. Sravana.*

for her valuable support throughout the conference.

My sincere thanks to the officials of *Technical Education Quality Improvement Program* (**TEQIP III**) for sponsoring this event. Without their help organization of this conference would not have been possible.

I would like to thank for Co-Conveners **Dr. Vaibhav Sapkal**, SMVDU-Jammu and **Mr. Padam Omar** for continuous support for conference under twinning activity

We have been very fortunate enough to be backed by a team of very motivated and dedicated experts of various committees in guiding us throughout the conference very meticulously. My sincere thanks to all the members of the Scientific and Advisory Committee, Technical Committee and Organizing Committee for their sincere advice and help from time to time.

I profusely thank all the Key note speakers, Chair persons and Co-chair persons of various technical sessions of conference have readily responded to our invitation to conduct the proceedings and to address the gathering and for their kind gesture in the conference.

My thanks are also due to various other Teaching and Non-teaching staff of Department of Civil Engineering who have cooperated on several occasions in organizing this Conference.

I sincerely thank M/s BS Publications for bringing out the souvenir and pre-conference proceedings well in advance.

My sincere thanks to my students **L. Ravi, B. Harish, K. Maruthi Rao**

for their continuous day and night support for this conference.

Finally, I thank all the people and organizations who are directly and indirectly involved in organizing the conference, but I could not mention their names due to paucity of space.

I thank one and all

K. ManjulaVani
Convener

Contents

Behaviour of Concrete Beams Strengthened with Basalt Fiber Reinforced Polymer Bars-A Review

Mallela ManojKumar[1], V. Rama Krishna[1] and V. Srinivasa Reddy[2]
[1]M. Tech Students, Department of Civil Engineering, GRIET Hyderabad.
[2]Professor of Civil Engineering, GRIET Hyderabad
vempada@gmail.com

Abstract

Concrete is good in resisting compressive forces and weak in resisting Tension. Hence to assist concrete in Tension we are reinforcing the concrete. Steel is the most widely used reinforcing material for concrete worldwide. To address the major disadvantage of Steel as reinforcement i.e. corrosion there is a requirement of alternative material. One such type of alternative which has gained prominence is the Fibre Reinforced Polymer (FRP) bars. They have high tensile strength, less weight, high thermal resistance and other advantageous properties over steel. This Paper presents the research work carried out on one such type of FRP which is Basalt Fiber Reinforced Polymer (BFRP) bars. The research works mainly aimed at bringing the BFRP bar as an alternative to steel as reinforcement in construction industry. Research works mainly focused on determining the Mechanical properties like Flexural behavior, Tensile Strength, Bond strength and Structural behavior when Basalt Fiber Reinforced Polymer (BFRP) bars are used as reinforcement in beams. Basalt FRPs can be a good alternative to Carbon and Glass FRPs and it may completely replace steel as reinforcement in the near future due to its varied properties.

Keywords: BFRP, FRP, Basalt Rebars, Basalt Fibre Reinforced Polymer Bar, Basalt Fibres.

Introduction

Corrosion of steel is the major cause for the loss of serviceability and durability in Reinforced Concrete structures. To address the matter of corrosion many alternative materials have emerged. One such type of alternative is replacement of Steel bars as reinforcement with Fiber Reinforced Polymer (FRP) bars. Glass FRP, Carbon FRP, Aramid FRP are some FRPs which have gained prominence due to their properties of high strength, modulus of elasticity equal to that of steel, light weight etc. Basalt Fiber Reinforced Polymer (BFRP) bar is one such type of FRP which is currently gaining importance in construction industry due to its varied advantages besides other available FRP bars. They are manufactured using Basalt rock which is a Mafic (rich in Mg and Fe) extrusive igneous rock. BFRP bars are manufactured by Gluing together of Continuous Basalt Fibers (CBF) in Epoxy Resin Matrix. CBFs are manufactured by melting crushed Basalt stone at a temperature of1500°c to 1800°c and drawing the molten lava into thin wires.[1]CBFs when chopped are used as Basalt Fibres. Paul Dhe of United States was granted patent for Baslat fibres in 1923. Initially they were used only for military purposes and aerospace applications. They usually have a diameter of 9-20µm. They are non-reactive to air and water and are not affected by chemicals. Hence Basalt Fibres are being used in reinforcing pipes which carry chemicals and life span of pipes reinforced with Basalt fibres is at least 50years without maintenance and repairs[2]. BFRP bars have high ultimate tensile strength (1000-1300MPa), Low density (2600kg/m^3), good thermal stability(the end use temperature range is -263 to 900°c), Easy and cheaper to produce than Glass fibers. FRP bars are non-magnetic and non-corrosive. They are Eco friendly since manufacture of basalt Fibers requires only 5KWh/Kg of Energy whereas steel requires 14KWh/Kg [3]. A good amount of research is being carried out around the world to give Basalt bars a permanent place in construction industry.

Fig. 1 Chopped Basalt strands and BFRP rebar

The Flexure and shear performances of concrete beams reinforced with BFRP rebars and stirrups are evaluated by Tomilsonet al. [4]. Nine 150 × 300 × 3,100-mm beams were tested in four-point bending to examine the effect of BFRP flexural reinforcement ratios. The stirrups used in beams were either BFRP or steel and some had no shear reinforcement. It was observed that the ultimate and service load increased with the Flexural Reinforcement and there was no

ISBN: 978-93-8830-599-0

effect of shear reinforcement on service load. The beams with BFRP flexural reinforcement and BFRP or steel stirrups had significantly higher strengths (almost 2.6–2.9 times) than control steel-reinforced specimens having the same reinforcement ratio. The failure of beams without stirrups was at 55-58% of ultimate flexural capacity whereas beams with BFRP stirrups has 90-96% of flexural capacity and both the specimen failed in shear whereasbeams with steel stirrups failed in flexure.

Marek Urbanski et al, [5] studied the effect of BFRP rebar as reinforcement in concrete beam and compared it with the beam with steel reinforcement. Three beams of size 180 x 140x 1200 mm reinforced at bottom with 3-8mm dia. bars were compared with three beams reinforced with steel bars of same percentage. It was noted that critical load for beams reinforced with BFRP bars (46.7KN) was much greater than the carrying capacity of beams with conventional steel reinforcement (37.6KN). The deflection was significantly higher in the beams reinforced with BFRP bars compared to steel bars. The failure of beams reinforced with BFRP was not sudden and this was due to the transformation of beam into a tie system as the Flexural BFRP rebar remain unbroken. Due to the relatively lower elasticity modulus of basalt rods(70-80GPa), compared to steel (200GPa), both the deflection and width of cracks can be a major factor in designing the BFRP reinforced concrete beams.

Thivanovitigala [6] studied the bond, flexure and shear behaviour of concrete beam reinforced with BFRP rebar and compared with steel reinforcement. In the pull-out bond test, 24specimens with varying bar diameters and bond lengths were tested and it was concluded that Twenty times the diameter of BFRP bar can be considered as the development length for reinforced flexural specimens and the bond stress increased with decrease in diameter of bar. In the flexure behaviour, 8 beams each of under reinforcement and over reinforcement of BFRP bars were compared with 6 beams reinforced with steel bars. Beams reinforced with BFRP bars showed higher ultimate load carrying capacity compared to steel reinforcement. When the beam was over reinforced using BFRP rebars it was observed that the beam failed by crushing of concrete. For beams reinforced with 3-10mm dia. and 3-25mm dia. BFRP rebars the ultimate moment carrying capacity was increased by 90% and deflection reduced by 60%. In the shear behaviour it was observed that the failure mode of the specimens depends on the BFRP reinforcement area and the span to depth ratio (a/d). The failure was more brittle, when the area of reinforcement was higher. When the span to depth ratio

increases, the failure mode changed from shear failure to flexural shear failure. This study also concluded that the equations given by ACI440.1R-06 gives satisfactory results for calculating the flexural capacity of beams reinforced with FRP bars and there is no necessity for the development of new equations for the design.

DawidPawłowski and Maciej Szumigała[7] made a study on 12 beams reinforced with various percentage of BFRP rebars and steel rebars. The main aim of this study was to evaluate the failure mechanisms, deflections and ductility of simply supportedBFRP RC beams depending on the reinforcement ratio. It was observed that the beams reinforced with 3-7 dia. and 3-9 dia. failed suddenly by rupture of rebar while beam reinforced with 5-9 dia. failed by crushing of concrete and the failure was not sudden in the last case (it showed some ductility). The Flexural capacity increased with the increase in reinforcement ratio. This study also concluded that with increase in reinforcement ratio deflection decreases.

Lijuan Li and Feng Liu [8] tested Flexural performance on ten basalt fibre bar-reinforced sea-sand concrete beams (150 x 250 x 2100mm) with two strength grades and four reinforcement ratios (2-6Ø,2-10Ø,2-14Ø,2-18Ø BFRP bars and 2-14Ø steel bars) by using the four-point bending method. They observed that compared to low grades of concrete, high concrete grades are good at using the tensile properties of BFRP. They observed two types of failure, tensile failure (2-6Ø. rebars) and crush failure (2-18Ø. rebars). There is good bonding between concrete and BFRP bar resulting in good force transfer which helps in perfect distribution of stresses. With increase in reinforcement the flexural capacity increased but it might make the structural failure brittle. Compared to the steel bar reinforced concrete beams withidentical reinforcement ratios, BFRP bar reinforced concretebeams exhibited a higher bearing capacity, largerdeformation, fewer cracks, wider crack spacing andgreater crack height.

Iyer[9] (2014) has investigated the mechanical properties of smart concrete made of chopped basalt fibre. Two different types of basalt fibres (bundles and filaments) were used. The basalt fibre specimens were cast using basalt fibres of varying length (12 mm, 36 mm, and 50 mm) and varying fibre dosage (4 kg/m3, 8 kg/m3,and 12 kg/m3). The results indicated that the 50 mm basalt bundled fibre at 8 kg/m3 was the optimum fibre length and fibre volume for basalt bundled fibres. It provided the optimum increase in flexural strength, compressive strength, and split tensile strength when compared with plain concrete.

ISBN: 978-93-8830-599-0

M. Naga Suresh [10] studied the effect of Basalt fibre content on compressive, split tensile and flexural strength of concrete made with optimum replacement of cement with fly ash and observed that all the properties increased at 2% Fiber content by volume of concrete. Similar study done by TehminaAyub [11] also showed the same thing but with different mineral admixture (silica fume, metakaolin) and also reported that strains corresponding to the maximum compressive strength as well as splitting tensile strengths were found to be increasing with the increasing fibre volume and this increment was significantly higher than control mix samples (containing no fibre), showing the ductile behavior of the concretes with the use of Basalt fibres.

TehminaAyub [12] analyzed the compressive stress-strain behavior of three mix types of high-strength fiber-reinforced concrete (HSFRC) having compressive strengths of 70–85 MPa and containing 1–3% volume fractions of basalt fibers. In the first mix of HSFRC, 100% cement content was utilized whereas 10% cement content was replaced by silica fume and metakaolin as replacement materials in the remaining two mixes. The addition of basalt fibers in the range 1–3% (by volume of concrete mix) did not increase compressive strength significantly. The inclusion of basalt fibers mainly increased the strain corresponding to the maximum compressive strength 4.85–12.24% in all series. The increase in toughness ratio for all HSFRC series was 3.8–47.15%, showing that the addition of basalt fibers enhanced concrete toughness.

Study carried by Ramakrishnan [13] was not in line with the previous mentioned papers i.e., author stated that there was no improvement in the compressive strength of concrete with the addition of basalt fibres. However, the study found that there was considerable improvement in the impact strength and flexural toughness of the BFRC specimens especially when large amounts (0.5% by volume) of fiber were added.

Conclusions

Based on the above studies and other observations we conclude that usage of BFRP bars as reinforcement increases the Flexural capacity of the member compared to steel reinforced members. They have good resistance to chemical attacks and are less affected by magnetic fields. Hence, they can be used in aggressive environments like sea waters, chemical factories and rooms containing MRI scanning missions (affected by magnetic fields). Usage of Basalt Fibers might not show significant improvement in the Compressive strength but there is notable improvement in the Split Tensile strength, ductility and toughness of concrete.

Continuous basalt fiber is more efficient and eco-friendlier than most glass & carbon fibresBasalt Fibres have wide variety of applications like manufacture of fire proofing clothes, reinforcement in pipes, conveyor belts, oil and gas supplying pipe lines, pipes supplying hot water, repair of cracks in buildings, bridges etc[2]. BFRP bars are highly resistant to corrosion and they result in reduction of concrete as cover material and also result in light weight structures (subjected to its low density). They could be a solution to construction in remote areas having low accessibility as they can be transported easily. However, it should be noted that they cannot be used as compression members since compressive modulus of FRP bars is lower than that of Tensile modulus [14].

Scope

As India has huge deposits of Basalt, more research has to take place on Basalt Fibers and Basalt Fiber Reinforced Polymer bars as a reinforcing material for concrete so that the Indian construction industry gets benefitted by its advantages.

Note: *Basalt fiber is (BF), known as "the green industrial material of the 21st century"* [2]

References

1. Novitskii A, Efremov M. "Some aspects of the manufacturing process for obtaining continuous basalt fiber."2011;67:361-5.
2. Piyush Sharma "An Introduction to Basalt Rock Fiber and Comparative Analysis of Engineering Properties of BRF and Other Natural Composites"- IJRASET volume 4 Issue 1, Jan-2016 ISSN 9321 9653
3. Fazio P. Basalt fibre: from earth an ancient material for innovative and modern application. *EAI* 2011;3: 89-96
4. Tomlinson, D., and Fam, A. (2014). "Performance of concrete beams reinforced with basalt FRP for flexureand shear"J. Compos. Constr.,10.1061/(ASCE)CC.1943-5614.0000491, 04014036.
5. Marek Urbanski, Andrzej Lapko, Andrzej Garbacz,"Investigation on Concrete Beams Reinforced with Basalt Rebars as an Effective Alternative of Conventional R/C Structures" 11th International Conference on Modern Building Materials, Structures and Techniques, MBMST 2013
6. ThilanOvitigala, "Structural Behaviour of Concrete Beams Reinforced with Basalt Fibre Reinforced Polymer (BFRP) bars". University of Illinois at Chicago, 2012.
7. DawidPawłowski and Maciej Szumigała, "Flexural behaviour of full-scale basalt FRP RC beams –

experimental and numerical studies" Procedia Engineering 108 (2015) 518 – 525

8. Li, Lu, Fang, Liu, Li "Flexural study of concrete beams with basalt fibre polymer bars" Structures and Buildings – ICE publishing

9. Padmanabhan Iyer, Sara Y. Kenno and Sreekanta Das "Mechanical Properties of Fiber-Reinforced Concrete Made with Basalt Filament Fibers" DOI: 10.1061/(ASCE)MT.1943-5533.0001272

10. M. Naga Suresh, K. Prafulla Devi (2017) "Experimental Investigation on Mechanical Properties of Basalt Fiber Reinforced Concrete with Partially Replacement of Cement with Fly Ash" ISSN: 2278 -7798International Journal of Science, Engineering and Technology Research, Volume 06, Issue 05, May 2017

11. TehminaAyub, Nasir Sharif, M.FadhilNuruddin "Mechanical Properties of High-Performance Concrete Reinforced with Basalt Fibers" Procedia Engineering 77 (2014) 131 – 139

12. TehminaAyub, S.M.ASCE ; Nasir Shafiq ; and Sadaqat Ullah Khan, "Compressive Stress-Strain Behavior of HSFRC Reinforced with Basalt Fibers".

13. V Ramakrishnan, N. Tolmare, 'Performance Evaluation of 3-D Basalt Fiber Reinforced Concrete & Basalt Rod Reinforced Concrete', IDEA Program Transportation Research Board National Research Council, November 1998.

14. ACI-440.1R-06 "Guide for the design and construction of structural concrete Reinforced with FRP bars"

ISBN: 978-93-8830-599-0

Strength Characteristics of Fibrous Self Curing Concrete Using Super Absorbent Polymer

Rathod Ravinder[1], Vivek Kumar C.[2], Akula Prakash[3], A.Vittalaiah[4] and P.V.V.S.S.R Krishna[5]

Assistant Professor, Department of Civil Engineering,

GokarajuRangaraju Institute of Engineering & Technology, Hyderabad, India.

rathod506ravinder@gmail.com, vivekkumarcr@gmail.com, akulprakash93@gmail.com,

vittal1559@gmail.com, siva.polinas@gmail.com

Abstract

In the present day, Water is the most required substance in the era. Across the world, water scarcity is a very critical issue that humankind is facing in contemporary times. The "scarcity" dimension remains as a significant "risk" to our very survival in this planet. Generally, Curing of concrete is maintaining moisture in the concrete during early ages specifically within 28 days of placing concrete, to develop desired properties. Appropriate curing of concrete is crucial to obtain maximum durability, especially if the concrete is exposed to serve conditions where the surface will be subjected to excessive wear, aggressive solutions and severe environmental conditions. Poor curing practices adversely affect the desirable properties of concrete which makes a major impact on the permeability of a given concrete. Unexpected shrinkage and temperature cracks can reduce the strength, durability and serviceability of the concrete. The surface zone will be seriously weakened by increased permeability due to poor curing. The development of concrete shrinkage is proportional to the rate of moisture loss in concrete. When concrete is properly cured, water retained in concrete would help continuous hydration and development of enough tensile strength to resist contraction stresses. The continuous development of strength reduces shrinkage and initial cracks or micro-cracks. As a part of this investigation of Fibrous Self Curing Concrete, proportion and addition of Polypropylene Fibre resulted in the formation of micro cracks in order to reduce the autogenous shrinkage and improvement of durability with a mix design of M40.

Keywords: Self-Curing Concrete , Fibrous Self-Curing Concrete, Super absorbent Polymer, Flexural Behavior, Finite Element Analysis , ANSYS 12.

INTRODUCTION

The development of high quality cement in the field of civil engineering is developing and it is gathering the enthusiasm during most recent few years. The curing of concrete structures in an effective manner is very important to make sure that concrete structures meet intentional performance and durability. The traditional external curing method could not achieve anticipated effect due to low permeability and high chemical reactions in conventional concrete.

Internal curing (IC), which has been widely investigated in thelast decade, shown to enhance hydration, diminishautogenous shrinkage, and mitigate early-age cracking. It also increases the internal porosity of concrete, however, which might reduce mechanical properties.After concrete is batched, water in freshly mixed concrete exclusive of absorbed water by the aggregatereacts with the cementitious materials and diverse chemical reactions.

A portion of the water combines to form calcium silicate hydrates (C-S-H), which give the primary strength phasein Portland cement concrete.As the cement hydration develops, the essential water to remain hydration is partially obtained from the capillarypores.

The emptying of the capillary pores will produce an escalation in surface tensions over the aggregates and thehydration products developed. As a result, adequate curing is essential for concrete to obtain advanced structural and durability properties and therefore is one of the most important requirements for optimum concrete performance in any environment or application. Curing techniques and Curing durability significantly affect curing efficiency. As per IS 456: 2000, Curing is the process of preventing the loss of moisture from the concrete whilst maintaining a satisfactory temperature regime.

REVIEW OF LITERATURE

Álvaro Paul and Mauricio Lopez, (2011),[1] internal curing (IC), which has been extensively investigated in the last decade, has been shown to enhance hydration, diminish autogenous shrinkage, and mitigate early-age cracking due to self-desiccation in high-performance concrete. It also increases the internal porosity of concrete, however, which might reduce mechanical properties.

Ambily and Rajamane, (2009),[2] studied the different aspects of achieving optimum cure of concrete

ISBN: 978-93-8830-599-0

without the need for applying external curing methods excessive evaporation of water (internal or external) from fresh concrete should be avoided, otherwise, the degree of cement hydration would get lowered and thereby concrete may develop unsatisfactory properties. Curing operations should ensure that adequate amount of water is available for cement hydration to occur.

Dale P.Bentz, (2007), [7] in the twenty-first century, most high-performance concretes, and many other ordinary concretes, are now based on blended cements that contain silica fume, slag, and/or fly ash additions. Because the chemical shrinkage accompanying the pozzolanic and hydraulic reactions of these mineral admixtures is generally much greater than that accompanying conventional Portland cement hydration, these blended cements may have an increased demand for additional curing water.

SCOPE AND OBJECTIVES

A comprehensive investigation has been undertaken to study the effects of self-curing agent like Super Absorbent Polymer (SAP) and the addition of Polypropylene Fiber on the mechanical properties of concrete and its durability. The test program was performed on concretes containing cement content, water cement ratio (0.4) cured in air 25°C and elevated temperature 50°C.

Special attention was given to the enhancement in self-curing concrete properties cured in normal 25°C and elevated temperature 50°C, as well as its durability and wet/dry cycles in water as affected by the type and doses of self-curing agent along with the addition of Super Absorbent Polymer (SAP).

PROPERTIES AND SIGNIFICANCE

A. Self Curing

The concept of self-curing is to reduce the water evaporation from concrete and hence increase the water retention capacity of the concrete compared to conventional concrete.

Internal Curing can be anticipated when cracking of concrete provides passageways resulting in deterioration of Reinforcing steel, Low early-age strength is a problem and Need for reduced construction time, Quicker turnaround time in precast plants, lower maintenance cost, Greater performance, predictability, Permeability/durability must be improved, Rheology of concrete mixture, modulus of elasticity of the finished product or durability of high fly-ash concretes are considerations

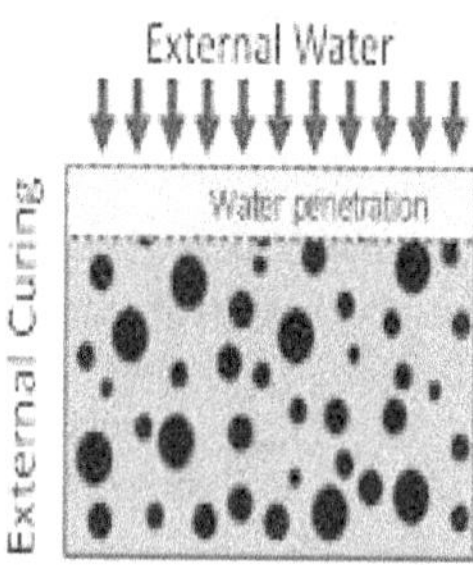
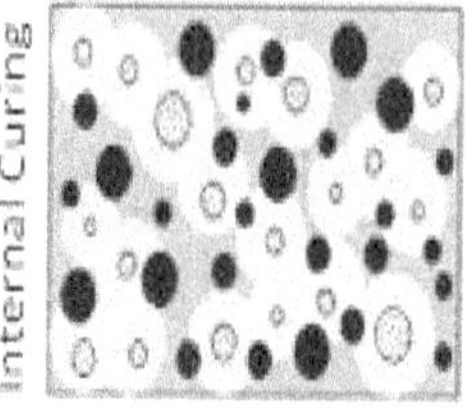

B. Effect of Fibres in Concrete

- Usually used in concrete to control cracking due to both plastic shrinkage and drying shrinkage.
- Reduce the permeability of concrete and thus reduce bleeding of water.
- Fibres produce greater impact, abrasion and shatter resistance in concrete.
- Fibres do not increase the flexural strength of concrete, and so cannot replace moment resisting or structural steel reinforcement
- The amount of fibres added to a concrete mix is expressed as a percentage of the total volume of the composite (concrete and fibres), termed volume fraction (Vf). Volume fraction typically ranges from 0.1 to 3%.

C. Benefits of Fibre Reinforced Concrete

- Improved impact & freeze-thaw resistance
- Improved resistance to explosive spalling in severe fire
- Increased resistance to plastic shrinkage during curing
- Improved structural strength
- Reduced steel reinforcement requirements & improved ductility
- Reduced crack widths and segregation

Table 1 Properties of material

Property of the Material	Values
Fineness of Cement	7.5%
Grade of Cement	43
Specific Gravity	3.15
Initial Setting time	28 min
Final Setting time	600 min
Specific Gravity of Fine aggregate	2.65
Fineness Modulus of Fine Aggregate	2. 25
Specific Gravity of Coarse Aggregate	2.77
Fineness Modulus of Coarse Aggregate	5.96

ISBN: 978-93-8830-599-0

Table 2 Properties of polypropylene fibre

Properties of Fibre	Values
Material	Virgin Homo polymer
Length	45 mm
Type/Shape	Macro/ monofilament
Colour	White
Specific Gravity	0.91
Acid and Salt resistance	High
Tensile Strength	620-758 MPa
Absorption	Nil
Fibre content	4-9 kg/m3 of concrete
Melting point	164° C (328° F)
Young's modulus	3.5 kN/mm2
Alkali resistance	Alkali Proof

Table 3 Properties of super absorbent polymer

Properties	Values
FORM – dry	Crystalline white powder / granules
FORM – wet	Transparent gel
Particle size	0 – 1 mm
Water absorption with distilled water	800 g for 1 g
Water absorbed with sea water	40 g for 1 g
pH of absorbed water	Neutral
Density	1.08
Bulk density	0.85
Hydration / Dehydration	Reversible
Stability of dry product	5 years
Stability of wet product	4 years
Decomposition in sun light	6 months
Available water	95% approx.

MIX DESIGN AND METHODOLOGY

Mix Design can be defined as the process of selecting ingredients of concrete and determine their relative proportions with the object of producing concrete of certain minimum strength and durability as economically. The object of any mix proportion method is to determine an economical combination of concrete constituents used for a first trail batch to produce a concrete that is close to that which can achieve a good balance between the various desired properties of concrete at the lowest possible cost. In this study mix design was done as per Indian Standard guidelines in IS: 10262-2009.

Table 4 Mix Ratio of Concrete – M40 (kg/m3)

Water (kg/m³)	Cement (kg/m³)	Fine Aggregate (kg/m³)	Coarse Aggregate ` (kg/m³)	Super plasticizer (kg/m³)
172	429	835	1004	4.3
0.4	1	1.94	2.34	0.01

Table 5 Mix Proportions of Fibrous Self-curing Concrete

MIX	Cem. kg/ cu.m	F.A kg/ cu.m	C.A kg/ cu. m	SAP kg/ cu. mm	Fiber kg/ cu.m	Water lit/ cu.m	S.P kg/ cu. m
CM SCC	429	835	1004	130	----	172	4.3
FSCC 0.1%	429	831	1004	130	0.9	172	4.3
FSCC 0.2%	429	829.5	1004	130	1.8	172	4.3
FSCC 0.3%	429	826.5	1004	130	2.7	172	4.3
FSCC 0.4%	429	822	1004	130	3.6	172	4.3

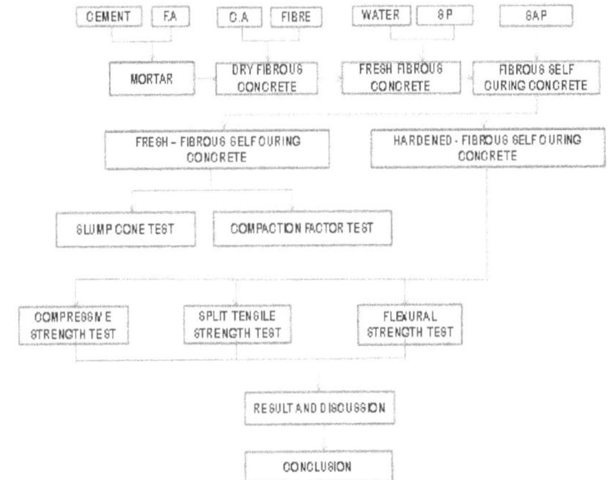

Fig. 1 Methodology of FSCC

Table 6 Cost Analysis of Fibrous Self-curing Concrete

Mix	C kg/m³	FA kg/m³	CA kg/m³	Water lit/m³	SP kg/m³	Fiber kg/m3	SAP kg/m³	Cost (Rs) Per m3
CM SCC	3003	417.5	402	43	232.4	-	58.5	4156
FSCC 0.1%	3003	415.5	402	43	232.4	32.4	58.5	4186
FSCC 0.2%	3003	414.7	402	43	232.4	64.8	58.5	4218
FSCC 0.3%	3003	413.2	402	43	232.4	97.2	58.5	4248
FSCC 0.4%	3003	411.0	402	43	232.4	129	58.5	4279

ISBN: 978-93-8830-599-0

TEST RESULTS AND DISCUSSIONS

The dosages of Super Absorbent Polymer were varied from 0.3% by weight of the cement was used as self-curing agent with that different dosages of PolypropyleneFibre varied from 0.1% - 0.4% by weight of the concrete to the self-curing concrete mix and compared the test result with the fibrous self-cured concrete as per Mix Design M40 grade (IS 10262:2009).

Table 7 Slump cone Values for various Mix Proportions

Mix Proportions	Slump Values in (mm)
CM SCC	110
FSCC 0.1%	120
FSCC 0.2%	125
FSCC 0.3%	115
FSCC 0.4%	95

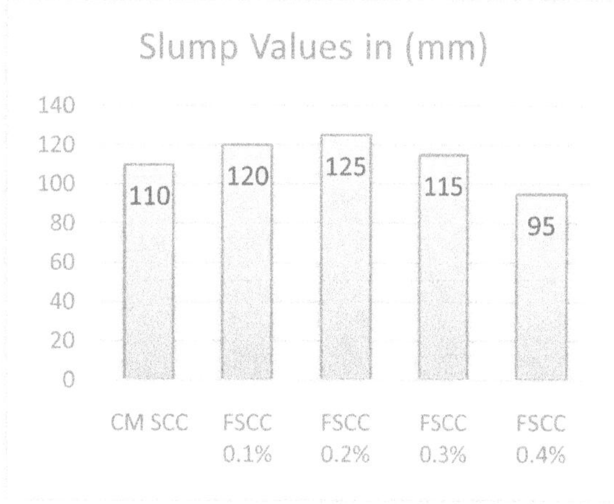

Fig. 2 Slump cone values for various Mix Proportions

Compressive Strength Test

The test is carried out on 150x150x150 mm size cubes, as per IS: 516-1959. The test specimens are marked and removed from the moulds and unless required for test within 24 hrs, immediately submerged in clean fresh water and kept there until taken out just prior to test. A 2000 KN capacity Compression Testing Machine (CTM) is used to conduct the test. The specimen is placed between the steel plates of the CTM and load is applied at the rate of 140 Kg/Cm2/min and the failure load in KN is observed from the load indicator of the CTM.

Compressive strength = Load / Area (MPa)

Table 8 Compressive Strength values for Various Mixes

Mix Proportions	7 Days Strength	14 Days Strength	28 Days Strength
CM SCC	18.11	35.35	43.11
FSCC 0.1%	18.27	35.68	43.51
FSCC 0.2%	19.11	36.80	44.88
FSCC 0.3%	19.22	38.91	45.77
FSCC 0.4%	18.77	33.45	43.7

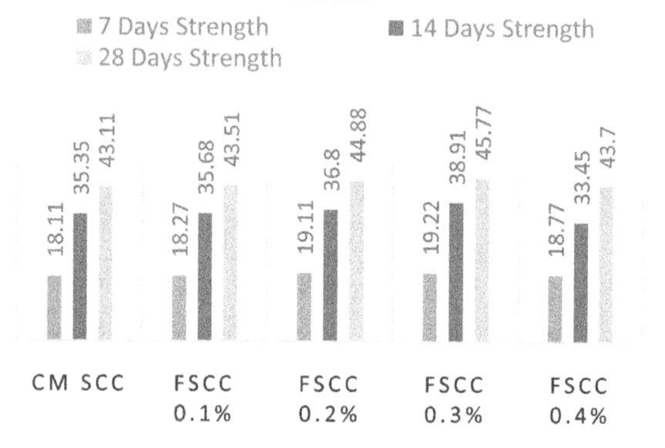

Fig. 3 Compressive Strength values for Various Mixes

Split Tensile Strength Test

The splitting tensile strength of concrete cylinder was determined based on IS 5816-1970 The load shall be applied nominal rate within the range 1.2 N/(mm2/min) to 2.4 /(mm2/min). The test was carried out on diameter of 150mm and length of 300mm size cylinder

$$\text{Split Tensile Strength } = \frac{2P}{\pi LD}$$

Table 9 Split Tensile Strength values for various Mix Proportions

Mix Proportions	7 Days Strength	14 Days Strength	28 Days Strength
CM SCC	1.67	3.25	4.92
FSCC 0.1%	1.68	3.29	4.95
FSCC 0.2%	1.64	3.17	4.81
FSCC 0.3%	1.83	3.77	5.37
FSCC 0.4%	1.64	3.30	4.81

ISBN: 978-93-8830-599-0

Fig. 4 Split Tensile Strength values for various Mix Proportions

CONCLUSION

The results of test on self-curing concrete was compared with Fibrous self-curing concrete. The Compressive strength, Split tensile strength and were conducted to investigate hardened properties of concrete.

1. Water retention for the concrete mixes incorporating self-curing agent is higher compared to conventional concrete mixes, as found by the weight loss with time.

2. The effectiveness of internal curing by means of SAP applied to concrete is the highest if 45 kg/ m3 water is added by means of 1 kg/m3 SAP.

3. In the study cubes were casted and kept for curing in room temperature about 25-30c.practical feasibility of fibrous self-cured member is need to be checked in hot regions

4. The Self-cured concrete using SAP was more economical than conventional cured concrete.

5. Performance of the self-curing agent will be affected by the mix proportions mainly the cement content and the w/c ratio

6. The optimum dosage is 0.3%.Addition of SAP leads to a significant increase of mechanical strength (Compressive and Splitting tensile).

7. Compressive strength for fibrous self-cured concrete for dosage of 0.3% was higher than self cured concrete.

8. Split tensile strength for fibrous self-curing concrete for dosage of 0.3% was higher than self cured concrete.

REFERENCES

1. Alvaro Paul and Mauricio Lopez, (2011), 'Assessing Lightweight Aggregate Efficiency for Maximizing Internal Curing Performance', ACI Materials Journal vol.108.

2. Ambily.P.S and Rajamane N.P, (2009), 'Self Curing Concrete – An Introduction', Concrete Composites Lab, Structural Engineering Research Centre, Chennai.

3. Balamuralikrishnan, R. and Antony Jeyasehar, C. (2009), 'Flexural Behavior of RC Beams Strengthened with Carbon Fiber Reinforced Polymer (CFRP) Fabrics' The Open Civil Engineering Journal vol.3, pp.102-109.

4. Bart Craeye, Matthew Geirnaerta and Geert De Schutter. (2011), 'Super absorbing polymers as an internal curing agent for mitigation of early-age cracking of high-performance concrete bridge decks', Construction and Building Materials 25, pp.1–13.

5. Bentz D.P. (2007), 'Internal curing of high-performance blended cement mortars', ACI materials Journal vol.104,No.4, pp.408–414.

6. Bentz, D.P and Lura, P. (2005), 'Mixture proportioning for internal curing', Concrete International Paper, pp.1–6.

7. Dale P.Bentz. (2007), 'Internal Curing of High-Performance Blended Cement Mortars' ACI Materials Journal, Vol.104.

8. Daniel Cusson and Ted Hoogeveen, (2008), 'Internal curing of high-performance concrete with pre-soaked fine light weight aggregate for prevention of autogenous shrinkage cracking',Cement and Concrete Research 38.

9. El-Dieb. A.S. (2006), 'Self – Curing concrete: Water retention, hydration and moisture transport', Construction and Building Materials vol.21, pp.1282 – 1287.

10. Ganesan.N, et al. (2007), 'Behaviour of Steel Fibre Reinforced High Performance Concrete Members under Flexure' IE(I) Journal–CV.

11. Gaston Espinoza-Hijazin and Mauricio Lopez, (2010), 'Extending internal curing to concrete mixtures with W/C higher than 0.42', Construction and Building Materials.

12. Geiker, M.R. Bent, D.P. and Jensen, O.M, (2004), 'Mitigating Autogeneous Shrinkage by Internal Curing' , American Concrete Institute vol.218, pp.143-148.

13. George C. Hoff, (2006), 'The use of lightweight fines for the internal curing of concrete', Northeast Solite Corporation.

14. IS 383 – 1970, (1993), 'Indian standard code of practice for Specification for Coarse and fine aggregates from natural sources for concrete', Bureau of Indian Standards,

15. IS 456 – 2000, (2002) 'Plain and Reinforced Concrete – Code of Practice', Bureau of Indian Standard, Bureau of Indian Standards, Fifth Reprint.

16. IS 516 - 1959 ,(1992) "Indian standard code of practice for methods of tests for strength of concrete", Bureau of Indian Standards.

17. IS 1199 -1959, (1999) " Indian standard code of practice for Methods of sampling and analysis of concrete", Bureau of Indian Standards.

18. IS 1489 - 1991 (Part - 1),(1993) 'Indian standard code of practice for Portland Pozzolana cement specification – Code of Practice', Bureau of Indian Standards.

19. IS 10262 – 2009,(2009), 'Concrete Mix Proportioning – Guidelines', Bureau of Indian Standards,(BIS 2009), First Revision.

20. John Roberts, (2006), 'High Performance Concrete Enhancement Through Internal Curing', Northeast Solite Corporation, pp.55-59.

21. Kumar, P.S , Mannan, M.A, Kurian, V.J and Achuytha, H (2007), 'Investigation on the flexural behaviour of high-performance reinforced concrete beams using sandstone aggregates', Construction and building materials Vol.21, pp. 1229-1237.

22. Ole Mejlhede Jensen and PietroLura, (2006), 'Techniques and materials for internal water curing of concrete', Materials and Structures Vol.39, pp.817–825.

23. PietroLura and Ole Mejlhede Jensen, (2006), 'Experimental observation of internal water curing of concrete', RILEM Technical Committee TC 196-ICC.

24. Ryan Henkensiefken, Javier Castro, Dale Bentz, and Jason Weiss. (2009), 'Internal Curing Improves Concrete Performance throughout its Life', pp.22-30.

A Perspective Study on
Engineering Properties of Fly Ash Bricks over Conventional Bricks

Shudhanshu Kumar[1], Rahul Kumar[1], Padam Jee Omar[2*] and Abhishek Kumar[2]
[1]Undergraduate Student, [2]Assistant Professor
Motihari College of Engineering, Motihari, Bihar, India
sss.padam.omar@gmail.com

Abstract

In this wild growing today's era development of new building growing with high rates on which the cutting of clay being rapidly increases under the conventional bricks to full fill the upcoming huge requirements that's not safe for our environment. Instead of conventional bricks, fly ash bricks are safer for our environment because of its manufacturing process. In fly ash bricks the maximum number of removal wastage of coal which left after burnt is directly came under use by mixing with sand/stone dust and cements with a particular ratio and percentage i.e. 1:3:6. The soundness test, hardness test, absorption test, efflorescence test and crushing strength test of fly ash bricks equate with conventional bricks that conveyed good results in various way even it is stronger, durable and yet more economical in every aspect. It also does not adapt the hydro properties of the bricks but fly ash bricks are lower in weight but higher in strength with comparison to conventional one. Fly ash bricks are eco welcoming as it safeguard our environment though conservation and shows dynamic role in the reduction of carbon dioxide and a injurious green house gas.

Keywords: Conventional Bricks, fertile agricultural, sand, fly ash, quarry dust.

I. Introduction

In present scenario in the construction industry, use of economic and environmental friendly material is of a great concern. The total play taken out from the agricultural fields for day was over 300 million tonnes for 10,000 crore bricks. At present, India has a production capabilities of over 10,000 crore bricks true around 45,000 local bhattas, in the unorganised sector so, the use of industrial waste products such a fly ash for making bricks is ecologically and economical advantages since apart from saving precious tap agricultural soil, it meets the social objective of disposing industrial waste that is fly ash which otherwise is a pollutant and a nuisance.

The ever increasing follow of fly ash quantities in the world has not been remotely matched by its utilisation. Australia is an example where such utilisation has been minimal. The most important and popular use of fly ash in Australia has been the partial replacement of Portland cement. Naturally seen that the bridge industries present a opportunity for the efficient utilization of the vast quantities of fly ash. Conservative attitude are among the factor that limited the use of fly ash in concrete to generally a maximum of 25% replacement of Portland cement. This conservative can be understood in context of concrete where the ash is mixed raw and the effects of high volume replacement are still subjected to research and sometimes controversy. It is however not quite justifiable that the bricks industry should take similar conservative attitude. Environmental concern has been raised in some parts of the world where coal is the main power generating resources and where bricks are also the main building material. Such concern have been resulted in legislation that applies the brick industry to incorporate at least 25% by weight of fly ash and/or bottom or pond ash in the brick making mixture if the industry is within 50 km from a coal power generation plant. Some successful ventures have been reported where fly ash was interrupted in the mixture at rate of 20 to 50%. Nevertheless, there is only little evidence that incorporation of fly ash in the breath mixer has exceeded than 30% volume even when the legislation was obeyed. Reason behind search reluctance is not clear. a most probable reason is the fear of change in mania small factories and the ingrained conservatism in the attitude of stake holders of the large producers. The possible incompatibility of the ash with the clay and shale during the various process of production including the crucial one of the firing may be legitimate difficulty. At high temperature the wind 1000 °C, the temperature and length of time of firing become very sensitive to the type of art and of course to the clay and shale of in the same mixture. Few attempts at manufacturing breeds from more than 80% as has been made. The author believes that fly as on its own cash be an excellent raw materials for brick making. This has now been proven and they pretend is taken for the manufacture of bricks from fly ash. the response of the ash firing temperature at 1000°C and beyond can be iterate li control it even in small factories. The potential savings with this approach are many. saving in production and transportation cost are producing breeds of superior qualities to those of standard clay bricks are in addition to the environmental solution that such venture may bring about.

ISBN: 978-93-8830-599-0

The objective of the present paper is to study the effect of fly ash bricks on the performance and the properties of bricks with the view to study the comparison between clay between clay bricks and a fly ash bricks because fly ash is enriched with silica is the main constituent for the conventional building material from the experiment it is for the desired to compare the strength of fly ash bricks by that of conventional clay bricks. The silent properties of bricks like crushing strength, water absorption, shape and size, soundness, hardness and efflorescence are too determined.

And to find out the optimum mix design for making brick so as to achieve the maximum compressive strength.

II. Used materials and Methodology

The details regarding the method and properties of materials used in this study.

Materials used

Fly ash + lime + gypsum + quarry dust = fly ash bricks

Properties of materials

Fly ash: It is finally divided reduce resulting from the combustion of powdered coal and transported by the flue gases and collected by the electrostatic precipitator.

ASTM broadly classified fly ash into two classes.

Class F: Fly ash normally produced by burning anthracite or bituminous coal, usually has less than 5% Cap. Class f fly ash has pozzolanic properties only.

Class C: Fly ash normally produced by burning lignite or sub bituminous coal. Some class C fly ash may have Cao content in excess of 10%. In addition to pozzolanic properties, class c fly ash also possesses cementious properties. Fly ash used is of type class C with a specific gravity of 2.19.

Lime: Lime is an important binding material in building construction. It is basically calcium oxide (Cao) in natural association with magnesium oxide (MgO). Lime reacts with fly ash at ordinary temperature and for a compound possessing cementious properties. After reaction between lime and fly ash calcium silicate hydrate are produced which are produced which are responsible for the high strength of the compound.

Gypsum: Gypsum is non hydraulic binder torturing naturally as a soft crystalline rod or sand gypsum have a valuable properties light small bulk density, incombustibility, good sound absorbing capacity, good fire resistance, rapid drawing and hardening with negligible shrinkage, superior surface finish etc.

In addition it can strength material or increase viscosity. It has a specific gravity of 2.31 gram per cubic centimetre. The density of gypsum powers is 2.8 to3 grams per cubic metres.

Quarry dust: First arriving mix percentage of fly ash bricks for fly ash (45 to 50%), lime (5 to 30%), gypsum (21%) tricks for each proportion 12 number of bricks are casting in that nine bricks are used to determine the compressive strength of brick in N/mm^2 at 7 days, 14 days and 21 days curing time and three bricks are used to determine the water absorption. Compressive stress is determined using compression testing machine (CTM) of 3000 kilo Newton capacity. The following flowchart describes the methodology of this study.

III. Experimental Study

In experimental study, following test was conducted under some constraints.

1. Soundness test

In soundness test we just take two bricks and collide them with each other such that they shouldn't break and clear sound should be produce. Those bricks will produce metallic sound that is of best quality.

Conventional bricks	Fly ash bricks
Produce metallic but with normal frequency.	It produces sound like metal. So, it is of best quality.

2. Shape and Size test

It should be in homogeneous shape and after test we found that both conventional and fly ash bricks are free from defects. But after the break, fly ash bricks are more in homogeneous shape as compare to conventional bricks.

3. Hardness test

In hardness test, we just check which is harder by scratching the brick with our nails on the surface of bricks and we found that both are almost same.

Conventional bricks	Fly ash bricks
No impact found after scratching with the help of nails.	No impact found after scratching with the help of nails.

4. Absorption test

In absorption test, check the percentage of water absorbed by bricks. After immersed the bricks in water for the period of 24 hours. The minimum percentage of water absorption is the maximum percentage of good quality bricks.

ISBN: 978-93-8830-599-0

<table>
<tr><td colspan="4" align="center">Conventional Bricks</td></tr>
<tr><td align="center">S. No.</td><td align="center">Weight of bricks, when it is in dry condition. W₁(kg)</td><td align="center">Weight of bricks, when it is in saturated condition. W₂(kg)</td><td align="center">Percentage of water absorption</td></tr>
<tr><td align="center">1.</td><td align="center">3.03</td><td align="center">3.31</td><td align="center">9.24%</td></tr>
<tr><td align="center">2.</td><td align="center">3.41</td><td align="center">3.84</td><td align="center">12.60%</td></tr>
<tr><td align="center">3.</td><td align="center">2.88</td><td align="center">3.25</td><td align="center">12.84%</td></tr>
<tr><td colspan="3" align="center">Average</td><td align="center">11.56%</td></tr>
</table>

<table>
<tr><td colspan="4" align="center">Fly ash Bricks</td></tr>
<tr><td align="center">S. No.</td><td align="center">Weight of bricks, when it is in dry condition. W₁(kg)</td><td align="center">Weight of bricks, when it is in saturated condition. W₂(kg)</td><td align="center">Percentage of water absorption</td></tr>
<tr><td align="center">1.</td><td align="center">3.27</td><td align="center">3.57</td><td align="center">9.17%</td></tr>
<tr><td align="center">2.</td><td align="center">2.99</td><td align="center">3.26</td><td align="center">9.03%</td></tr>
<tr><td align="center">3.</td><td align="center">3.00</td><td align="center">3.27</td><td align="center">9%</td></tr>
<tr><td colspan="3" align="center">Average</td><td align="center">9.066%</td></tr>
</table>

5. Efflorescence test

This test is use to check the appearance of white patches on the bricks after immersed the bricks in water for 24 hours in 25 ml of water. This activity was done twice and we found that more efflorescence is appeared in the conventional bricks as compare to fly ash bricks:-

Conventional bricks	Fly ash bricks
White patches appear up to 40% of bricks-	White patches appear less than 15 % of bricks

6. Crushing strength test

It is the test by which we can find the strength of bricks by using compression testing machine. The minimum strength should not be less than 35 N/m^2 at any cost for the good quality of bricks.

Conclusion

1. In the case of hardness conventional bricks and fly ash bricks are almost same.
2. In efflorescence test the percentage of white patches are less in fly ash bricks so it is good in quality.
3. The metallic sound in fly ash bricks is much better than conventional bricks.
4. The structure are found more homogeneous and compact in fly ash bricks rather than conventional bricks.
5. The average of absorbed moisture in fly ash bricks in 9.066% even the percentage of conventional bricks is 11.56%. That's why fly ash bricks are better than conventional bricks.

References

1. Butterworth B., "Bricks made with pulverized Fuel Ash.", Trans. Brit. Ceram.Soc' ENGLAND, 53 (5), 1954, 293- 313
2. Gupta P.C, Ray S. C. ," Commercialization of Fly ash", The Indian Concrete Journal, 167, 1993, 554-560.
3. Gupta, R.L., Bhagwan, Gautam, D.K., Garg, S.P., "A Press for Sand Lime Bricks Research and Industry", 22(40), 1977, 246-249.
4. I.S Code 12894-1990 for Fly ash – Lime Bricks-Specification
5. I.S Code: 8112-1989, Ordinary Portland Cement, 43 Grade- Specification
6. IS: 3495 (P-II) -1976, Determination of Water Absorption (Second Revision).
7. IS: 3495 (P-II)-1976, Determination of Effloresce (Second Revision).
8. IS: 3495(P-I)-1976, Determination of Compressive Strength (Second Revision).
9. Parul, R. Patel, "Use of Fly ash in Brick Manufacturing", National Conference On Advances in Construction Materials, " AICM-India, 2004, 53-56.
10. S.C Rangwala, "Engineering Material", Charotar Publications, 15th Edition, 1991, 72-112.
11. Thorne, D.L. , Watt , J.D. , "Composition and Properties of pulverized fuel ashes" Part II, Journal of Applied Chemistry, London,15,1965, 595-604.

ISBN: 978-93-8830-599-0

Experimental Investigation on the Effect of
Binder Index on Compressive Strength of Geopolymer Concrete

R. Shankaraiah[1*], B. Sesha Sreenivas[2] and D. Rama Seshu[3]

[1] E E & Research Scholar, Kakatiya Univ., Warangal, India.
[2] Professor and Principal, University College of Engg, Kothagudem, Kakatiya Univ., India.
[3] Professor of Civil Engineering, National Institute of Technology, Warangal, India.
shankarracharla@gmail.com

Abstract

The experimental study consisted of determination of the Compressive strength of Geopolymer concrete (GPC) by casting and testing of cubes of size 150 X 150 X 150 mm size. Fly Ash and GGBS are used as Binders. The Robo sand (RS) also called as manufactured sand produced by stone crushing was used as fine aggregate and crushed granite of 20 mm nominal size are used as coarse aggregate. Sodium hydroxide solution and Sodium Silicate solution are used as Alkaline solution. The testing of specimens was carried out at the end of 7 days (7D) and 28 days (28D) of outdoor curing. A total of 108 cubes representing 6 different GGBS / FA ratios (0.25, 0.43, 067. 1.0, 1.5 & 2.3), 3 different molarities (6, 8, & 10) of Alkaline solution, 2 different curing periods (7 & 28 days), 3 identical specimens for each variation were cast and tested. The compressive strength GPC for different GGBS to Fly Ash ratio and for different molarity of Alkaline activator and the variation of 7 days and 28 days strength is carried out and presented.

Keywords: Geopolymer concrete (GPC), Compressive strength, Fly ash (FA), Ground Granulated Blast Furnace Slag (GGBS), alkaline molar activators (Molarity), Robo Sand (RS), Binder Index (BI), Sodium hydroxide (NaoH), Sodium Silicate.

1.0 INTRODUCTION

The growth in infrastructure development and boom in the Housing sector has put a lot of demand on cement. Due to environmental concerns of cement industry, there arises a strong need to make use of alternate sustainable technology. The development of alkali activated binders with promising engineering properties and longer durability has emerged as an alternative to OPC. In this context the development of Geopolymer concrete (GPC) is being viewed as an emerging class of concrete material and could be the next generation concrete for applications in civil engineering infrastructure. The Geopolymer concrete is an environmental friendly material in sense that it uses the industrial byproducts such as Ground Granulated Blast furnace Slag (GGBS) and Fly Ash (FA) along with alkaline activated solutions. The commonly used combination of alkaline activator solution is Sodium Hydroxide and Sodium Silicate. These rich in silica byproducts form with alkaline solution a binder matrix to bound aggregate in the mixture and to produce the hardened concrete.

[1].Geopolymer concrete developed by Prof. Joseph Davidovits, is a new class of concrete that is attracting growing interest around the world due to its environmental and performance benefits compared to conventional Portland cement concrete. Davidovits (1970) reported the use of waste material like Fly Ash (FA) and Ground Granulated Blast furnace Slag (GGBS) and high alkaline solution as activators. Gluckovsky was the first which used the name of this binder "geopolymers" which caused the critical discussions among Russian Scientists working in cement chemistry. Hardjito et al concluded that a combination of Sodium hydroxide and Sodium silicate solutions can be a good application for activators and higher concentration of Sodium hydroxide solution and curing temperature enable the concrete compressive strength to be higher keeping in view of the previous research studies, the present research is aimed at study on the effect of binder index on compressive strength of Geopolymer concrete by casting and testing of cubes of 150 X 150 X 150 mm size. Fly Ash and GGBS are used as Binders. The Robo sand (RS) also called as manufactured sand produced by stone crushing was used as fine aggregate and crushed granite of 20 mm nominal size are used as coarse aggregate. Sodium hydroxide solution and Sodium Silicate solution are used as Alkaline solution. A total of 108 cubes representing 6 different GGBS / FA ratios (0.25, 0.43, 067. 1.0, 1.5 & 2.3), 3 different molarities (6, 8 & 10) of Alkaline solution, 2 different curing periods (7 & 28 days), 3 identical specimens for each variation were cast and tested. The compressive strength GPC for different GGBS to Fly Ash ratio and for different molarity of Alkaline activator and the variation of 7 days and 28 days strength is carried out and presented.

ISBN: 978-93-8830-599-0

2.0 EXPERIMENTAL PROGRAM

The experimental program consisted of finding the compressive strength of GGBS and FA based Geopolymer concrete by casting and testing of cubes of size 150 mm. A total of 108 cubes representing 6 different GGBS / FA ratios (0.25, 0.43, 067. 1.0, 1.5 & 2.3), 3 different molarities (6, 8 & 10) of Alkaline solution, 2 different curing periods (7 & 28 days), 3 identical specimens for each variation were cast and tested.

2.1 MATERIALS

Fly Ash and GGBS are used as binders in this work. GGBS is obtained from Toshali Cement Pvt Ltd, Bayyavaram near Vizag. Fly Ash is obtained from NTPC Ramagundam – Telangana Specific gravity of GGBS and Fly Ash are 2.90 and 2.17 respectively. As fine aggregate, the Robo Sand (RS) produced by stone crushing is used. This specific gravity and bulk density are 2.65 g/cm3 and 1.45 g/cm3 respectively.Crushed granite of 20 mm normal size obtained from a local crushing unit is used as Coarse Aggregate. This specific gravity and bulk density are 2.80 g/cm3 and 1.5 g/cm3 respectively. Potable water is used for preparation of alkaline solution.The Combination of Sodium Hydroxide and Sodium Silicate is used as alkaline solution. Chemical details of Fly Ash and GGBS are shown in table 1. The alkaline molar activators of sodium hydroxide solution used are 6,8 & 10. The Sodium hydroxide pellets used for preparation of NaOH solution is given in table 2. The ratio of Sodium silicate solution to NaOH solution is fixed as 2.5. The NaOH solution and Na_2SiO_3 solution are mixed and stored for 24 hours at room temperature before casting. Conplast Sp-430 is used as Super Plasticizer to obtain the desired workability.

Table 1 Chemical composition of Flyash and GGBS percentage by mass.

Material	SiO_2	Al_2O_3	Fe_2O_3	SO_3	CaO	MgO	Na_2O	LOI
Fly ash	60.12	26.63	4.22	0.32	4.1	1.21	0.2	0.85
GGBS	34.16	20.1	0.81	0.88	32.8	7.69	nd	.

Table 2 Materials used for NaOH solution preparation.

	6 moles/L	8 moles/L	10 moles/L
Sodium hydroxide pellets , (grams)	200	255	306
Potable Water (grams)	800	745	694

2.2 COMPOSITION OF GPC

The GPC mix proportions are shown in table 3.

Table 3 GPC mix proportion.

FA : GGBS	GGBS / FA Ratio	Coarse Aggregate	Robo Sand	Fly Ash	GGBS	NaOH Solution	Sodium Silicate	Super Plasticizer (2% of the Binder)
80: 20	0.25	18.00	13.20	7.85	2.00	1.80	4.50	
70:30	0.43	18.00	13.20	6.90	2.95	1.80	4.50	
60:40	0.67	18.00	13.20	5.91	3.94	1.80	4.50	
50:50	1.00	18.00	13.20	4.93	4.92	1.80	4.50	
40:60	1.50	18.00	13.20	3.94	5.91	1.80	4.50	
30:70	2.30	18.00	13.20	2.95	6.90	1.80	4.50	

3.0   CASTING OF GPC SPECIMENS

The dry materials were weighed and mixed using a rotating drum type pan mixer.The alkaline liquids and super plasticizer of optimum dosage were added Homogenous mixing was obtained by continues mixing for about 5 minutes. After mixing, the fresh GPC was casted in cube moulds followed by compaction. After compaction, the top surface was leveled with a trowel. The specimens were demoulded after 24 hours of casting & during this period the hydrated in the air and cured outdoor.For outdoor curing, specimens were left out air dried for a period of 28 days at temperature of about 20 degrees centigrade. The specimens were casted for 6M, 8M & 10M of GPC

3.1   TESTING FOR COMPRESSIVE STRENGTH

The cube specimens were tested on compression testing machine of capacity 2000 KN. The testing of specimens was carried out at the end of 7 days (7D) and 28 days (28D) of outdoor curing.A total of (108) cubes representing 6 different GGBS/FA ratios , 3 different molarities (6, 8 & 10) of alkaline solution, 2 different curing periods (7 & 28 days) 3 identical specimens for each variation were casted and tested. The compression strength of GPC for different GGBS to Fly Ash ratio and for different molarity of alkaline solution is shown in table 4.

Binder Index is taken as the product of molarity of alkaline activator and binder's ratio as given below.

Binder Index (BI) = [GGBS / Fly Ash] x Molarity

The Binder Index (BI) Vs Compressive Strength of GPC is shown in table 5.

Table 4 Compressive strength of GPC

Sl.	GGBS/FA	6moles/L		8 moles/L		10 moles/L	
		7D	28D	7D	28D	7D	28D
1	0.25	12.0	16.3	12.4	18.9	13.8	22.1
2	0.43	12.8	17.8	13.4	23.0	14.8	25.5
3	0.67	20.9	24.5	22.0	29.6	23.7	36.7
4	1.00	29.3	37.1	29.4	37.8	29.7	38.8
5	1.50	34.9	40.9	37.0	41.9	39.6	43.0
6	2.30	42.1	44.8	46.1	48.4	48.8	52.9

Table 5 Binder Index Vs Compressive Strength

Sl.No	Binder Index (BI)	7 D Strength (Mpa)	28 D Strength (Mpa)	7D / 28 D strength
1	1.5	12.0	16.3	0.736
2	2.0	12.4	18.9	0.656
3	2.5	13.8	22.1	0.624
4	2.58	12.8	17.8	0.719
5	3.44	13.4	23.0	0.583
6	4.3	14.8	25.5	0.580
7	4.02	20.9	24.8	0.843
8	5.36	22.0	29.6	0.743
9	6.7	23.7	36.7	0.646
10	6.0	29.3	37.1	0.790
11	8.0	29.4	37.8	0.778
12	10.0	29.7	38.8	0.765
13	9.0	34.9	40.9	0.853
14	12.0	37.0	41.9	0.883
15	15.0	39.6	43.0	0.921
16	13.8	42.1	44.8	0.940
17	18.4	46.1	48.4	0.952
18	23.0	48.8	52.9	0.922

ISBN: 978-93-8830-599-0

4. 0 DISCUSSIONS

4.1 The effect of GGBS to Fly Ash ratio on compressive strength of GPC for a particular

molarity of alkaline activator is shown in figure 1 to 4.

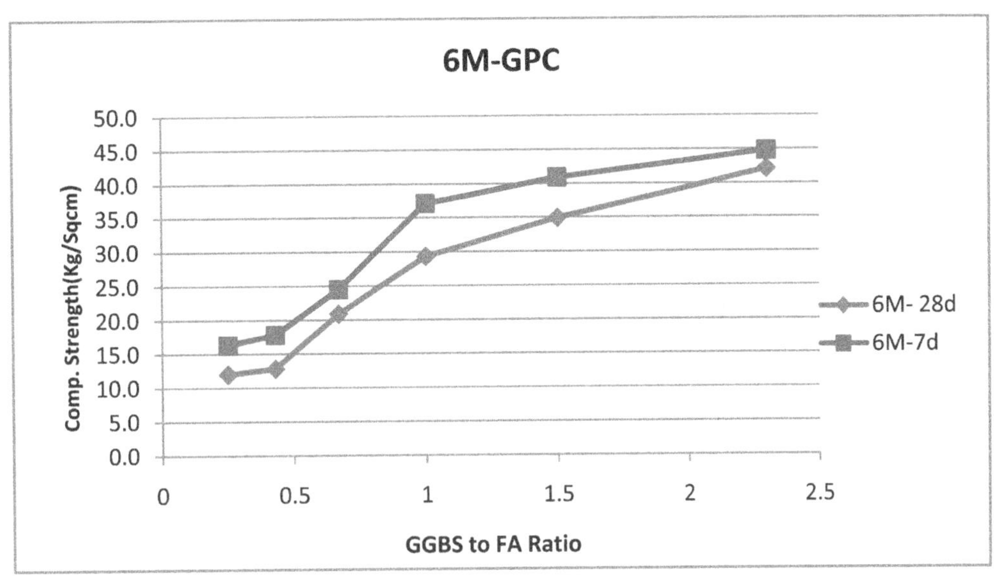

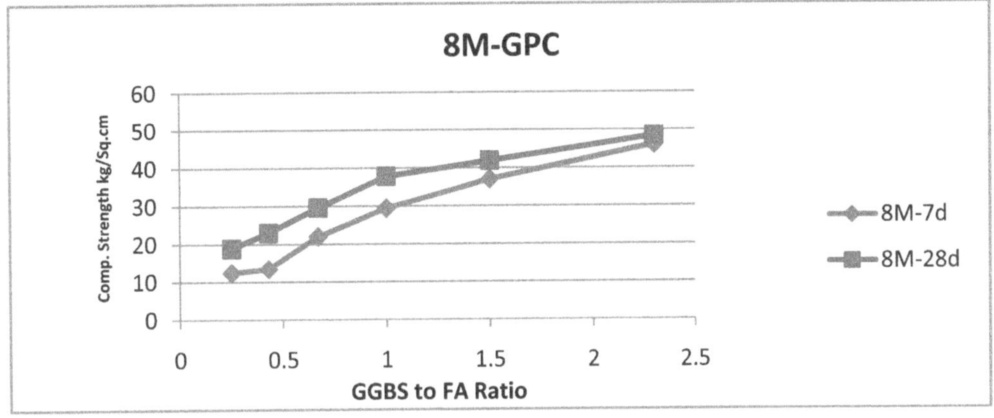

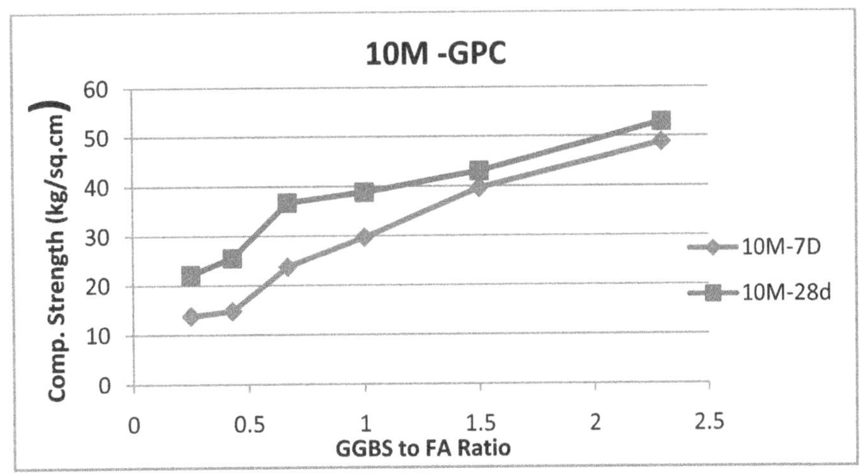

From the above figures, it can be observed that the compressive strength of GPC has increased with increase in GGBS to FA ratio.

4.2 Effect of molarity of alkaline activator on compressive strength of GPC

In general as the molarity is increasing, the 28 day compressive strength of GPC also as increased.

4.3 Effect of binder index on the compressive strength of GPC

The values of compressive strength of GPC at 7 days and 28 days of air curing with the level of binder index are given in table 5. This variation indicates that both 7days and 28 days compressive strength of GPC as increased with increase in binder index.

CONCLUSIONS

The following conclusions can be made from the experimental analysis done.

1. The Compressive strength of GPC has increased with increase of GGBS to FA ratio for a particular molarity of activator used.
2. The rate of compressive strength is higher for GGBS to FA ratios less than 1.0
3. As the molarity of activator was increasing the 28-days compressive strength is not in proporation to the increase in activator solution molarity.
4. A new parameter called "Binder Index" which combines the effects of both GGBS to FA ratio and molar concentration of activator solution can be used to control the compressive strength of geopolymer concrete.
5. Both 7-days and 28-days compressive strength of GPC has increased with increase in Binder Index.
6. There is a non-linear variation between Binder Index and compressive strength of GPC.
7. Fly Ash and GGBS Combination can be used for the production of GPC without the need of heat curing.

REFERENCES

1. Anuradha R. Sreevidya V. Venkatasubramani R. Rangan BV, "Modified guidelines for gepolymer concrete mix design using Indian Standard", Asian J Civil Eng (Build Hous), 13(3):353-364 (2012).
2. Hordijito D, Wallah SE, Surnajouw DMJ, and Rangan BV, "On the development of fly ash – based geopolymer concrete", ACI Mater J., 101(6):467-72(2004).
3. IS: 516-1956 (Reaffirmed 1999), Indian Standard Methods of Tests for Strength of Concrete"
4. J. Davidovits, "Synthetic mineral polymer compound of the silicoaluminate family and preparation process", US patent 4472199, (1978).
5. Rangan BV, "Mix design and productio of fly ash based geopolymer concrete", Indian Concr J, 82: 7-15 (2008).
6. Wang SD, Pu XC, Scrivener KL, Pratt PL, "Alkali – activated slag cement and concrete: a review of properties and problems", Adv Cement Res, 27:93-102 (1995).

ISBN: 978-93-8830-599-0

Transparent Concrete: A Glowing Future

Durga Raghavi T[#1]

[#] Gajalakshmi Associates, Hyderabad, India
[1] durgaraghavi3@gmail.com

Abstract

Concrete plays a very vital role in construction and has been used since roman times for development of infrastructure and housing. Due to rapid urbanization in 1960, concrete was often misconceived and disfavoured. But since that time, considerable progress has been made in concrete, not only in scientific terms but also in aesthetic terms. Nowadays, the space between buildings is reduced due to globalization and the construction of high-rise buildings, this leads to increase in the use of non- renewable energy sources and there is a requirement of new construction technique like green building and indoor thermal system. In 2001, the Hungarian architect Aron Losonczi invented the concept of transparent concrete and the first transparent concrete block was favourably produced by adding large amount of glass fibre into concrete. In 2003, it is named as LiTraCon. The transparent concrete mainly concentrates on transparency and its aim is to attain the artistic finish and to generate the green technology. It has major strength characteristics no way less than conventional concrete. Though it has major advantages, but it is still in developmental stages.

Keywords: Transparent concrete, Plastic optical fibre, Transmitting concrete, LiTraCon.

I. INTRODUCTION

Construction of high rise buildings and sky scrapers leads to the obstruction of natural light in buildings. Due to this problem, use of artificial light has increased in large amount. Thus, it becomes necessary to reduce artificial light consumption in structure. This leads to the introduction of innovative concrete called transparent concrete. The Transparent concrete is fibre reinforced concrete which is used for aesthetical application by inserting the optical fibres in concrete. Both natural as well as artificial light passes through the transparent concrete due to optical fibres. The study is not only constrained with the decorative purpose but the effect of fibre application in strength aspect is also discussed. This concrete can be used in interiors of hall, lobby and ceiling to glow in dark by external lighting source and in day time it glows by the light transmission from natural resources. By switching the ingredients of traditional concrete with transparent ones or embedding fibre optics, transparent concrete has become a reality. It is the brightest building material developed in recent years. It is one of the newest, most functional and revolutionary element in green construction material. However, this innovative new material, while still in development stages, is being used in a variety of applications in architecture and promises vast opportunities in the future.

II. TRANSPARENT CONCRETE

Translucent concrete (Transparent concrete) is an innovative concrete which is dissimilar from normal concrete. Transparent concrete permits light to pass through it and are light-weight as compared to normal concrete. In 2003, it is named as LiTraCon. Joel S. and Sergio O.G. developed a transparent concrete material, which can allow 80% light through and only 30% of weight of common concrete. Due to economic development and space utilization requirements, high-rise buildings and skyscrapers are mostly built downtown in metropolitan areas around the world, especially countries with great populations. Those buildings are isolated biosphere only based on man-made lights to maintain people's optical activities. It is considered to be one of the best sensor materials available and has been used widely since the 1990s. It can also be developed with plastic optical fibre (POF) or polymers (resins) in order to use solar energy and reduce the energy used by other resources as these are transparent and allow the light passing through them. Compared with a traditional electric lighting system, illuminating the indoors with daylight also creates a more appealing and healthy environment for building occupants. The characteristics of transparent concrete are presented in Table 1 below.

Table 1 Characteristics of Transparent concrete

Form	Prefabricated blocks
Ingredients	96% concrete, 4% optical fiber
Density	2100-2400 Kg/m2
Color	White, grey or black
Compressive strength	50 N/mm2
Bending tensile strength	7N/mm2

A. How Transparent concrete is made

The following are the steps for preparation of transparent concrete:

ISBN: 978-93-8830-599-0

- Strands of optical fibres are cast by the thousands into concrete to transmit light, either natural or artificial, into all spaces surrounding the resulting translucent panels. Light-transmitting concrete is produced by adding 2.5% to 4% optical fibres (by volume) into the concrete mixture. Holes are drilled on the opposite sides of the mould that is selected. The fibres run parallel to each other, transmitting light between two surfaces of the concrete element in which they are embedded. Thickness of the optical fibres can be varied between 2 μm and 2 mm to suit the particular requirements of light transmission. Wooden and PVC specimens were prepared and cast as shown in Figs.1 to 5 for transparent concrete.

- Originally, the fibre filaments were placed individually in the concrete, making production time-consuming and costly. Newer, semi-automatic production processes use woven fibre fabric instead of single filaments. Fabric and concrete are alternately inserted into moulds at intervals of approximately 2 mm to 5 mm. Smaller or thinner layers allow an increased amount of light to pass through the concrete. Following casting, the material is cut into panels or blocks of the specified thickness and the surface is then typically polished, resulting in finishes ranging from semi-gloss to high gloss.

- The concrete mixture is made from fine materials only: it contains no coarse aggregate. The compressive strength of greater than 70 MPa (over 10,000 psi) is comparable to that of high-strength concretes.

Fig. 1 Wooden Cube specimen for Transparent concrete

Fig. 2 Wooden prism specimen for Transparent concrete

Fig. 3 Cylinder specimen made with Polyvinyl Chloride (PVC) pipes for Transparent concrete

Fig. 4 Transparent concrete cube transmitting light where flash light is source of light from back

ISBN: 978-93-8830-599-0

Fig. 5 Transparent concrete specimens transmitting light where sunlight is source from back

III. LITERATURE REVIEW

Satish Kumar and Suresh (2015) produced the concrete specimen by reinforcing optical fibres with different proportion based on the volume of the cube by 0.15%, 0.25%, 0.35% to compare the strength and intensity of light passing through it. The cube and cylinder mould used in the project is of standard size 150mm×150mm and150mm×300mm respectively. Different test was carried out on the specimen like Compressive strength test, Split-Tensile strength, Intensity of light passing through it, etc. They have observed that the reinforcing of optical fibre will transmit light and also eventually increases the strength of the concrete as compared to conventional concrete. Compressive strength of the concrete is increased by 22.99% of the normal concrete for 0.25% of optical fibre. The tensile strength of the concrete is increased by 83.95% for 0.25% of optical fibre, which clearly indicates that transparent concrete transmits light without affecting the strength of concrete.

Bhagyasri et al (2016) reduced the quantity of manufacturing of cement by partially replacing with glass powder and avoided the disposal problem of waste glass material. The specimens are cubes, prisms and cylinders were casted by partially replacing cement by glass powder with 10%, 20%, 30% and 40%. They used finely ground glass material to increase the strength without using the strength increasing admixtures. They even studied the role of glass powder in mechanical properties of concrete. Initial and final setting time of cement, specific gravity test, workability test, UPV test, compressive strength test, and flexural strength test has been performed. The study concludes that the setting time of concrete increases as there is increase in the percentage of glass powder. Workability

of concrete decreases with increase in the glass powder content. From UPV test it is concluded that the quality of concrete increases with increase in the cement replacement with glass powder. At the ages of 7 and 28 days maximum compressive strength occurs at 20% of glass powder. Water absorption of concrete reduces by increasing the glass powder content. There is increased in the flexural strength as the glass powder content increases.

IV. PLASTIC OPTICAL FIBRE

An Optical Fibre is a flexible, transparent fibre made of glass (silica) or plastic, slightly thicker than a human hair as shown in Fig.6. It functions as a waveguide or light pipe, to transmit light between the two ends of the fibre. The field of applied science and engineering concerned with the design and application of optical fibres is known as fibre optics. Optical fibres are widely used in fibre-optic communications, which permits transmission over longer distances and at higher bandwidths (data rates) than other forms of communication. Fibres are used instead of metal wires because signals travel along them with less loss and are also immune to electromagnetic interference. Fibres are also used for illumination, and are wrapped in bundles so that they may be used to carry images, thus allowing viewing in confined spaces. Specially designed fibres are used for a variety of other applications, including sensors and fibre lasers.

Optical fibres typically include a transparent core surrounded by a transparent cladding material with a lower index of refraction. Light is kept in the core by total internal reflection. This causes the fibre to act as a waveguide. Fibres that support many propagation paths or transverse modes are called multi-mode fibres (MMF), while those that only support a single mode are called single-mode fibres (SMF). Multi-mode fibres generally have a wider core diameter, and are used for short-distance communication links and for applications where high power must be transmitted. Single-mode fibres are used for most communication links longer than 1,050 meters (3,440 ft.).

Optical Fibre Elements: The following are the elements of plastic optical fibre:

Core - The thin glass centre of the fibre where the light travels.

Cladding - The outer optical material surrounding the core that reflects the light back into the core. To confine the reflection in the core, the refractive index of the core must be greater than that of the cladding.

ISBN: 978-93-8830-599-0

Coating - Plastic coating that protects the fibre from damage and moisture. Our solution used the same principle. For our translucent concrete panel, we needed a core material such as acrylic that will transmit light continuously into the inside of the building, a white cladding layer that reflects the light back into the core and concrete as the protective coating. The elements of plastic optical fibre can be referred in Fig.7 below.

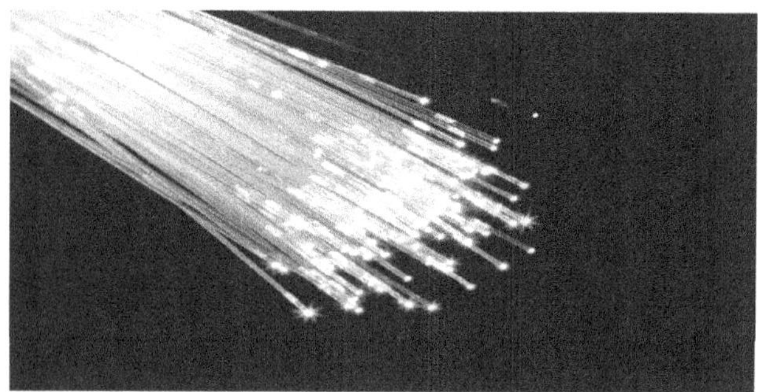

Fig. 6 Light end plastic optical fibre

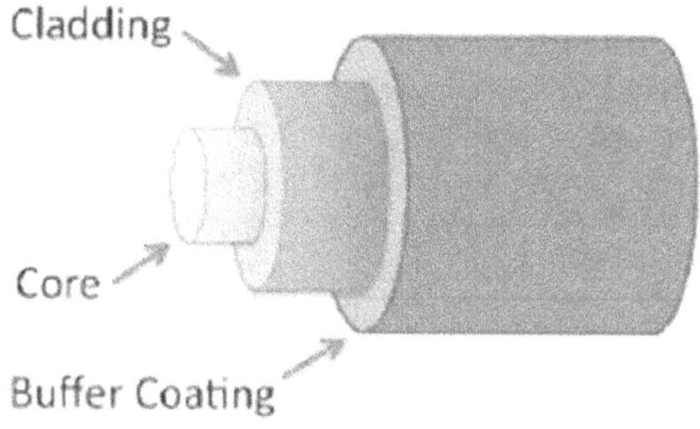

Fig. 7 Elements of optical fibre

Optical Fibre Types:

There are 3 basic types of optical fibres:
  (i)   Multi-mode graded-index fibre
  (ii) Multi-mode step-index fibre and
  (iii) Single mode step-index fibres.

A multimode fibre can propagate hundreds of light modes at one time while single mode fibres only propagate one mode as shown in Fig.8 below.

The difference between graded-index and step-index fibres is that in a graded-index fibre it has a core whose refractive index varies with the distance from the fibre axis, while the step-index has core with the same refractive index throughout the fibre. Since the single-mode fibres propagate light in one clearly defined path, intermodal dispersion effects is not present, allowing the fibre to operate at larger bandwidths than multimode fibre. On the other hand, multimode fibres have large intermodal dispersion effects due to the many light modes of propagations it handles at one time. Because of this multimode fibres operate at lower bandwidths, however they are typically used for enterprise systems such as offices, buildings, universities since they are more cost effective than single mode ones. In this experimental work the intensity of light passing through concrete blocks provided with optical fibre is measured in lux unit for measurement of light. Optical fibres of 0.75 mm were used in the present thesis work. The properties of an optical fibre are presented in Table 2 below.

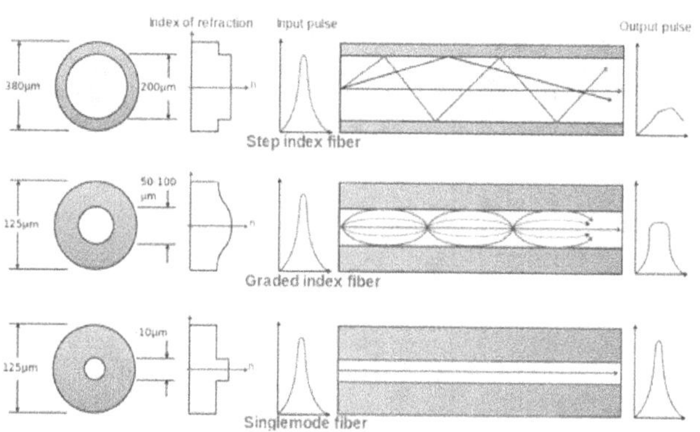

Fig. 8 Types of plastic optical fibre

Table 2 Properties of plastic optical fibre

Product	Fibre Diameter (mm)	Fibre Diameter (inches)	Attenuation db/m (650nm)	Numerical Aperture	Allowable bending radius
FOF .25	0.25	1/100th	Under .35		
FOF .50	0.50	1.25/64th	Under .25		
FOF .75	0.75	1.9/64th			> 9 mm
FOF 1.0	1.0	1.25/32nd		0.5/0.6	
FOF 1.5	1.5	1/16th	Under .20		
FOF 2.0	2.0	5/64th			> 20 mm
FOF 3.0	3.0	1/8th			

V.   ADVANTAGES

1. The main advantage is that on large scale objects the texture is still visible.
2. When a solid wall is imbued with the ability to transmit light, it means that a home can use fewer lights in their house during daylight hours.
3. It has very good architectural properties for giving good aesthetical view to the building.
4. Where light is not able to come properly at that place transparent concrete can be used.
5. Energy saving can be done by utilization of transparent concrete in buildings.
6. It is a green building material.
7. It Increases the visibility in dark subway stations.

ISBN: 978-93-8830-599-0

8. It provides high UV resistance, and structurally stable when compared to conventional concrete resulting in major applications.

VI. DISADVANTAGES

The following are the disadvantages of transparent concrete:
1. The concrete is costly due to plastic optical fibres.
2. Casting of transparent concrete block is difficult for the labour so skilled persons are required.
3. Natural or any artificial light passes through in the direction of fibres laid; only at this condition shadow appears on the darker side.

VII. APPLICATIONS

The applications of transparent concrete are defined below:
1. Transparent concrete blocks suitable for floors, pavements and load-bearing walls.
2. Facades, interior wall cladding and dividing walls based on thin panels, partition walls and it can be used where the sunlight does not reach properly.
3. It can be used in furniture for the decorative and aesthetic purpose.
4. It can be used as speed bumps in roads.
5. It can be used as light sidewalks at night.
6. Decorative lamps
7. Decorative tiles, stairs, floors, etc.

VIII. ENERGY EFFIECIENCY OF TRANSPARENT CONCRETE

In total domestic usage of electricity 23% of electricity is used for lightening purpose as shown in Fig.9, so it is necessary to utilize natural light for illuminating interior of building.

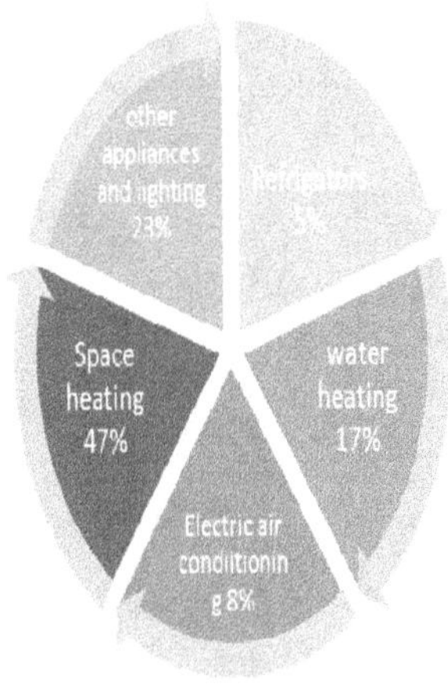

Fig. 9 Domestic electric consumption

A. *Cost Analysis:*

The cost analysis and comparison of cost is made between conventional concrete and transparent concrete with 2.5% and 4% plastic optical fibre and presented in Tables 3 and IV below.

Table 3 Cost Analysis of M25 grade (Conventional and Transparent concrete) for 1cum

	Cost		Total Cost(Rs.)
	Concrete	Optical Fibres	
Conventional concrete	4198.53	0.00	4199.00
Concrete with 2.5% plastic optical fibre	1018.2	5369.5	6387.7
Concrete with 4% plastic optical fibre	3051.35	7116.0	10167.35

Table 4 Comparison of cost

S.No	Number of cubes	Volume in Cum	Cost(Rs.)	
			Conventional Blocks	Optical concrete blocks
1	20	0.068	285.50	3876.35
2	16	0.054	226.72	3101.08
3	12	0.041	172.14	2325.81

1. *Power consumption by artificial lighting*: Power consumption when one 60 Watt light bulb is used for illumination for 30days for 8 hours $= 60 \times 30 \times 8 = 14400 = 14.4$ Units.

Table 5 Tariff/Charges for power supply for residential and commercial buildings

Residential	Public
Rs.4.85	Rs.7.92

2. Payback Period:

   (a) For 0.36 Sq.m area (16 Blocks): *With reference to comparison of cost Table IV, difference in initial cost for 16 No. of Cubes = Rs. 3101.08 - Rs. 226.72 = Rs. 2874.36 /-*

Energy saving in residential room in one year

14.4 X 4.85 X 12 = Rs 838.08 /-

ISBN: 978-93-8830-599-0

- Period require to recover extra amount for transparent block $= \dfrac{2874.36}{838.08} = 3.42$ years.

- Energy saving in commercial/Industrial room in one year = 14.4 X 7.9 X 12 = Rs. 1368.58/

- Period require to recover extra amount for OFRC block $= \dfrac{2874.36}{1368.38} = 2.1$ Years.

From above cost analysis and payback period calculation it was confirmed that, the recovery period for light transmitting block is 3.42 years for residential use and it is 2.1years for commercial use. This payback period is too less as compared to benefits of light transmitting concrete. Similarly Payback period for other areas are presented in Table 6 below.

Table 6 Payback period

Area in Sq.m	No. of Blocks	Payback for Residential	Payback for commercial
0.45	20	4.3	2..7
0.36	16	3.5	2.1
0.27	12	2..6	1.6

IX. CONCLUSIONS

The main aim of transparent concrete is to use sunlight as source of light in spite of using electrical energy in order to minimize the use of non-renewable sources. This technique results in to energy saving. Optical fibres are a detecting or transmission element, to reduce the use of non-natural light. The normal concrete is swapped by translucent concrete, which has natural lighting and art design. By introducing the concrete with optical glass fibres, light travels from outside in or inside out.

- An innovative material, light transmitting concrete, or LiTraCon, is still partially in the development stage. Yet, it is beginning to be used in a variety of applications in architecture, and holds tremendous promise for the future. There are few manufacturers of translucent concrete. These include LiTraCon, Lucon and Lucem Lichbeton. In India, till today there is no availability.

- Energy efficiency of transparent concrete is observed to be good and more compared to conventional concrete. Hence, structural elements made with transparent concrete are cost saving. Thus transparent concrete is considered to be a green building material as it is energy efficient and cost effective.

- Further study on replacing optical fibre with acrylic rods, waste glass, etc. can be carried out.

- Replacing with different types of cement like white cement and also with various types of recyclable materials which are suitable and strength and durability tests can be done and results can be found out accordingly.

- By introducing various percentages of optical fibre in design procedure, studies can be carried on.

- Studies can be carried out on high strength concrete made with recycled aggregates.

REFERENCES

1. M.N.V. Padma Bhushan et.al "Optical fibre in the modelling of translucent concrete blocks", ISSN: 2248-9622. Vol.3, Issue 3, May-June 2013.

2. Soumyajit Paul and Avik Dutta "Translucent Concrete" IJSRP, vol. 3, Issue 10, October 2013.

3. Prachi Sharma et al "A Review of the Development in the Field of Fiber Optic Communication Systems", IETAE, Volume 3, Issue 5, May 2013. [9] P. S. Mane Deshmukh and R. Y. Mane Deshmukh "Comparative Study of Waste Glass Powder Utilized in concrete", IJSR, Volume 3, Issue 12, December 2014.

4. P. M. Shanmugavadivu et al "An Experimental study on Light Transmitting concrete", IJJRET Volume:03, Special issue:11, June 2014.

5. Sameer Shaikh et al "Effective Utilisation of Waste Glass in concrete", IJERA, Volume 5, Issue 12, (Part-4), December 2015.

6. Jadhav G. S et al "Partially Replacement of Waste Toughen Glass as Fine Aggregates in concrete", IJIFR, Volume 3, Issue 9, May 2016.

7. T. Bhagyasri et al "Role of Glass Powder in Mechanical strength of concrete", IRF International Conference, March 2016.

8. Shetty, M.S "Concrete technology" Chand S and Co.Ltd, India (2009).

9. Plain and Reinforced Concrete Code of Practice (Fourth Revision), IS 456:2000, Bureau of Indian Standards, New Delhi.

An Experimental Study on Strength of Geopolymer Concrete (GPC) Curing With Ambient Temperature

J. Srinivas[1], B. Sesha Sreenivas[2] and D. Rama Seshu[3]

[1]Research Scholar, [2]Professor of Civil Engineering& Principal, Kakatiya University, Warangal, India
[3]Professor of Civil Engineering, NIT, Warangal, India

Abstract

This research article presents the Experimental investigation on compressive strength(fck), Split tensile strength(fct) and Modulus of Rupture(fcr) of Geopolymer concrete. This article mainly aims at the study of effect of fly ash (FA) and Ground Granulated Blast Furnace slag (GGBS) on the properties of Geo Polymer Concrete(GPC) as a replacement for OPC, using 8Molarity Sodium Hydroxide (NaoH) and Sodium Silicate (Na2SiO3) solutions as alkaline activators. The proportions of Fly Ash to GGBS used are (80:20),(60:40),(40:60) and (20:80). Alkaline liquid content to Fly ash ratio taken as 0.36 and fine aggregate to total aggregate ratio is taken as 32 percentage[1]. The ratio of Sodium Silicate (Na2Sio3) solution to Sodium Hydroxide solution is kept constant as 2.5[1]. Three identical specimens for each of variation were cast and tested after 7days and 28days of ambient curingThe test results of Compressive Strength (fck), Split Tensile Strength (fct) and Modulus of Rupture (f_{cr}) of Geopolymer concrete are presented.

Keywords: Geopolymer Concrete, GGBS, Fly ash, Sodium Silicate, Sodium Hydroxide, Compressive Strength, Split Tensile Strength, Modulus of Rupture.

I. INTRODUCTION

Concrete is most widely used construction material.Cement is the main component for making concrete.It is estimate that production of cement is increase 3% annually. As cement is a key element in the construction, but during the manufacturing of cement it emits some toxic gases like co_2 in the excess amount .To overcome this problem scientist has done a lot of research on it and finally a french scientist davidovits found complete replacing agent called Geopolymer .On the other scenario huge quantiy of fly ash generate all around the world from the thermal power plant and lead to waste management problem. Fly ash is a by-product produced from thermal power stations in large quantities, the safe disposal of which has become very difficult. On the other hand, the abundance and availability of fly ash and GGBS worldwide created an opportunity to utilize these by-products, as partial replacement for OPC. In 1978, Davidovits [3] developed a composite material called geo-polymer to describe an alternative cementitious material which has ceramic-like properties. Geo-polymer concrete is a new technology considered to reduce the use of Portland cement in concrete. Geopolymer is environmental friendly materials that do not emit green house gases during polymerization process. Geopolymer can be produced by combining a pozzolanic compound or alumino silicate source material with highly alkaline solutions[4]. Geopolymers are made from source materials with silicon (Si) and Aluminium (Al) content and thus cement can be completely replaced.

II. EXPERIMENTAL PROGRAM

The present investigation consisted of determining the Compressive Strength (fck), Split Tensile Strength (fct) and Modulus of Rupture of the GPC, by casting and testing cubes (100mm size), cylinders (100mm X 200mm) and prisms (100mm X100 X500mm) respectively.

A. Materials

Low calcium, Class F dry fly ash, conforming to IS 3812(part 1:2003)[7], is obtained from Kothagudem Thermal power station, Bhadradri Kothagudem Dist, Telangana, India . Ground Granulated Blast furnace slag conforming to IS 12089:1987 is obtained from Blue way exports supplier, from Vijayawada, Andhra Pradesh, India. Specific gravity of fly ash and GGBS are 2.17 and 2.90 respectively. Chemical composition details are shown in Table 1. Natural river sand was used as fine aggregate. The bulk specific gravity in oven dry condition and water absorption of the sand as per IS 2386 (Part III, 1963)[9]were 2.45 and 1% respectively. The gradation of the sand was determined by sieve analysis as per IS 383 (1970)[8.] Fineness modulus of sand was found to be 2.50. Crushed granite stones of size 12 mm and 10 mm were used as coarse aggregate. The bulk specific gravity in oven dry condition and water absorption of the coarse aggregate 12 mm and 10mm as per IS 2386 (Part III, 1963)[9] were 2.35 and 0.28% respectively. Super Plasticizer Conplast Sp-430 was used to obtain the desired workability. Potable water was used in the experimental work for preparation of alkaline solution.

ISBN: 978-93-8830-599-0

Table 1 Showing the chemical composition of Fly ash and GGBS, percentage by mass

Material	SiO2	Al2O3	Fe2O3	SO3	CaO	MgO	Na2O	LOI
Fly ash	60.12	26.63	4.22	0.32	4.1	1.21	0.2	0.85
GGBS	34.16	20.1	0.81	0.88	32.8	7.69	nd	.

B. Alkaline liquid: 8 molarity of sodium hydroxide solution is used in the present investigation. The ratio of sodium silicate solution to sodium hydroxide solution is 2.5[2]. NaOH solution was prepared by dissolving 262 grams of NaOH pellets in potable water of 738 grams. The NaOH solution thus prepared is mixed with Na_2SiO_3 solution. The mixture was stored for 24 hours at room temperature before casting.

Table 2 GPC mix proportions [5]

FA:GGBS	GGBS/FA	Molarity	Materials in Kg/m^3							
			Coarse Agg	Fine Agg	Fly ash	GGBS	NaOH Solution	Sodium Silicate	Super Plasticizer	Extra water (10% of the binder)
0:100	--	8M	1100	517.45	0	575.2	59.10	148.25	11.50	57.52
20:80	4	8M	1100	517.45	115.04	460.16	59.10	148.25	11.50	57.52
40:60	1.5	8M	1100	517.45	230.08	345.12	59.10	148.25	11.50	57.52
60:40	0.67	8M	1100	517.45	345.12	230.08	59.10	148.25	11.50	57.52
80:20	0.25	8M	1100	517.45	460.16	115.04	59.10	148.25	11.50	57.52
100:0	0	8M	1100	517.45	575.2	0	59.10	148.25	11.50	57.52

The GPC specimens for Compressive strength, split tensile strength and modulus of rupture were tested on universal testing machine of capacity 1000KN. The load was increased gradually at constant rate until failure. The prisms were tested under two point loading for modulus of rupture. Three identical specimens with each variation were cast and tested after 7 days & 28 days of ambient curing. A total of 30 cubes, 30 cylinders & 30 prisms using different GGBS/FA ratios (0, 0.25, 0.66, 1.5, and 4.0) and 8M alkaline activator were cast and tested after 7days & 28 days of ambient curing. The maximum loads applied on various specimens were recorded as per IS 516-1956. The test results are given in table 3.

Table 3 Compressive Strength, split tensile strength and modulus of rupture values for GPC

FA:GGBS	GGBS/FA	Compressive strength (N/mm^2)			Split tensile strength (N/mm^2)			Modulus of rupture (N/mm^2)		
		fck	fck	7D/28D	fct	fct	7D/28D	fcr	fcr	7D/28D
		7D	28D		7D	28D		7D	28D	
80:20	0.25	13.26	19.06	0.69	0.906	1.49	0.608	1.25	1.95	0.64
60:40	0.67	29.83	36.20	0.82	1.869	2.067	0.904	1.6	3.0	0.53
40:60	1.5	40.36	48.63	0.83	2.09	2.219	0.941	1.7	3.2	0.53
20:80	4	43.45	51.12	0.85	2.19	2.35	0.986	1.9	3.65	0.52

III. RESULTS & DISCUSSIONS

The effect of GGBS on compressive strength, split tensile strength and modulus of rupture for GPC is shown in figures 1,3 and 5. From figures 1,3and 5 it is observed that the Compressive Strength, Split Tensile Strength and Modulus of rupture of Geopolymer concrete has increased with increase in GGBS percentage for 8M.

The effect of GGBS to FA ratio on Compressive Strength(fck), Split Tensile Strength(fct) and Modulus

ISBN: 978-93-8830-599-0

of Rupture(fcr) for GPC is shown in figures 2,4and 6. From figures 2,4 and 6 it is observed that the Compressive Strength(fck), Split Tensile Strength(fct) and Modulus of Rupture (fcr) of GPC has increased with increase in GGBS to FA ratio for 8M.

A. Effect of Compressive strength:

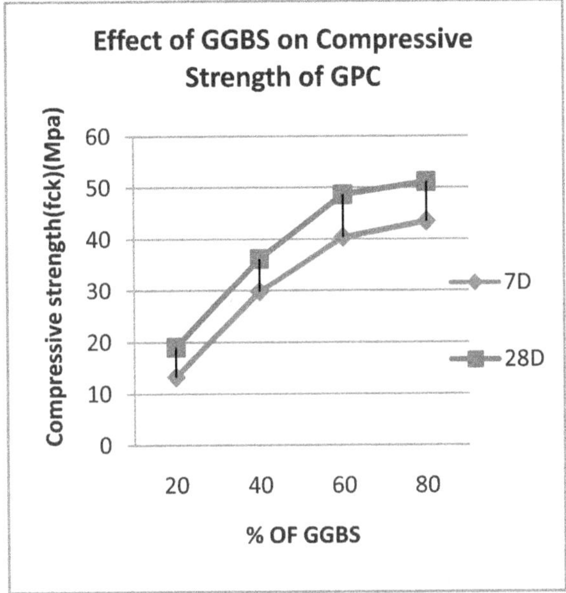

Fig. 1 Effect of replacement of fly ash with GGBS on Compressive Strength of GPC

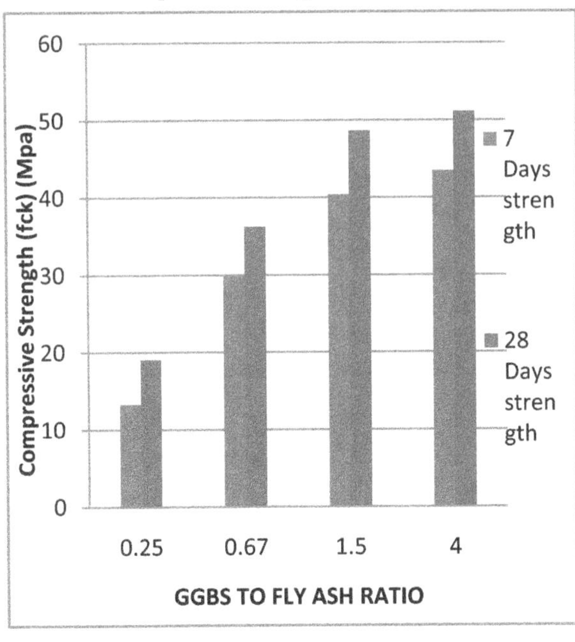

Fig. 2 Effect of GGBS to Fly Ash Ratio on Compressive Strength of GPC (With an age 7days and 28days)

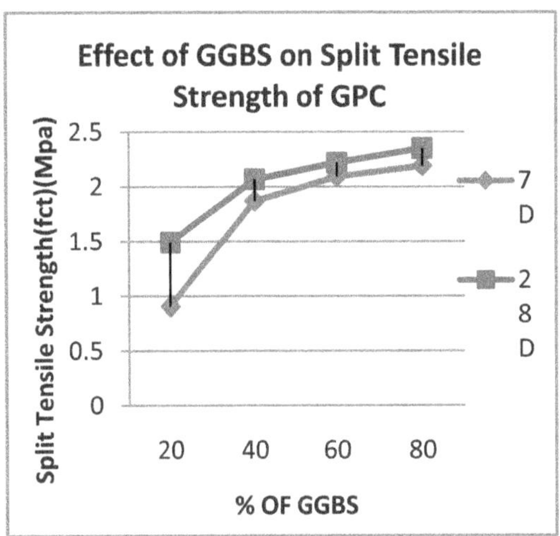

Fig. 3 Effect of replacement of Fly ash with GGBs on Split tensile Strength of GPC

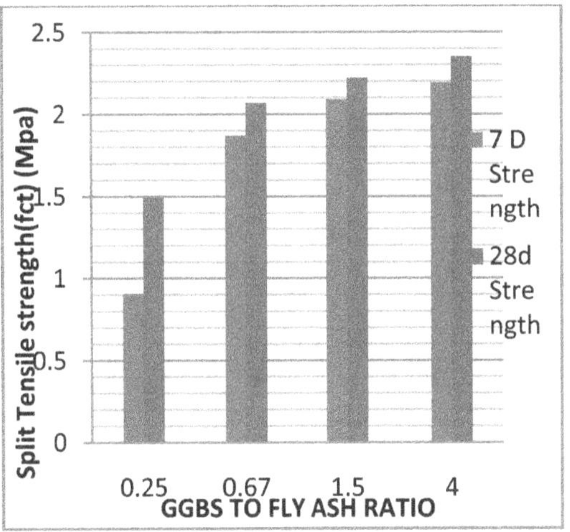

Fig. 4 Effect of GGBS to Fly Ash Ratio on on Split tensile Strength of GPC (With an age of 7days and 28days)

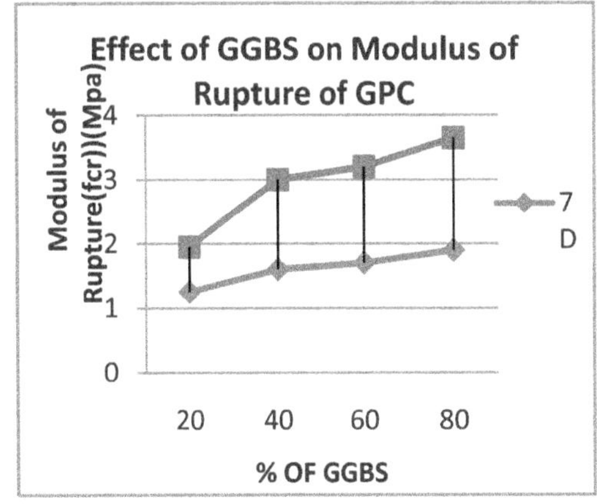

Fig. 5 Effect of Replacement of Fly ash with GGBs on Modulus of Rupture of GPC

ISBN: 978-93-8830-599-0

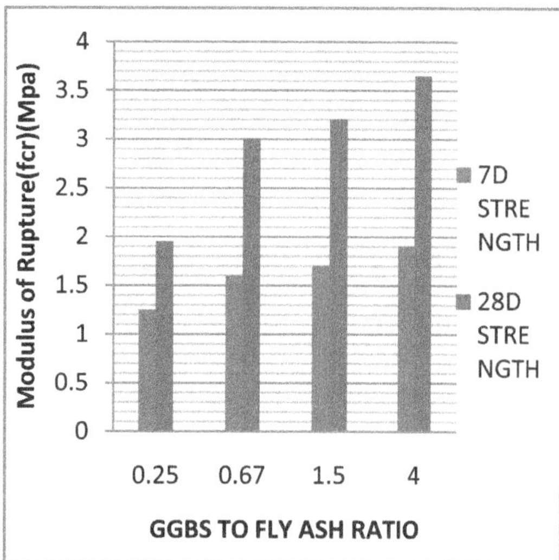

Fig. 6 Effect of GGBS to Fly Ash Ratio on Modulus of Rupture of GPC (With an age of 7days and 28days)

IV. CONCLUSIONS

Based on the experimental investigation carried out on geopolymer concrete following conclusions were drawn

1. The compressive strength, Split tensile strength and modulus of rupture at 7 days and 28 days of ambient curing has increased with increase in percentage of GGBS.

2. The compressive strength, Split tensile strength and modulus of rupture at 7D & 28D have increased with increase in percentage of GGBS to Fly ash ratios.

3. The combination of GGBS TO Fly ash can be conveniently used for production ofGPC avoiding heat curing.

4. Due to Geopolymer Concrete the onsumption of cement, emission of corbon di-oxide and green house effect are reduced.

V. REFERENCES

1. B.V.Rangan, Mix design and production of Fly Ash based Geopolymer concrete, Indian Concrete journal, 82,7(2008).

2. K.Vijai, R.Kurnutha and B.G.Vishnuram "Effect of Inclusion of Steel fibres on the propertie concrete Composites."International Journal of Society for Information Display September-2011.

3. J. Davidovits, Synthetic mineral polymer compound of the silico alumi-nate family and Preparation process, US patent 4472199, 1978.

4. R. Anuradha, V. Sreevidya, R. Venkatasubramani, B. V. Rangan, Modified guidelines for Geopolymer concrete mix design using Indian standard, Asian J. Civil. Eng., Build. House. **13** 3, 353(2012).

5. D. Hardjito, S. E. Wallah, D. M. J. Sumajouw, B. V. Rangan, on the evelopment of fly ash-based geopolymer concrete, ACI Mater. J., **101**, 6, 467 (2004).

6. B. V. Rangan, Mix design and production of fly ash based geopolymer concrete, Indian ConcreteJ., **82**, 7 (2008).

7. Is 3812 Part-1 2003,(pulverized fuel Ash-Specifications for use as pozzolana on cement Mortar and concrete.

8. Is 383-1970, Specification for coarse and fine Aggregates from natural sources for Concrete.Bureau of Indian Standards, New Delhi

9. Is 2386-, "Methods testing for aggregate for concrete".

10. IS: 516–1956 (Reaffirm rmed 1999), Indian Standard Methods of Tests for Strength of Concrete

11. Is 1199-1959, Indian standard Methods of sampling and analysis of concrete.Burau of Indian standards, New Delhi, 1959.

12. K.Sandeep Dult, K.vinay Kumar, l.siva kishor, Ch. Mallika chowdary , "A case study on fly ash based geopolymer concrete" international journal of engineering trends and technology vol -34, Nov-2 April -2016

13. B.V.Rangan, Fly ash based Geopolymer concrete, Research Report GC 4 Engineering Faculty Curtin University of Technology Perth, Australia.

14. M.I.Abdul Aleem, P.D Arumairaj, "Geopolymer Concrete – A review" Vol 1, issue 2, pp118-122 IJESET,Feb- 2012.

15. Adanagouda, Pampapathi G S, A.Varun, Ramya Madagiri, Merugu Keerthan "Experimental Study on Fly Ash based Geopolymer Concrete with Replacement of Sand by GBS" Adanagouda.et.al. Int. Journal of Engineering Research and Application ISSN : 2248-9622, Vol. 7,Issue 7, (Part -2) July 2017, pp.57-6

Detection and Prevention of Leakages in Dams

R. Prathyusha[1] and M. Sirisha[2]

UG, Department of Civil Engineering, JNTUH-CEH

prathyushar98@gmail.com & sirishgoud100@gmail.com

Abstract

Dams are very important and huge structures in all around the world. As the dam is continuously exposed to the water, it to should be constructed cautiously. A small damage in dam may lead to huge failure. Repairing these damages needs costly remedies, so it is better to detect the damages as early as possible to make it economical.

Leakages are the huge problems in dams. These should be detected as soon as possible and relative measures should be taken to repair these damages. It is very important for engineers to know about the techniques to detect and prevent the leakages of dams. In the past there are some hydrological techniques to detect the leakages.

This paper explains all the techniques to detect damages (leakages) and how to repair those damages which includes evaluation, design and mitigation of these leakages.

INTRODUCTION

There are many types of dams which many causes for failure of dams. Mostly earth dam structures are easily prone to failure. Earth dams have many problems compared to gravity dams. Some of the problems of earth dams are overcomed in gravity dams. But still there some problems in gravity dams, which are leading to failure. Some of the causes for failure of dam structures are

1. Erosion of soil near to the foundation causing settlement
2. The quality of the material used is very poor
3. Rock foundation subjected to horizontal shear.
4. Inadequate grouting action of the contraction joints causing cracks in the structure.

Most common cause of failure of dams is due to leakages and infiltration of water in the reservoirs. The main purpose of reservoirs to store more water without wasting, so infiltration in reservoirs should be reduced. To reduce the infiltration, infiltration zones should be detected. The need to reduce the risk of failure or control water loss has led to costly remedial repairs that are planned and executed without a complete understanding of the problem. A lack of appropriate leakage investigation and monitoring can result in repairs that are unsuccessful in controlling or reducing leakage.

To detect the leakages and infiltration zones there many methods in last few decades. These methods are very useful in detecting the main causes of failures in dams. Some the methods are discussed in this paper with some examples.

METHODS TO DETECT LEAKAGES IN DAMS

Leakage and preferential flow paths are often controlled by the geology of the site. Therefore, any leakage study should always include obtaining detailed geological and hydrogeological information as the first step. A single technique is not always sufficient to investigate leakage in dams. Often a combination of techniques results in successful leakage detection, because an key information can be obtained from any one of the techniques.

Relationship between Reservoir Level and Flow Rate:
The relationship between the water level in the reservoir and the flow rate at the leak provides valuable information regarding the elevation interval within the reservoir where the infiltration is taking place. Utilization of this technique requires frequent measurements of the flow at the leak(s) and at different reservoir elevations. When more than one leak is present, separate flow-rate readings should be taken in each leak.

It can be seen from below Figure that when all the water emerging from a leak comes from the same reservoir elevation, the points tend to align in a straight line. The flow rate in the leaks is proportional to the reservoir water level.

In other cases, the straight-line relating flow rate at the leaks and reservoir levels exhibit a slope change as shown in Figure. The slope change results from the activation of another infiltrating zone feeding that leak.

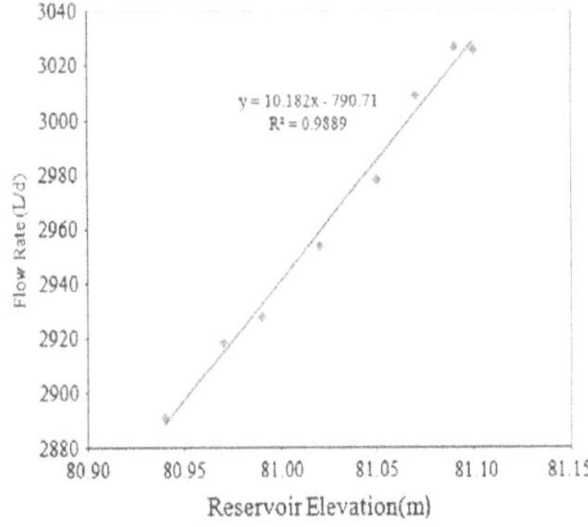

ISBN: 978-93-8830-599-0

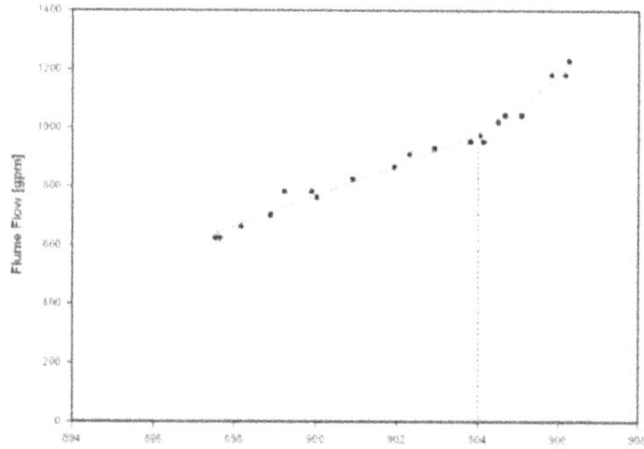

Fig. 1 Relationship between Reservoir level and flow rate at west flume in boney falls, Michigan

Data in above Figures do not include information on the time response of the flow rate at the leak and the reservoir level. This in fact can provide further information on the characteristics of the ground along the flow path between the reservoir and the leak.

Natural Water Tracers: Water has natural tracers, which can be used to obtain information on the origin and location of the infiltrating points in the reservoir.

Water temperature: When the transit time of the water between the reservoir and the boreholes or leaks is less than a few days, the temperature of the water does not change significantly, as it flows through the ground. This is because the specific heat capacity of water is much higher than most natural materials found in the ground. As a result, the water associated with the flow path tends to maintain the same temperature as the reservoir at the infiltration zone.

Conductivity: Similar to water temperature, conductivity is a good natural tracer that can provide valuable information when evaluating leaks in dams and reservoirs. This is because, as in the case of temperature, reservoirs also develop saline stratification and therefore deep waters exhibit higher salinity contents than water in upper layers. Measurements of conductivity in the reservoir, boreholes, and leaks are strongly recommended during the first stages of dam leakage investigations. It is common to obtain temperature and conductivity measurements simultaneously during the same campaign because both parameters are usually measured using the same instrument. The thermo conductivity probe is typically used to measure both the temperature and conductivity.

Conductivity of the reservoir waters shows great variation with time and space. These variations can be of interest in the investigation of hydraulic connections between reservoir, boreholes, and leaks

downstream of the dam. This is obtained by correlating the peaks in conductivity in the reservoir, boreholes, and leaks. If important changes in conductivity are found between water samples collected in the reservoir, boreholes, and leaks, the next step is to perform chemical analyses to identify the ions responsible for those changes. Chemical analyses can provide valuable information regarding the change of the conductivity.

Chemical Analysis: The results obtained from temperature and conductivity measurements are improved by measuring the major chemical components of water.

Chemical composition of water provides additional information on its origin and geochemical evolution, facilitating the differentiation between reservoir and aquifer waters or a mixture of those waters. Chemical composition testing is often complemented with tools such stable isotopes.

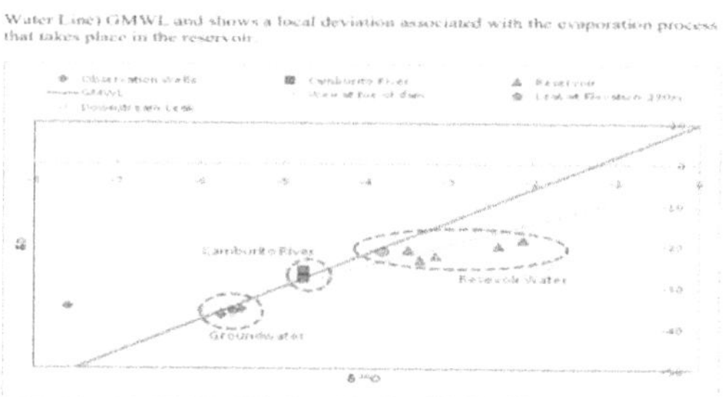

Fig. 2 Relationship between δD and $\delta^{18}O$ at Borde Seco dam in Venezuela

Artificial Tracers: During investigation of leaks from dams and reservoirs, it is usually necessary to use artificial tracers. They are typically used after the chemical composition and the other natural tracers have been used. The techniques that utilize artificial tracers have had limited application in the past due to complexity of the hydrological framework, lack of information on the utilization of artificial tracers, technological limitations for detecting low concentrations, and the difficult interpretation of the results.

Suspended Solids: Suspended solids can be applied only for investigation of groundwater flows through well-developed fractures, solution channels and cavities such as karstic terrain. In porous media, these tracers will not travel long distances from the injection point.

Fluorescent Tracers: Fluorescents are the most common tracers used in dam leak investigations. Fluorescent traces emit a characteristic fluorescence when they are excited by a light beam of certain wavelength. The wavelength of the excited light beam and the emitted fluorescence are characteristic for each product. The technology on the instruments used to measure fluorescence has improved significantly.

ISBN: 978-93-8830-599-0

Interconnection Tests: Interconnection tests are an integral portion of leak investigations from dams and reservoirs. They provide proof of hydraulic connection between the point of injection and the measuring point(s). The test consists of the injection of a given tracer amount in the reservoir or boreholes and its subsequent measurement of arrival time and concentrations at boreholes or downstream leaks

The transit time, obtained from the passing curve and the flow rate measured at the leak can be used to estimate the minimum volume of cavities associated with the flow path between the entry and exit points.

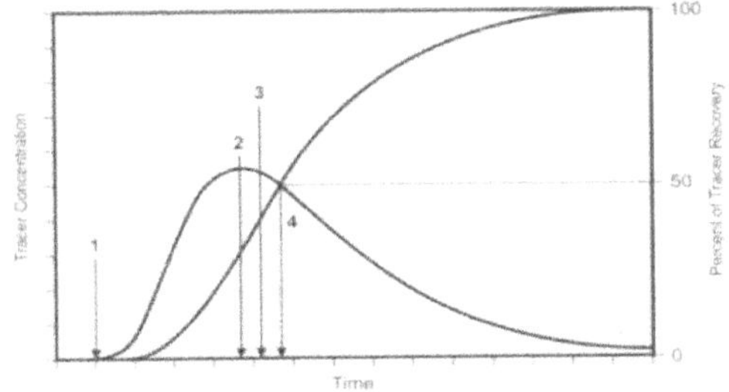

Fig. 3 Ideal passing curve for a point injection test

Interconnection test at the Boney Falls reservoir in Michigan: The tracer used during the test was Rhodamine WT and the measuring point was the West Flume, which corresponds to a leak on the west side of the dam. It can be seen from Figure that the tracer was detected only on the West Flume, which indicates no connection between the release point and Barney Creek and West Weir. The tracer arrival time was 30 minutes. The concentration then rises sharply up to 2,000 ppb indicating that the recharge point(s) is associated with karstic media. After reaching the peak, the concentration drops sharply with another smaller peak, indicating a rather concentrated flow condition. The shape of the passing curve suggests flow through a karstic media with relatively short path.h. The second peak suggests the presence of a second smaller recharge point sharing the same flow path.

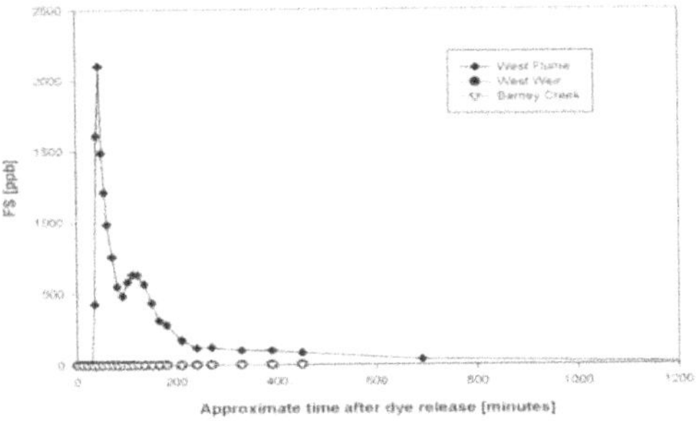

Fig. 4 Passing curve measured at west flume in boney falls, Michigan

Tests performed within the Reservoir: The main objective of the tests performed within the reservoir is identifying the point or area of infiltration. In most leakage studies, lowering the reservoir to search for the infiltration point or area is not an option. The methods used for location of the infiltration zones at the bottom of the reservoir include:

- use of float drogues
- tracing of the reservoir water
- migration of tracer cloud
- filter tube
- absorbed tracers
- direct infiltration measurements.

The authors are most familiar with a direct infiltration method using a single point dilution. This method is relatively simple and does not require sophisticated equipment.

The method consists of injecting a pulse of brine solution through a hose from the boat at the surface of the reservoir. The end of the hose is encased in a diffuser connected close to a conductivity meter. The diffuser is located up near the bottom of the reservoir. After injection of the brine pulse, measurements of the conductivity with time are performed. The relative time for the brine to move out of the diffuser provides an indication of the water velocity at the bottom of the reservoir. The tests are performed on a grid and then contours of velocities at the bottom of the reservoir can be developed. This information is then used to identify the zone of infiltration.

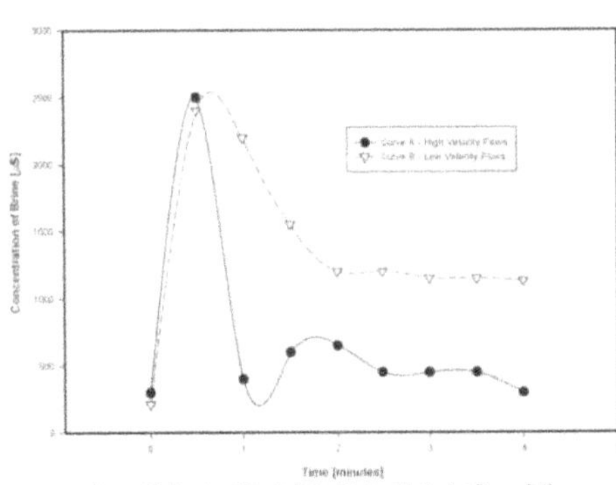

Figure 12 Results of Single Point Dilution Method at Boney Falls

Fig. 5 Results of single point dilution method at boney falls

It can be seen from above Figure that in a high velocity zone the concentration curve moves out of the diffuser in about 1 minute. On the other hand, in a low velocity zone the concentration curve do not move as fast and takes more than 4 minutes to move out.

ISBN: 978-93-8830-599-0

GPR method: The GPR (ground penetrating radar) technology is one of the advanced and effective methods to detect hidden hazards in dams due to its features of quickness, accuracy, high resolution, intutive image and less constraints on geological conditions of the field. It is a non destructive method.

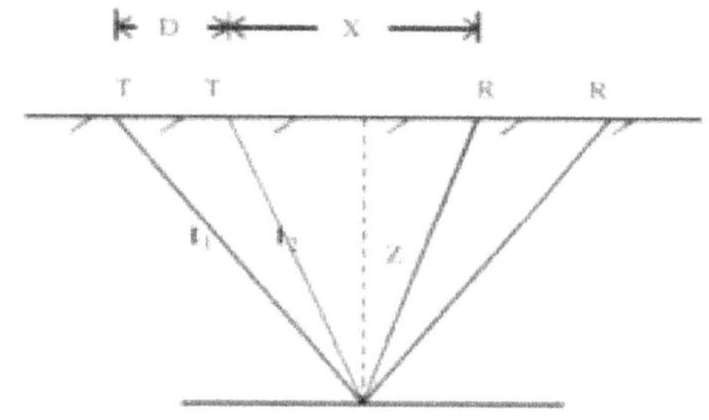

Fig. 6 Schematic diagram of GPR principle

Fig. 7 Schematic diagram of the wide-angle method to determine the velocity

Principle of GPR detection: GPR transmits high frequency electromagnetic waves (10 mhz to 2ghz) from the surface to the ground in the form of broad band short pulses through the transmitting antenna. The receiving antenna receives electromagnetic waves that have been reflected. Based on two way travel time, amplitude and wave form if received wave, image processing and data analysis are used to distinguish internal surface of underground medium or location of invisible object.

Methods and Techniques: There are generally three GPR testing methods, which are profile, wide angle and transmission methods.

Profile method is the most common detection method. The measuring can be showed by time profile image, in which the horizontal axis is the position of the antenna on the surface and the vertical axis is the reflected wave two way trip. It can also be converted into the depth of the geological slope.

Wide-Angle method applies to the situation in which underground medium is relatively homogenous and reflects interface is close to the horizontal.

Transmission method mainly used for state detection ancient buildings, bridges and so on and close quarter inter-porous observation.

PREVENTION METHODS

- Small cracks in concrete structures should be sealed with a quality polyurethane chalk.
- Paints and coatings on reinforcing steel and metal structures.
- Apply concrete sealers or top coats to protect slightly weathered concrete.
- Poorly sealed joints in conduits should be sealed with expandable rubber gaskets or by injecting hydrophobic urethane sealant.
- Materials like high performance concrete, cementitious mortar, epoxy compounds, fibre reinforced concrete, various types of grouts, Resurfacing materials etc. are used.

CONCLUSION

Leakage from dams and reservoirs is a major safety issue that, if left unattended, may result in dam failure by different mechanisms. These techniques allow identification of recharge zones, preferential paths, and transit times, which aid in monitoring and mitigating the dam leakage. A specific procedure describing each step of leaks investigations in dams is not available.

REFERENCES

1. https://pdfs.semanticscholar.org
2. https://www.researchgate.net

ISBN: 978-93-8830-599-0

Experimental Studies on Concrete with Percentage Replacement of Cement with Rice Husk Ash and Aluminum Powder

T. Siva Sankar Reddy[1], K. Ramreddy[2], K. Ruchitha[3] and K. Karan Kumar[4]

Ibrahimpatnam, UG Student of Civil Engineering Guru Nanak Institutions Technical Campus

Sreekanthreddy899@gmail.com, ramreddy83933@gmail.com, kushigumpularuchitha@gmail.com

karankumar.karra@gmail.com

Abstract

Cement is widely noted to be most expensive constituents of concrete. The entire construction industry is in search of a suitable and effective the waste product that would considerably minimize the use of cements and ultimately reduces the construction cost. Rice husk ash (RHA) which has the pozzolanic properties is a way forward. A comparative study on effects of concrete properties when OPC of varying grades was partially replaced by RHA is discussed in this paper. Percentage replacement of OPC with RHA was 0, 10, 20, 30 and 40% respectively. The compressive strength, water absorption, shrinkage and durability of concrete were mainly studied. The study suggests that up to 20% replacement of OPC with RHA has the potential to be used as partial cement replacement, having good compressive strength performance and durability.Percentage replacement of OPC with Aluminum powder is 0.5% which is kept constant. We are using M sand instead of normal sand to decrease the cost of construction and increases quality and durability concrete. The study suggests the replacement of cement with RHA and Aluminum powder increases the compressive strength, durability, and decreases the cost of construction.

Keywords: Rice Husk Ash, ordinary Portland cement (OPC), M sand.

INTRODUCTION

In progress to develop the construction industry all over the world, many attempts have also been made by various researchers to reduce the cost of the constituents and hence total construction cost by investigating and ascertaining the waste material which could be classified as local materials.

In this project we preferred aluminum powder, quarry dust and rice husk ash as partial replacement of cement. Because during the consumption of cement, the amount of $CO2$ is produced which causes pollution problems. Quarry dust waste is generated during the crushing process of rock. The waste can cause land disposal, health and environmental problems. To reduce these problems alternative technique is replacing the cement by aluminum powder, quarry dust and rice husk ash. Not only reducing the problems but also reducing the cost of cement in construction industry

A comparatively good strength is expected when cement is replaced partially or fully with or without concrete admixtures and chemical admixtures. It is proposed to study the possibility of replacing cement with locally available waste without sacrificing the strength and workability of concrete.

LITERATURE SURVEY

Anoj et al 2017: The aim of this journal is to investigate the partial replacement of cement as quarry dust and M30 grade concrete cubes were casted for finding the compressive strength. The analysis of this experiment gives partial replacement 25% quarry dust with cement gives the compressive strength of 41.5N/mm2 and after that it goes on decreasing when compared to conventional without any replacement of quarry dust. It is clearly observed that addition of quarry dust up to certain extent will increase the compressive strength of concrete.

Selvaraj.R (2015): According to his study per capita consumption of concrete is approximately two tones; the usage includes construction of all civil engineering structures. Gas concrete is one category of concrete family falls under light weight concrete. Volume and void increase in mortar is studied by adding aluminum powder to cement mortar of proportion 1:3 with and without alkali solutions. An effort is made to optimize the percentage of aluminum powder and alkalinity of mixing solution. Various properties of concrete such as sorptivity, water absorption, micro structure, density etc. are examined for gas concrete.

Ahsan Habib, et.al., (2015): In this experiment, generation method of hydrogen gas was used for the aeration process. For various percentages of OPC, as described in the gasification method, aluminum powder is added to the slurry. To evaluate the effect of aluminum powder on concrete various tests such as density, water absorption and compressive strength test were carried out. In the case of aerated concrete, 0.15% aluminum powder helps in gaining strength.

Aruova Lyazat. Dr (2014): Aerated concretes belong to the most effective materials for fencing structures of buildings of different purpose. The component made out of aerated concrete has high strength, freeze resistance,

ISBN: 978-93-8830-599-0

low average density, fire safety, high air and vapor transmission while giving comfortable living conditions inside the building.

Ismail and waliuddin (1996): Had worked on effect of rice husk ash on high strength concrete. They studied the effect the rice husk ash (RHA) passing 200- and 325-micron sieves with 10-30% replacement of cement on strength of HSC. Test result indicated that strength of HSC decreased when cement was partially replaced by RHA for maintaining same value of workability. They observed that optimum replacement of cement by RHA was 10-20%.

MATERIALS AND METHODOLOGY

Cement: Ordinary Portland cement (OPC) of 53 grade was used in which the composition and properties follows the Indian standard organization. Cement can be defined as the bonding material having cohesive & adhesive properties which makes it capable to unite the different construction materials and form the compacted assembly.

Fine Aggregate: Fine aggregate are basically sands won from the land or the marine environment. Those particles passing through 4.75mm sieve and predominantly retained on the 75 μm (no.200) sieve are called fine aggregates. For increased workability and economy as reflected by use of less cement, the fine aggregate should have a rounded shape.

Coarse Aggregate: The maximum size aggregate used may be dependent upon some conditions. In general, 40mm size aggregate used for normal strengths and 20mm size is used for high strength concrete.

Rise Husk Ash: Rice husk ash is produced by burning the outer shell of the paddy that comes out as a waste product during milling of rice. In most of the cases, the husk produced processing of the rice is either burnt or dumped as waste material. Rice husk ash contains 90%-95% of reactive silica.

Aluminum Powder: Fine, uniform, smooth metallic powder free from aggregates available from market is used in this research and it has an atomic weight of 26.98. The aluminum powder of grade was used in this project. It had a density of 0.55g/cm^3.

MIX DESIGN

Mix proportions=1: 2.25: 1.86(cube)

Mix proportions=1:5.32:4.42(cylinders)

RESULTS AND DISCUSSIONS

COMPRESSION TEST

Compressive strength tests are carried out on cubes of size 150 mm as specified by IS: 516– 1959. All the cubes are tested under dry condition, after drying the surface of the specimens containing moisture in them. For each mix proportion (trial mix), three cubes are tested at 7 days, 14 days and 28 days using compression testing machine, the specimen properly placed and centered in the testing machine. Some Al dross specimens were cured and tested for compressive strength. The ultimate load (P) be tested immediately on removal from the water and will be noted down.

Compressive strength=load/area Fc=p/A

Fig. 1 Compression test by using UTM

Table 1 Compression test results

Sl No	Specimen No	% replacement Of Al	% replacement Of RHA	Load At 14 Days In KN	Compressive Strength on 14 days	Load At 28 Days in KN	Compressive Strength on 28 days	Avg Of 14 days	Avg Of 28 days
1	01	0%	0%	583	25.91	900	40.00	26.0	40.0
	02			588	26.30	905	40.22		
	03			583	25.91	901	40.04		
2	01	0.5%	0%	440	19.55	810	36.10	20.2	36.2
	02			480	21.33	821	36.48		
	03			450	20.00	800	35.55		
3	01	0.5%	10%	520	23.11	820	36.40	23.9	36.5
	02			560	24.80	810	36.00		
	03			537	23.86	840	37.30		
4	01	0.5%	20%	520	23.11	825	36.66	23.4	37.1
	02			510	22.66	850	37.70		
	03			527	23.42	833	37.02		
5	01	0.5%	30%	506	22.48	812	36.08	21.9	35.6
	02			487	21.64	796	35.37		
	03			492	21.86	791	35.15		

SPLIT TENSILE STRENGTH TEST

Splitting tensile strength test on concrete cylinder is a method to determine the tensile strength of concrete. The concrete is very weak in tension due to its brittle nature and is not expected to resist the direct tension. The concrete develops cracks when subjected to tensile forces.

Fig. 2 Split tensile strength test by using UTM

Table 2 Split tensile strength results

Sl No	Specimen No	% replacement Of Al	% replacement Of RHA	Load At 14 Days In KN	Tensile Strength on 14 days	Load At 28 Days in KN	Tensile Strength on 28 days	Avg Of 14 days	Avg Of 28 Days
1	01	0%	0%	143	2.02	440	6.20	2.01	5.29
	02			135	1.90	350	4.95		
	03			150	2.12	335	4.73		
2	01	0.5%	0%	160	2.26	200	2.82	2.10	2.99
	02			152	2.15	220	3.11		
	03			150	2.12	215	3.04		
3	01	0.5%	10%	160	2.26	220	311	2.40	3.18
	02			170	2.40	225	3.18		
	03			180	2.4	230	3.25		
4	01	0.5%	20%	181	2.56	235	3.32	2.52	3.24
	02			180	2.54	232	3.28		
	03			177	2.5	222	3.14		
5	01	0.5%	30%	139	1.96	211	2.98	2.03	3.00
	02			145	2.05	209	2.95		
	03			149	2.1	218	3.08		

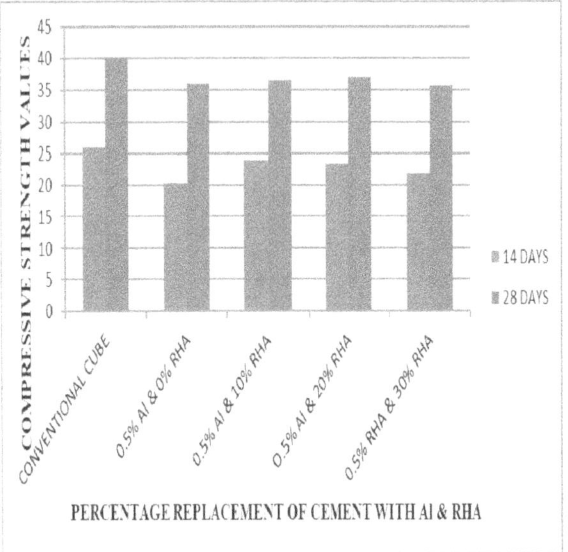

Graph 1 Comparison of compressive strength of cubes

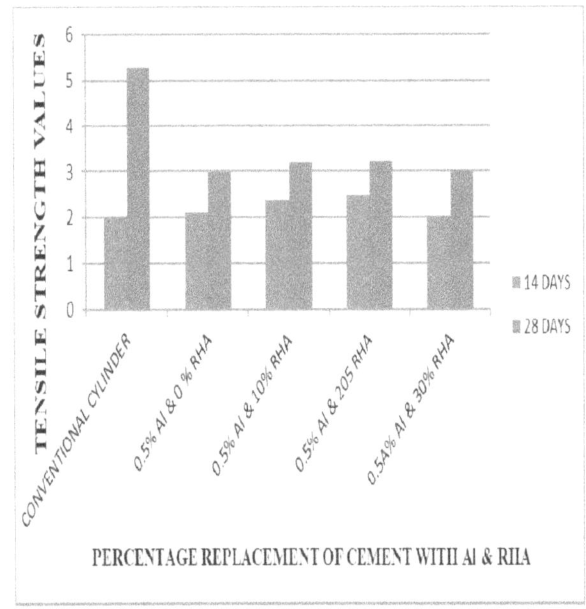

Graph 2 Comparison of split tensile strength values

By increasing of Rice husk ash percentage and with different volume fractions and keeping the aluminum powder constant compressive, split tensile, graphs are plotted. The Load carrying capacity was found to be increasing by increase in percentage of Rice husk ash up to certain extent.

Compressive strength, split tensile strength of 0%,10%,20%,30% addition of Rice husk ash and 0.5% aluminum powder is compared with conventional cubes and cylinders respectively. From the above discussion it is observed that with the increase in the

ISBN: 978-93-8830-599-0

percentage of Rice husk ash the workability of the self-compacting concrete increased.

It is concluded that aluminum powder and rice husk ash can be used as partial replacement of cement in concrete up to 20% RHA & 0.5% Al powder.

CONCLUSIONS

1. In compressive strength graph all the alumina & Rice husk ash percentages 0.5% Al & 20% RHA will give the more strength.
2. In Split tensile strength this Alumina & Rise Husk ash percentages 0.5% Al & 20% RHA will give the more strength.
3. It is observe d that if we increase the percentage of rice husk ash the workability of concrete will be increase.
4. From the above research if increase the rice husk percentage up to 20% the compressive strength will increase after that increase the rice husk percentage the strength is decrease.

REFERENCES

1. T. Subramani, A. Fizoor Rahman, " An Experimental Study on The Properties of Pet Fibre Reinforced Concrete ", International Journal of Application or Innovation in Engineering & Management (IJAIEM), Volume 6, Issue 3, March 2017, pp. 058-066, ISSN 2319 – 484
2. Manoj et.al, -(2017)-International journal of research in engineering and technology (IJRET)-" Partial Replacement of Cement by Quarry Dust" Anoj et al 2017, The aim of this journal is to investigate the partial replacement of cement as quarry dust and M30 grade concrete cubes were casted for finding the compressive strength
3. Akshay S. Pachor et al April 2017, this paper summarizes the research work on the properties of Rice Husk Ash
4. Aruova Lyazat. Dr (2014), Aerated concretes belong to the most effective materials for fencing structures of buildings of different purpose.
5. AL Khalaf and A. Yusuf (1984), Have investigated the effect of rice husk on pozzolanic behavior of rice husk ash

Strength Conversion Factors for Concrete based on Specimen Geometry, Aggregate Size and Direction of Loading

B. Sunil[1], B. Akhil Reddy[1], S. Madhav[1] and V. Srinivasa Reddy[2]
[1]B.Tech Students, [2]Professor
Department of Civil Engineering, GRIET Hyderabad.
vempada@gmail.com

Abstract

The main goal of this study is to find out the effect of effect of specimen shape and size, aggregate size and directions of loading and placement on the compressive strength of M20, M40, M60 and M80 grades of concrete. During the experimental study, different shaped and sized concrete specimens of different concrete mix designs were tested for compressive strength at 28 days. For casting the concrete samples, totally four different moulds were utilized, which were two different sizes of cubes and two different sizes of cylinders. The cubic moulds were 100 and 150 mm. The cylindrical moulds were 150×300 and 100×200 mm. So the relationship between size and shape effect on compressive strength of concrete samples is evaluated. Casted cubes and cylinders are tested for the compressive strength under axial compression on completion of 28 days as per IS: 516-1999.In this study, the effect of specimen sizes, specimen shapes, and placement directions on concrete compressive strengths for various grades widely used is evaluated. In addition, correlations between compressive strengths with size, shape, and placement direction of the specimen are investigated. It was found that with the increase of the size of the concrete specimen, compressive strength tends to decrease. The effect of grade of concrete on the shape effect of the compressive strength decreases as the specimen size increases regardless of strength level. Conversion factors of 0.80 to 0.90 were suggested for converting compressive strength of cylinders to compressive strength of cubes. For cubes, when the placement direction is parallel to the loading direction, the compressive strength is higher than the normal case. As aggregate size increases, compressive strength is found to be increasing.

Keywords: specimen shape and size, Conversion factors, strength correction factors, aggregate size, loading direction.

I. INTRODUCTION

Determining the compressive strength of concrete is one of the most crucial stages of construction works as the mechanical and durability properties of the concrete depends on the compressive strength of the concrete. To maintain a control on the quality of concrete, concrete samples are casted in various moulds according to various standards adopted in various countries. On the other hand, it is known that different shapes and sizes of concrete samples can cause variations in results of compressive strength. This paper is focused to study the effect of specimen sizes and shapes on compressive strength of concrete. Many previous studies and experimental investigations have been conducted in order to find out how changing specimen shape and size could influence the results. In this study conversion factors are proposed to convert the compressive strength results of different specimens based on size, shape, aggregate size and placement direction. For most countries, sizes and shapes of test specimens to determine the compressive strength of the concrete are different. However, commonly used specimens are cylinders and cubes. Due to the differences in the shape, height/diameter ratio, and end restraint due to the machine platens, cylinder and cube strengths obtained from the same batch of concrete could differ.

II. OBJECTIVES OF THE PRESENT WORK

1. To evaluate the effect of specimen shape and size on the compressive strength of M20, M40, M60 and M80 grade concretes.
2. To suggest strength conversion factors for M20, M40, M60 and M80 grade concrete specimens of different shapes and sizes.
3. To compare the compressive strengths of concrete cubes of M20, M40, M60 and M80 grades when subjected to loading parallel and perpendicular to the placement of concrete.
4. To evaluate the effect of aggregate size on the compressive strength of M20, M40, M60 and M80 grade concrete specimens of different shapes and sizes.

III. MIX PROPORTIONS

The mix proportions of M20 and M40 grade concrete are designed using IS: 10262-2009 and mix proportions for M60 and M80 grade concrete are designed using Entroy and Shacklock's empirical graphs. The mix proportions and materials required for one cubic meter are given in table 1 below.

ISBN: 978-93-8830-599-0

Table 1 Mix proportions and quantities per 1 cu. M. Of m20, m40, m60 and m80 grade concretes

Grade of the Concrete	Mix proportions	Cement kg/m³	Micro-silica kg/m³	Fly ash	Fine Aggregate kg/m³	Coarse Aggregate kg/m³	Water L
M20	1: 2.90: 3.14: 0.54	320.4	-		927.3	1005.4	173 L
M40	1: 2.25: 2.56: 0.42	390.7	-		876	999.7	164.1 L
M60	1:1.60:1.84:0.26 (Micro Silica - 6% bwc*) (Superplasticizer(SP) – 1% bwc*)	436	27.8	50.34	819.9	948	130.6 L
M80	1:1.12:1.40:0.23 (Micro Silica - 10% bwc*) (Superplasticizer(SP) – 1.2% bwc*)	451.8	50.2	155.2	732.3	914.1	115.5 L

*bwc-by weight of cement

IV. LATERAL SURFACE AREA/VOLUME RATIO

The table 2 presents the lateral surface/volume ratio and load application area for different concrete specimen shapes and sizes.

V . EFFECT OF SPECIMEN SHAPE AND SIZE ON COMPRESSIVE STRENGTH OF CONCRETE

The figure 1 presents variation of compressive strength of various grades of cylindrical and cube concrete specimens.

Table 2 Lateral surface/volume ratio and load application area for different specimens

Sample typeand size in mm	Lateral surface area/volume ratio (mm⁻¹)	Area on which load applied mm²
Cyl.100×200	0.050	7850
Cyl.150×300	0.033	17662.5
Cube 100	0.060	10000
Cube 150	0.040	22500

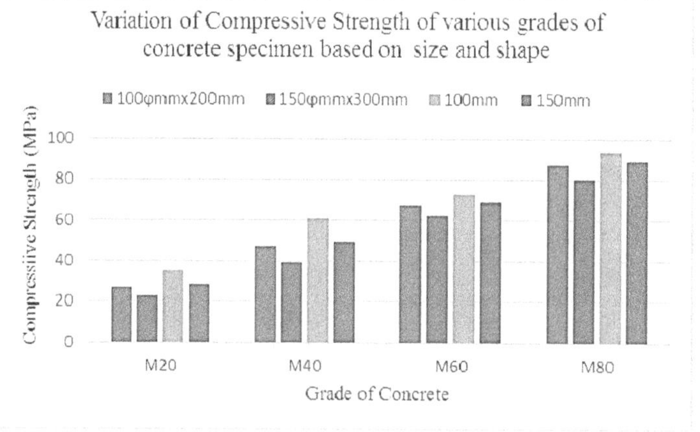

Fig. 1 Variation of compressive strength of various grades of cylindrical and cube concrete specimens.

ISBN: 978-93-8830-599-0

VI. COMPRESSIVE STRENGTH CONVERSION

FACTORS

The table 3 presents the compressive strength conversion factors based on various shapes and sizes of specimens of M20 and M40 grade.

The table 4 presents the compressive strength conversion factors of cylindrical and cube concrete specimens of M60 and M80 grade.

To convert compressive strength of the specimen to the strength of other specimens (with different cross sectional area), the specimen's compressive strength should be multiplied by the values which were obtained from the table 5.

Table 3 Compressive Strength Conversion Factors of Cylindrical and Cube Concrete Specimens Of M20 And M40 Grade

		Cylinder		Cube	
		100☐mmx200mm	150☐mmx300mm	100mm	150mm
Cylinder	100☐mmx 200mm	1.00	1.16	0.84	0.95
	150☐mmx 300mm	0.86	1.00	0.70	0.80
Cube	100mm	1.19	1.43	1.00	1.24
	150mm	1.05	1.25	0.80	1.00

Table 4 Compressive Strength Conversion Factors of Cylindrical and Cube Concrete Specimens of M60 And M80 Grade

		Cylinder		Cube	
		100☐mmx 200mm	150☐mmx 300mm	100mm	150mm
Cylinder	100☐mmx 200mm	1.00	1.05	0.90	0.98
	150☐mmx 300mm	0.95	1.00	0.89	0.90
Cube	100mm	1.11	1.12	1.00	1.05
	150mm	1.02	1.11	0.95	1.00

Table 5 Compressive Strength Multiplication Factors of Cylindrical and Cube Concrete Specimens of Different Grade Concretes

		Cylinder		Cube	
		100☐mmx200mm	150☐mmx300mm	100mm	150mm
Compressive Strength	M20	0.95	0.80	1.24	1
	M40	0.95	0.80	1.24	1
	M60	0.98	0.90	1.05	1
	M80	0.98	0.90	1.05	1

VII. EFFECT OF LOADING AND PLACEMENT DIRECTIONS ON COMPRESSIVE STRENGTHS

The table 6 presents the compressive strengths based on loading and placement directions.

Table 6 Compressive Strengths Based On Loading and Placement Directions For Different Grade Concretes Of Various Specimen Sizes

Specimen shape	Specimen size (mm)	Grade of the concrete	Compressive Strength (MPa)	
			Placement direction parallel to loading direction	Placement direction perpendicular to loading direction
Cube	100	M20	36.90	35.14
		M40	64.17	61.11
		M60	83.35	72.48
		M80	107.61	93.57
	150	M20	29.76	28.34
		M40	51.74	49.28
		M60	79.38	69.03
		M80	102.48	89.11

ISBN: 978-93-8830-599-0

VIII. EFFECT OF COARSE AGGREGATE SIZE ON COMPRESSIVE STRENGTHS OF CONCRETE

The table 7 presents the compressive strengths of various grades of concrete specimens of different size and shape made with 20mm and 10mm coarse aggregate.

Table 7 Compressive of Various Grades of Concrete Specimens of Different Size and Shape Made with 20mm and 10mm Coarse Aggregate

Specimen shape	Specimen size (mm)	Grade of the concrete	(20mm CA) Compressive Strength (MPa)	(10mm CA) Compressive Strength (MPa)
Cylinder	100×200 (diameter ×length)	M20	26.92	24.23
		M40	46.82	42.14
		M60	67.65	60.89
		M80	87.33	78.60
	150×300 (diameter ×length)	M20	22.67	20.40
		M40	39.42	35.48
		M60	62.13	55.92
		M80	80.20	72.18
Cube	100	M20	35.14	31.63
		M40	61.11	55.00
		M60	72.48	65.23
		M80	93.57	84.21
	150	M20	28.34	25.51
		M40	49.28	44.35
		M60	69.03	62.13
		M80	89.11	80.20

IX. DISCUSSIONS

The cube with a 100mm size had the highest compressive strength of all. For this reason, that size of test cube should be more representative, and any cube size below 100 mm cube seems to be unsuitable for testing the compressive strength of concrete.

The ratio of the cube strength to cylinder strength with increasing compressive strength of concrete decreases progressively from 1.24 to 1.05. Above-mentioned 1.24 and 1.05 are the ratios corresponding to the cylinder compressive strengths of 20-40 MPa and 60- 80MPa, respectively

The compressive strength was insignificantly affected by changing l/d ratio as the strength of concrete increases. For high-strength concrete (HSC), the influence of specimen shape is not so significant.

For cubes, size effect difference with placement direction is insignificant. This is because the failure of cubes occurs not due to lateral expansion but due to crushing and the lateral expansion is restrained due to the end restraint occurred by the machine platens.

For cubes, when the placement direction is parallel to the loading direction, the compressive strength is higher than the normal case.

HSC shows more brittle behavior than NSC. This means that the size of fracture process zone and the size effect of HSC, respectively, are smaller and more apparent than NSC.

For cubes, because the face of casting is rough compared to the other faces. When load acted on it, it doesn't share the load uniformly and hence the face other than the face of casting is tested. Casting face with

ISBN: 978-93-8830-599-0

the load uneven may cause the cracks prior giving the low compressive strength.

In the uniaxial compression test, smaller contact area between the specimen surface and steel platen of test machine resulting in lower friction-based shear forces in small specimens.It has been observed that the restraining effect due to friction between the specimen and the platens extends over the entire height of the cube. But in the case of the cylinder, it remains unaffected. Hence, it is clear that the total stress that will be created in the cube will be higher compared with the cylinder specimen. This will result in a higher value of the compressive strength in cubes than the cylinder even by employing the same concrete mix.

X. CONCLUSIONS

The current work investigated a narrative approach of finding the effect of specimen shape and size, aggregate size and directions of loading and placement on the compressive strength. Based on the results reported in this work and key findings during the experimental investigations, the following conclusions are drawn:

1. With increase of the size of the concrete specimen, compressive strength tends to decrease. The cube with a 100mm size had the highest compressive strength of all.

2. The effect of grade of concrete on the shape effect of the compressive strength decreases as the specimen size increases regardless of strength level. More specifically, for M60 and M80, the difference of compressive strengths between cylinders and cubes is more rapidly disappeared than that of M20 and M40 grades of concrete.

3. Conversion factors of 0.80 to 0.90 were suggested for converting compressive strength of cube to compressive strength of cylinder. The ratio of the cube strength to cylinder strength with increasing compressive strength of concrete decreases progressively from 1.24 to 1.05.Abovementioned 1.24 and 1.05 are the ratios corresponding to the cylinder compressive strengths of 20-40 MPa and 60- 80MPa, respectively

4. For cubes, when the placement direction is parallel to the loading direction, the compressive strength is higher than the normal case. For normal strength concrete (NSC) (M20 and M40 grades), the size effect difference with displacement direction is not distinct compared to high strength concrete (HSC) (M60 and M80 grades).

5. To study the wall effect, compressive strength results of different specimens have been plotted against surface/volume ratio of samples. By increasing the ratio of lateral surface to volume, compressive strength has a steady decreasing trend. Increasing the ratio of surface/volume, in fact means that for each cubic meter of the specimens, there is an increasing amount of surface area. As a result, specimens with higher ratio of surface area/volume are willing to have less compressive strength.

6. Compressive strength results were strongly influenced by their specimen sizes and wall effect. When the concrete bonds are weaker, the results of compressive strengths seem to be more uniform, since the compressive strength of the samples are controlled by their bonds' strength.

7. By decreasing the ratio of lateral surface/volume of specimens, the compressive strength increases.

8. The cylinder specimen indicated a small change in compressive strength when the l/d ratio changed from 1.0 to 2.0; compared to cubes. In fact, compressive strength for cylinder increased as l/d increased; whereas for cube, there was a reduction in compressive strength.

9. As aggregate size increases, compressive strength is found to be increasing.

REFERENCES

1. Arioz, O., Ramyar, K., Tuncan, M., Tuncan, A., & Cil, I. (2007). Some factors influencing effect of core diameter on measured concrete compressive strength. ACI Materials Journal , 291-296.

2. Bazant, Z.P., 1984. Size effect in blunt fracture; concrete, rock, metal. J. Eng. Mech. ASCE 110 (4), 518–535.

3. Bazant, Z.P., 1987. Fracture energy of heterogeneous material and similitude. In: SEM-RILEM International Conference on Fracture of Concrete and Rock, pp. 390–402.

4. Bazant, Z.P., 1993. Size effect in tensile and compressive quasibrittle failures. In: JCI International Workshop on Size Effect in Concrete Structures, pp. 141–160.

5. Bazant, Z.P., Xiang, Y., 1997. Size effect in compression fracture: splitting crack band propagation. J. Eng. Mech. ASCE 123 (2), 162–172.

6. Benjamin, J.R., Cornell, C.A., 1970. Probability, Statistics, and Decision for Civil Engineers. McGraw-Hill Publishing Company, New York (Section 4.3). CEB-FIP Model Code, 1990, 1993. Comite Euro-International du Beton (CEB), Bulletin D'Information No. 203/205, Lausanne, 437 pp.

7. BESHR, H., A. A. ALMUSALIAM and M. MAsLEHUDDIN. 2003. Effect of coarse aggregate quality on the mechanical properties of high strength concrete. Construction and Buikling Materials 17: 97-103.

8. Carpinteri, A., Chiaia, B., and Ferro, G. (1998). Size effect on nominal tensile strength of concrete structures: Multifractality of material ligaments and dimensional transition from order to disorder. Materials and Structures, 28:311–317.

9. Chin, M.S., Mansur, M.A., Wee, T.H., 1997. Effects of shape, size, and casting direction of specimens on stress–strain curves of high-strength concrete. ACI Mater. J. 94 (3), 209–219.

10. Del Viso, J., Carmona, J., & Ruiz, G. (2008). Shape and size effects on the compressive strength of high-strength concrete. Cement and Concrete Research , 386–395.

11. Del Viso, J., J. Carmona and G. Ruiz, 2008. Shape and size effects on the compressive strength of high strength concrete. Cement Concrete Res., 38(3): 386-395.

12. Del Viso, J.R., Carmona, J.R. and Ruiz, G. "Shape and size effects on the compressive strength of highstrength concrete," Cement and Concrete Research, 38, 2008, 386-395.

13. Elwet, D.J. and G. Fu, 1995. Compression Testing of Concrete: Cylinders vs. Cubes. Transportation Research and Development Bureau, New York State Department of Transportation, Albany.

14. Gonnerman, H.F., 1925. Effect of size and shape of test specimen on compressive strength of concrete. Proc. ASTM, 25: 237-250.

15. Gonnerman, H.F., 1925. Effect of size and shape of test specimen on compressive strength of concrete. ASTM Proc. 25, 237–250.

16. Gyengo, T., 1938. Effect of type of test specimen and gradation of aggregate on compressive strength of concrete. J. ACI 33, 269–283.

17. IMSL, 2003. Library, 8th ed. IMSL, Inc. Jansen, D.C., Shah, S.P., 1997. Effect of length on compressive strain softening of concrete. J. Eng. Mech. ASCE 123 (1), 25–35.

18. J.R. Del Viso, J.R. Carmona, G. Ruiz, Shape and Size Effects on The Compressive Strength Of High-Strength Concrete, Cement And Concrete Research 3 (2008) 386-395.

19. Kim, J.H.J., Yi, S.T., Kim, J.K., 2004. Size effect of concrete members applied with flexural compressive stresses. Int. J. Fracture 126 (1), 79–102.

20. Kim, J.K., Eo, S.H., 1990. Size effect in concrete specimens with dissimilar initial cracks. Mag. Concrete Res. 42 (153), 233–238.

21. Kim, J.K., Yi, S.T., Kim, J.H.J., 2001. Effect of specimen sizes on flexural compressive strength of concrete. ACI Struct. J. 98 (3), 416–424.

22. Kim, J.K., Yi, S.T., Park, C.K., Eo, S.H., 1999. Size effect on compressive strength of plain and spirally reinforced concrete cylinders. ACI Struct. J. 96 (1), 88–94.

23. Kim, J.K., Yi, S.T., Yang, E.I., 2000. Size effect on flexural compressive strength of concrete specimens. ACI Struct. J. 97 (2), 291–296.

24. M. Tokyay, M. Ozdemir, Specimen Shape and Size Effect On the Compressive Strength Of Higher Strength Concrete. Cement And Concrete Research, 27 (8) (1997) 1281-1289

25. Malaikah, A. S. (2009). Effect of Specimen Size and Shape on the Compressive Strength of High Strength Concrete. Pertanika Journal of Science & Technology , 87-96.

26. Malaikah, A.S., 2009. Effect of specimen size and shape on the compressive strength of high strength concrete. Pertanika J. Sci. Technol., 13(1): 87-96.

27. Mansur, M.A., Islam, M.M., 2002. Interpretation of concrete strength for nonstandard specimens. J. Mater. Civil Eng. ASCE 14 (2), 151–155.

28. Markeset, G., 1995. A compressive softening model for concrete. In:Wittmann, F.H. (Ed.), Fracture Mechanics of Concrete Structures, FRAMCOS-2, AEDIFICATIO Publishers, pp. 435–443.

29. Markeset, G., Hillerborg, A., 1995. Softening of concrete in compression localization and size effects. Cement Concrete Res. 25 (4), 702–708.

30. Murdock, J.W., Kesler, C.E., 1957. Effect of length to diameter ratio of specimen on the apparent compressive strength of concrete. ASTM Bull. 221, 68–73.

31. S.T. Yi, E.I. Yang, J.C. Choi, Effect Of Specimen Sizes, Specimen Shapes and Placement Directions on Compressive Strength of Concrete. Nuclear Engineering And Design 236 (2006) 115-127.

32. Tokyay, M. and M. Ozdemir, 1997. Specimen shape and size effect on the compressive strength of higher strength concrete. Cement Concrete Res., 27(8): 1281-1288.

33. Tokyay, M., & Ozdemir, M. (1997). Specimen Shape and Size Effect on the Compressive Strength of Higher Strength Concrete. Cement and Concrete Research , 1281-1288.

34. Turkel, A., & Ozkul, M. H. (2010). size and wall effects on compressive strength of concrete. ACI Materials Journal , 372-379.

35. Turkel, A., & Ozkul, M. H. (2010). Size and Wall Effects on Compressive Strength of Concretes. ACI Materials Journal , 372- 379.

36. Yi, S.T., Kim, J.H.J., Kim, J.K., 2002. Effect of specimen sizes on ACI rectangular stress block for concrete flexural members. ACI Struct. J. 99 (5), 701–708.

37. Zheng, J.J. and C.Q. Li, 2002. Three-dimensional aggregate density in concrete with wall effect. ACI Mater. J., 99(6): 568-575.

ISBN: 978-93-8830-599-0

Experimental Investigation on Mechanical Properties of Geopolymer Concrete made with Partial Replacement of Coarse Aggregate by Recycled Aggregate

T. Srinivas[1], G. Sravani[2], D. Divya[3] and A. Mounika[4]

[1]Professor, [2, 3 &4]B.Tech Students, Department of Civil Engineering, GRIET, Hyderabad.

Abstract

Concrete is the world's most versatile, durable and reliable construction material. There are many ecological issues connected with the manufacture of OPC, at the same time availability of natural coarse aggregate is also becoming scarcity. Hence, it is inevitable to find an alternative material to the existing most resource consuming Portland cement and natural aggregates. Geopolymer concrete is an innovative construction material which shall be produced by the chemical action of inorganic molecules and made up of fly ash, GGBS, fine aggregate, coarse aggregate, and an alkaline solution of sodium hydroxide and sodium silicate, plays a significant role in its environmental control of greenhouse effects. The main objective of this paper is to study the mechanical properties of geopolymer concrete of grade G40 when natural coarse aggregate is replaced with recycled aggregate in different proportions such as 10%, 20%, 30%, 40% and 50% and also to compare the results of geopolymer concrete made with recycled coarse aggregates with geopolymer concrete of natural coarse aggregate and controlled concrete of respective grade. It has been observed that the mechanical properties are enhanced in geopolymer concrete, both in natural coarse aggregate and recycled coarse aggregate up to 30% replacement when it is compared with the same grade of controlled concrete.

Keywords: Geopolymer Concrete, Recycled Aggregate, Alkaline Solutions, controlled concrete.

1.   INTRODUCTION

Global warming and climate change are increasingly important issues. Carbon dioxide is one of the most detrimental greenhouse gases with 65 percent of global warming caused by carbon dioxide. Portland cement contributes significantly to greenhouse gas emissions approximately 0.8-1 tonne of carbon dioxide per tonne of cement.

Increased focus is also being placed on recycling as the world's natural resources are being depleted and the amount of waste being disposed of into landfill is increasing globally. Therefore, as the industrial development process continues, the re-use of construction and demolition waste is becoming increasingly important and various solutions have been researched for high-volume use of recycled concrete.

Geopolymer is a promising alternative binder to Portland cement. It is produced mostly from by-product materials such as fly ash and blast furnace slag. There are many efforts are being made to reduce the use of Portland cement in concrete by finding alternative binders to Portland cement this include the utilization of supplementary cementing materials fly ash, granulated blast furnace slag, rice-husk ash and metakaolin. In 1972, Joseph Davidovits coined the name ''geopolymers'' to describe the zeolite like polymers. Geopolymers are the alumina-silicate polymers which consist of amorphous and three dimensional structures formed from the geopolymerisation of alumina-silicate monomers in alkaline solution. Investigations have been carried out on calcined clays (e.g., metakaolin) or industrial wastes (e.g., fly ash or metallurgical slag). A reaction pathway involving the polycondensation of orthosialiate ions (hypothetical monomer) is proposed by Davidovits.

This paper presents experimental results of geopolymer concrete of grade G40 when natural coarse aggregate is replaced with recycled aggregate in different proportions such as 10%, 20%, 30%, 40% and 50% and also to compare the results of geopolymer concrete made with recycled coarse aggregates with geopolymer concrete of natural coarse aggregate and controlled concrete of respective grade.

2.   MATERIALS

2.1   Ordinary Portland Cement

In the experimental investigations, 53-grade of ordinary Portland cement is used. The cement thus procured was tested for physical properties in accordance with the IS: 4031-1968 and found to be conforming various specifications of IS 12629-1987.

2.2   Fine Aggregate

In the present investigation, fine aggregate used is obtained from local sources. The sand is made free from clay matter, silt, and organic impurities and sieved on 4.75mm IS sieve. The physical properties of fine aggregate like specific gravity, bulk density, gradation and fineness modulus are tested in accordance with IS: 2386. Grain size distribution of sand shows it is close to Zone II of IS 383-1970.

2.3   Coarse Aggregate

The crushed angular aggregate of 20mm maximum size obtained from the local crushing plants is used as coarse aggregate in the present study. The physical properties

ISBN: 978-93-8830-599-0

of coarse aggregate such as specific gravity, bulk density, flakiness and elongation index are tested in accordance with IS: 2386-1963.

2.4   Fly Ash

In the present study of work, the Class F-fly ash is used, which is obtained from Vijayawada thermal power station in Andhra Pradesh.

2.5   Ground Granulated Blast Furnace Slag

Ground Granulated Blast Furnace Slag (GGBS) is a byproduct of the steel industry. Blast furnace slag is defined as "the non-metallic product consisting essentially of calcium silicates and other bases that is developed in a molten condition simultaneously with iron in a blast furnace". About 15% by mass of binders was replaced with GGBS.

2.6   Recycled Aggregate Concrete (RAC)

Construction and demolition waste contributes up to 40 percent of all waste generated worldwide. The majority of recycled aggregate that is used in Australia is recycled concrete aggregate (RCA) produced form construction and demolition waste, as it is the most suitable replacement of natural coarse aggregate. Utilizing recycled aggregate can result in around 60 percent less waste and 50 percentages less mineral depletion per cubic metre of concrete produced. The strength of ordinary Portland cement concrete utilizing recycled aggregate depends largely on the percentage of recycled aggregate used.

Fig. 1 Recycled Coarse Aggregate (Source: Laboratory)

2.7   Water

Water free from chemicals, oils and other forms of impurities is used for mixing of concrete as per IS: 456:2000.

2.8   Sodium Hydroxide

Sodium Hydroxide is one of the major ingredients of geopolymer concrete. The following are the specifications of Sodium hydroxide pellets and this material is procured from the local laboratory chemical

vendors in Hyderabad. Specifications are tabulated in table 1 as given by the suppliers.

Table 1 Shows Physical properties of NaOH

Molar mass	40 gm/mol
Appearance	White solid
Density	2.1 gr/cc
Melting point	318°C
Boiling point	1390°C
Amount of heat liberated when dissolved in water	266 cal/gr

2.9   Sodium Silicate Solution

Sodium silicate solution is a type of alkaline liquid plays an important role in the polymerisation process. This material is procured from the local laboratory chemical vendors in Hyderabad. Specifications are tabulated in table 2 as given by the suppliers.

Table 2 Properties of Na_2SiO_3 Solution

Specific gravity	1.57
Molar mass	122.06 gm/mol
Na_2O (by mass)	14.35%
SiO_2 (by mass)	30.00%
Water (by mass)	55.00%
Weight ratio (SiO_2 to Na_2O)	2.09
Molarity ratio	0.97

2.10   Super Plasticizer

Super plasticizer GLENIUM B233 of Fosroc chemical India Ltd. was used as water reducing admixture, it increases workability.

3.   EXPERIMENTAL INVESTIGATION

3.1   General

An objective of this paper is to study the mechanical properties of geopolymer concrete of grade G40 when natural coarse aggregate is replaced with recycled aggregate in different proportions such as 10%, 20%, 30%, 40% and 50% and also to compare the results of geopolymer concrete made with recycled coarse aggregates with geopolymer concrete of natural coarse aggregate and controlled concrete of respective grade. The cubes of size 100mm×100mm×100mm were cast and after one day rest period, the specimens were cured in an oven at 60°C for 24 hours and the remaining period cured in sun light until the testing is done.

3.2   Mixing and Casting of Geopolymer Concrete

Geopolymer concrete is prepared by using the same procedure whatever is used in the conventional concrete. In the laboratory, the fly ash and the aggregates were mixed together in dry by using a pan mixer for about two minutes, then the alkaline liquid was mixed with the super plasticizer and extra water if any. The liquid component of the mixture was then added to the dry material and the mixing continued usually for another two minutes. The fresh concrete was cast and compacted by the usual methods used in the case of conventional concrete. The workability of the fresh concrete was measured by means of the conventional slump test.

4. TEST RESULTS

4.1 Compressive Strength of CC and GPC

Compressive strength test is the direct measure of the strength of concrete. From the results given in table 3, it is evident that the strength of Geopolymer concrete with Natural Aggregate is slightly more than that of Conventional concrete with Natural Aggregate.

Table 3 Compressive strength of CC & GPC

	CC (N/mm^2)	GPC (N/mm^2)
3 days	28.75	46.71
7 days	40.25	53.91
28 days	56.72	58.30

4.2 Compressive Strength of Conventional Concrete with Recycled Aggregate (RAC)

Compressive strength test is performed with recycled aggregate of varying percentages i.e., 10%, 20%, 30%, 40% & 50%. In this regard, the compressive strength value has been slightly decreasing up to 30% of replacement with recycled aggregate and thereon it started to decrease drastically. From the results given in table 4, it can be inferred that, 30% replacement is the optimum percentage of utilization of recycled aggregate in the concrete, because at this percentage the strength is higher than target mean strength.

Table 4 Compressive strength of Conventional Concrete with varying percentages of Recycled aggregate

	0%	10%	20%	30%	40%	50%
3 days(N/mm^2)	28.45	26.80	24.91	24.56	23.26	21.98
7 days(N/mm^2)	39.50	37.66	35.50	34.65	32.70	31.50
28 days(N/mm^2)	56.72	53.81	51.04	49.75	46.72	45.08

4.3 Compressive Strength of Geopolymer Concrete with Recycled Aggregate (RAGPC)

Compressive strength test is performed with recycled aggregate of varying percentages i.e., 10%, 20%, 30%, 40% & 50%. In this regard, the compressive strength value has been slightly decreasing up to 30% of replacement with recycled aggregate and thereon it started to decrease drastically. From the results given in table 5, it can be inferred that, 30% replacement is the optimum percentage of utilization of recycled aggregate in the concrete, because at this percentage the strength is higher than target mean strength.

Table 5 Compressive strength of Geopolymer Concrete with varying percentages of Recycled aggregate

	0%	10%	20%	30%	40%	50%
3 days(N/mm^2)	46.71	45.56	43.65	40.77	38.24	38.02
7 days(N/mm^2)	53.91	51.18	48.44	46.31	43.98	42.72
28 days(N/mm^2)	58.30	56.25	53.24	50.34	47.81	46.95

4.4 Comparision of Test Results

From Table 6, it is evident that the strength of Geopolymer concrete is more than that of Conventional concrete. When the conventional and geopolymer concrete replaced with recycled aggregate the strengths have been slightly decreased up to 30% and thereon deceased drastically. In recycled aggregate concrete and recycled aggregate geopolymer concrete, the strength has been decreased slightly but still it is more than target mean strength, so optimum strength results at 30% replacement of recycled aggregate geopolymer concrete can be considered.

Table 6 Comparison of Test Results

	CC	GPC	RAC	RAGPC
3 days (N/mm^2)	28.75	4671	24.17	40.15
7 days (N/mm^2)	40.25	53.91	34.52	46.31
28 days (N/mm^2)	56.72	58.30	49.75	50.34

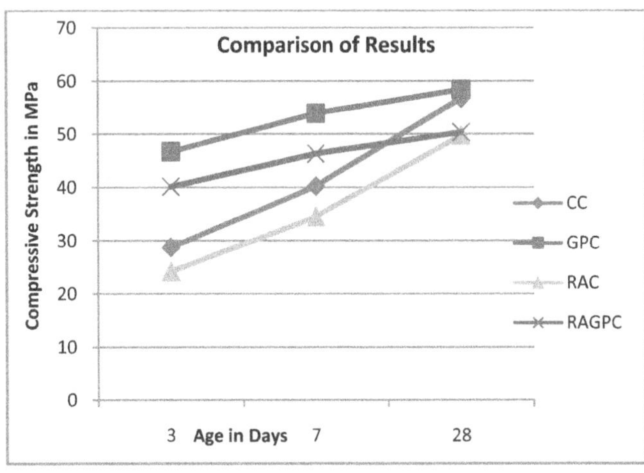

Fig. 2 Comparison of Test Results

5. CONCLUSIONS

The following specific conclusions can be drawn from the present experimental investigation:

(i)   From the results, it is concluded that the strength of Geopolymer concrete with Natural Aggregate is 2.79% more than that of Conventional concrete with Natural Aggregate.

(ii) It is observed that the strength of Conventional concrete with Recycled Aggregate is 12.29% less than that of Conventional concrete with Natural Aggregate at 30% replacement, but still it is higher than target mean strength so it can be considerable.

(iii) The strength of Geopolymer concrete with Recycled Aggregate is 13.65% less than that of Geopolymer concrete with Natural Aggregate at 30% replacement, but still it is higher than target mean strength so it can be considerable.

(iv) The replacement of recycled aggregate in concrete gives better results till 30% replacement, so it is an alternative solution for disposal of C & D waste.

(v)   The geopolymer concrete gives an early strength when cured in an oven, so it can be preferred wherever early strength is required.

REFERENCES

1. Chamila Gunasekara, Sujeeva Setunge (2018) "Engineering Properties of Geopolymer Aggregate Concrete", Journal of Materials in Civil Engineering, 30(11), pp.040182991-0401829911.

2. D.V. Reddy, Jean-Baptiste Edouard(2013) "Durability of Flyash based Geopolymer Structural Concrete in the Marine Environment", Journal of Materials in Civil Engineering, 25(6), pp.781-787.

3. Gaurav Nagalia, Yeonho Park(2016) "Compressive Strength and Micro structural Properties of Flyash-Based Geopolymer Concrete", Journal of Materials in Civil Engineering, 28(12), pp.040161441-0401614411.

4. Kolli. Ramujee, M. PothaRaju(2017) "Mechanical Properties of Geopolymer Concrete Composites", Materials Today: Proceedings 4, pp.2937-2945.

5. Kunal Kupwade-Patil, Erez N. Allouche(2013) "Impact of Alkali Silica Reaction on Flyash-Based Geopolymer Concrete", Journal of Materials in Civil Engineering, 25(1), pp.131-139.

6. M.S. Laskar, S. Talukdar(2017) "Influence of Super plasticizer and Alkali Activator Concentration on Slag-Flyash based Geopolymer", Urbanization Challenges in Emerging Economies, pp.330-337.

7. M. Elchalakani, M. Dong, A. Karrech(2018) "Development of Flyash-and Slag- Based Geopolymer Concrete with Calcium Carbonate or Micro silica", Journal of Materials in Civil Engineering, 30(12), pp.040183251-0401832514.

8. Majidi, B.(2009) "Geopolymer Technology, from Fundamentals to Advanced Applications: A Review." Material Technology, 24(2), pp.79–87.

9. Mohamed Elchalakani, Hakan Basarir(2017) "Green Concrete with High Volume Flyash and Slag with Recycled Aggregate and Recycled Water to build Future Sustainable Cities", Journal of Materials in Civil Engineering, 29(2), pp.040162191-0401621912.

10. Mostafa Vafaei, Ali Allahverdi(2018) "Acid-Resistant Geopolymer based on Flyash-Calcium Aluminate Cement", Journal of Materials in Civil Engineering, 30(7), pp.040181431-0401814311.

11. Najif Ismail, Hilal El-Hassan(2018) "Development and Characterization of Flyash-Slag blended Geopolymer Mortar and Lightweight Concrete", Journal of Materials in Civil Engineering, 30(4), pp.040180291-0401802914.

12. Nella Shiva Kumar, Gone Punneshwar(2018) "Strength Properties of the Geopolymer Concrete as Partial Replacement of Recycled Aggregate", International journal of emerging technologies in engineering research, 6(2), pp.63-68.

13. O. Sanusi, B. Tempest and V. O. Ogunro(2011) "Mitigating Leachability Flyash based Geopolymer Concrete using Recycled Coarse Aggregate", Geo-Frontiers, pp.1315-1324.

14. P. Chindaprasirt, T. Chareerat(2011) "High-Strength Geopolymer using Fine High-Calcium Flyash", Journal of Materials in Civil Engineering, 23(3), pp.264- 270.

15. Pawan Anand Khanna, Durga Kelkar(2017) "Study on the Compressive Strength of Flyash based Geopolymer Concrete", Journal of Materials Science and Engineering, 263, pp.1-5.

16. Rangan, B. V., Hardjito, D., Wallah, S. E., and Sumajouw, D. M.(2005) "Studies on Flyash-Based Geopolymer Concrete", Proc., World Congress Geopolymer, Geopolymer Institute, Saint Quentin, France, pp.133–137.

17. Rao, A., Jha, K. N., and Misra, S.(2007) "Use of Aggregates from Recycled Construction and Demolition Waste in Concrete", Resource. Conserv.Recycl, 50(1), pp.71–81.

ISBN: 978-93-8830-599-0

Compative Study on Acid Resistance of Geopolymer Concrete to Conventional Concrete

P. Sai Shraddha[1] and M. Madhuri[2]

[1,2]Assistant Professor, Department of Civil Engineering, Vignan Institute of Technology and science, Deshmukhi, Hyderabad

Abstract

Growth of Environmental pollution is a serious issue to be addressed in the global scenario. Among the Major industries contributing to the Pollution, a Cement industry contributes to around 5% of Carbon dioxide emissions globally. Hence there is a need to use substitute material for cement which reduces this impact on Environment. Geopolymer concrete is made utilizing binder materials like Flyash, Silica fume, rice husk ash, GGBS, Metakaolin etc., and Alkali Activator solutions. This paper deals with the study on chemical resistance of Geopolymer concrete when compared to conventional concrete. Sulphuric acid is chosen for attack in this study since it can cause both acid and sulphuric attack on concrete matrix. The test results of present study compares the resistance offered by ordinary Portland cement concrete and geo-polymer concrete.

Keywords: Geopolymer, Flyash, Acid resistance, Sulphuric acid.

I. INTRODUCTION

Among the several vital properties of hardened concrete, Durability plays a major role as it influences the life of structures. Reaction and resistance of concrete to the aggressive environment quantifies durability of hardened concrete. Durability of hardened concrete can be characterized as its capacity to resist Chemical attack, weathering action, wear and tear while retaining its designed engineering properties. Chemical attack on concrete can be explained under acid attack, Chloride attack, Alkali aggregate reaction, Sulphate attack, carbonation etc., Sulphur attack can occur due to association between concrete and sewage water or sea water or Ground water containing sulphates. Soil contains Sulphates in many forms mainly as magnesium sulphate, calcium sulphate, Ammonium sulphate, Potassium sulphate and Sodium sulphate. Generally sulphates present in solid form does not cause any significant damage to concrete but when present in dissolved form reacts with hydration products of concrete. Calcium sulphate causes unsubstantial deterioration due its low soluble nature while magnesium sulphate causes greater deterioration by reaction with calcium hydroxide and hydrated calcium aluminates. This reaction causes deterioration of cement paste volume in the concrete matrix and converts concrete in to granulated and powdered mass.

Even though Ordinary Portland Cement is the broadly utilized component in construction industry, its resistance to Chemical attacks is a noteworthy concern. In the recent studies binders of Geopolymer concrete has been found to be more effectively resisting the chemical attack apart from environmental friendliness. Geopolymer concrete is made utilizing binder materials like Flyash, Silica fume, rice husk ash, GGBS, Metakaolin etc., and Alkali Activator solutions. Generally either Sodium Hydroxide, Sodium silicate solutions or Potassium hydroxide , Potassium silicate solutions can be used as Alkali activator Solutions. In this study Flyash based Geopolymer concrete is made using Sodium hydroxide and Sodium silicate solutions. Coal based thermal power plants produce Flyash. It is a powdered residue originating from oxidization of powdered coal. Amorphous silica in flyash triggers the Pozzolanic activity. Geo-polymer concretes have alumina-silicates as binder and they generally do not have free lime. Therefore their resistance to acids can be estimated to be superior to the Portland cement based concrete.

II. MATERIAL CHARACTERISTICS

Tests on materials used for preparation of conventional concrete and Geopolymer concrete are conducted in order to find the material properties.

(a) Cement: Ordinary Portland Cement (OPC) of 53 grade is used for preparation of conventional concrete of M30 grade. Specific gravity of cement is tested and found to be 3.15 using Density bottle method. Different properties of cement are tested and the results are discussed in Table 1.

Table 1 Properties of OPC 53 grade cement

S.No	Material property	Test result	Specifications (IS 12269:2013)
1	Specific gravity	3.15	3.0-3.2
2	Soundness using Le chatelier's method	4	Should be less than 10mm
3	Normal consistency	33%	—

Contd…

ISBN: 978-93-8830-599-0

S.No	Material property	Test result	Specifications (IS 12269:2013)
4	Initial setting time	30 min	Should not be greater than 30min
5	Final setting time	560 min	Should not be greater than 600min
6	Percentage of cement retained on sieve no.9	5%	Should not be greater than 10%

(b) Coarse aggregate: Durability against chemical attack of a concrete depends not only upon the cement characteristics but also on the properties of Aggregate. Specific gravity of the Coarse aggregate is found as 2.67 using density basket apparatus. Coarse aggregate size used for casting of both conventional and Geopolymer concrete is 20mm.

(c) Fine aggregate: Locally available sand passing through 4.75mm IS sieve is used for casting. Specific gravity of sand is found as 2.67 using Pycnometer bottle method. The zone of sand is confirmed to be Zone III sand in accordance with IS 383-1970.

(d) Flyash: Fly ash is generally available as Class C and Class F type. Active compounds in addition to calcium alumina-silicate glass that may be present in Class C fly ash are free lime, an hydrate, tricalcium aluminate, calcium sulfo-aluminate, and rarely, calcium silicates. Low calcium (Class F) fly ash characteristically contain a large proportion of silicate glass of high silica content plus crystalline phases of low reactivity. Hence as Class F flyash contains low or no free lime content it is selected as binder in this study. Composition of flyash used in this study is discussed in Table 2.

Table 2 Composition of Class F Flyash

Constituent	Composition/Percentage
Cao	0.72-3.6
SiO_2	49-67
Al_2O_3	16-28
Fe_2O_3	4-10
MgO	0.32-2.6
SO_3	0.1-1.9

(e) Alkali Activator Solutions: Alkaline environment is essential for the initiation of geopolymer binder reaction. Sodium hydroxide solution in combination with Sodium silicate solution or Potassium hydroxide solution in combination with Potassium silicate solution are generally used as Alkali Activator Solutions.

Sodium hydroxide solution: Sodium hydroxide is available in the form of pellets, flakes, sticks or chips and in solutions of different concentrations and purities. Sodium hydroxide solution in combination with Sodium silicate solution is most commonly used because Sodium hydroxide is abundantly available at low cost.

Sodium Silicate solution: The preparation of sodium meta silicate is done for 33.3% concentration by mixing Sodium meta silicate powder in water.

III. EXPERIMENTAL STUDY

Mix design: The grade of concrete used in this study is M 30. Conventional concrete mix design is adopted according to IS 456:2000. Mix proportions of Geopolymer concrete used in this study are presented in table 3.

Table 3 Mix proportion of Geopolymer Concrete

Ingredients of Geo-polymer concrete	Quantity (Kg/m^3)
Fly ash	405
NaOH	70.88
Na_2SiO_3	70.88
sand	683.13
Coarse aggregate	1268.66
Total water	108.35
Extra water	29.46

Mixing and Casting: The preparation of sodium hydroxide solution is done for 13 M by mixing 520grams of sodium hydroxide pellets in 1 litre of distilled water. The preparation of sodium hydroxide solution is an exothermic in nature. The preparation of sodium silicate solution is done by mixing 590 grams of sodium meta silicate powder in 1litre of water. 97% purity of sodium meta silicate powder is used. Alkali Activator solutions are prepared 24 hours prior to mixing of Geopolymer Concrete.

Weigh Batching is adopted for accounting the materials. Concrete mixer machine is used for mixing the materials. Flyash, Coarse aggregate and Fine aggregate are thoroughly mixed in dry state. After attaining a homogenous dry mixture alkali activator solutions and free water content are added. A decent mixing is carried out before placing the fresh Geopolymer Concrete mix in moulds. Conventional concrete is mixed and cast according to IS 10262:2009. Cube specimen moulds of size 150x150x150mm and Cylinder specimens of diameter 300mm and height 150mm are used for casting the specimens. A Thin

ISBN: 978-93-8830-599-0

layer of grease or waxing agent is applied to the mould for easy demoulding of specimens.

Fig. 1 Casting of Geopolymer specimens

The specimens are demoulded after 24hours of cast. Conventional concrete specimens are Water cured for 28 days. Geopolymer concrete specimens oven cured for 24 hours at 60^0C.

Fig. 2 Demoulded concrete specimens

After 24 hours of oven curing the specimens are air cured for 28 days and tested for compressive and Split tensile strengths.

Fig. 3 Testing of Concrete Specimens

Concentrated sulphuric acid (98% and density of 1.84 g/cc) was used to prepare the diluted sulphuric acid of 10%concentration.Conventional concrete and Geopolymer Specimens are immersed in Sulphuric acid after 28 days for a week period and tested for the compressive and split tensile strength.

Fig. 4 Concrete specimens immersed in Sulphuric acid

IV. RESULTS

Conventional concrete cubes are removed from curing tank before 4-5 hours of testing and surface dry condition is ensured. Cubes and Cylinder specimens are placed in Compression Testing Machine and tested. Compressive strength Results are presented in table 4.

Table 4 Cube strength of conventional concrete specimens before and after Acid attack

Cube	Cube Strength (N/mm^2)	Cube Strength after immersion in acid (N/mm^2)
C1	30.22	19.55
C2	30.22	20
C3	30.67	20.44
C4	31.11	20.88
C5	31.11	20.88
C6	30.22	20
C7	30.67	20.88
C8	31.11	20.88
C9	30.67	21.33
C10	30.67	21.77

Table 5 Split Tensile strength of conventional concrete specimens before and after Acid attack

Cylinder	Cylinder strength(N/mm^2)	Cylinder Strength after immersion in acid (N/mm^2)
CY1	3.8	2.26
CY2	3.53	2.12
CY3	3.39	2.41

ISBN: 978-93-8830-599-0

Geopolymer Concrete Cubes has lost their texture and appearance, revealing the aggregates in them after exposing to the acid environment.

Table 6 Cube strength of Geopolymer concrete specimens before and after Acid attack

Cube	Cube Strength(N/mm^2)	Cube Strength after immersion in acid (N/mm^2)
GC1	30.67	22.22
GC2	30.22	21.33
GC3	31.11	22.22
GC4	31.56	22.22
GC5	31.11	21.78
GC6	30.67	22.22
GC7	31.11	22.22
GC8	31.11	21.78
GC9	31.11	21.33
GC10	31.11	22.22

Table 7 Split Tensile strength of Geopolymer concrete specimens before and after Acid attack

Cylinder	Cylinder strength(N/mm^2)	Cylinder Strength after immersion in acid (N/mm^2)
GCY1	3.54	2.83
GCY2	3.54	2.69
GCY3	3.68	2.97

Comparision of Conventional concrete and Geopolymer concrete specimens are graphically presented below.

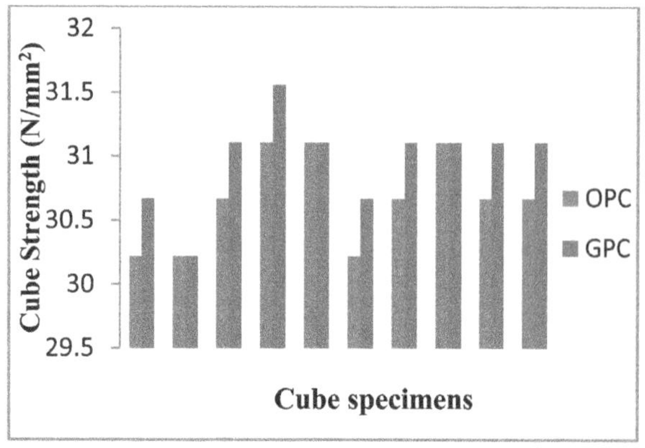

Fig. 5 Compressive Strength Comparison of cubes before Acid attack

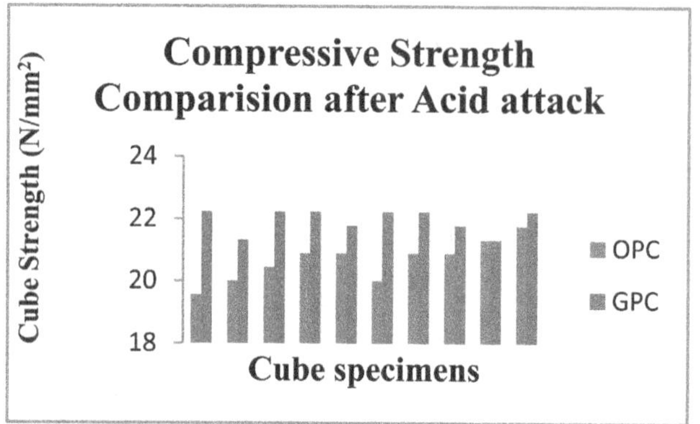

Fig. 6 Compressive Strength Comparison of cubes after Acid attack

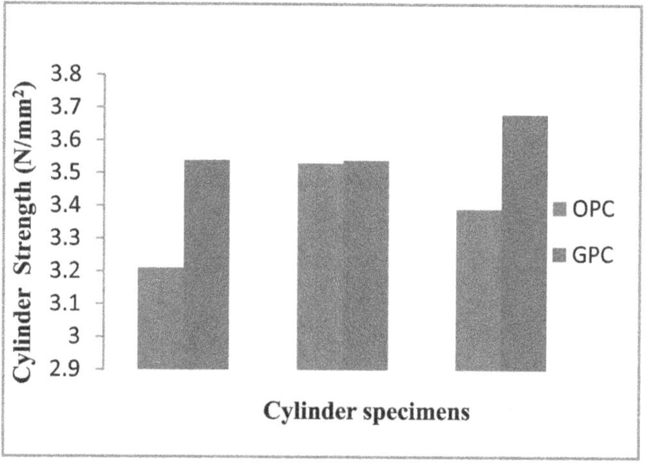

Fig. 7 Split tensile Strength Comparison of cubes Before Acid attack

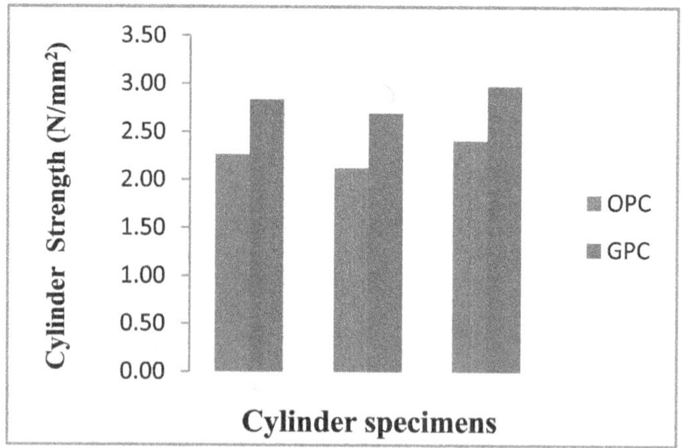

Fig. 8 Split tensile Strength Comparison of cubes after Acid attack

V. CONCLUSIONS

The average of 28days compression test of GPCs were in the range of 30.2 to 31.11 MPa. The corresponding values for OPCCs were 29.33 to 30.67 MPa

ISBN: 978-93-8830-599-0

respectively. It may be noted that this strength levels are quite adequate when compared with the minimum structural grade recommended in IS456-2000 for extreme exposure condition.

GPC specimens has maintained their shapes without any signs of severe external deteriorations but the GPCs had strength losses of 28.49% which is quite considerable. OPC concrete specimens had almost lost their shape with many locations of severe externally visual deteriorations; the strength losses were very high i.e; 35%. The OPC concrete specimens has exposed coarse aggregates, the surfaces were rough and whitish in color. Expansive chemical reaction should have happened in the Conventional Concrete specimens since a perceptible increase in the diameter of the specimens was noticed. These specimens could be considered as having reached their ultimate level of integrity. In contrast the GPC specimens were intact even though significant strength losses have occurred. From the above results, it is clear that GPC mixes have comparatively high resistance to Sulphuric acid attack. This can be ascribed to the fact that the Geopolymer Concrete do not have free lime content in its components and Geopolymer themselves are not easily attacked by the acids. Conventional concrete do contain free lime content in its components and are easily attacked by the acids.

REFERENCES

1. Steenie E. Wallah, Djwantoro Hardjito "Performance of Fly Ash-Based Geopolymer Concrete under Sulfate and Acid Exposure".
2. P Abhilash, C Sashidhar and I V Ramana Reddy "Evaluation of performance of Geopolymer Concrete in acid environment".
3. Salmabanu Luhar "Durability Studies of Fly Ash Based Geopolymer Concrete".
4. Sanjukta Sahoo, B. B. Das, A. K. Rath and B. B. Kar "Acid, Alkali and Chloride Resistance of High Volume Fly Ash Concrete"
5. S. Kumaravel[l], K. Girija "acid and salt resistance of geopolymer concrete with varying concentration of NaoH"
6. Mr. K. Madhan Gopal, Mr. B. Naga Kiran "Investigation on Behaviour of Fly Ash Based Geopolymer Concrete in Acidic Environment".
7. T Srinivas, N V Ramana Rao "studies on acid attack resistance of low calcium flyash and slag based geopolymer concrete"
8. Davidovits, J., (1994), Properties of geopolymer cements, Proceedings of first International conference on alkaline cements and concretes.
9. P. Chindaprasirt, U. Rattanasak, S. Taebuanhuad, Resistance to acid and sulfate solutions of microwave-assisted high calcium fly ash geopolymer.
10. Davidovits, J.,(2002), 30 years of successes and failures in geopolymer applications-Market trends and potential breakthroughs, Proceedings of the geopolymer.
11. Suresh Thokchom, Dr. Partha Gosh and Dr. Somnath Gosh, (2009), Acid resistance of fly ash based geopolymer mortars, International Journal of Recent Trends in Engineering.
12. Mittal A, Kaisar MB, Shetti RK. Experimental study on use of fly ash in concrete. 2005; 1–22.
13. Naik TR, Singh SS, Hossain MM. Properties of high-performance concrete systems incorporating large amounts of high-lime fly ash. Construction and Building Materials. 1995 Aug; 9(4):195–204.
14. Naik TR, Sivasundaram V, Singh SS. Use of high volume class F fly ash for structural grade concrete. 1991; 23–34.
15. Poon CS, Lam L, Wong YL. A study on high strength-concrete prepared with large volumes of low calcium fly ash. Cement and Concrete Research. 2000 Mar.
16. Thirugnanam GS. An experimental investigation on the mechanical properties of bottom ash concrete. Indian Journal of Science and Technology. 2015May.
17. Reddy BSK, Varaprasad J, Reddy KNK. Strength and -workability of low lime fly-ash based geopolymer -concrete. Indian Journal of Science and Technology.
18. Allahverdi Ali, Skavara Frantisek. 2006. sulfuric acid attack on hardened paste of geopolymer cements. Part Corrosion mechanism at mild and relatively low concentrations.
19. Song X.J, Marosszeky M, Brungs M, Munn R. 2005. Durability of fly ash based geopolymer concrete against sulphuric acid attack. 10 DBMC International Conference on Durability of Building Materials and Components.
20. Wallah S.E, Rangan B.V. 2006. Low Calcium Fly ash based Geopolymer Concrete: Long term properties.

Research Report GC 2, Curtin University of Technology, Australia.

21. Davidovits, J., Geopolymers: Inorganic Polymeric New Materials. Journal of Thermal Analysis,1991. 37: p. 1633-1656.

22. Davidovits, J. Properties of Geopolymer Cements. Presented at the First International Conference on Alkaline Cements and Concretes. 1994. Kiev, Ukraine.

23. Davidovits, J. High-Alkali Cements for 21st Century Concretes. Presented at the V. Mohan Malhotra Symposium on Concrete Technology: Past, Present And Future. 1994. University of California, Berkeley: ACI.

24. van Jaarsveld, J.G.S., van Deventer, J.S.J., and Lukey, G.C., The effect of composition and temperature on the properties of fly ash- and kaolinite-based geopolymers. Chemical Engineering Journal, 2002. Article in press: p. 1-11.

ISBN: 978-93-8830-599-0

Influence of Copper Slag on the Fresh Properties of Self Compacting Concrete

Rahul Sharma[1], Bharat Bhushan Jindal[2] and Vaibhav Sapkal[3]

[1,2,3]Shri Mata Vaishno Devi University, Katra, J&K, India.

Abstract

The present study involves incorporation of copper slag (CS) as fine aggregates to assess the fresh properties of self-compacting concrete (SCC).The fresh properties evaluated include slump flow values, T_{500} time, V-funnel time and L box ratio. The six mixes were prepared as replacement of sand by CS upto 100% with constant w/b ratio of 0.45. The replacement level of sand for each mix with CS was kept 20% with different dosages of superplastizicer. The results revealed that slump flow value increased with increment in the CS content. The maximum slump flow value was observed for 100% replacement of sand whereas T_{500} time decline with increase in the CS content. The maximum T_{500} time was noticed for control concrete and minimum value exhibited for mix containing 100% CS. Similar trend was noticed for V-funnel time. The L box ratio increased as the replacement content of sand increased and maximum value was observed for mix with 100% CS.

Keywords: Copper slag, Self Compacting Concrete (SCC), Fresh Properties

INTRODUCTION

Self-compacting concrete (SCC) also known as self-consolidating concrete is a special type of concrete which can flow in the dense reinforcement and narrow formwork by its own weight without any need of vibrator. SCC came into origin in early 80 in the country of Japan to overcome the premature deterioration of reinforced concrete structures and the cause of deterioration was found to be inadequate compaction and unskilled labour (Okamura et al; 2003). As a solution to the problem, SCC was developed in the University of Tokyo. SCC mainly differs from conventional concrete in the composition of fine aggregates and high cementitious materials along with superplasticizer and viscosity modifying agent if required.

Now a days, industrial byproduct is increasing which can be utilized in the construction sector to overcome the scarcity of the materials in the industry of construction. One such industrial byproduct is copper slag (CS) which is obtained from copper industry during the melting and smelting process of copper metal. CS is highly rich in the oxides of iron, silica and alumina which constitutes about 80-85% (Gorai et al; 2003, Shi et al; 2008). This CS can be utilized as fine aggregates to develop SCC. The aim of the present investigation is to develop and assess the fresh properties of the SCC which includes slump flow values, T_{500} time, V-funnel time and L box ratio

MATERIALS

Cement

OPC 43 Grade cement was used throughout the investigation satisfying IS 8112(IS 8112 1989). The FA of class F was used as SCMs. FA was procured from thermal power plant located in Ropar, Punjab, India. The physical properties of OPC and FA are shown in Table 1 and their chemical composition is shown in Table 2.

Table 1 Physical properties of OPC and FA

Characteristic Properties	OPC	FA
Standard consistency (%)	32	-
Initial Setting time (minutes)	62	-
Final setting time (minutes)	270	-
Soundness by Le-Chat Expansion (mm)	1.0	-
Compressive strength (MPa)		-
3 days	24.6	
7-days	34.3	
28-days	45.2	
Specific gravity	3.15	2.1
Loss on ignition (%)	2.04	3.3

Note: OPC- Ordinary Portland Cement, FA- Fly Ash

Table 2 Chemical composition of OPC, FA, SF and CS

Component (%)	OPC	FA	CS
SiO_2	20.99	57.6	30.53
Al_2O_3	5.98	30.5	2.80
Fe_2O_3	4.10	3.72	57.82
CaO	60.78	1.10	1.60
MgO	0.96	0.38	1.48
SO_3	2.86	0.22	1.59
K_2O	1.18	1.35	0.71
Na_2O	0.86	0.10	0.34
TiO_2	0.25	1.72	0.26
CuO	-	0.017	0.64

Locally available sand was utilized in the SCC mixes conforming IS 383(IS 383 1970). The water absorption of sand was 0.80%, fineness modulus of

ISBN: 978-93-8830-599-0

2.79 and specific gravity of 2.6. CS was used as a replacement of sand from 0 to 100% at an interval of 20%. CS had fineness modulus of 3.33, water absorption 0.36% and specific gravity of 3.51.

Locally available coarse aggregate of nominal size 12.5 mm and10 mm was used in the proportion of 40% and 60%. The water absorption of coarse aggregate was 0.68%, fineness modulus of 6.93 and specific gravity of 2.64. Master Glenium SKY 8765 as per IS 9103 (IS 9103 1999)was used in different dosages by weight of binder to achieve desirable properties of SCC.

MIXTURES

A total six mixes were cast with different proportions of CS from 0% to 100% at an interval of 20% with constant w/b ratio of 0.45, total coarse aggregates of 700 kg/m^3 and water content of 247.5 kg/m^3. The control concrete consists of 100% fine aggregates with 330 kg/m^3 of OPC and 220 kg/m^3 of FA. Table 3 illustrates the different mixes of the SCC. The total cementitious content was kept constant i.e. 550 kg/m^3 in each mix and fine aggregates were replaced by CS through equivalent volume approach. Initially coarse aggregates, fine aggregates, and CS were mixed for 1 minute in hand operated laboratory concrete mixer. Then cementitious materials were added in the mixer and rotated for 2 more minutes to obtain a homogeneous mix. In the next stage, 70 % water was discharge into the mixer and rotated for 3 minutes. At last, 30 % left over water was combined with required

dosage of superplasticizer and added into the mixer then rotated for 2 more minutes. SCC was poured into the well-oiled moulds and demoulded after 24 hours. Samples were water cured in temperature controlled curing tank at 27±2□ till the age of testing.

TEST METHODS

The fresh properties of SCC were evaluated by conducting tests for filling ability, viscosity and passing ability. The filling ability was assessed by using slump flow whereas viscosity and passing ability was examined by T500, V-funnel and L-box test. The fresh properties were conducted as per European Federation of National Association Representing Producers and Applications of Special Building Products of Concrete (EFNARC)(EFNARC 2005).

RESULTS AND DISCUSSIONS

The results of fresh properties i.e. Slump flow, T$_{500}$, V-funnel, L-box ratio of all SCC mixes are shown in Fig. 1 to Fig. 4. The slump flow value of SCC mixes increased with increment of CS in the concrete. This may be attributed to the non porous and low absorption characteristics of CS particle which increases the free water content. The highest slump flow value of 735 mm was obtained for OF-CS100 without any bleeding and segregation whereas lowest value of 705 mm was recorded for control mix OF-CS0. According to the ENFARC, the slump flow values of whole mixes fall in the category classified as SF2 is shown in Fig 1.

Table 3 Detail of Material Proportions of SCC Mixes in kg/m^3

Mix Description	OPC	FA	CS	Fine Aggregates	Coarse Aggregates		Water	SP
					10 mm	12.5 mm		
OF-CS0 (cc)	330	220	-	1020	420	280	247.5	4.40
OF-CS20	330	220	284.4	816	420	280	247.5	4.40
OF-CS40	330	220	568	612	420	280	247.5	3.30
OF-CS60	330	220	853.2	408	420	280	247.5	2.75
OF-CS80	330	220	1137.6	204	420	280	247.5	2.20
OF-CS100	330	220	1422	-	420	280	247.5	2.20

ISBN: 978-93-8830-599-0

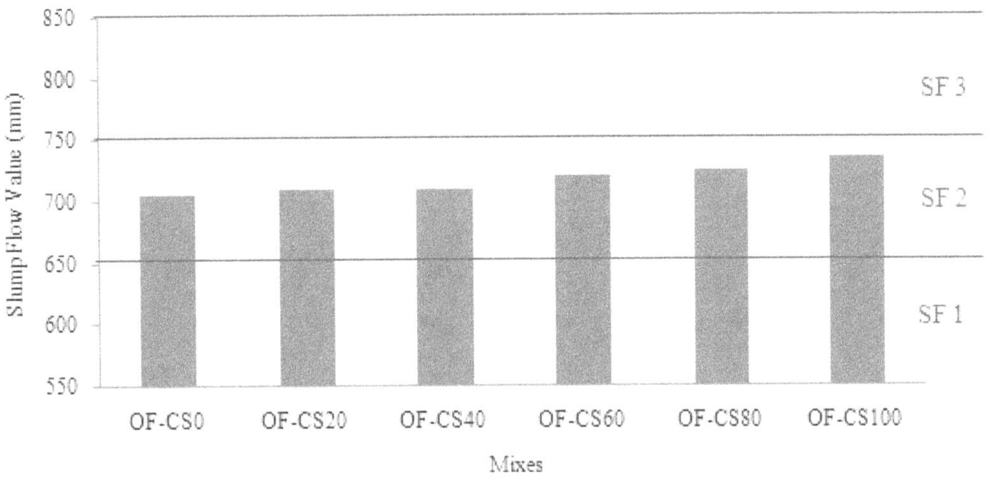

Fig. 1 Slump Flow Values of SCC Mixes

It is evident from the Fig 2 that T_{500} times for all SCC mixes were within the permissible limits as suggested by ENFARC. The results indicate that T_{500} time varied from 2.18 s to 3.8 s and all mixes were in the class VS2 on the basis of viscosity. In addition, on the basis of viscosity, V-funnel time for two mixes i.e. OF-CS0 and OF-CS20 with the values of 8.55 s and 8.12 s respectively were in the upper limit of viscosity class VF1. The value of viscosity class VF1 is equal to and less than 8 s, whereas the value of VF2 is between 9 to 25 s. The other four mixes were within the range of class VF1 as shown in Fig 3. It can be noticed from Fig 4 that blocking ratio varied from 0.87 to 0.96 for entire mixes. As passing ability test conducted include 3 rebars with blocking ratio over 0.8 were classified as PA2. The inclusion of 100 % CS in the concrete mix reduced the requirement of superplasticizer to half of 0 % CS substitution for achieving the desired consistency properties of fresh SCC at similar w/b ratio.

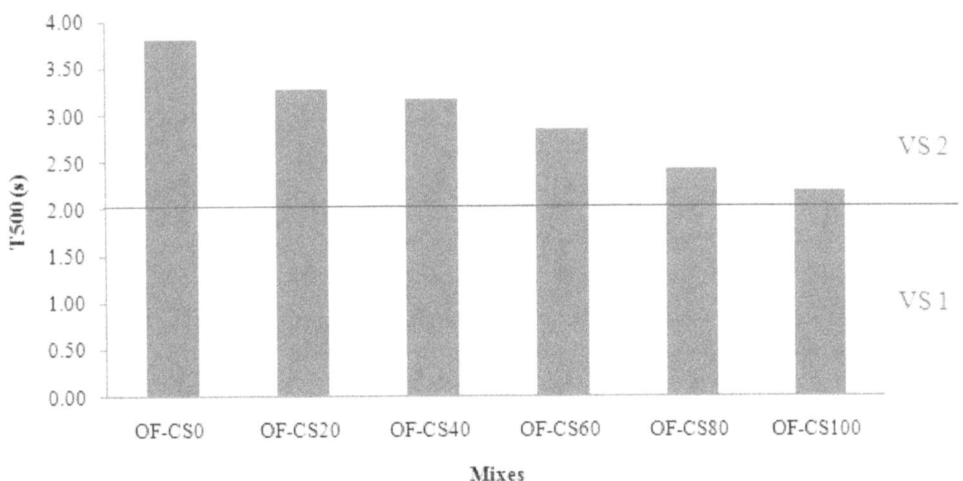

Fig. 2 T_{500} Values of SCC Mixes

ISBN: 978-93-8830-599-0

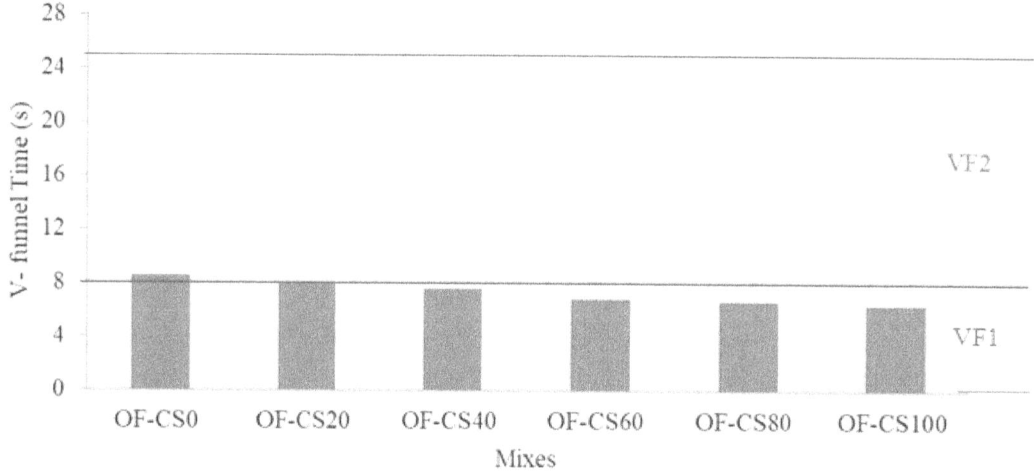

Fig. 3 V-funnel Values of SCC Mixes

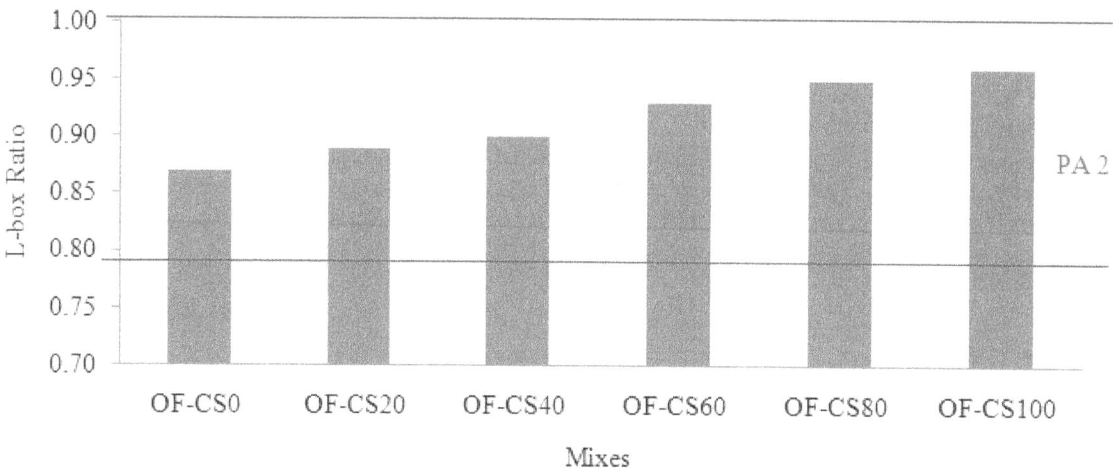

Fig. 4 L-box Ratio of SCC Mixes

CONCLUSIONS

The fresh properties enhanced with increase in the CS content. The maximum slump flow value was observed for 100% CS substitution whereas lowest value was noticed for control concrete. The T_{500} time and V-funnel was minimum for 100% CS whereas maximum value exhibited by control concrete. Moreover dosage of superplasticizer decline with increase in the content of CS to achieve the desired fresh properties of the SCC. For 100% CS substitution, dosage of superplasticizer was half of the control concrete. The main reason was found to be the low water absorption characteristics of CS which increased the free water content inside the mixes and resulting into the enhancement of the fresh properties with less consumption of superplasticizer.

This investigation shows that CS can be utilised as substitute to fine aggregates for the development of SCC.

REFERENCES

1. EFNARC. (2005). The European Guidelines for Self-Compacting Concrete. The European Guidelines for Self Compacting Concrete.

2. Gorai, B., Jana, R. K., and Premchand. (2003). "Characteristics and utilisation of copper slag - A review." Resources, Conservation and Recycling, 39(4), 299–313.

ISBN: 978-93-8830-599-0

3. IS 383.(1970). Specification for coarse and fine aggregates from natural sources for concrete. Bureau of Indian Standards, New Delhi.

4. IS 8112.(1989). 43 Grade Ordinary Portland Cement-Specification. Bureau of Indian Standards, New Delhi.

5. IS 9103.(1999). Concrete Admixtures-Specification. Bureau of Indian Standards, New Delhi.

6. Okamura, H; and Ouchi, M. (2003)."Self-compacting concrete."Journal of Advance Concrete Technology, 1, 5-15.

7. Shi, C., Meyer, C., and Behnod, A. (2008). "Utilization of copper slag in cement and concrete." Resources, Conservation and Recycling.

ISBN: 978-93-8830-599-0

Experimental Investigation on Split Tensile Strength of PolypropyleneFiber Reinforced Geopolymer Concrete

Hathi Ram Gugulothu[1*] B. Sesha Sreenivas[2] and D. Rama Seshu[3]

[1]Research Scholar (Reg.No15016HD0704), katiya Univ., Warangal, India,
[2]Professor and Principal, University College of Engg, Kothagudem, Kakatiya Univ., India.
[3]Professor of Civil Engineering, National Institute of Technology, Warangal, India.
knrecs@gmail.com

Abstract

The experimental study consisted of determination of the Split Tensile Strength of Polypropylene fiber reinforced Geopolymer concrete (PPFRGPC) by casting and testing cylinders of size 100 mm diameter and 200 mm long. Polypropylene fibers (Recron 3S) were used in different volume fractions 0.2, 0.4, 0.6, 0.8 and 1 (S2, S4, S6, S8 and S10). For preparing PPFRGPC two different Ground Granulated Blast Furnace Slag to Fly Ash ratios (60:40, 40:60), three different alkaline molar activators 8, 10 and 12 were used as binders. Three identical specimens for each variation were cast and tested after 7 days and 28 days ambient curing. Two Parameters called **Binder Index** and **Modified Binder Index** are introduced to quantify the effect of alkaline molar activator, Ground Granulated Blast FurnaceSlag, fly ash, fiber volume fraction and fiber tensile strength on Split Tensile Strength of PPFRGPC. The variation of 7 days and 28 days Split tensile strength of PPFRGPC with different fiber volume proportions and Modified Binder Index are presented.

Keywords: Geopolymer concrete (GPC), Split tensile strength, Polypropylene fiber reinforced Geopolymer concrete (PPFRGPC), soft fibers, Fly ash (FA), Ground Granulated Blast Furnace Slag (GGBS), Ambient temperature, alkaline molar activators (Molarity).

1.0 INTRODUCTION

The cement production is highly energy intensive (at 1.3 kWh/kg of cement) next only to steel and aluminium, also consumes significant amount of non-renewable natural resources such as lime stone deposits, coal etc [1].Geopolymer concrete developed by Prof. Joseph Davidovits, is a new class of concrete that is attracting growing interest around the world due to its environmental and performance benefits compared to conventional Portland cement concrete. The manufacture of Geopolymer concrete is carried out using the usual concrete technology methods .As in the case of Ordinary Portland cement concrete, the aggregates occupy about 75-80 % by mass, in Geopolymer concrete. The silicon and the aluminium in the fly ash react with an alkaline liquid that is a combination of sodium silicate and sodium hydroxide solutions to form the Geopolymer paste that binds the aggregates and other un-reacted materials. However, like most ceramics, Geopolymer concrete suffers from quasi-brittle characters, deficiency of low flexural strength and sudden failure [2–5]. Fiber-reinforced Geopolymer composites were first investigated by Davidovits with the aim of fabricating molding tools and Patterns for the plastics processing industry [6]. Studies on reinforcement of Geopolymers with different type of fibers were carried on through organic fiber like cotton fiber [7] and protein-based fibers [8], carbon fibers [9, 10], polypropylene fiber [11] to overcome the brittleness. Keeping in view of the previous research studies , present research is aimed at determination of the Split Tensile Strength of Polypropylene fiber reinforced Geopolymer concrete (PPFRGPC) by casting and testing cylinders of size 100 mm dia and 200 mm long. Polypropylene fibers (recron 3S) were used in different volume fractions 0.2,0.4,0.6, 0.8 and 1 (S2, S4, S6, S8 and S10). For making PPFRGPC two different Ground Granulated Blast Furnace Slag to Fly Ash ratios (60:40, 40:60), three different alkaline molar activators 8, 10 and 12 are used as binders. Three identical specimens for each variation were cast and tested after 7 days and 28 days ambient curing. Two Parameters called **Binder Index** and **Modified Binder Index** are introduced to quantify the effect of alkaline molar activator, Ground Granulated Blast Furnace Slag, fly ash, fiber volume fraction and fiber tensile strength on Split Tensile Strength of PPFRGPC. The variation of 7 days and 28 days Split tensile strength of PPFRGPC with different fiber volume proportions and **Modified Binder Index** are presented.

2.0 EXPERIMENTAL PROGRAM

The experimental program consisted of determination of the Split tensile strength of PPFRGPC by casting and testing cylinders of size 100 mm dia and 200 mm long. Fly ash to GGBS proportions 60:40 and 40:60 are used with alkaline molar activators 8, 10 and 12 as binders. Polypropylene fibers (recron 3S) were used in different volume fractions 0.2,0.4,0.6, 0.8 and 1 (S2, S4, S6, S8 and S10). Three identical specimens for each variation were cast and tested for 7 days and 28 days ambient curing.

ISBN: 978-93-8830-599-0

2.1 MATERIALS

Fly ash is obtained from Kothagudem Thermal Power Station, Bhadradri Kothagudem Dist, Telangana, India. GGBS is obtained from Blue way exports supplier from Vijayawada, Andhra Pradesh, India. Specific gravity of Fly ash and GGBS are 2.17 and 2.90 respectively. Chemical composition details of fly ash and GGBS are shown in Table 1. Natural river sand confirming to grading zone II of IS 383:1970 was used. Specific gravity and fineness modulus of sand used were 2.32 and 2.81 respectively. Coarse aggregate of maximum size 12 mm from local source was used. Polypropylene fiber (Recron 3S) of length 12mm and diameter 20 microns with tensile Strength 490.3 Mpa is used. The density of soft fibers is 946 Kg/m^3. The alkaline molar activators of sodium hydroxide solution used are 8, 10 and 12. The sodium hydroxide pellets used for preparation of NaOH solution is given in table 2. The NaOH solution thus prepared is mixed with Na_2SiO_3 solution. The ratio of sodium silicate solution to sodium hydroxide solution is fixed as 2.5[12, 13, 14]. The mixture was stored for 24 hours at room temperature before casting. Super Plasticizer Conplast Sp-430 is used to obtain the desired workability.

2.2 MIX PROPORTIONS

The unit weight of PPFRGPC is taken as 2400Kg/m^3.

The PPFRGPC mix proportions are shown in Table 3 and 4

Table 1 Chemical composition of Flyash and GGBS percentage by mass.

Material	SiO_2	Al_2O_3	Fe_2O_3	SO_3	CaO	MgO	Na_2O	LOI
Fly ash	60.12	26.63	4.22	0.32	4.1	1.21	0.2	0.85
GGBS	34.16	20.1	0.81	0.88	32.8	7.69	nd	.

Table 2 Materials used for NaOH solution preparation.

	8 moles/L	10 moles/L	12 moles/L
Sodium hydroxide pellets, (grams)	262	314	361
Potable Water (grams)	738	686	639

Table 3 GPC mix proportion.

FA:GGBS	Molarity(M)	GPC mix proportions (Kg/m^3)							
		Coarse Aggregate	Fine Aggregate	Fly Ash	GGBS	NaOH Solution	Sodium Silicate	Super Plasticizer (2% of the Binder)	Extra water (7.5% of the Binder)
60:40	8	1100	517.45	345.10	230.10	59.10	148.25	11.50	43.15
60:40	10	1100	517.45	345.10	230.10	59.10	148.25	11.50	43.15
60:40	12	1100	517.45	345.10	230.10	59.10	148.25	11.50	43.15
40:60	8	1100	517.45	230.10	345.10	59.10	148.25	11.50	43.15
40:60	10	1100	517.45	230.10	345.10	59.10	148.25	11.50	43.15
40:60	12	1100	517.45	230.10	345.10	59.10	148.25	11.50	43.15

Table 4 Polypropylene Fiber mix proportion.

Polypropylene fiber mix proportions		
@fiber designation.	Soft fiber volume fraction (%)	Soft fiber weight (Kg/m^3)
S2	0.20	1.9
S4	0.40	3.8
S6	0.60	5.7
S8	0.80	7.6
S10	1	9.5

@ 1st letter indicates the Soft fiber designation (S),
2nd letter indicates the volume fraction percentage for Soft fiber (0.2, 0.4, 0.6, 0.8 &1).

ISBN: 978-93-8830-599-0

2.3 CASTING OF PPFRGPC SPECIMENS

The solids constituents of the PPFRGPC i.e. the aggregates, fly ash, GGBS and fibers were dry mixed for about three minutes. The liquid part of the mixtures i.e. the alkaline solution, added water and the super plasticizer were premixed then added to the solids. The wet mixing usually continued for another four minutes. The fresh PPFRGPC concrete was dark in colour and shiny in appearance. The mixtures were usually very cohesive. The workability of the fresh concrete was measured by means of the conventional slump test. Compaction of fresh concrete in the cylindrical moulds was done in three equal layers followed by compaction on a vibration table for ten seconds. The demoulding was done after 24 hours and kept for ambient curing.

2.4 SPLIT TENSILE STRENGTH

The Split tensile strength tests on hardened PPFRGPC were performed on a 1000 kN capacity universal testing machine in accordance to the relevant Indian Standard code IS 516[15]. Three identical 100 mm dia and 200 mm long PPFRGPC cylinders for each variation were cast and tested for their average Split Tensile strength. The results given in the various figures and tables are the mean of these values. To quantify the effect of alkaline molar activator, Ground Granulated Blast Furnace Slag, fly ash, fiber volume fraction and fiber tensile strength on Split Tensile Strength of PPFRGPC, **Binder Index** and **Modified Binder Index** are introduced. Binder index is taken as the product of molarity of alkaline activator and binder's ratio as given below [16].

The variation of 7 days and 28 days Split tensile strength of PPFRGPC with different fiber volume proportions and **Modified Binder Index** are presented.

**Binder Index =
Molarity × [GGBS /(GGBS + Fly Ash)]   .....(1)**

Modified Binder Index: To know the effect of binder index, fiber volume fraction and fiber tensile strength on Split Tensile Strength of PPFRGPC modified binder index is introduced [17, 18, 19].

Modified binder index

$$(P) = \quad B_i \; X \; (\sqrt{pf_{ef}}) \qquad ..... (2)$$

Where B_i = Binder Index, pf_{ef} = Polypropylene fiber effect

$$f_{ef} = (F_{tp} \; X \; V_{pf}) \qquad .....(3)$$

F_{tp} = Tensile strength of Polypropylene fiber = 490.33Mpa, V_{pf} = Volume fraction of Polypropylene fiber.

Fiber effect and Modified binder index for Polypropylene fibers is calculated and tabulated in table 7.

The variation of Split tensile strength of PPFRGPC with different volume proportions of Polypropylene fibers is Shown in fig 1. to fig 6.

Table 5 Split Tensile strength values of PPFRGPC

FA:GGBS	Molarity (M)	Binder Index (Bi)	Split Tensile strength values of PPFRGPC (Mpa)									
			S2		S4		S6		S8		S10	
			7D	28D	7D	28D	7D	28D	7D	28D	7D	28D
60:40	8	3.2	1.75	2.13	2.07	2.8	3.18	3.2	3.18	3.98	3.34	4.14
60:40	10	4	2.04	2.39	2.55	3.03	3.3	3.5	3.3	4.15	3.66	4.24
60:40	12	4.8	2.45	3.25	2.93	3.66	3.5	4.0	3.5	4.5	4.14	4.62
40:60	8	4.8	2.1	2.68	2.55	3.03	3.38	3.66	3.38	3.85	3.41	4.30
40:60	10	6	2.39	3.15	2.74	3.54	4.11	4.50	4.11	4.68	3.79	4.52
40:60	12	7.2	2.74	3.66	3.34	3.89	4.5	4.71	4.5	4.87	4.78	5.38

Table 6 Polypropylene fiber effect & Modified Binder Index.

Polypropylene Fiber effect (pf_{ef})					Modified binder index for Polypropylene fiber (P)						
Fiber designatation	F_{tp}	V_{pf}	pf_{ef}	$\sqrt{pf_{ef}}$	Molarity	Binder Index	Modified binder index (P) = B_i X ($\sqrt{pf_{ef}}$)				
							S2	S4	S6	S8	S10
S2	490.33	0.002	0.98	0.98	8	3.2	3.2	4.5	5.5	6.4	7.1
S4	490.33	0.004	1.96	1.4	10	4	3.9	5.6	6.9	7.9	8.8
S6	490.33	0.006	2.94	1.72	12	4.8	4.7	6.7	8.3	9.5	10.6
S8	490.33	0.008	3.92	1.98	8	4.8	4.7	6.7	8.3	9.5	10.6
S10	490.33	0.01	4.90	2.21	10	6	5.9	8.4	10.3	11.9	13.3
					12	7.2	7.1	10.1	12.4	14.3	15.9

ISBN: 978-93-8830-599-0

Table 7 Binder Index Vs Split tensile strength of PPFRGPC

S. NO.	Binder Index	Modified binder index	Split tensile strength of CFRGPC		Ratio of 7 day strength to 28 day strength of CFRGPC
			7 days	28 days	7D/28D
1	3.2	3.2	1.75	2.13	0.822
2	4	3.9	2.04	2.39	0.854
3	4.8	4.7	2.45	3.25	0.754
4	4.8	4.7	2.1	2.68	0.784
5	6	5.9	2.39	3.15	0.759
6	7.2	7.1	2.74	3.66	0.749
7	3.2	4.5	2.07	2.8	0.739
8	4	5.6	2.55	3.03	0.842
9	4.8	6.7	2.93	3.66	0.801
10	4.8	6.7	2.55	3.03	0.842
11	6	8.4	2.74	3.54	0.774
12	7.2	10.1	3.34	3.89	0.859
13	3.2	5.5	3.18	3.35	0.949
14	4	6.9	3.3	3.5	0.943
15	4.8	8.3	3.5	4.0	0.875
16	4.8	8.3	3.38	3.66	0.923
17	6	10.3	4.11	4.50	0.913
18	7.2	12.4	4.5	4.71	0.955
19	3.2	6.4	3.25	3.98	0.816
20	4	7.9	3.5	4.15	0.843
21	4.8	9.5	3.80	4.5	0.845
22	4.8	9.5	3.38	3.85	0.878
23	6	11.9	4.3	4.68	0.878
24	7.2	14.3	4.68	4.87	0.918
25	3.2	7.1	3.34	4.14	0.807
26	4	8.8	3.66	4.24	0.863
27	4.8	10.6	4.14	4.62	0.896
28	4.8	10.6	3.41	4.30	0.793
29	6	13.3	4.45	4.90	0.908
30	7.2	15.9	4.78	5.38	0.888

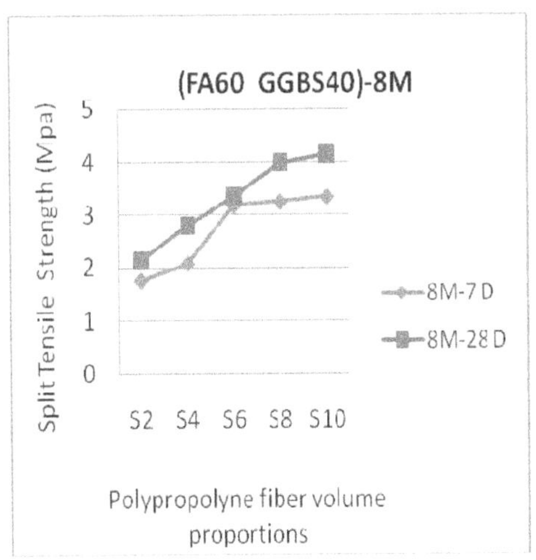

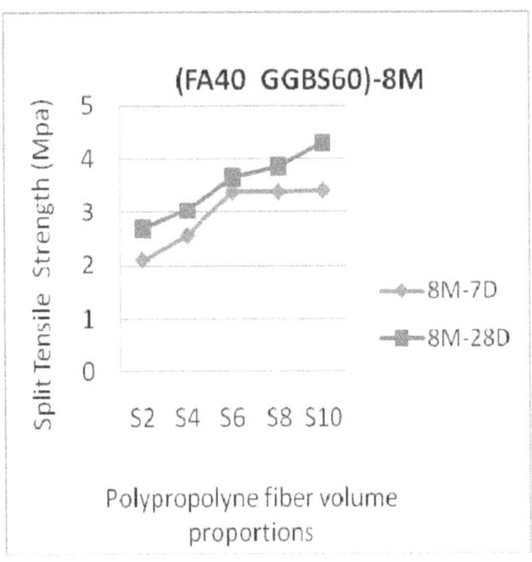

Fig.1 Polypropylene fiber effect on split tensile strength of PPFRGPC

Fig. 2 Polypropylene fiber effect on split tensile strength of PPFRGPC

ISBN: 978-93-8830-599-0

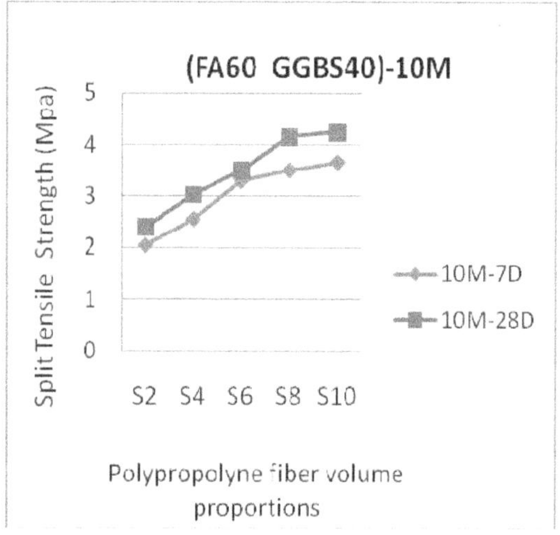

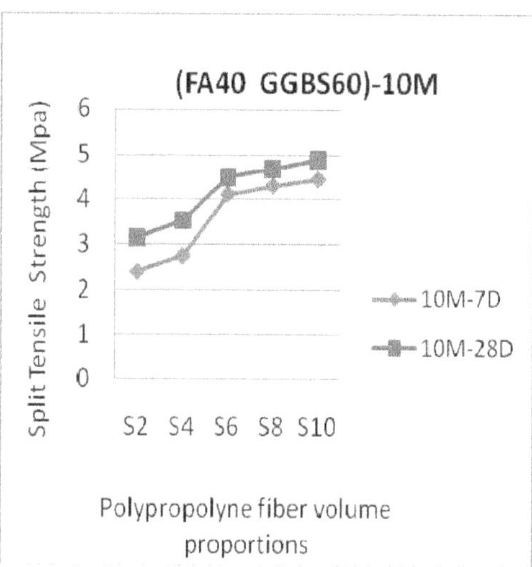

Fig. 3 Polypropylene fiber effect on split tensile strength of PPFRGPC

Fig. 4 Polypropylene fiber effect on split tensile strength of PPFRGPC

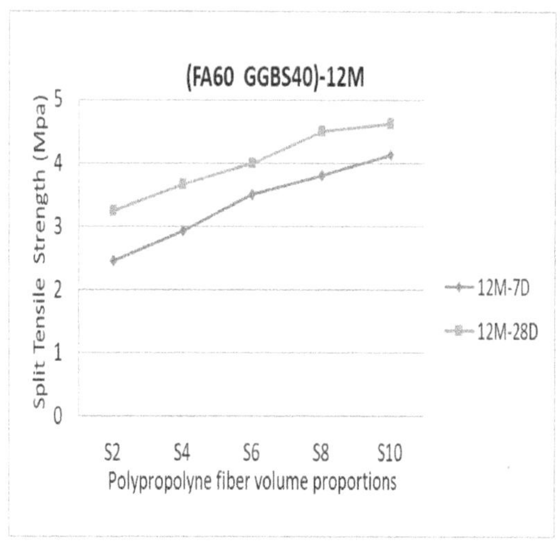

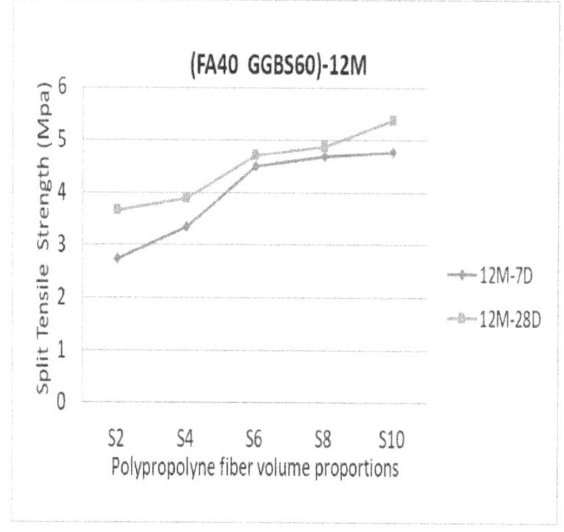

Fig. 5 Polypropylene fiber effect on   split tensile strength of PPFRGPC

Fig. 6 Polypropylene fiber effect on split tensile strength of PPFRGPC

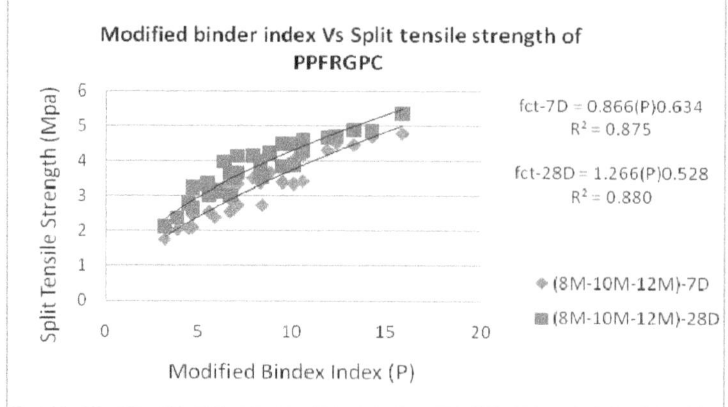

Fig. 7 Modified Binder index effect on Split tensile strength of Polypropylene fiber reinforced geopolymer concrete (PPFRGPC).

ISBN: 978-93-8830-599-0

3.0 DISCUSSIONS

3.1 Effect of Polypropylene fiber on Split tensile strength of PPFRGPC

From fig 1 to fig 6, it is observed that for any choosen molarity of alkaline activator, fly ash to GGBS proportion, the 7 days and 28 days split tensile strength of PPFRGPC increased with increase in Polypropylene fiber proportions.

3.2 Effect of Molarity of alkaline activator on Split tensile strength of PPFRGPC

The effect of molarity of alkaline activator for different fiber proportion , fly ash to GGBS proportion is shown in fig 1 to fig 6. In general as the molarity of alkanine activator   increased   the   Split   tensile   strength   of PPFRGPC increased.

3.3 Effect of fly ash to GGBS proportion on Split tensile strength of PPFRGPC

From fig 1 to fig 6, it is observed that for any chosen molarity of alkaline activator, Polypropylene fiber proportion, the 7 days and 28 days split tensile strength of PPFRGPC increased with increase in GGBS proportion.

3.4 Effect of binder index on Split tensile strength of PPFRGPC.

From table 5, it is observed that the split tensile strength of PPFRGPC increased with the increase in binder index values.

3.5 Effect of Modified binder index on Split tensile strength of PPFRGPC

From table 8, it is observed that the split tensile strength of PPFRGPC increased with the increase in Modified binder index values. Also From fig 7, it is observed that the proposed modified binder index combine the effect of binder index, molarity , fiber volume fraction and fiber tensile strength is found reasonably well in predicting the split tensile strength of PPFRGPC. The following best fit equations give the relation between the Spli tensile strength at 7 days and 28 days of PPFRGPC with modified binder index along with the correlation coefficient (R^2).

$$f_{ct}\text{-7D}=0.866(P)^{0.634} \qquad R^2 = 0.875$$

$$f_{ct}\text{-28D}=1.266(P)^{0.528} \qquad R^2 = 0.880$$

Where

f_{ct}-7D is Split tensile strength of PPFRGPC at

7 days

f_{ct}-28D is Split tensile strength of PPFRGPC at 28 days

P is modified binder index for Polypropylene fibers.

R^2 correlation coefficient

4.0 CONCLUSIONS

The following conclusions can be made from the experimental analysis done.

1. The 7 days and 28 days Split tensile strength of PPFRGPC   increased with increase in with increase in Polypropylene fiber proportions

2. The 7 days and 28 days Split tensile strength of PPFRGPC increased with increase in molarity of alkaline activator .

3. The 7 days and 28 days split tensile strength of PPFRGPC increased with the increase GGBS proportion.

4. The 7 days and 28 days split tensile strength of PPFRGPC increased with the increase Binder index values.

5. The proposed Modified binder index combine the effect of binder index, molarity , fiber volume fraction and fiber tensile strength is found reasonably well in predicting the split tensile strength of PPFRGPC.

6. Fly ash and GGBS combination can be used for the production of GPC without the need of heat curing.

5.0 REFERENCES

1. Geopolymer Concrete – An Ecofriendly Concrete, N. P. Rajamanea et al.concrete composites, The Master Builder, November 2009 vol.11, No.11, Pg no. 200 to 2006.

2. Ranjbar N, Talebian S, Mehrali M, Kuenzel C, Metselaar HSC, Jumaat MZ. Mechanisms of interfacial Bond in steel and polypropylene fiber reinforced geopolymer composites. Composites Science and Technology. 2016; 122:73–81.

3. Alomayri T, Shaikh FUA, Low IM. Synthesis and mechanical properties of cotton fabric reinforced Geopolymer composites. Composites Part B: Engineering. 2014; 60(0): 36–42.

4. Dias DP, Thaumaturgo C. Fracture toughness of geopolymeric concretes reinforced with basalt fibers.Cement and Concrete Composites. 2005; 27(1):49–54.

5. Pan Z, Sanjayan JG, Rangan BV. Fracture properties of geopolymer paste and concrete. Magazine of concrete research. 2011; 63(10):763–71.

6. Davidovits J. Geopolymers: Inorganic polymeric new materials. Journal of Thermal Analysis. 1991; 37:1633–56.

ISBN: 978-93-8830-599-0

7. Alomayri T, Shaikh FUA, Low IM. Synthesis and mechanical properties of cotton fabric reinforced geopolymer composites. Composites Part B: Engineering. 2014; 60(0):36–42.

8. Alzeer M, MacKenzie KJ. Synthesis and mechanical properties of new fibre-reinforced composites of inorganic polymers with natural wool fibres. J Mater Sci. 2012; 47(19):6958–65.

9. Ranjbar N, Mehrali M, Mehrali M, Alengaram UJ, Jumaat MZ. Graphene nanoplatelet-fly ash based geopolymer composites. Cement and Concrete Research. 2015; 76(0):222–31.

10. He P, Jia D, Lin T, Wang M, Zhou Y. Effects of high-temperature heat treatment on the mechanical properties of unidirectional carbon fiber reinforced geopolymer composites. Ceramics International. 2010; 36(4):1447–53.

11. Navid Ranjbar, Mehdi Mehrali, Arash Behnia, Alireza Javadi Pordsari, A Comprehensive Study of the Polypropylene Fiber Reinforced Fly Ash Based Geopolymer, PLOS ONE | DOI:10.1371/journal.pone.0147546 January 25, 2016.

12. R.Anuradha, et al , Modified Guide lines for Geopolymer concrete mix design using Indian Standard, Asian journal of Civil Engineering (Building and Housing), 133, 353(2012).

13. D. Hardjito, S. E. Wallah, D. M. J. Sumajouw, B. V. Rangan, on the development of fly ash- based geopolymer concrete, ACI Mater. J., 101, 6, 467 (2004).

14. B. V. Rangan, Mix design and production of fly ash based geopolymer concrete, Indian Concrete J., 82, 7 (2008).

15. IS: 516–1956 (Reaffi rmed 1999), Indian Standard Methods of Tests for Strength of Concrete.

16. D.Rama seshu , R.Shankaraiah, B.Sesha Srenivas, (2017), A study on the effect of Binder index on compressive strength of Geopolymer concrete, CWB -3/2017. Pages 211-215.

17. G.Hathiram, B.Seshasreenivas, D.Rama seshu , (2018), Experimental investigation on compressive strength of combined fiber reinforced geopolymer , Journal of Emerging technologies and Innovative Research.(ISSN 2349-5162), volume 4, Issue 6, Pages 530-536.

18. G.Hathiram, B.Seshasreenivas, D.Rama seshu , (2018), Experimental investigation on compressive strength of fiber reinforced geopolymer concrete, International journal of Civil Engineering and Technology(IJCIET) , ISSN 0976-6316, volume 9, Issue 12, Pages 1162-1173.

19. G.Hathiram, B.Seshasreenivas, D.Rama seshu , (2019), Experimental evaluation of the split tensile strength of combined fiber reinforced geopolymer concrete, International Journal of Technical Innovation in Modern Engineering & Science (IJTIMES), Impact Factor: 5.22 (SJIF-2017), e-ISSN: 2455-2585 Volume 5, Issue 01, January-2019, Pages 142-149.

ISBN: 978-93-8830-599-0

Fiber Reinforced Nano Concrete using Recycled Steel Fibers – A Study

V. V. Sai Akhil[1] and K. Manjula Vani[2]
[1]PG Student, JNTUH College of Engineering, Hyderabad
[2]Professor of Civil Engineering, JNTUH College of Engineering Hyderabad

Abstract

Since ancient Rome, Concrete has been widely used as a man-made stone due to its strong mechanical properties. There is a need for improving the performance of the concrete and concern for the environmental impact arising from the continually increasing demand. Tire production is increasing every year due to the increase of vehicle sales. The generation and disposal of waste are inherent to life itself and have presented very serious problems to the human community in world. Recently, some research has been devoted to the use of recycled steel tire fibers (RTWF) in concrete. Nowadays, the application of Nano materials has received numerous attentions to enhance the conventional concrete properties. Eventually, the introduction of Nano materials in concrete is to increase its strength and durability. Nano Silica is found to be a suitable material to replace cement in concrete as it has got high pozzolanic activity. The aim of this study is to enhance the strength of concrete in which cement is partially replaced with Nano-silica by reinforcing it with recycled steel fibers which is obtained from scrap tires. Concrete reinforced with recycled steel fibers has got very good mechanical properties and the behavior of such fibers in concrete is found to be almost similar to that of commercial fibers. In this study, cement is replaced with optimum 3% Nano silica and is reinforced with Recycled steel fiber by 0.25%, 0.5%, 0.75%, 1%, 1.25%, 1.5% weight of concrete. Compressive strength, Split Tensile strength and Flexural strength was tested. From the test results, an optimum percentage at which the strength of concrete is maximum was found.

Keywords: Recycled Tire wire, Recycled Tire wire fiber, Nano-silica, Nano Concrete, Fiber Reinforced Concrete.

INTRODUCTION

Currently, more than 152 million tires are produced in country each year along with existing 200 million tires stockpiled throughout the country. With the rapid development, higher living standards and higher population growth, the quantities of scrap tires are expected to be sharply increased. The total number of abandoned scrap tires will be expected to reach 300 million tires in 2019. In order to properly dispose of these millions of tires, the use of innovative techniques to recycle them is important. Without the proper disposal of these waste tires, the resulting stockpiles would cause major health risks for the public and the environment.

In **2000, Youjiang Wang et al.** studied and reviewed some of the work on concrete reinforcement using recycled fibers. They found out that fiber reinforcement using virgin fibers are effective in improving the toughness, shrinkage, and durability characteristics of concrete. They concluded that the use of recycled fibers from industrial or postconsumer waste could offer additional advantages of waste reduction, resources conservation and low-cost waste fiber for concrete reinforcement could lead to improved infrastructure with better durability and reliability.

In **2007, Job Thomas et al.** studied the influence of the addition of fibers on mechanical properties of concrete. In this study models derived based on the regression analysis of 60 test data for various mechanical properties of steel fiberreinforced concrete were presented. The study concluded that the fiber matrix interaction contributes significantly to enhancement of mechanical properties caused by the introduction of fibers, which is at variance with both existing models and formulations based on the law of mixtures.

Ana Baricevic et al. (2017) developed an environmentally and financially sustainable hybrid fiber–reinforced concrete (SHFRC), obtained by combining traditional components [manufactured steel fibers (M)] and unsorted RTSF. For the first time it is demonstrated that by determining the optimal amount and ratio of fibers (M and RTSF), sustainable hybrid fiber–reinforced concrete (SHFRC) with equal or better properties compared to ordinary fiber-reinforced concrete with only M fibers can be prepared. At the same time, it is also shown that in the optimum quantity and correct ratio, synergy between RTSF and M fibers can be used even for preparation of SHFRC with deflection-hardening behaviour.

In **2017, Amin Yaghoobi et al.** designed fourteen reinforced concrete beams of 6 x 10 in. (150 x 254 mm) in cross section with a combination of multiscale fibers including nanoscale carbon, microscale polyvinyl alcohol, and two types of macroscale polypropylene/ steel fibers to experimentally investigate the structural performance under impact loading. It is concluded that fiber-reinforced concrete (FRC) members singly reinforced with nanofibers alone in conventional concrete-fiber mixtures do not significantly change the

ISBN: 978-93-8830-599-0"

fracture-dominant failure mode, despite excellent nanomaterial properties. In contrast, nanoscale materials provided in combination with macroscale and microscale fibers positively alter the impact energy absorption mechanism.

In **2018, X. F Wang et al.** used 1%, 2% and 3% dosage Nano-SiO2 in lightweight aggregate concrete (LWAC) to investigate the influence of nano-SiO2 on the compressive strength, long-term shrinkage and early cracking of LWAC.

Results revealed that the incorporation of 3% Nano-SiO2 increased compressive strength of LWAC significantly while the influence of Nano-SiO2 on the long-term shrinkage of LWAC was not significant.

In **2018, Filip Grzymski et al.** In this research the effectiveness of recycled steel fibres (obtained from machining process discards) in increasing the ductility of concrete is examined. The effect of a fibre addition was examined for three groups of test specimens differing in their fibres. A series made of plain concrete was prepared as the reference. In the other two groups the concrete matrix was reinforced with identical volume fractions of two types of steel fibres.

EXPERIMENTAL PROGRAM

An experimental Investigation has been conducted on 12 full scale test specimens prepared using Control Concrete (CC), Nano Concrete (NC) and Fiber Reinforced Nano Concrete (FRNC). Recycled Fibers from Tire Wire with aspect ratio 40 are used in producing FRNC. Flexural strength test was conducted on the beam specimen. Split tensile strength test was conducted on the cylinder specimen and compressive strength was conducted on the cube specimen.

MATERIALS AND MIX PROPORTIONS

Research was performed using OPC 53 grade Cement, fine aggregate (FA) confirming to Zone II according to IS: 383-1970, coarse aggregate (CA) confirming to IS: 383-1970, Nano Silica (NS) of specific gravity 2.28, recycled tire wire fiber of aspect ratio 40. The mix proportions of concretes studied are given in table 1.

Mixes were grouped as Control Concrete (CC), Nano Concrete (NC), Recycled Tire Wire Fiber Reinforced Nano Concrete (RTFNC) with varying percentages of fibers.

The identifier (ID) for each RTFNC mix denoted amount of RTWF in percentage of concrete. For example, the mix denoted as 0.25% RTFNC contains 0.25% of RTWF by weight of concrete.

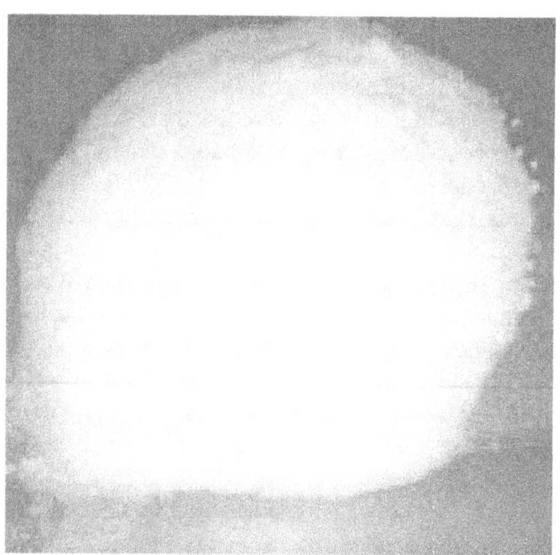

Fig. 1 Nano-Silica

Table 1 Composition of Concrete for 1m3 of volume.

S.NO	MIX IDENTIFIER	Cement(kg)	NS (kg)	RTF(Kg)	FA (kg)	CA (kg)	Water(kg)
1	CC	370	0	0	774	1135	162.8
2	NC	358.9	11.1	0	774	1135	162.8
3	0.25% RTFNC	358.9	11.1	6.09	774	1135	162.8
4	0.5% RTFNC	358.9	11.1	12.19	774	1135	162.8
5	0.75% RTFNC	358.9	11.1	18.29	774	1135	162.8
6	1.0% RTFNC	358.9	11.1	24.38	774	1135	162.8
7	1.25% RTFNC	358.9	11.1	30.48	774	1135	162.8
8	1.5% RTFNC	358.9	11.1	37.1	774	1135	162.8

ISBN: 978-93-8830-599-0

(a) Recycled Tire Wire

(b) Recycled Tire Wire Fiber

Fig. 2 Recycled Tire Wire & Fibers

RESULTS AND DISCUSSIONS

Compressive Strength: The results of the compressive tests performed on CC, NC and RTFNC with different percentages of RTWF tested on the 7th day and 28th day are shown in Fig. 1. As seen in Fig. 1 the compressive strength of concrete increases with the increase of RTWF content up to 0.5% RTFNC (0.5% RTWF in NC) and thereafter decreases with increase in RTWF content. The Optimum percentage of RTWF is 0.5% as the compressive strength obtained is maximum

when RTWF content is 0.5% in Nano-concrete. As the fiber content increases in concrete compressive strength will get reduced due to the formation of air voids, hence when fiber content is increased beyond optimum % i.e. 0.5% reduction in strength is observed. However further research is to be carried out by increasing Nano-silica percentage in Nano-concrete (NC) as the Nano materials acts as filler materials. When RTWF is added at optimum the improvement in compressive strength is 40% more than control concrete (CC) and 7% more than Nano-concrete (NC).

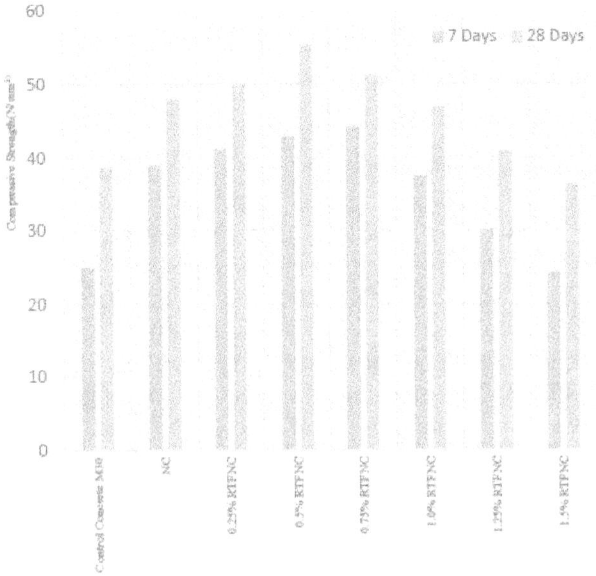

Fig. 3 Compressive strength (N/mm2) at 7 and 28 days

Flexural Strength: The results of the flexural strength tests performed on RTFNC at different percentages of RTWF content is shown in Fig. 2. From Fig. 2, it can be seen that the flexural strength increases with increase in RTWF content upto 0.5% RTFNC and thereafter decreases with increase in RTWF content. The presence of RTWF improves the concrete flexural strength. The highest flexural strength is obtained at 0.5% RTFNC (0.5% RTWF in NC) and can be considered as optimum for Flexural Strength. At optimum the improvement in flexural strength is 50% more than CC and 40% more than NC. The increases of the concrete flexural strength with the presence of RTWF might be due to the influence of concrete bonds
with added materials in the form of RTWF. The maximum flexural strength obtained is 6.62 Mpa.

ISBN: 978-93-8830-599-0

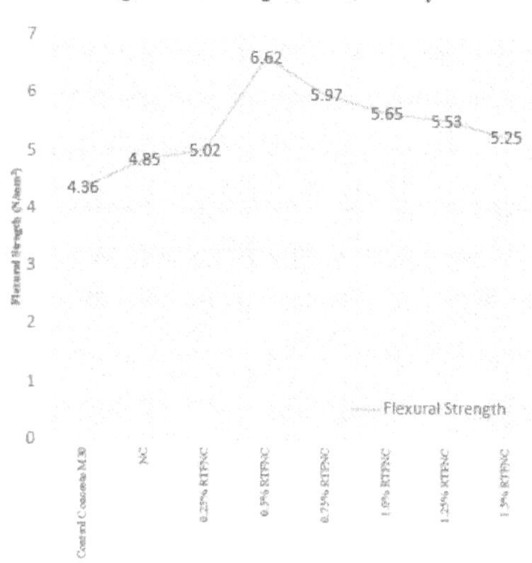

Fig. 4 Flexural Strength (N/mm2) at 28 days

Split Tensile Strength: The splitting tensile strength test results RTFNC with different contents of RTWF are shown in Fig. 4. The that the tensile splitting strength increases as the RTWF content increases RTFNC and thereafter decreases with increase in RTWF content. This consequence of the strong mechanical interlocking force in the concrete presence of RTWF. It is found that the maximum splitting tensile obtained by adding 1.0% RTWF content in NC and obtained strength which is 65% more than CC and 58% more than NC.

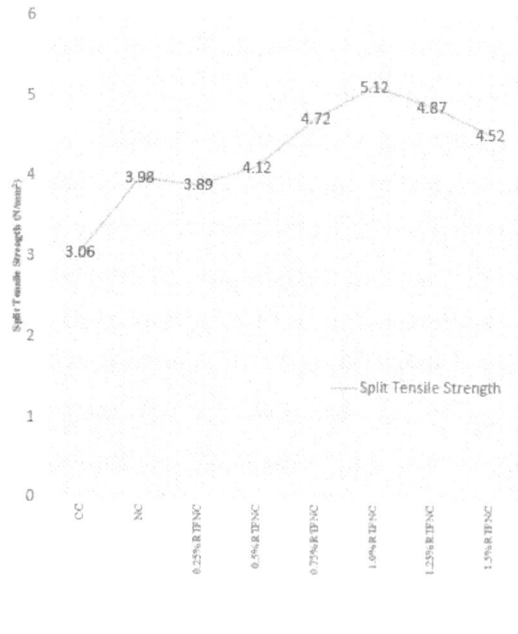

Fig. 5 Split Tensile Strength at 28 days

Table 2 Compressive Strength

SNO	MIX	7 DAYS COMPRESSIVE STRENGTH (N/mm²)	28 DAYS COMPRESSIVE STRENGTH (N/mm²)
1	CC	25	38.8
2	NC	39	48.1
3	0.25% RTFNC	42.1	50.1
4	0.5% RTFNC	47.5	55.4
5	0.75% RTFNC	42.8	51.3
6	1.0% RTFNC	37.5	46.9
7	1.25% RTFNC	30.4	40.9
8	1.5% RTFNC	24.3	36.3

Table 3 Flexural And Split Tensile Strength

SNO	MIX	FLEXURAL STRENGTH (N/mm²)	SPLIT TENSILE STRENGTH (N/mm²)
1	CC	4.36	3.06
2	NC	4.85	3.98
3	0.25% RTFNC	5.02	3.89
4	0.5% RTFNC	6.62	4.12
5	0.75% RTFNC	5.97	4.72
6	1.0% RTFNC	5.65	5.12
7	1.25% RTFNC	5.53	4.87
8	1.5% RTFNC	5.25	4.52

CONCLUSIONS

1. The Optimum Recycled Tire Wire Fibers (RTWF) in Nano Concrete (NC) with 3% optimum Nano Silica is 0.5%.

   a) The corresponding Mechanical Properties are

   b) Compressive Strength: 55.4 N/mm2

   c) Flexural Strength: 6.62 N/mm2

   d) Split Tensile Strength: 4.12 N/mm2.

2. When 0.5% Recycled Tire Wire Fiber (RTWF) are added to Nano Concrete (NC) the improvement in Compressive strength is 40% more than Control Concrete (CC). 7% more than Nano Concrete (NC).

3. When 0.5% Recycled Tire Wire Fiber (RTWF) are added to Nano Concrete (NC) the improvement in Flexural strength is 50% more than Control Concrete (CC). 40% more than Nano Concrete (NC).

ISBN: 978-93-8830-599-0

4. When 1.0% Recycled Tire Wire Fiber (RTWF) are added to Nano Concrete (NC) the improvement in Split Tensile strength is 65% more than Control Concrete (CC). 58% more than Nano Concrete (NC).

5. Considering the environmental and economic sustainability, from the obtained results it can be concluded that using Recycled tire wire as fibers in concrete is highly effective. The cost of a Recycled Tire Wire fiber is 10% of the ordinary steel fibers and the strength properties obtained are almost similar to that of regular steel fibers when used in Nano concrete.

FUTURE SCOPE

- In this study tire wire fibers from passenger cars are only used, hence further research is to be carried out by using tire wire from different types of vehicle as the strength of wires used in vehicles varies.

- Further research is to be carried out by varying Aspect Ratio of fibers as we know that aspect ratio influences strength properties.

- In this study Optimum Nano-silica is chosen for Nano-concrete (NC), further research is to be carried out by varying Nano-silica content.

REFERENCES

1. Ana Baricevic, Dubravka Bjegovic, et al. (2017), "Hybrid fiber–reinforced concrete with unsorted recycled-tire steel fibers." Journal of Materials in Civil Engineering / Volume 29 Issue 6 - June 2017(ASCE)

2. Amin Yaghoobi, Mi G. Chorzepa, Mahadi Masud and Hua Jiang "Multiscale fiber-reinforced concrete including polypropylene and steel fibers" ACI Structural Journal/November-December 2017 (ACI)

3. Filip Grzymski, Michał Musiał, Tomasz Trapko "mechanical properties of fibre reinforced concrete with recycled fibres." Construction and Building Materials 198 (2019) 321–331 (ELSEVIER)

4. Job Thomas and Ananth Ramaswamy (2007), "Mechanical properties of steel fiber-reinforced concrete." Journal of Materials in Civil Engineering / Volume 19 Issue 5 - May 2007 (ASCE)

5. Youjiang Wang, H. C. Wu, et al. (2000), "Concrete reinforcement with recycled fibers." Journal of Materials in Civil Engineering / Volume 12 Issue 4 - November 2000 (ASCE).

6. X.F. Wang, Y.J. Huang, et al. (2018), "Effect of nano-sio2 on strength, shrinkage and cracking sensitivity of lightweight aggregate concrete." Construction and Building Materials 175 (2018) 115–125 (ELSEVIER)

ISBN: 978-93-8830-599-0

An Experimental Study is to Assess the Utility and Efficiency of Coconut Shell as a Partial Replacement with Coarse Aggregate in Concrete

A. Laxminarayana[1], G. Abinash[2] and Uday Baskar[3]

[1,2]Department of Civil Engineering, JNTUH college of Engineering (Autonomous), Hyderabad, India

[3S]Associate Professor Department of Civil Engineering, MRIT, Hyderabad, India

laxminarayanachary@gmail.com, abhinashnaik143i@gmail.com and uday.mendu@gmail.com

Abstract

This paper presents an objective of this study is to assess the utility and efficacy of CS as a coarse aggregate as an alternative to natural aggregate in concrete. CS has not been tried as aggregate in structural concrete. The properties of CS have to be known before it can be used as a coarse aggregate in concrete. This study focuses on coconut shell as a structural material.

If structural LWC can be developed from CS, which is locally available in abundance, it would be a milestone achievement for the local construction industries. Thereforle, the main objective of this research is to determine the feasibility of using solid waste CS as coarse aggregate for structural LWC. The objectives of the study are briefly summarized below.

- To study the workability of CSAC for both M_{20} and M_{25} concrete mixes.

To determine the compressive strength of CSAC for M_{20} and M_{25} grade concrete mixes..

Keywords: Coconut shell (CS), Concrete, Workability, compressive strength, light weight concrete (LWC), Coconut shell Aggregate concrete (CSAC).

1. INTRODUCTION

Concrete, artificial engineering material made from a mixture of Portland cement, water, fine and coarse aggregates and a small amount of air. It is the most widely used construction material in the world. Concrete is the only major building material that can be delivered to the job site in a plastic state. This unique quality makes concrete desirable as a building material because it can be molded to virtually to any form or a shape. Concrete provides wide latitude in surface textures and colors and can be used to construct a wide variety of structures such as highways and streets, bridges, dams, large buildings, airport runways, irrigation structure, break waters, piers and docks, sidewalks, soils and farm building homes and even barges and ship. Other desirable qualities of concrete as a building material are its strength, economy and durability. Depending on the mixture of materials used, concrete will support, in compression, 700 or more kg/sq cm, (10,000 or more 1b/sq cm). The tensile strength of concrete is much lower when compared to compressive strength of concrete, but by using properly designed steel reinforcing, the structural members can be made that are as strong as in compression. The durability of concrete is evidenced by the fact that concrete columns built by the Egyptians more than 3600 years ago are still standing.

Concrete is the premier construction material around the world and is most widely used in all types of construction works, including infrastructure, low and high-rise buildings, and domestic developments. It is a man-made product, essentially consisting of a mixture of cement, aggregates, water and admixture(s). Inert granular materials such as sand, crushed stone or gravel form the major part of the aggregates. Traditionally aggregates have been readily available at economic prices and of qualities to suit all purposes. But, the continued extensive extraction use of aggregates from natural resources has been questioned because of the depletion of quality primary aggregates and greater awareness of environmental protection. In light of this, the non-availability of natural resources to future generations has also been realized. Concrete, the structural material, is a mixture of binder and filler, the binders being the Portland cement and water, the filler being the aggregate coarse and fine.

2. METHODOLOGY

The methodology here follows An attempt was made to utilize coconut shell as a substitution of natural coarse aggregate for making concrete and to verify its strength properties. For this a mix design for M_{20} and M_{25} grades concrete were prepared with natural coarse aggregate with target mean strengths of 26.6 and 31.6 N/mm^2 and then natural coarse aggregate was substituted with coconut shell as a 10%, 20%, 30% of natural coarse aggregate (NCA).The mechanical properties such as compressive strength and workability tests were found out and the test values are compared with different code values.

The test results shows that concrete using coconut shell aggregate up to 30 % replacement with natural coarse aggregate has resulted in acceptable strength

ISBN: 978-93-8830-599-0

required for structural lightweight concrete. It is concluded that the lightweight concrete developed from CSA can be used for both structural and non-structural applications.

In this part of the paper, the experimental results of cs concrete mixes related are workability tests such as slump cone, compaction factor test and compressive strength test is conducted on hardened concrete.

4. RESULTS &DISCUSSIONS

MATERIALS AND MIX PROPORTIONS

Tests conducted on cement

Normal consistency	Setting time Initial - Final	Fineness of cement	Specific gravity of cement	Compressive strength(Mpa) on cement on the day of		
				7days	14days	28days
34%	32mins-8hr	7%	3.1	24	36	54.2

Tests conducted on aggrigates

Fineness Modlr		Spacific gravity			Bulk density (gm/cm$^{3)}$)			Water absorption		
FA	CA	Sand	Gravel	Coco shell	Sand	Gravel	Coco shell	Sand	Gravel	Coco shell
2.43	-	2.5	2.7	1.43	1.63	1.56	0.56	0.1%	0.35%,	0.6%

COCONUT SHELL AGGREGATE

Coconuts are referred to as "man's most useful trees", "king of the tropical flora" and "tree of life". Coconuts or its scientific name cocos nucifera are the most important of cultivated palms and the most widely distributed of all palms. Coconut shell is available as a waste product from the coconut industry obtained during the breaking operation of coconuts.

CSAC, which is produced using CS aggregates, was the main concrete studied in this investigation. CS is discarded at coconut industries as half-shell rounds. CS was collected from the local coconut oil mills to analyze the properties of CS in this study. CS has maximum thickness in range of 2-8 mm. The available coconut shells were hammered and crushed into smaller pieces in the range of 3-12 mm in length as shown in the fig3.4.The crushed materials were washed with

clean water for several times and soaked in water for 24hrs and then dried in sun, made saturated and then required quantity was taken for casting.

To prevent water absorption occurred during the mixing process, soak CS aggregates at SSD condition based on 24hr submersion in potable water. When these aggregates are used in concrete, it is believed that the cementitious hydration can be enhanced due to the process of internal curing from the internal reservoir of water absorbed by the LWA. In addition, the high absorption capacity of CS aggregates may also be beneficial in the sense that it can strengthen the contact zone between the cement paste and the CS through the formation of new hydrated compounds at the interface zone. The water stored within the CS aggregate may also help reduce the heat generated during cement hydration.

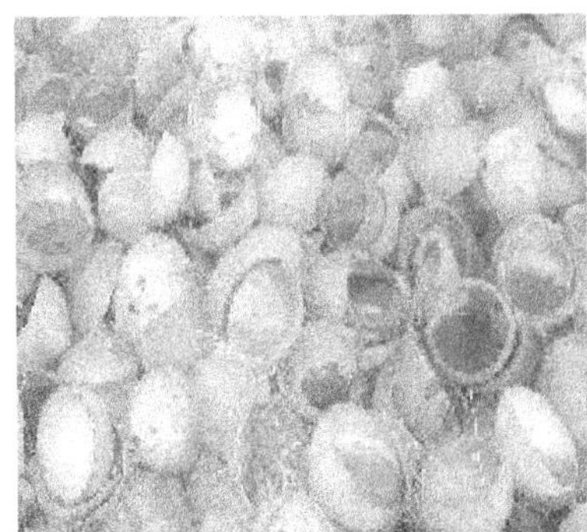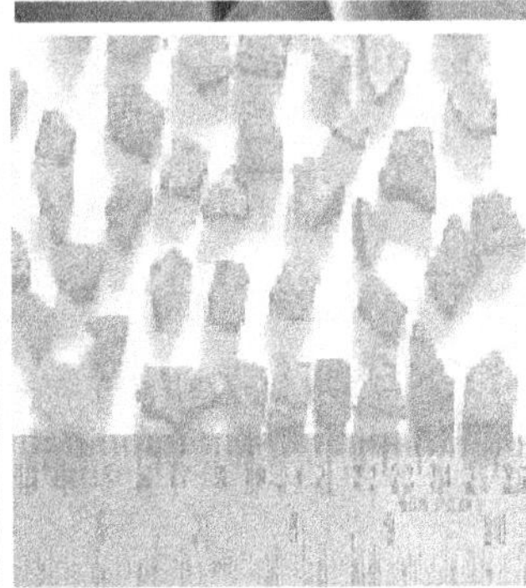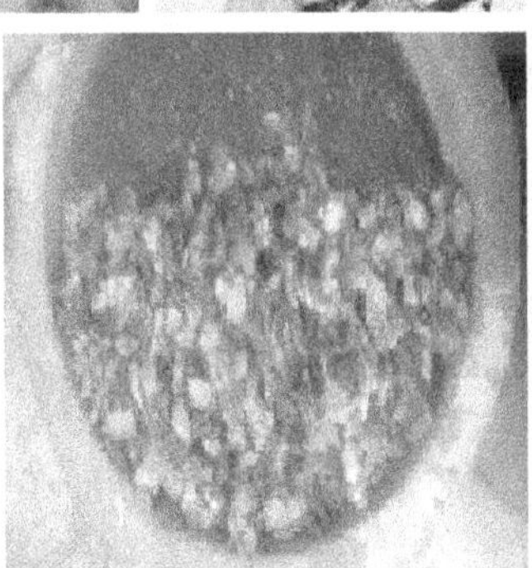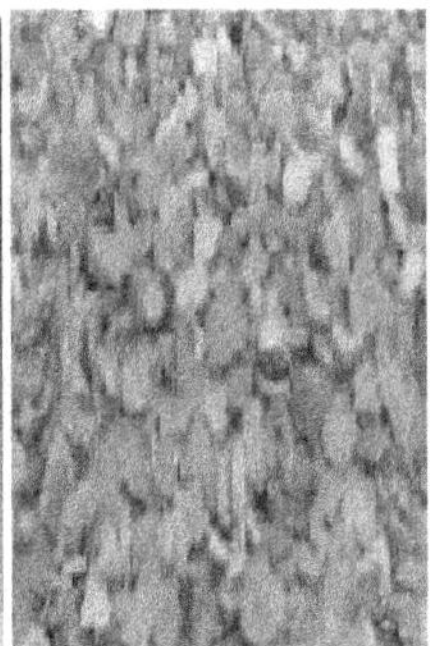

Fig. 1 Process involved in preparing the coconut shell aggregates

ISBN: 978-93-8830-599-0

MIX PROPORTIONS

Indian standard recommended method IS 10262-82

In this study Indian standard recommended method is used for designing the mix and the procedure of this method is as follows.

Table Assumed standard deviation as per IS 456: 2000

Grade of Concrete	M 10, M 15	M 20, M 25	M 30, M 35, M 40, M 45, M 50
Assumed Standard Deviation N/mm²	3.5	4.0	5.0

1. Obtain the water cement ratio for the desired mean target using the empirical relationship between compressive strength and water cement ratio so chosen is checked against the limiting water cement ratio. The water cement ratio so chosen is checked against the limiting water cement ratio for the requirements of durability given in table and adopts the lower of the two values.

2. Estimate the amount of entrapped air for maximum nominal size of the aggregate from the table.

Table Approximate entrapped air content

Maximum size of aggregate (mm)	10	20	40
Entrapped air, as % of volume of concrete	3.0	2.0	1.0

3. Select the water content, for the required workability and maximum size of aggregates (for aggregates in saturated surface dry condition) from graphs shown in figure 2.

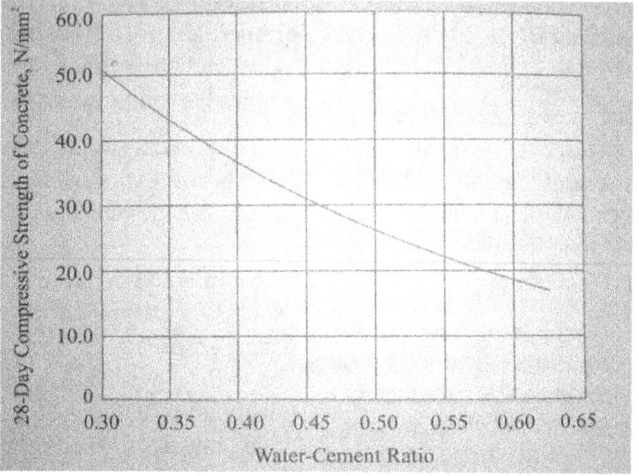

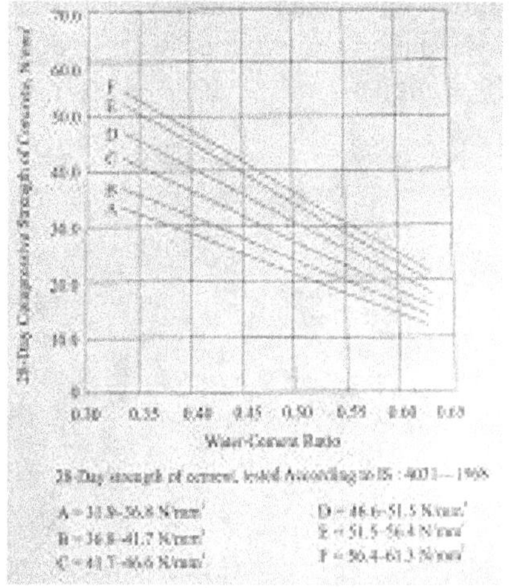

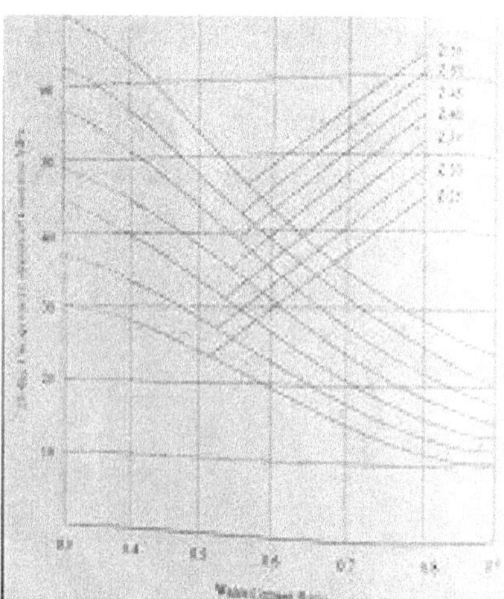

Fig. 2 Relationship between W/C ratio and compressive strength

ISBN: 978-93-8830-599-0

4. Determine the percentage of fine aggregate in total aggregate by absolute volume from table for the concrete using crushed coarse aggregate.

5. Adjust the values of water content and percentage of sand as provided in the table for any difference in workability, water cement ratio, grading of fine aggregate and for rounded aggregate the values are given in table.

Table Approximate sand and water contents per cubic meter of concrete

Maximum size of aggregate (mm)	Water content including surface water, per cubic meter of concrete (kg)	Sand as % of total aggregate by absolute volume
10	200	40
20	186	35
40	165	30

Table Adjustment of values in water content and sand percentage for other conditions

Change in conditions stipulated for tables	Adjustments required in water content	Adjustment required in % sand in total aggregate
For sand conforming to grading Zone 1,Zone 3or Zone 4 of table 4, IS:383-1979	0	+1.5% for zone 1 -1.5%for zone 2 -3.0%for zone 4

Design

Target strength	cement content	volume of coarse agg (IS:10262:2009 table 3)		volume of aggregates	W/C ratio	water content
26.6 N/mm^2	383.2kg/m^3	C Agg 0.64m^3	Fine ag 0.36m^3	CA=1182kg/m^3 FA=616kg/m^3	**0.5**	191.6 kg/m^3

3. Determination of water content

Table 4.4 Water content for different sizes of aggregates

Size (mm)	10	20	40
Water (lts)	206	186	168

Calculation of Mix Proportion

Table 4 Mix design for 0% replacement of coarse aggregate with CS aggregates in M$_{20}$ grade concrete

Change in conditions stipulated for tables	Adjustments required in water content	Adjustment required in % sand in total aggregate
Increase or decrease in the value of compacting factor by 0.1	±3%	0
Each 0.05 increase or decrease in water cement ratio	0	±1%
For rounded aggregate	-15kg	-7

Determine the concrete mix proportions for the first trial mix As per is code provision.

4.1 Mix Design for M20 Grade Concrete

In this study Indian standard recommended method IS 10262-82 is used for designing the mix of M$_{20}$ grade concrete.

Parameters required

1. Slump   -   75mm
2. Exposure condition   -   mild
3. Grade of concrete   -   M$_{20}$
4. Specific gravity

Cement	Fine aggregate	Coarse aggregate	Coconut shell aggregate
3.1	2.5 (zone3)	2.7 (20mm size)	1.43

Cement	Fine Aggregate	Coarse Aggregate	Water
383kg/m^3	616 kg/m^3	1182 kg/m^3	191.6 kg/m^3
1	1.6	3.08	0.5

There fore the mix design of M$_{20}$ grade concrete = **1:1.6:3.08.**

In this study the coarse aggregate is replaced with coconut shells in 10%, 20%, 30% so the mix design for these replacements are as follows

Table coarse aggregate is replaced with coconut shells in 10%, 20%, 30%

% replacement of CAgg with coco shell	cement	fine Agg	Course Agg	Coconut Shell	water
0%	383kg/m^3	616 kg/m^3	1182 kg/m^3	0	191.6 kg/m^3
10%	383kg/m^3	617 kg/m^3	1136 kg/m^3	62.5 kg/m^3	191.6 kg/m^4
20%	383kg/m^4	618 kg/m^3	945.56kg/m^3	125.19kg/m^3	191.6 kg/m^5
30%	383kg/m^5	619 kg/m^3	827.36kg/m^3	187.79kg/m^3	191.6 kg/m^6

ISBN: 978-93-8830-599-0

4.2 MIX DESIGN FOR M25 GRADE CONCRETE

In this study Indian standard recommended method IS 10262-82 is used for designing the mix of M_{25} grade concrete.

Design Procedure

Parameters required

Slump	Exposure condition	Grade of concrete	Specific gravity
100 mm	mild	M_{25}	As per above

Target strength	cement content	volume of aggregates	volume of aggregates	W/C ratio	water content
$31.6N/mm^2$	$419kg/m^3$	Ca = $0.646\,m^3$ Fa = $0.354m^3$	CA=$1164kg/m^3$ FA=$590kg/m^3$	0.5	$197\ kg/m^3$

Table Coarse aggregate is replaced with coconut shells in 10%, 20%, 30%

% replacement of CAgg with coco shell	cement	fine Agg	C Agg	Coconut Shell	water
0	$419kg/m^3$	$590kg/m^3$	$1164\ kg/m^3$	0	191.6 kg/m3
10%	$419kg/m^4$	590 kg/m^3	1047 kg/m^3	61.6 kg/m^3	191.6 kg/m4
20%	$419kg/m^5$	591 kg/m^3	930.7kg/m	122.1kg/m^3	191.6 kg/m5
30%	$419kg/m^6$	$197kg/m^3$	814.36kg/m	184.84kg/m^3	191.6 kg/m6

5. TEST RESULTS (TESTS ON CONCRETE)

5.1 Workability Tests on Fresh Concrete

Slump value for different percentage replacements in **M20** grade concrete

Replacement in%	0%	10%	20%	30%
Slump Value in mm	50	30	20	10

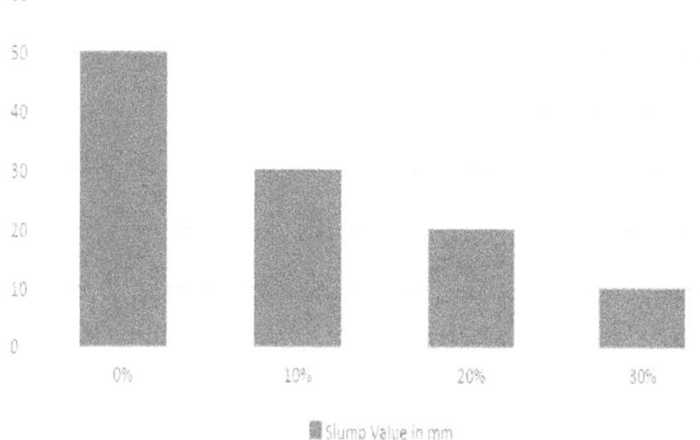

Slump value for different percentage replacements in M_{20} grade concrete

Slump value for different percentage replacements in **M25** grade concrete

Replacement in%	0	10	20	30
Slump Value in mm	70	50	35	20

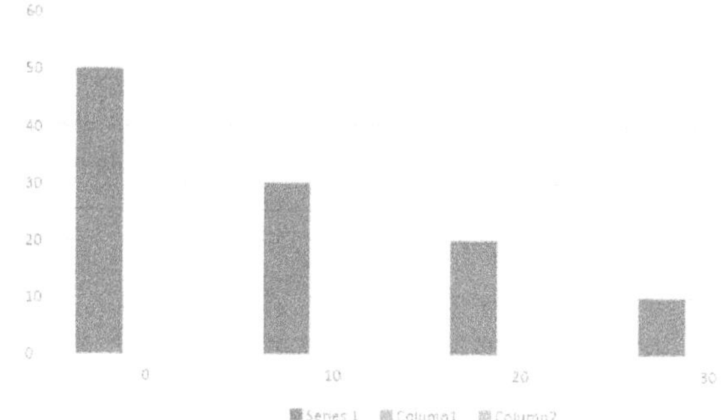

Slump value for different percentage replacements in M_{25} grade concrete

5.2 Compaction Factor Test

Compaction factor values for different cs percentage replacements in **M 20** grade concrete

Replacement in%	0%	10%	20%	30%
Compaction Factor	0.9	0.87	0.86	0.85

Compaction factor values for different cs percentage replacements in **M 25** grade concrete

Replacement in%	0%	10%	20%	30%
Compaction Factor	0.86	0.85	0.84	0.83

From the above tables we can see that as the % replacement of coarse aggregate with coconut shells increases the value of compaction factor value decreases. The compaction factor value is directly

ISBN: 978-93-8830-599-0

proportional to workability, so we can say that as the coconut shells % in concrete increases the workability decreases.

5.3 Compressive Strength Test on Hardened Concrete

The main objective of this test is to determine the compressive strength of concrete for cube samples. Compression test develops a rather more complex system of stresses. Due to compression load, the cube undergoes lateral expansion owing to Poisson's ratio effect. The steel plates do not undergo lateral expansion to some extent of concrete, with the result that steel restrains the expansion tendency of concrete in the lateral direction. This includes a tangential force between the end surfaces of concrete specimen and adjacent steel plates of testing machine. Due to this platen restrains the lateral expansion of the concrete in the parts of the specimen near its end.

Compressive strength of **M20** grade concrete:

	7DAYS				28DAYS			
Coconut Shell (%)	0%	10%	20%	30%	0%	10%	20%	30%
Load Crushing in (KN)	182	169	151	139	246	227	219	206
Compressive Strength (N /mm^2)	**18.2**	**16.9**	**15.1**	**13.9**	**24.6**	**22.7**	**21.9**	**20.6**

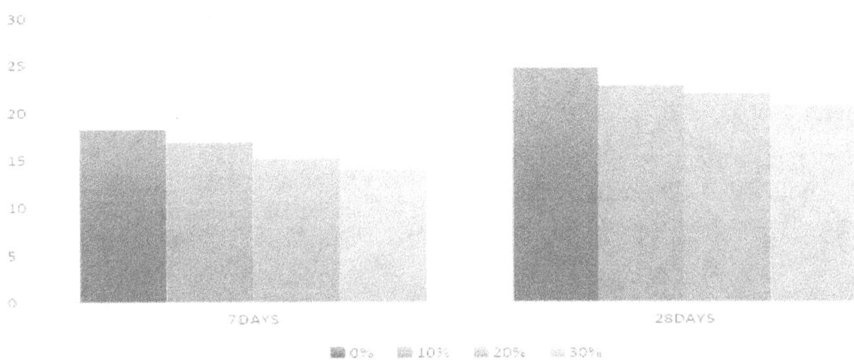

Compressive strength of **M25** grade concrete

	7DAYS				28DAYS			
Coconut Shell (%)	0%	10%	20%	30%	0%	10%	20%	30%
Load Crushing in (KN)	232	216	192	176	293	274	262	254
Compressive Strength (N /mm^2)	**23.2**	**21.6**	**19.2**	**17.6**	**29.3**	**27.4**	**26.2**	**25.4**

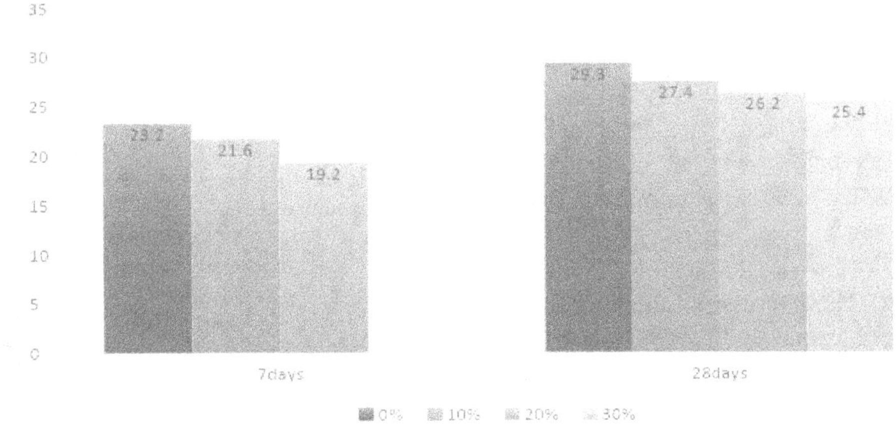

ISBN: 978-93-8830-599-0

5. CONCLUTION AND FUTURE SCOPE

5.1 Conclusion

It was concluded that the CS's were more suitable as low strength-giving lightweight aggregate when used to replace common coarse aggregate in concrete production.

The present project work gives the following:

- Results of experiments on compressive strength and workability for different CS replaced concretes have been presented with those of control concrete. The data shows that CS aggregate can be used in place of normal aggregate, however, performance of CS aggregate concrete is little lower than normal aggregate concrete.

- With CS percentage increase the 7 day strength gain also increased with corresponding 28 day curing strength. However, the overall strength decreased with CS replacement when compared to control concrete.

- Increase in CS replacement permeable voids also increased. With 30% CS replacement the permeable were 40 percent higher than normal concrete.

- The fresh concrete density and hardened concrete density after 28 days using coconut shell for M_{20} and M_{25} grade concrete was found to be in the range of 1970-2450 kg/m^3 and 2110-2360 kg/m^3.

- The 28 days compressive strength of coconut shell concrete for M_{20} grade was found to be 24.6, 22.7, 21.9 and 20.6 N/mm^2 for 0%, 10%, 20% and 30% replacement by coconut shell aggregate under full water curing and it satisfies the requirement for structural lightweight concrete.

- The 28 days compressive strength of coconut shell concrete for M_{25} grade was found to be 29.3, 27.4, 26.2 and 25.4 N/mm^2 for 0%, 10%, 20% and 30% replacement by coconut shell aggregate under full water curing and it satisfies the requirement for structural lightweight concrete.

- If we use M_{20} grade concrete, the strength is greater than 20 N/mm^2 up to 30 % replacement of coarse aggregate with CS, similarly for M_{25} grade concrete the strength is greater than 25 N/mm^2 up to 30 % replacement of coarse aggregate.

- So, we can replace the coarse aggregate with CS up to 30 %.

- Maximum attempt is made to maintain economy, strength and serviceability of concrete.

5.2 Future Scope

- From the study as the percentage replacement of CS is increased the workability decreases, to increase the workability, we can add plasticizers and any other material to increase the workability.

- As the percentage of CS in concrete increases the strength decreases, so to increase the strength of concrete, we can add 10 % of other ingredients like fly ash, rock powder etc to CS concrete and the experiments should be done.

- Experiments should be done on split tensile strength and flexural strength.

- Coconut shell may exhibit more or less resistance against crushing, impact and abrasion, compared to crushed granite aggregate.

6. REFERENCES

1. IS 456:2000 - Plain and reinforced concrete
2. IS 383:1970 - Specification for coarse and fine aggregate
3. IS 516: 1959 - Methods of tests of concrete
4. IS 3025 - Water quality parameters for concrete
5. IS 299 - Properties of cement
6. IS 2387 - Tests of aggregate
7. IS 10262: 2009 - Concrete mix deign proportioning guidelines
8. Properties of concrete - A.M Neville (1997)
9. Concrete Technology by M.L.Gambhir 5th edition, 2013 and by M.S.Shetty
10. Light Weight Aggregate Concrete – Science, Technology and Applications by Satish Chandra & Leif Berntsson
11. Mechanical and Bond Properties of Cocnut Shell Concrete - K. Gunasekaran, P.S. Kumar, M. Lakshmipathy
12. Use of Recycled Materials in Building and Structures - Cyr M, Aubert JE, Husson B, Clastres P
13. Developing Lightweight Concrete Using Agricultural and Industrial Solid Wastes - Gunasekaran K, Kumar PS, (2008).

ISBN: 978-93-8830-599-0

Experimental Study on Fiber Reinforced Nano Concrete using Recycled Glass Fibers

Diwakar Prasad[1] and K. Manjula Vani[2]
[1]PG Student, [2]Professor, JNTUH, Hyderabad.

Abstract

Nowadays, numerous attentions have been paid to applying nano materials to enhance convectional concrete properties. The introduction of nano materials in concrete ultimately increases its strength and durability. Nano Silica is a material suitable for replacing cement in concrete because it has a high pozzolanic activity. As concrete is rich in compression and poor in tension, Fibers are tensile in nature, so Recycled Glass Fibers are added to the concrete mix to improve the Tensile Strength. This study has been conducted for understanding the mechanical behavior of Nano-Concrete replacing 3% cement with Nano silica and adding recycled glass fibers by weight of concrete. Studies were conducted on the compressive, flexural and split tensile strengths of concrete by varying the fiber percentage from 0 to 1.25% by weight of concrete, and 3% Nano silica in cement. The obtained results then compared with M30 control concrete and optimum percentage at which the strength of concrete is maximum was found.

keywords: Recycled glass fiber; Nano-Silica; Nano-concrete; Mechanical properties.

INTRODUCTION

Over the last three decades, sustainable cementitious materials have been rapidly developed, and the use of waste materials in cement based composites has simultaneously achieved particular importance. Using waste materials reduces negative environmental impacts and construction costs **(Meddah et al. 2009)**. Therefore, various studies have been carried out to assess the feasibility of using the waste materials obtained from different sources in concrete. In this way, many efforts have been made to improve the mechanical properties of concrete through using recycled fibers in reinforced concrete.

In **2018, X. F Wang et al.** investigated that, 1%, 2% and 3% dosage Nano-SiO$_2$ was used in lightweight aggregate concrete (LWAC) to investigate the influence of nano-SiO$_2$ on the compressive strength, long-term shrinkage and early cracking of LWAC. Results revealed that the incorporation of 3% Nano-SiO$_2$ increased compressive strength of LWAC significantly while the influence of Nano-SiO$_2$ on the long-term shrinkage of LWAC was not significant.

In **2017, Karam Mahmoud et al**. Presented the development of nano-modified fiber reinforced cementitious composite (FRCC) mixtures. In this study they used the Nano silica, Blast furnace pellets with different lengths, Glass fiber-reinforced polymer (GFRP) bars. The overall results indicated that 2.5% by volume of 36mm long Basalt Fiber (BF) pellets to the mixture achieved the best performance like acceptable rate of hardening, benefit on the post cracking behavior, enhanced ductility and bonding to GFRP bars and enhanced the flexural toughness of the cementitious composites.

The influence of nanolimestone/nanoCaCO$_3$ (NC) on the properties of ultrahigh-performance concrete (UPHC) cured at standard and heat conditions was experimentally investigated. The NC was used at ratios of 1, 2 and 3% as partial replacement for cement. Test results shows that replacing cement with NC achieved the higher autogenous shrinkage of UPHC mixture than that of control mixture without NC. Through incorporating NC, the UHPCs have benefits of accelerated early-age hydration and enhanced particle packing density, while achieving enhanced mechanical strength compared with that of the control mixture without NC. It has also been concluded that replacing cement with NC resulted in lower flow ability of UPHCs mixtures with that of the control mixture (**Wengui Li et al. 2016).**

This research program explored the influence of recycled glass fiber reinforced polymer (GFRP) on compressive strength, splitting tensile strength and drying shrinkage in concrete and alkali silica reaction (ASR) expansion in accelerated mortar beam and concrete prism tests. Compressive strengths and drying shrinkage were not improve but splitting tensile strength was improved by recycled GFRP additions at a substitution level of 5% weight of coarse aggregate. Negligible expansion was observed from the ASR testing **(Alireza Dehghan et al. 2017).**

In this experimental study, the effects of recycled glass fiber on the drying shrinkage and mechanical properties of the self compacting mortar and fly ash-slag geopolymer were investigated. The rate of drying shrinkage was much lower in SCMs than in FSGMs. The experimental results showed that adding recycled

ISBN: 978-93-8830-599-0

glass fiber has much greater effects on increasing flexural strength and decreasing drying shrinkage rate in fly ash – slag geopolymer mortars when compared to self compacting mortars **(Z. Abdollahnejad et al. 2017).**

In this paper, the quantities of cement, nano-silica, aggregates, and water for each batch were determined. Mix design for M30 in batches is designed. A total of 84 test specimens are casted. Fibers added to concrete during mixing stage manually, by dispersing them uniformly and randomly. Recycled Glass fibers added 0%, 0.25%, 0.5%, 0.75%, 1% and 1.25% by weight of concrete and 3% Nano-Silica replacement to cement. Compressive Strength Test at 7 days & 28 days, Split Tensile Strength and Flexural Strength test at 28 days are carried out.

EXPERIMENTAL PROGRAM

An experimental Investigation has been conducted; a total of 84 test specimens were prepared for Control Concrete (CC), Nano Concrete (NC) and Fiber Reinforced Nano Concrete (FRNC). Recycled Glass Fibers of length 12mm, diameter 14µm and having aspect ratio of 857 are used in producing FRNC. Flexural strength test was conducted on the beam specimen. Split tensile strength test was conducted on the cylinder specimen and compressive strength was conducted on the cube specimen.

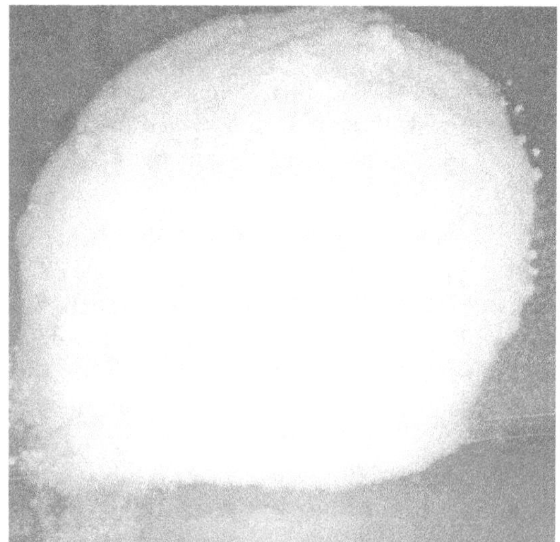

Fig.1 Nano-Silica

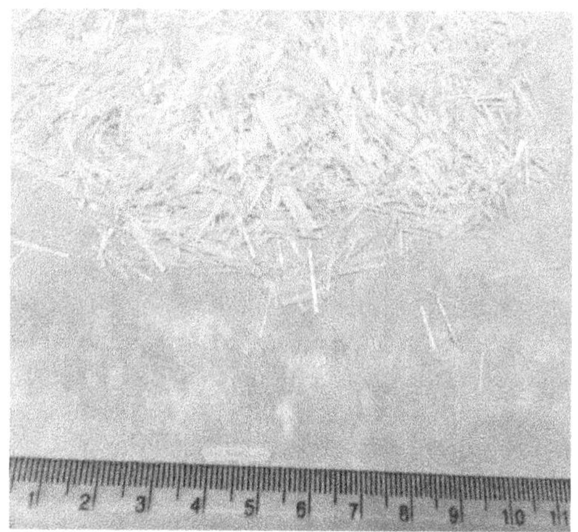

Fig. 2 Recycled Glass Fibers

MATERIALS AND MIX PROPORTIONS

Research was performed using OPC 53 grade Cement, fine aggregate (FA) confirming to Zone II according to IS: 383-1970, coarse aggregate (CA) confirming to IS: 383-1970, Nano Silica (NS) of specific gravity 2.28, recycled glass fiber (RGF) of aspect ratio 857. The mix proportions of concretes studied are given in table 1, Mixes were grouped as Control Concrete (CC), Nano Concrete (NC), and Recycled Glass Fiber Reinforced Nano Concrete (RGFRNC) with varying percentages of fibers. The identifier (ID) for each RGFRNC mix denoted amount of RGF in percentage of concrete. For example, the mix denoted as 0.25% RGFRNC contains 0.25% of RGF by weight of concrete.

Table 1 Mix Proportion for 1m^3 of volume for M30 Grade.

S. No.	Mix identifier	Cement (kg)	NS (kg)	RGF (Kg)	FA (kg)	CA (kg)	Water (kg)
1	CC	370	0	0	774	1135	162.8
2	NC	358.9	11.1	0	774	1135	162.8
3	0.25% RGFRNC	358.9	11.1	6.09	774	1135	162.8
4	0.50% RGFRNC	358.9	11.1	12.19	774	1135	162.8
5	0.75% RGFRNC	358.9	11.1	18.29	774	1135	162.8
6	1.00% RGFRNC	358.9	11.1	24.38	774	1135	162.8
7	1.25% RGFRNC	358.9	11.1	30.48	774	1135	162.8

RESULTS AND DISCUSSIONS

Table 2 Compressive Strength, Flexural Strength and Split Tensile Strength details for M30 mix

S. No.	Concrete Mix	Percentage of Nano-Silica (%)	Percentage of Recycled Glass Fibers (%)	Compressive Strength (N/mm²)		Flexural Strength for 28 Days (N/mm²)	Split Tensile Strength for 28 Days (N/mm²)
				7 Days	28 Days		
1	CC	0	0	25	38.8	4.36	3.01
2	NC	3	0	39	48.1	4.85	3.98
3	0.25% RGFRNC	3	0.25	40.6	48.9	4.92	4.10
4	0.50% RGFRNC	3	0.50	42.8	51.2	5.11	4.43
5	0.75% RGFRNC	3	0.75	41.3	47.4	5.75	4.91
6	1.00% RGFRNC	3	1.00	40.1	46.1	5.43	4.61
7	1.25% RGFRNC	3	1.25	39.2	42.8	5.03	4.20

CC – Control Concrete

NC – Nano-Concrete

RGFRNC – Recycled Glass Fiber Reinforced Nano-Concrete

Compressive Strength: By adding recycled glass fibers by weight of concrete mix by 0%, 0.25%, 0.5%, 0.75%, 1% & 1.25% and replacing 3% nano silica to cement. From Fig.1, the maximum optimum dosage of fiber in nano-concrete by compressive test on cube is obtained at 0.5% recycled glass fiber and 3% Nano silica. The average 7 days compressive strength is 42.8 N/mm^2 and 28 days compressive stre ngth is 51.2 N/mm^2. Up to 0.5% recycled glass fiber and 3% nano silica; there is a gradual increase in Strength at 28 days similar to the compressive strength of 7 days. And then there is a gradual decrease of compressive strength. This may be due to increase of fibrous material in concrete. From the test results it indicates that there is an increase of about 23.9% of compressive strength by utilization of 0.5% recycled glass fiber and 3% nano silica.

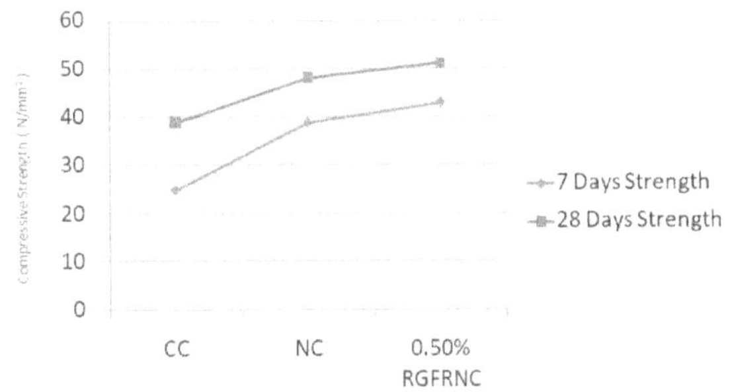

Fig. 4 Compressive Strength (N/mm^2) for 7 Days and 28 Days

Fig.4 shows the comparison of Control Concrete (CC), Nano-Concrete (NC) and optimum RGF percentage i.e.   0.5% RGFRNC. Nano-Concrete is increased upto 24% compared to Control Concrete. Recycled Glass Fiber reinforced Nano-Concrete increased to 32% compared Control Concrete.

Flexural Strength: By adding recycled glass fibers by weight of concrete mix by 0%, 0.25%, 0.5%, 0.75%, 1% & 1.25% and replacing 3% Nano Silica to cement. From Fig.2, the maximum optimum dosage of fiber in nano-concrete by flexural test on beam is obtained at 0.75% recycled glass fiber and 3% Nano Silica. The average 28 days flexural strength is 5.75 N/mm^2. From the test results it indicates that there is an increase of about 32% of flexural strength by utilization of 0.75% recycled glass fiber and 3% Nano Silica.

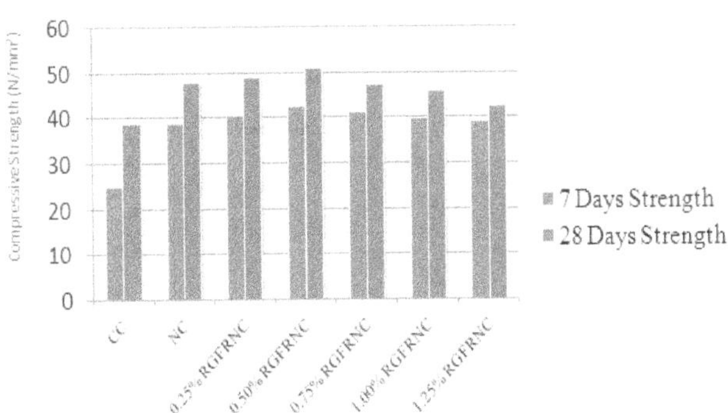

Fig. 3 Compressive Strength (N/mm^2) for 7 Days and 28 Days

ISBN: 978-93-8830-599-0

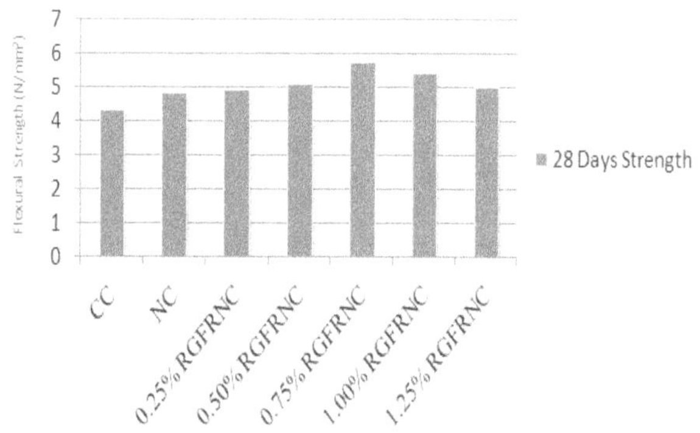

Fig. 5 Flexural Strength (N/mm^2) for 28 Days

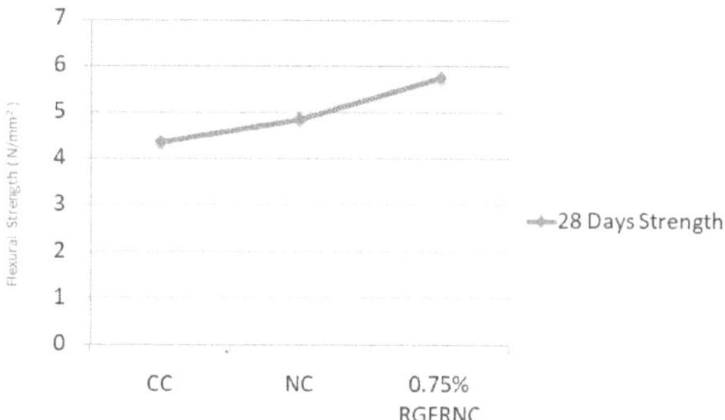

Fig. 6 Flexural Strength (N/mm^2) for 28 Days

Fig. 6 shows the comparison of Control Concrete (CC), Nano-Concrete (NC) and optimum RGF percentage i.e. 0.75% RGFRNC. Nano-Concrete is increased upto 11% compared to Control Concrete. Recycled Glass Fiber reinforced Nano-Concrete increased to 31% compared Control Concrete.

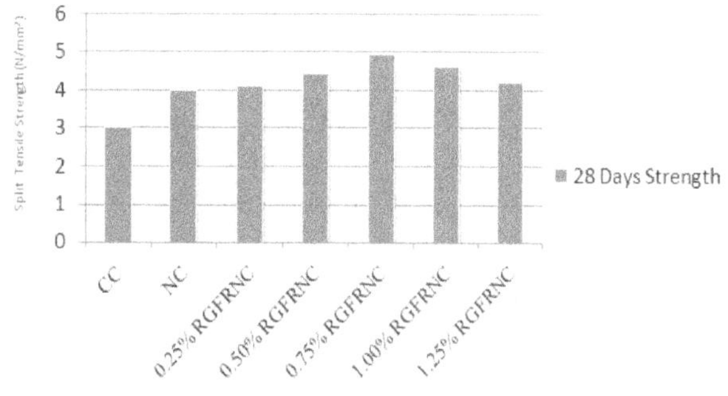

Fig. 7 Split Tensile Strength (N/mm^2) for 28 Days

Split Tensile Strength: By adding recycled glass fibers by weight of concrete mix by 0%, 0.25%, 0.5%, 0.75%, 1% & 1.25% and replacing 3% nano silica to cement. The maximum optimum dosage of fiber in nano-concrete by split tensile test on cylinder is obtained at 0.75% recycled glass fiber and 3% Nano silica. The average 28 days split tensile strength is 4.91 N/mm^2.

From the test results it indicates that there is an increase of about 63% of tensile strength by utilization of 0.75% recycled glass fiber and 3% Nano Silica.

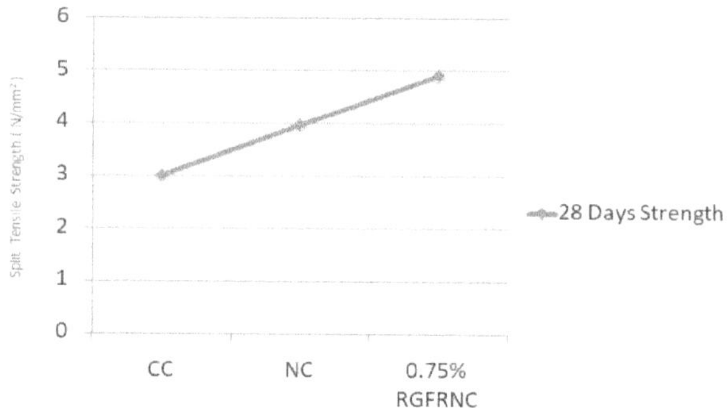

Fig. 8 Split Tensile Strength (N/mm^2) for 28 Days

Fig.8 shows the comparison of Control Concrete (CC), Nano-Concrete (NC) and optimum RGF percentage i.e. 0.75% RGFRNC. Nano-Concrete is increased upto 32% compared to Control Concrete. Recycled Glass Fiber reinforced Nano-Concrete increased to 63% compared Control Concrete.

CONCLUSIONS

1. The Properties of M30 grade Control Concrete for Compressive Strength, Flexural Strength and Split Tensile Strength are **38.8 N/mm^2, 4.36 N/mm^2 and 3.06 N/mm^2** respectively.

2. The Properties of M30 grade Nano-Concrete with optimum of 3% Nano-Silica for Compressive Strength, Flexural Strength and Split Tensile Strength are **48.1 N/mm^2, 4.85 N/mm^2 and 3.98 N/mm^2** respectively.

3. The Optimum % of Recycled Glass Fiber in Nano-Contrete for Compressive Strength is **0.5** and corresponding compressive strength is **51.2 N/mm^2.**

4. The Optimum % of Recycled Glass Fiber in Nano-Contrete for Flexural Strength is **0.75** and corresponding Flexural Strength is **5.75 N/mm^2.**

5. The Optimum % of Recycled Glass Fiber in Nano-Contrete for Split Tensile Strength is **0.75** and corresponding Split Tensile Strength is **4.91 N/mm^2.**

6. The Optimum Recycled Glass Fiber in Nano-Concrete with 3% Nano Silica is 0.5%. The corresponding Mechanical Properties for Compressive Strength, Flexural Strength and Split Tensile Strength are **51.2 N/mm^2, 5.11 N/mm^2 and 4.43 N/mm^2** respectively.

7. When **0.5% Recycled Glass Fiber (RGF)** is added to Nano-Concrete the improvement in

ISBN: 978-93-8830-599-0

Compressive Strength is **32%** more than Control Concrete (CC) and **6%** more than Nano-Concrete (NC).

8. When **0.75%** Recycled Glass Fiber (RGF) is added to Nano-Concrete the improvement in Flexural Strength is **31%** more than Control Concrete (CC) and **19%** more than Nano-Concrete (NC).

9. When **0.75%** Recycled Glass Fiber (RGF) is added to Nano-Concrete the improvement in Split Tensile Strength is **63%** more than Control Concrete (CC) and **23%** more than Nano-Concrete (NC).

REFERENCES

1. Alireza Dehghan, Karl Peterson, Asia Shvarzman (2017). "Recycled glass fiber reinforced polymer additions to Portland cement concrete." *Construction and Building Materials 146 (2017) 238–250 (ELSEVIER)*

2. Karam Mahmoud, Ahmed Ghazy, Mohamed T. Bassuoni and Ehab El-Salakawy (2017). "Properties of Nanomodified Fiber-Reinforced Cementitious Composites." *J. Mater. Civ. Eng., 2017, 29(10): 04017173 (ASCE)*

3. Meddah, M. S., and Bencheikh, M. (2009). "Properties of concrete reinforced with different kinds of industrial waste fibre materials." *Constr. Build. Mater., 23(10), 3196–3205 (ELSEVIER)*

4. Wengui Li, M. ASCE; Zhengyu Huang; Tianyu Zu; Caijun Shi (2016) . "Influence of Nano limestone on the Hydration, Mechanical Strength, and Autogenous Shrinkage of Ultrahigh-Performance Concrete." *J. Mater. Civ. Eng., 2016, 28(1): 04015068 (ASCE)*

5. X.F. Wang, Y.J. Huang, G.Y. Wu, C. Fang, D.W. Li, N.X. Han, F. Xing (2018). "Effect of nano-SiO2 on strength, shrinkage and cracking sensitivity of lightweight aggregate concrete." *Construction and Building Materials 175 (2018) 115–125 (ELSEVIER)*

6. Z. Abdollahnejad, M. Mastali, M. Mastali and A. Dalvand (2017). "Comparative Study on the Effects of Recycled Glass–Fiber on Drying Shrinkage Rate and Mechanical Properties of the Self-Compacting Mortar and Fly Ash–Slag Geopolymer Mortar." *J. Mater. Civ. Eng., 2017, 29(8): 04017076 (ASCE)*

Experimental Study on High Strength Concrete with Copper Slag and Carbon Fiber

Hima Bindu. K[1] and Lalitha. G[2]

[1]M Tech Student, Civil Engineering Department, VNRVJIET, Telangana, India

[2]Assistant Professor, Civil Engineering Department, VNRVJIET Telangana, India,

himabindu453@gmail.com[1] and lalitha_g@vnrvjiet.in[2]

Abstract

As day by day the infrastructure development leads to the demand of sustainable materials in construction and to protect natural materials like river sand and aggregates. In recent years in India, sand mining has become widespread and extraction of river sand causes dredging of river beds shown a great impact on structures near by. To decrease the exploit of natural sand as it created a great problem to the environment and human life.In my present Research work fine aggregate was partially replaced with copper slag (in 0%, 10%. 20%, 30%, 40%, 50%) proportions. Copper slag is an industrial waste from Copper industry. By using copper slag in construction work we can reduce the open land dumping there by protection of imprudent place as well as natural resources .Along with copper slag 0.25% carbon fibers were added as admixture constantly for all 5 mixes as they resist against cracks with high durable and high tensile nature

Main aim of my present research work is to show the effective usage of copper slag and to study the Mechanical properties ,tests like compressive strength, split tensile strength, flexural strength test are conducted and test values are obtained after 7, 14, 28 days of curing. A highest compressive strength of 114.33MPa, split tensile strength of 6.5MPa ,flexural strength of 14.5MPa was obtained by replacing 30% of fine aggregate with copper slag and 0.25% carbon fiber admixture while compared to nominal mix having compressive strength of 59.4MPa,split tensile strength of 3.8MPa and flexural strength of 5.36MPa.

Keywords: Fine aggregate, copper slag, Carbon fiber, compressive strength, split tensile strength, Flexure strength

1. INTRODUCTION

Concrete is a melded material composed mainly of cement, fine aggregate, and coarse aggregate. Fine aggregate (Sand) is an important material for the preparation of concrete. copper slag is a by-product obtained during copper smelting and refining process [1]. As refineries draw metal out of copper ore, they promote a large volume of non-metallic dust, soot, and rock inclusively , these materials make up slag, which can be used for several applications in the building and industrial fields [4].The copper slag serves as binding agent, which helps to clutch the larger gravel particles with in the concrete together then used in this manner, the slag helps to improve the properties of the concrete.

It has been estimated that approximately 24.6 million ton of slag is generated from world copper industry [3].

There are many stalwartness and endurance characteristics that qualify copper slag to be used in the concrete as partial substitute of aggregates. This copper slag usage will serve two purposes; first, it will be environment friendly, second, it can be utilized as a resource in place of precious and relatively costlier natural resources. There is huge potential of using copper slag in the concrete construction sector [2]. The reduced drying shrinkage of copper slag concrete could be attributed to its reduced permeability relative to normal concrete, which is apparent in the moisture absorption rate. In order to Reduce the consumption of river sand the waste copper slag is used by partially replacing fine aggregate in various (0% 10%, 20%, 30%, 40% and 50%) proportions with carbon fiber admixture by utilizing carbon fiber ,the randomly oriented carbon fibers starts arresting both the randomly oriented micro cracking and its proportion improving strength and durable characteristics and carbon fiber wrap is applied as glue to the defected areas in concrete construction instead of steel as it is more durable [6]. To acquire the strength parameters compressive, split tensile, flexural strength test have been conducted.

2. OBJECTIVES

There was very high demand of natural sand as it is deficient material now a days for infrastructure construction. In order to limit the usage of natural sand the one and only material with similar characteristics and properties is copper slag can be used as a replacing agent. Main aim is to examine the workability and strength properties of concrete like workability, compressive strength, Split tensile strength and Flexural strength test, with partial replacement of fine aggregate with copper slag and carbon fiber admixture in concrete.

3. METHODOLOGY

3.1 Sieve Analysis

A sieve analysis was performed for determination of particle size distribution. The proper sizes of sieves are taken and samples are pretended to be taken in sieve

ISBN: 978-93-8830-599-0

and permitted to rotate in clock wise direction [1]. The aggregates were retained in sieves. According to IS: 2386 (Part-3)-1963 sieve analysis test was performed. The Table 1 shows the sieve analysis.

Table 1 Ssieve analysis of copper slag

S. No	Size of sieve(mm)	% retained
1	4.75	0
2	2.36	3.68
3	1.18	29.47
4	0.6	71.57
5	0.3	63.4
6	0.15	96.4
7	<=0.07	100

3.2 Mix Design

The preference given in selecting distinct materials in order to accomplish the goal of strength and workability characteristics. In my present investigation mix design of M60 was prepared as per ACI 211-1993 code-5 different mix proportions are prepared i.e., 0%, 10%. 20%, 30%, 40%, 50% as partial replacement of fine aggregate with copper slag. Table (2) shows the quantities of material required for one cubic meter of concrete. Figure (1) shows the combination of all materials at various proportions by replacement of fine aggregate with copper slag.

4. MATERIALS

4.1 Cement

Cement is a binder that is formed by heating of limestone, shells, chalk or marl combined with shale, clay, slate, blast furnace slag, silica, sand, iron ore are heated at high temperature form a hard rock like substance that is ground into fine powder called cement. In my present research work ultratech cement of ordinary portland cement 53 grade was used for all concrete mixes.

Table 2 Mix proportions for cubic meter of concrete

S. No.	Particulars	Quantity (kg/m³)
1	Ordinary Portland cement 53 grade	518.42
2	Fine aggregate	620
3	Water	172.83
4	Coarse aggregate	1095
5	Carbon fiber	1.29
6	Super plasticizer	9.07

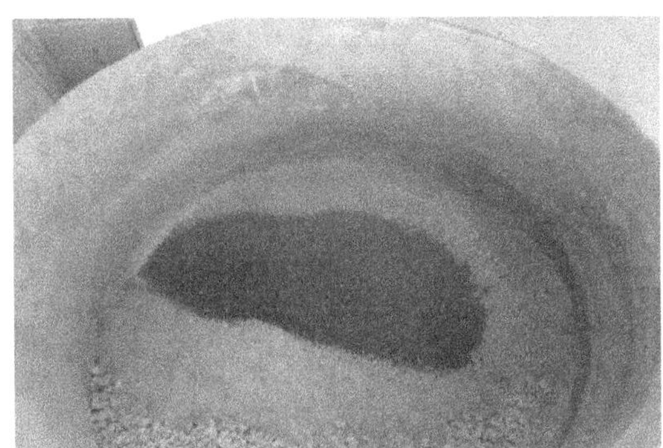

Fig. 1 Mixing of materials

4.2 Fine aggregate

The mixture of small particles of grainswhich pass through 4.75mm sieve comes under fine aggregate. The selection fine aggregate must be such that it should be free from all inorganic material. In my present work locally available river sand belongs to zone-2 with fine characteristics the Figure-2 shows fine aggregate and Table- 3 shows properties of fine aggregate.

Fig. 2 Fine aggregate

Table 3 Physical properties of Fine Aggregate

S. No.	Property	Values
1	Specific gravity	2.61
2	Fineness modulus	2.67
3	Bulk density	3.0

4.3 Coarse Aggregate

Gravel plays a key role in formation of solid mass matrix concrete. The size of coarse particles Greater than 4.75mm is termed to be coarse aggregate and in my research work coarse aggregate of size 12.5mm is used which was bought from quarrying site located in Hyderabad. Figure-3 shows coarse aggregate and Table-4 specifies properties of coarse aggregate.

ISBN: 978-93-8830-599-0

Figure 3 Coarse aggregate

Table 4 Properties of coarse aggregate

S. No	Property	values
1	Specific gravity	2.75
2	Bulk Density(kg/m³)	1680
3	Fineness modulus	7.34
4	Water Absorption	0.30

4) Copper Slag

Copper slag is an irregular, black, glassy, and granular in nature and its properties are similar to river sand. Now in my research work copper slag is obtained from Vivekananda fertilizers located in kukatpally.The chemicals present in copper slag does not show any health effects And most worthy advantage is it is having having similar properties that river sand have. Figure-4 indicate copper slag and Table-5 shows physical properties of copper slag.

Fig. 4 copper slag

Table 5 Physical properties of copper slag

S. No.	Property	Values
1	Specific gravity	3.72
2	Fineness modulus	3.2
3	Bulk density	3.0

4.5 Carbon fiber

Chopped carbon fiber of size 6mm with 7 micrometer diameter was used as an admixture which was obtained from vruksha composites, tenali. Figure-5 point out the carbon fibers. Table-6 specifies physical properties of carbon fiber.

Fig. 5 Carbon Fiber

Table 6 Physical properties of carbon fiber

S. No.	Properties	Values	Units
1	Chopped carbon fiber	6	mm
2	Carbon content	95	%
3	Electrical Resistivity	$1.6*10^{-3}$	Ωcm
4	Elongation	1.8	%
5	Tensile Strength	4250	MPa
6	Tensile Modulus	230	GPa
7	Bulk Density	350	g/liter
8	Filament Die	7	μm

4.6 Super plasticizer

Conplast SP-430 super plasticizer was utilized as a plasticizing gent which was obtained from Vivekananda chemicals, Jagadgirigutta, Hyderabad. Figure-6 Show superplasticizer.

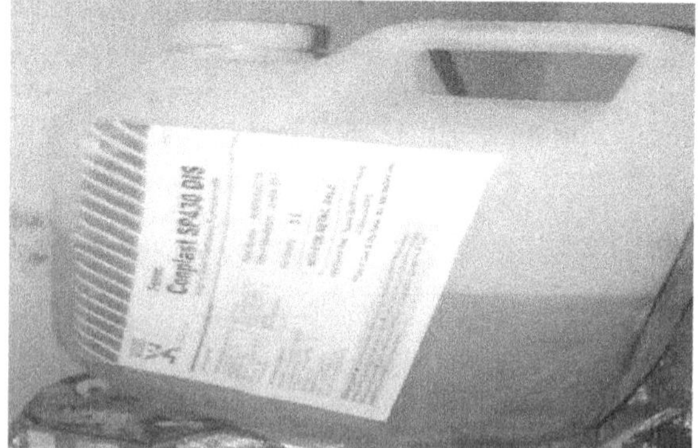

Fig. 6 Super Plasticizer

4.7 Water

Potable tap water is used which is free from pathogenic and other deleterious materials .water is crucial element in construction industry which starts bonding with every element during mixing and helps in hydration of mix[9].

ISBN: 978-93-8830-599-0

5. RESULTS AND DISCUSSIONS

5.1 Workability

The freshly mixed concrete that can be easily handled, consolidated with minimum loss of homogeneity with minimum effort is termed to be noteworthy workable.

a) Slump cone test: slump indicates the characteristics of the concrete and by slump test results we can recognize whether the concrete is effected by segregation figure-9 shows the slump cone test and figure-7 specifies the graph between %replacement of copper slag and amount of slump.

Fig. 7 Slump cone

Amount of slump

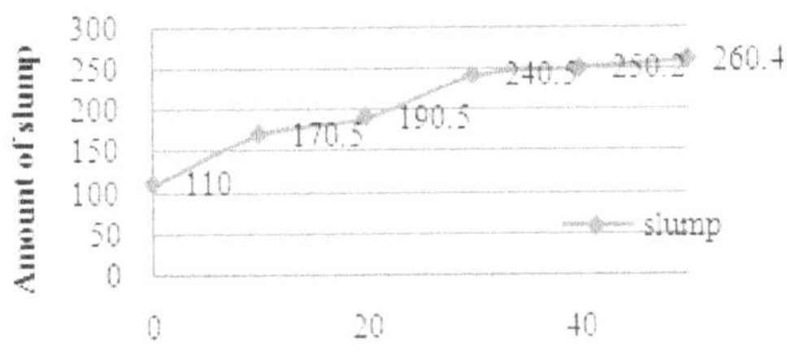

Graph 1 slump test

5.1 Compressive Strength

Compressive testing machine of 2000 KN Capacity is used to determine the compressive strength of concrete cube specimen of size 150mm×150mm×150mm that was cured after 7,14,28 days . And after curing, the surface of the cubes are dried in order to prevent the prevent the moisture. concrete cubes were cast with 0%,10%,20%30%,40%,50% of different proportions of copper slag with fine aggregate replacement with constant 0.25% percentage of carbon fiber. The whole material failure is observed When the ultimate strength reaches the uniaxial compressive stress and the compressive strength is calculated and units of compressive strength is measured in N/mm².figure 8 shows the compressive strength of the cubes.

Fig. 8 Compressive strength testing machine

The graph-2 resembles the results obtained from compressive strength. At later stages the strength has been decreased with increase in content of copper slag. The maximum compressive strength was obtained at 30% replacement of fine aggregate with copper slag. And slowly at later replacements there was a gradual reduction in strength parameters.

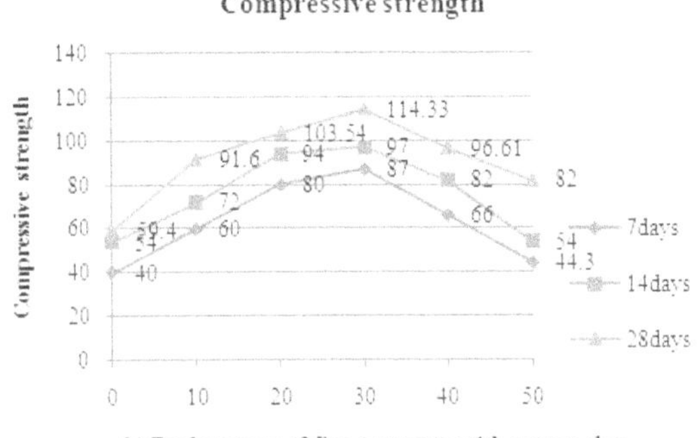

Graph 2 Compressive strength

5.2 Split Tensile Strength

Cylinders of size 200mm×100mm were cast for every mix. After casting of cubes the period of 24 hours is required to harden the specimens and allowed to cure for 7,14,28 days and has been tested 3 cylindrical specimens for every 3 different curing periods. And tensile strength is obtained when the maximum stress that a material can withstand while being stretched before breakage. And units of split tensile strength is N/mm². The figure-9 designates the split tensile strength of cylinders. Graphs are drawn for various strength Verses replacements.

Fig. 9 Split tensile strength test machine

ISBN: 978-93-8830-599-0

Graph-3 demonstrate the results of split tensile strength of the cylinders at later stages with increase in copper slag replacement with fine aggregate i.e., Above 40% replacement there was a reduction in tensile strength parameters and the maximum split tensile strength value is obtained at 30% replacement of fine aggregate with copper slag along with 0.25% carbon fiber.

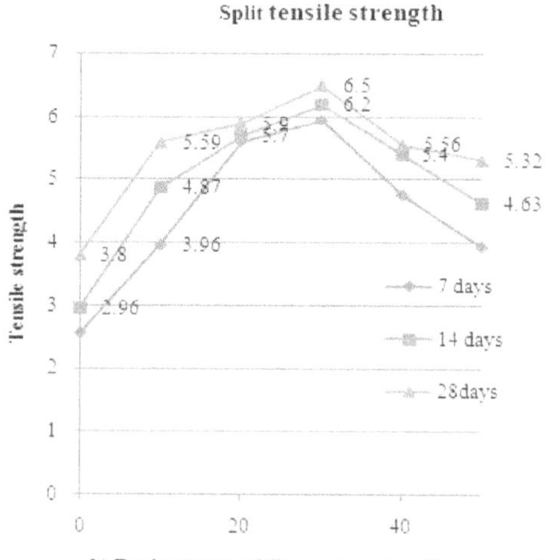

Graph 3 Split tensile strength

5.3 Flexural Strength

prisms of size 500mm×100mm×100mm were cast .After casting of cubes the period of 24 hours is required to harden the specimens and allowed to cure for 7,14,28 days. And has been tested prisms for every 3 different curing periods. The flexural strength is also termed as modulus of rupture, fracture strength. And the flexural strength represents the highest stress experienced within the material at its moment fracture. The flexural strength test values are obtained from the maximum applied load that is required to make the beam to break at the middle third portion of the beam specimen. And flexural strength is calculated in N/mm². Figure-10 specifies the flexural strength of the prism.

Fig. 10 Flexural strength test machine

Graph-4 point out the results of flexural strength of the prisms. At initial stages strength has been increased but at later stages strength values are declined due to the increase in percentage of copper slag with fine aggregate replacement beyond 30%.

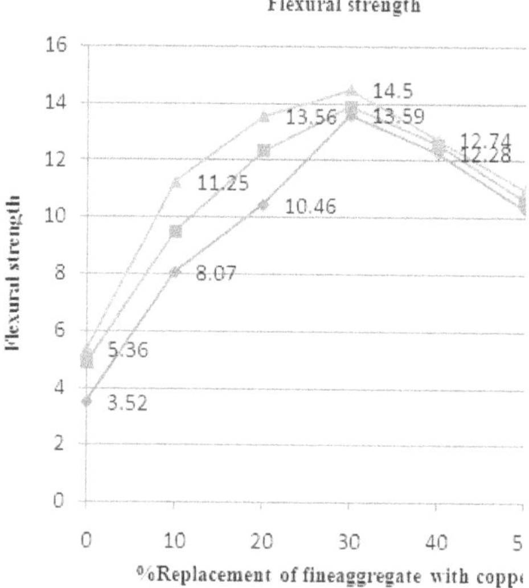

Graph 4 Flexural strength

6) CONCLUSIONS

➤ The increment propornality of copper slag as a partial substitute of fine aggregate hikes the workability.

➤ The utilization of copper slag depicted many benefits such as high strength, toughness, as a good binder of all substances, and acts a resistor to chemicals and temperature and exposure conditions with environment friendly nature.

➤ There was an abrupt increase in compressive and flexural strength values with increase in copper slag content up to 30% and maximum compressive strength of 114.33N/mm²and flexural strength of 14.5 N/mm²was observed and beyond that 30% there was a reduction of change in strength parameter.

➤ By usage of carbon fiber admixture at constant 0.25% for all mixes acts as a good stiffening and high temperature tolerant with its low weight shown its impact on increase in split tensile strength of cylindrical specimen. And maximum of 6.5 N/mm² tensile strength is observed at 30% replacement of fine aggregate with copper slag.

By my research work I come to know by utilization of copper slag as a resource from copper manufacturing industry can reduce digging of river sand and having much more benefits in construction industry with its favourable characteristics.

ISBN: 978-93-8830-599-0

7) REFERENCES

1. Ch. Sai Bhavagana (2017)"Experimental study on concrete(M30) by partial replacement of fine aggregate with copper slag", *in international journal of science and technology(IJRET),* ISSN:0976-6316, **Volume-8**, issue1, January-2017, pp.1031-1038.

2. S.M. kinayekar (2007)"The effect of addition of carbon fibers on mechanical properties of high strength concrete",*in international journal of innovative research in science in engineering and technology (IJIRSET),* ISSN:231753, Volume3, issue1, January 2014, ISO 3297: 2007.

3. Ch. srinivas (2018)" Experimental study on Mechanical properties of concrete" *In inter-nationational journal of applied engineering research(IJAER),* ISSN 0973-4562 **Volume-13,** Number(2017)PP.5328-5331.

4. Tamil selvi "Experimental study on concrete using copper slag as a a replacement of fine aggregate"*In an open access journal,* **volume 4** issue-5 1000156, ISSN: 2165-784X JCEE.

5. PankajThakur and khushpreet Singh "Effect of carbonfibre on different mixes on concrete" a Review paper *in International research Journal of Engineering And Technology (IRJET),* **Volume-5.** Issue-03, March 2011,e-ISSN: 2395-0056, p-ISSN:2395-0072. 7 "Methods of test for strength of concrete", IS 516:1959, NewDelhi, 2002.

6. Reported by ACI Committee 211,"Guide for selecting proportions for high strength concrete with Portland cement "ACI 211.4R-93,pp.1-13,1998.

7. M.N.V Dayanara Reddy (2017) "Mechanical properties of High strength concrete with Copper slag as Fine aggregate and Partially Replacing cement with Nano silica*"in international journal of innovative research in science engineering and technology,* **Volume-6,** Issue 2, Febraury 2017, ISSN: 2347-6710.

ISBN: 978-93-8830-599-0

Innovative Decorative Material Using Plastic Waste in Civil Constructions and Eco-Friendly

Md Jalal Uddin

Asst professor, M.E.(OSMANIA UNIVERSITY)

The project elucidates about the use of plastic in civil construction. The components used include everything from plastic screws and hangers to bigger plastic parts that are used in decoration, electric wiring, flooring, wall covering and waterproofing.

Plastic use in road construction that have shown same hope in terms of using plastic waste in road construction i.e., plastic roads. Plastic roads mainly use plastic carry bags, disposable cups and PET bottles that are collected from garbage dumps as important ingredients of the construction materials.

By using plastic waste as modifier, we can reduce the quantity of cement and sand by their weight, hence decreasing the overall cost of construction. At 5% optimum modifier content, strength of modified concrete we found to see the times greater than the plain cement concrete.

Using plastic poisons our food chain under the plastic affects human health. By the disposable plastics is the main source of plastic. For these plastic pollution is not only the ocean also in desert.

Plastic will increase the melting point of the bitumen. Rain water will not seep through because of the plastic in the tar. So, this technology will result in lesser road repairs.

Keywords: M_2O plain cement concrete, waste plastic.

1.1 INTRODUCTION

Nowadays, human apply all of its potentiality to consume more. The result of this high consumption is nothing unless reducing the initial resources and increasing the landfill. In recent times, human from the one hand is always seeking broader sources with lower price and from the other hand is following the way to get rid of the wastes. The waste today can be produced wherever humans footprints be existed, and remind him that they have not chosen the appropriate method for exploitation of the nature.

This paper introduces the development and low cost housing in India. At the present time, the possibility of utilizing the renewable resources such as solar, geothermal has been provided for us more than before, and development of the renewable and alternative energies is making progress. Plastic have become an essential part of our day to day life since their introduction over hundred years ago

The only way to reduce the hazards of plastic is reduce and reuse

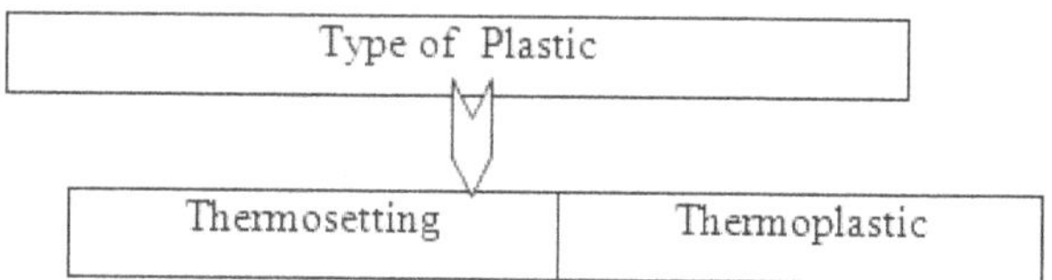

Fig. 1 Types of plastic

1.2 PET PLASTIC

Table 1 Introduction of PET

Full form	Polyethylene Terephthalate
Molecular formula	C10H8O4
Structure Composition	Polyester of Terephthalic acid and ethylene glycol

1.3 INTRODUCTION OF PET

PET is used for high impact resistant container for packaging of soda, edible oils and Peanut butter. Used for cereal box liners, Microwave food trays. Used in medicine for plastic vessels and for Implantation. Plasticis heat resistant and chemically stable. PET is resistant to acid, base, some solvents, oils, fats. PET is difficult to melt and transparent and other properties are,

Table 2 Properties of Plastic

Colour	White or light cream Material
Density of plastic	1.33220 gm/cm3
Melting point	255 to 265 °C
Solubility	Insoluble in water

1.4 PROPERTIES OF PLASTIC

Plastic have many good characteristics which include versatility, light-ness, hardness, and resistant to chemicals, water and impact. Plastic is one of the most disposable materials in the modern world. It makes upmuch of the street side litter in urban and rural areas. It is rapidly filling up landfills as choking water bodies.

ISBN: 978-93-8830-599-0

Plastic bottles make up approximately 11% of the content landfills, causing serious environmental consequences

Due to the consequences some of the plastic facts are as follow:

- More than 20,000 plastic bottles are needed to obtain one ton of plastic.
- It is estimated that 100 million tons of plastic are produced each year.
- The average European throws away 36 kg. of plastics each year.
- Some plastic waste sacks are made from 64% recycled plastic.
- Plastics packaging totals 42% of total consumption and every year little of this is recycled. According to ENSO Bottles, in the 1960's plastic bottle production has been negligible but over the years there was an alarming increase in bottles produced and sold but the rate of recycling is still very low

Plastics are produced from the oil that is considered as non-renewable resource. Because plastic has the insolubility about 300 years in the nature, it is considered as a sustainable waste and environmental pollutant. So reusing or recycling of it can be effectual in mitigation of environmental impacts relating to it. It has been proven that the use of plastic bottles as innovative materials for building can be a proper solution for replacement of conventional materials. The use of this material has been considered not only for exterior walls but also for the ceiling of the building.

The objective of this paper is to investigate the key and positive characteristics of this product and the benefits obtained by using it in building. It also intends to compare the characteristics of some construction materials such as brick, ceramic and concrete block with bottle. One can use solar bomb (bottle filled with bleaching powder solution) will be fitted on the roof for light source.

PET scales Recycled/Beverages sale

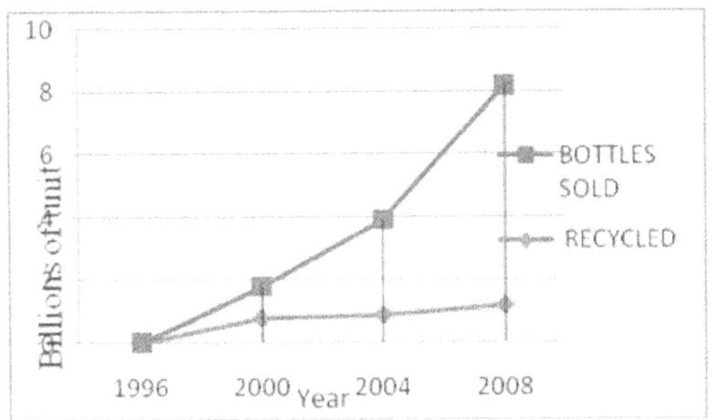

Fig. 2 PET Bottle Sales/ Recycled

DISADVANTAGES OF PLASTIC

2.1 INTRODUCTION

Plastic bottles are certainly ubiquitous they bring us everything from house hold cleaners to soft dryings to things so readily available as water these bottles, while convenient, do have disadvantages when used on as wide a scale while most of these disadvantages are environmental in nature the consequences could have widespread economical consequences in the long-term.

Disadvantages of Plastic Bottles:
1. decomposition
2. non renewable
3. hard to use
4. difficult to recycle
5. decomposition

The main disadvantages of plastic bottles is the shear amount of time they take to decompose he averages plastic bottle takes 500 years plastics decomposition can be agented by various factors, such as the types of plastic, the climate and acids in the landfill; plastic still lasts a long time, filling landfills for an indefinite period.

2.3 NON-RENEWABLE

Plastic is manufactured using oil by products and natural gas material that could be used in numerous other applications or conserved were plastic usage lower. natural gas for example, can be used to heat houses and cook food. using plastic in the volume we currently do reduces the availability of these resources, which are gone forever when used up.

2.4 HARD TO USE

The standard disposable plastic bottle is meant for one use, not many. recycled plastic bottles are not refilled in mass they glass beer bottles are , and flimsy plastic bottles do not lead themselves well to at home re-usage. water bottles, for example, are often reused in the home but become less and less sturdy over time and are ultimately thrown away.

2.5 DIFFICULT TO REYCLE

Glass bottles can be meted and easily reused as can tin cans. recycling plastic is not so simple. much of the plastic placed in recycling boxes is not recycled at all, as most plastic cannot be recycled those bottles that are recycled are not used to make new bottles. instead, recycled plastic bottles are used to make non-recyclable products, such as t-shirts, lactic lumber or parking lot bumers this means more raw materials need to be used

ISBN: 978-93-8830-599-0

to create new plastic bottles than is the case with easily recycled material, such as glass or tin.

BASIC CONSTRUCTION MATERIALS AND PROPERTIES

3.1 INTRODUCTION

This construction require some of the basic materials which ensures a stable, eco friendly structure and also results in cheap construction as compared to brick wall. Materials uses for Bottle wall masonry construction are:

1. Soil
2. Plastic bottles
3. Cement
4. Nylon rope
5. Water

3.2 SOIL

Soil is the basic element in any construction project so before using it in our project we have to study the basic properties of the soil and go through different tests, so as to check whether the soil sample selected is suitable for the given project.

Table 3 Diameter of Soil Particles

Soil Particle	Diameter(mm)
Gravel	>2.0
Sand	0.05-2.0
Silt	0.002-0.05
Clay	<0.002

3.2.1 PROPERTIES OF SOIL

- **Soil Texture:** Soil texture can have a profound effect on many other properties and is considered among the most important physical properties. Texture is the proportion of three mineral particles, sand, silt and clay, in a soil. These particles are distinguished by size, and make up the fine mineral fraction.
- **Soil Colloids:** Soil colloids refer to the finest clay in a soil. Colloids are an important soil fraction due to properties that make them the location of most physical and chemical activity in the soil. One such property is their large surface area. Smaller particles have more surface area for a given volume or mass of particles than larger particles. Thus, there is increased contact with other colloids and with the soil solution. This results in the formation of strong friction and cohesive bonds between colloid particles and soil water, and is why a clay soil holds together better than a sandy soil when wet.
- **Soil Structure:** Soil structure is the arrangement and binding together of soil particles into larger clusters, called aggregates or pads. Aggregation is important for increasing stability against erosion, for maintaining porosity and soil water movement, and for improving fertility and carbon sequestration in the soil. Granular structure consists of loosely packed spherical pads that are glued together mostly by organic substances.
- **Soil Porosity:** Many important soil processes take place in soil pores (air or water-filled spaces between particles). Soil texture and structure influence porosity by determining the size, number and interconnection of pores. Coarse textured soils have many large (macro) pores because of the loose arrangement of larger particles with one another. Fine-textured soils are more tightly arranged and have more small (micro) pores. Macro pores in fine textured soils exist between aggregates. Because fine-textured soils have both macro- and micro pores, they generally have a greater total porosity, or sum of all pores, than coarse-textured soils.

3.3 PLASTIC BOTTLE

In this paper plastic bottles are used as a fundamental element, so we have gone through every property of the PETE bottles so as to ensure a stable structure.

3.3.1 PROPERTIES OF PETE BOTTLE

Polyethylene Terephthalate Ethylene (PETE) bottles is thermoplastic materials. This type of plastic are polymers and with or without cross linking and branching, and they soften on the application of heat, with or without pressure and require cooling to be set to a shape. Following are properties of plastic bottle:

1. Wax like in appearance, translucent, odorless and one of the lightest plastics.
2. Flexible over a wide temperature.
3. Heat resistance.
4. Chemically stable.
5. Do not absorb moisture.
6. Transparent.

3.4 CEMENT

Cement is the important binding material. In these paper it is use to bind the plastic bottles to make the masonry wall more durable so that the quality of cement is check by following properties.

3.4.1 PROPERTIES OF CEMENT

- **Fineness:** Fineness or particle size of Portland cement affects Hydration rate and thus the rate of strength gain. The smaller particle size, and the greater the surface area-to-volume ratio so that the more area available for water cement interaction per unit volume. The effects of greater fineness on strength are generally seen during the first seven days.

- **Soundness:** Soundness is defined as the volume stability of the cement paste.
- **Strength:** Cement paste strength is typically defined in three ways: compressive, tensile and flexural. These strengths can be affected by a number of items including: water cement ratio, cement-fine aggregate ratio, type and grading of fine aggregate, curing conditions, size and shape of specimen, loading conditions and age.
- **Setting Time:** The initial setting time is defined as the length of time between the penetration of the paste and the time when the needle penetrates 25mm into the cement paste.

3.5 NYLON ROPE

Nylon rope has a very high tensile strength so that it is use as the main binder for PETE bottles masonry.

3.5.1 Properties of Nylon Rope

Nylon rope is gotten from coal, Petroleum, air and water. It is a polyamide thermoplastic produced by series of condensation reaction between an amine and organic acids. the properties of nylon as follow:

1. Good abrasion resistance.
2. Tough and strong but flexible too.
3. High impact strength.
4. Absorb water which causes reduction in strength and impact properties
5. Resistant to most of the solvents and chemicals
6. High softening temperatures and thus molding becomes difficult.

3.6 WATER

Water is in a similar way like cement, an active component in mortar. For cement-sand mortar, without water no hydration can be attained, hence no strength can be achieved. Water is responsible for the workability of a fresh mortar. 20% of the overall weight of the cement and soil was used to determine the quantity of water to be used in the mix. A slump test and a flow test were conducted to evaluate the consistency of the fresh mortar.

BACKGROUND

INTRODUCTION

The first bottle house was constructed in 1902 by William F.Peck in Tonopah; Nevda. The house was built with 10,000 bottles of beer, which were 90%

alcohol and 10% opium. The Peck house was demolished in the early 1980's. The use of empty vessels in construction dates back to ancient Rome, which had structures with amphorae embedded in concrete. This was not done for aesthetic reasons, but to lighten the load of upper levels of structures empty, and to reduce concrete usage. The first plastic bottle construction project in Africa was pioneered in Uganda by Butakoola Village Association for Development (BUVAD) in 2010 in Cayuga district. The idea followed a BUVAD community survey in 2009 that revealed that many farmers in Kayunga were experiencing low crop yields due to poor soil fertility, which was a result of the presence of waste plastics, such as bottles and polythene bags, in the soil.

Fig. 3 First bottle house construction

Dec 2nd, 2010 proving that there are all kinds of uses for recycled PET plastic, Taiwan-based Er. Arthur Huang processed 1.8 million used plastic bottles into honey comb-shaped bricks for a boat-shaped exhibition hall called the Eco-ARK. Built for Taipei's flower show, Eco-ARK was constructed for just one-third the cost of a conventional structure. Once locked together, the bricks are extremely strong.

Fig. 4 Eco-ARC built for taipei's

ISBN: 978-93-8830-599-0

May 2011 Samarpan Foundation has constructed a children school in New Delhi, using hundreds of used PET bottles instead of conventional bricks.

Fig. 5 Samarpan foundation

METHEDOLOGY

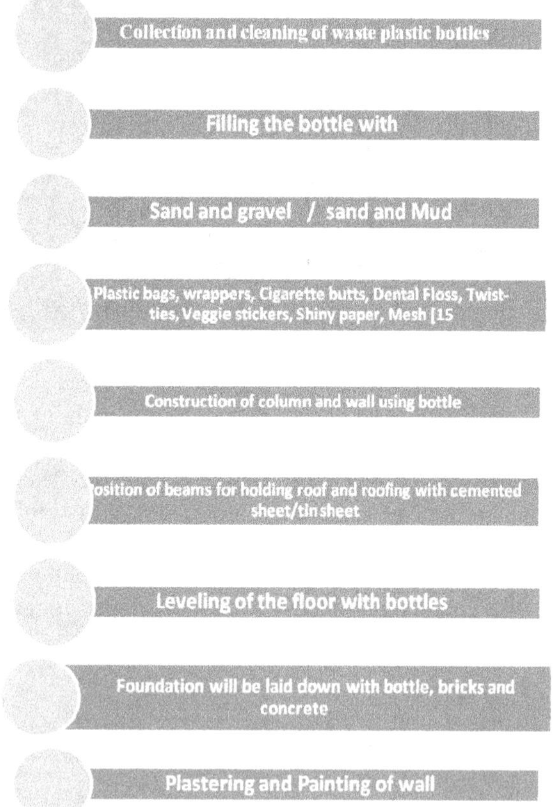

Methodology for construction of plastic bottle House

6.2 MORTAR CALCULATION

Table 4 Cost Estimation of Brick Wall Masonry

Sr. no	Material	Quantity	Rate	Per	Amount(Rs.)
1	Brick	1150 nos.	5	1 no.	5750
2	Cement	5.45	300	1 bag	1635
3	Sand	0.237	250	1 m3	59.25
				Total	7444.25

6.4 NO. OF LABOUR CALCULATION

One labour can made 400 bottles per day (filling soil in bottles).

Total no. of bottles = 1572

Numbers of labour needed = (1572/400) = 4 nos.

Table 5 Cost Estimation of Plastic Bottle Wall Masonry

Sr. no	Material	Quantity	Rate	Per	Amount(Rs.)
1	Plastic bottle	1572 nos.	0.5	1 no.	786
2	Cement	5.45	300	1 bag	1635
3	Sand	0.237	250	1m3	59.25
4	Soil	1.99	100	1m3	199
5	Labour work	4	300	1 person	1200
				Total	3879.25

6.5 COMPARASION BETWEEN THE WALLS BY PLASTIC BOTTLE WALL AND BRICK WALL

For construction Time and speed of Execution for 5 persons team-one working day for plastic wall is 15% faster and for brick wall 120 m2. Material and equipment cost for plastic bottle wall is less as compared to brick wall. Transportation cost for plastic bottle wall construction is less than brick wall. Plastic bottle wall construction require less manpower as compare to brick wall and require high cost. Strength and load capacity for plastic bottle wall construction is 20 times more than brick wall construction.

ISBN: 978-93-8830-599-0

Table 6 Comparison between the Wall by Plastic bottle and Brick

Sr. No	Factors	Considerations	Plastic Bottle Wall	Brick Wall
1	Time and speed of Execution	5 persons team oneworkingday	15% faster	120 m2
2	Material and equipment costs	Implementation and installation of materials and equipment	Saving in cement, water, grinder and fitting	More weight, more materials
3	Transportation Costs	Displacement in the building	Lighter and higher volume, easy and cheap displacement	Greater weight and less volume, hard and costly displacement
4	Execution cost	Using calculations of panel	Less manpower and indigenous	More human resources- the higher cost
5	Strength and load Capacity	-	20 times more than brick	Greater wall thickness, lower strength
6	Resistance to Earthquake	Earthquake hasa direct relationship with the weight of each structure	Low and Integrated weight without falling debris	High weight and loss of material
7	Cleanness and beauty of work		Very clean execution, no construction waste	High volume of construction waste
8	Flexibility		Very clean execution, no construction waste	Low flexibility
9	Material waste		No wastage	High and unusable

BENEFITS OF PLASTIC BOTTLE MASONRY WALL

The most important benefits of these alternative innovative materials compared to conventional materials such as brick can include:

7.1 GOOD CONSTRUCTION ABILITY

The walls built by these bottles are lighter than the walls built by brick and block, and that makes these buildings to show a good response against earthquake. Due to the compaction of filling materials in each bottle, resistance of each bottle against the load is 20 times higher compared to brick. And these compressed filling materials, makes the plastic bottle to be prevented from passing the shot that makes the building as a bulletproof shelter.

7.2 LOW COST

Constructing a house by plastic bottles used for the walls, joist ceiling and concrete column offers us 45% diminution in the final cost. Separation of various components of cost shows that the use of local manpower in making bottle walls can lead to cost reduction up to 75% compared to building the walls using the brick and concrete block. It must be noted that the sophisticated manpower can lead to reducing the construction time and the relative costs also become lower.

7.3 NON-BRITTE CHARACTERSTIC

Using the non-brittle materials can reduce construction waste. Unlike brick, plastic bottle is non-brittle. So due to the frangibility property, the percentage of producing construction waste in brick is more than plastic bottles.

7.4 ABSORBS ABRUPT SHOCK LOADS

Flexibility is a characteristic which makes the buildings performance higher against the unexpected load. Since the plastic bottles are not fragile, they can be flexible and tolerates sudden loads without failure. This characteristic can also increase the buildings bearing capacity against the earthquake.

7.5 GREEN CONSTRUCTION

Plastic bottles can cause the green construction by saving energy and resources, recycling materials, minimizing the emission, having significant operational savings and increasing work place productivity

These 13 plastic bottle vertical garden ideas will interest you if you are a creative person, DIY lover and love to grow plants.

This way you can use plastic bottles to make something amazing out of them. Repurpose those old bottles, which you usually throw away to grow your favorite plants either indoor or outdoor and help to save our environment.

ISBN: 978-93-8830-599-0

Here are 13 inspiring plastic bottle vertical garden ideas to make a vertical soda bottle garden.

1. Window Farm

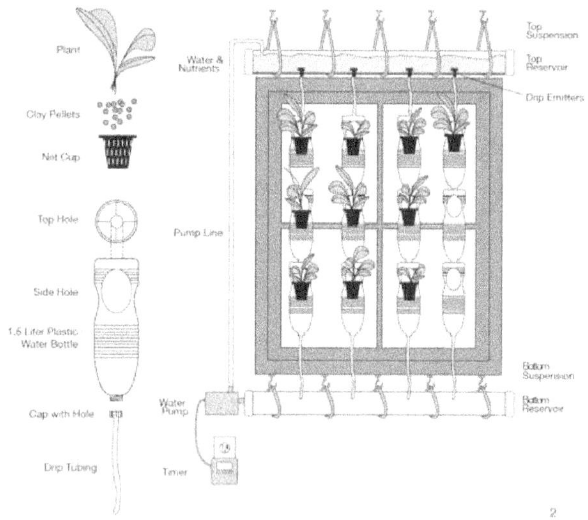

If you love DIY ideas and you have a green thumb then starting a window farm is a smart idea. A windowfarm will let you do a lot with the little amount of space you have. The indoor windowfarms allows the crops to take full advantage of the light and vertical space available at the windows. Here in this PDF, you will find all the instructions on how to build a Window Farm.

2. Plastic Bottles on Walls

Follow this amazing idea for growing small leafy vegetables, such as lettuce, fenugreek and spinach, herbs and medicinal plants. This **plastic bottle vertical garden** is made of by stringing the bottles horizontally in a grid along an interior wall, which then filled up by substrate and herbs.

3. Plastic Bottle Tower Garden

A remarkable kitchen garden with plastic bottles with minimal means and efforts. It can be set up easily and does not require regular watering. Here is the tutorial with more images of it.

4. Growing Cactus in Hanging Plastic Bottles

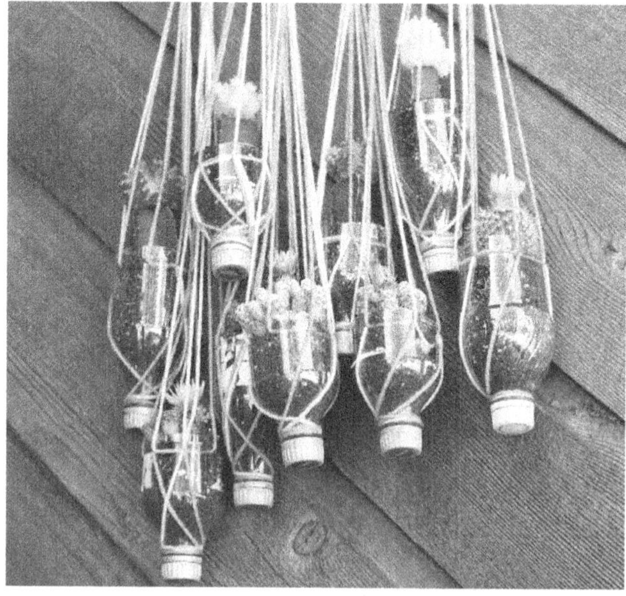

Do you want to create a low maintenance vertical soda bottle garden? Follow this idea. All that is required is bottles cut in half, cactus plants or succulents, and many colorful threads to get a really cool decorative effect.

ISBN: 978-93-8830-599-0

5. Half Plastic Bottle Vertical Garden on Wooden Frame

Use two-liter soda bottles, cut them in half and use the neck side. Turn them upside down. Adhere the bottles to a wooden frame and arrange them in such a way so that the open neck of the bottle will drain out the water into the bottle below it.

Do not use the word "ess6. Green Soda Bottle Vertical Garden

Here's an another idea to create a vertical garden using the plastic bottles. It is a great way to reuse old plastic bottles and to introduce some greenery to a small urban space.

7. Another Vertical Garden

One more wonderful idea to make use of plastic bottles, more useful if you don't have much space on the ground.

8. Bottles Hanging on String

A hanging plastic bottle garden to make full use of vertical space. In this post, which we found on

Source: Container Gardening

ISBN: 978-93-8830-599-0

9. Plastic Bottles Hanging on Net

Another useful idea on using plastic bottles vertically.

10. Inspiring Plastic Bottle Garden

Plastic bottles are mounted on the wall for utilizing the vertical space. Bushy and trailing plants like lettuces and strawberries hide the structure, creating a nice 'green wall' effect.

11. Hanging Soda Bottle Garden

Another innovative and great looking plastic bottle vertical garden. Bottles are hanging horizontally, attached through the strings.

12. Vertical Plastic Bottle Herb Garden

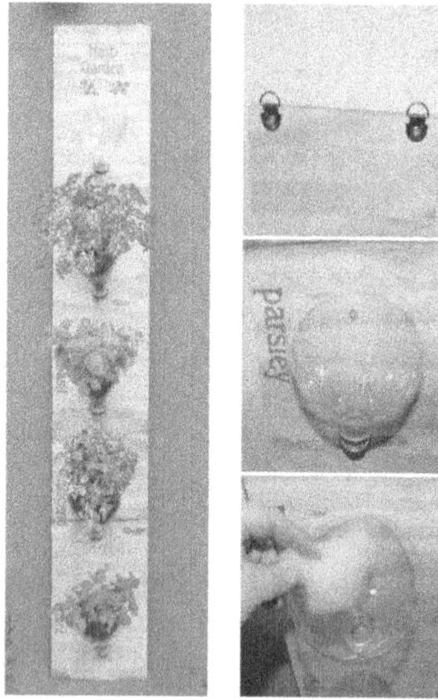

Want to grow herbs but you don't have space? Well, even a wooden plank is enough. All you need is some plastic bottles, hooks, nails and hammer and you're all set to grow your own herbs. Be sure to check out our post on balcony herb garden ideas to find out more ideas like that.

ISBN: 978-93-8830-599-0

13. Pyramid Plastic Bottle Garden

A vertical pyramid garden made of plastic bottles.

CONCLUSION

1. Use of innovative materials with sustainable application such as plastic bottles can have considerable benefits including finding the best optimization in energy consumption of the region, reducing environmental degradation.
2. Generally the bottle houses are bio-climatic in design, which means that when it is cold outside is warm inside and vice versa.
3. Re-using the plastic bottles as the building materials can have substantial effects on saving the building embodied energy by using them instead of bricks in walls and reducing the CO_2 emission in manufacturing the cement by reducing the percentage of cement used.
4. Plastic bottles can cause the green construction by saving energy and resources, recycling materials, minimizing the emission, having significant operational savings and increasing work place productivity.
5. Cost compression between bottles wall is roughly half than conventional brick masonry. i.e., Total cost of10 m2 Brick masonry wall is Rs. 7444.25 and total cost of 10 m2 Bottle masonry wall is Rs. 3879.2
6. Use of innovative materials with sustainable application such as plastic bottles can have considerable benefits including finding the best optimization in energy consumption of the region, reducing environmental degradation.
7. Plastic bottles can cause the green construction by saving energy and resources, recycling materials, minimizing the emission, having significant operational savings and increasing work place productivity.

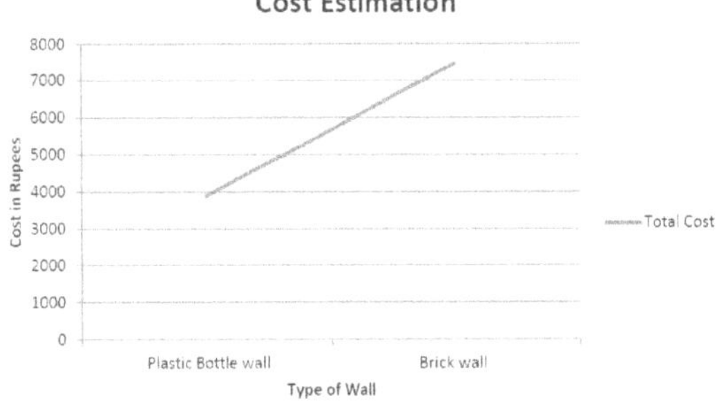

REFERENCES

1. Mojtaba Valinejad Shoubi., Azin Shakiba Barough.; 'Investigating the Application of Plastic Bottle as aSustainable Material in the Building Construction', International Journal of Science, Engineering andTechnology Research (IJSETR) Volume 2, Issue 1, January 2013 ISSN: 2278 – 7798.

2. Shilpi Saxena., Monika Singh.; 'Eco-Architecture: PET Bottle Houses', International Journal of Scientific Engineering and Technology Volume No.2, Issue No.12, pp: 1243-1246 1 Dec 2013, ISSN: 2277-1581.

3. Puttaraj Mallikarjun Hiremath., Shanmukha shettya.; 'Utilization Of Waste Plastic In Manufacturing Of Plastic-Soil Bricks', International journal of technology enhancements and emerging engineering research, volume 2, issue 4, ISSN 2347-4289.

4. Pratima Patel., Akash Shah.; 'Sub stainable development using waste PET bottles as construction element'www.wastebottleconstruction.com.

5. Arulmalar Ramaraj., Jothilakshmy Nagammal.; '30th INTERNATIONAL PLEA CONFERENCE' 16-18December 2014, CEPT University, Ahmedabad

6. Vikram Pakrashi.; 'Experimental Characterization of Polyethylene Terephthalate 1 (PET) Bottle Eco-Bricks'.

7. Andreas Froese (2001), 'Plastic bottles in construction who is the founder of ECO-TEC',

8. Yahaya Ahmed, of Nigeria's Development Association for Renewable Energies,

9. www.Throughthesandglass.typepad.com the sandglass construction material.

10. Seltzer D.J. (2000) bottle houses. www.agilitynut.com.

11. Job Bwirea., Arithea Nakiwala.; 'Cut costs with a plastic bottle house', NEW VISION: Uganda's leadingdaily Publish Date: Feb 11, 2013.

12. Samarpan foundation, 'House construction with plastic bottles, New Delhi', Available form:www.samarpanfoundation.org

13. Rajputa, R. K. (2007), "Engineering materials: including construction materials" 3rd Ed. S. Chand &company, New York.

14. K Jayaprakash (2008), "Treasure from the trash" in Indian Express 15th December 2008.

15. K Jayaprakash (2008), News Article of Indian express: "Treasure from the trash", Published: 15thDecember 2008 03:34 AM Last Updated: 14th May 2012 05:20 PM.

16. How-To Make a Bottle Brick www.earthbench.org

17. Ms. K. Ramadevi.; 'Experimental Investigation on the Properties of Concrete With Plastic PET (Bottle)Fibers as Fine Aggregates' International Journal of Emerging Technology and Advanced Engineering, ISSN2250-2459, Volume 2, Issue 6, June 2012

18. Lauren Vork, 'How to Build a Plastic Bottle Wall', How Contributor.

Study on Micro Structural behaviour of Silica Fumes and Non-Bio Degradable Fibers in High Performance Concrete

K. Manjula Vani[1], G. Srinivasa Rao[2] and B. Ashritha[3]
[1]Professor, Civil Engineering Dept., JNTU College of Engineering, Hyderabad, Telangana
[2]General Manager, Constructions, Hyderabad, Telangana
[3]Academic Assistant, Civil engineering department, JNTUH

Abstract

Silica fume is a by product of producing silicon metal or ferrosilicon alloys. One of the most beneficial uses for silica fume is to improve density in concrete .In recent years high performance concrete plays a major role in construction industry. The high performance concrete demand has increased significantly due to use in high rise towers mainly for columns. High-density polyethylene fibers coming out of Non-Bio Degradable waste are added to improve the tensile strength of concrete. To transform the brittleness to ductility the high density polyethylene fibers added as an additive in the fiber reinforced concrete .The M40 grade of concrete is prepared by adding mineral admixtures silica fume and an additive high density polyethylene fibers to the control concrete .The optimum dosage fixed for replacement of cement as 15% by weight of cement for mineral admixtures and 3% by volume of concrete for an additive. The samples prepared to investigate the formation and effects of agglomerates, calcium silicate hydrate (C-S-H) gel, porosity, portlandite, ettringite, etc., on the selected concretes. The laboratory tests energy dispersive spectroscopy, X-ray diffraction and scanning electron microscopy are conducted on specimens.

Keywords: silica fumes, X-ray diffraction, scanning electron microscopy, Energy Dispersive X-Ray Spectroscopy

1. INTRODUCTION

In recent years the construction of high rise towers is increased tremendously throughout the world. To reduce the cross section of structural elements the demand for using of HPC is increased from the constructions industry. Finally, the environmental agents can harm the fiber reinforced polymer surface which allows moisture or other environmental agents to enter into the concrete core [1]. The use of mineral admixtures such as SF, MK, fly ash, slag, etc., consumption for the replacement of cement is very high due to its pozzolana action, high reactive, economical, early age strength gaining, etc. By adding of MK as a replacement of cement from 5 to 20% increased the high range water reducer admixtures demand, mixture viscosity, passing and flowing ability, segregation resistance and compressive strength [2]. The usage of plastic waste is increasing day to day. The Goa department of tourism announced that more than 40.58

Lakhs tourists were visited in the year 2014 [39]. The reuse of plastic waste as building material in mass constructions is to reduce the landfill space and to save the earth. The Polyethylene *Teraphthalate* FRC had high deformation capacity and is a good alternative in repairing or strengthening applications for seismic safety [3]. Alkali resistant glass was the most cost effective and available textile reinforcement for FRC [4]. The safe disposal of MSW is a big challenge to the municipalities. When design the new landfill cells the barrier system should be reflect the type of waste such as MSW, aluminium production waste, incinerator ash and mode of operation like dry cells, *leachate* recirculation, bioreactor [5]. The sonicated SF show higher reactivity allied with increased consumption of portlandite during curing, relative to pastes together with densified SF [6]. For the HPC, it is proposed to maintain ultimate concrete strain at a constant value of 0.003 instead of 0.002 [7]. The high strength concretes are very useful for early removing of supports and applying loads in constructions. The ultra HPC specimen was failure by distributed small cracking, without any cover spalling or crushing, in contrast to conventional reinforced concrete [8]. The percentage of mineral admixtures for replacement of cement is playing a vital role in properties of concrete. The optimum mix of 25% by weight MK provides a distinct improvement in the hydration materials, a relatively low porosity, leading to an enhancement in the mechanical properties and microstructure of the blended samples [9]. The water binder ratio should be maintained as low as possible for HPC to enhance the properties of concretes. The high w/c ratio MK blended concretes were more susceptible to deterioration from aggressive sulphate ions when air curing and subjected to drying-immersion cyclic exposure [10]. The fibers size, shape, type of fibers, orientation, etc., are significantly influence the mechanical properties of fiber reinforced concretes. The fiber orientation was significantly effect on flexural strength of the panels made with ultra-HPFRC [11]. For the HPC, if the higher uniaxial strength occurs then the smaller increase in normalized multi-axial strength happens [12]. The replacements of cement by MK with 5 or 10% will significant improvement in compressive strength of

ISBN: 978-93-8830-599-0
"

mortar [13]. The MK blended concrete shows a clear reduction in compressive strength during the early stages of hydration at low temperature curing [14]. The type of curing is also governing the early age strength of concrete. The steam-cured SF blended concrete samples after 365 days displayed significantly higher compressive strengths than the 28 days strengths [15]. The authors concluded that the use of UHPFRC which made of cementitious materials in structural elements leads to higher stiffness and increased resistance [16]. The addition of polymer dispersion with an amount of 10 or 20% by weight to the concrete results in increases the workability, tensile strength and all other properties [17]. The high performance light weight concrete is a viable alternative for bridge constructions and it was proved by deep investigations [18]. The authors found that the fiber strips length of 50.8mm produced optimum results under normal condition tests [19]. The steel micro fibers in very HPFRC improve the flexural behaviour of beams in terms of strength and stiffness with enhanced crack bridging effect [20]. The complete stress-deformation curves for mineral admixtures blended young concrete under uniaxial tension show less ductility as compared to those under uniaxial compression [21]. To reduce the concrete crushing, the minimum amounts of flexural reinforcement are provided, which is also protect the structural elements from large cracks developing due to the concrete tensile stress is exceeded [22]. The experimental results are proved that the replacement of natural coarse aggregate with sintered sludge pellets in concrete [23]. A series of laboratory tests are conducted on the selected concretes to study the improvement in compressive strength, workability, density, chemical changes, microstructure and formation of agglomerates, ettringite, portlandite, C-S-H gel, etc., in this paper.

2. EXPERIMENTAL PROGRAM

The strength of concrete is mainly depended on the type of materials used and the compaction of concrete. All the laboratory tests were conducted on materials and concrete specimens as per Bureau of Indian Standards (BIS) procedure. The results are recorded and discussed in this paper.

2.1. Materials and Mix Design

The ingredients used throughout study are cement, fine aggregate, coarse aggregate, water, SF, MK and HDPE fibers. Normally available ordinary Portland cement (OPC) of 53 Grade was used and confirmed to Indian Standard Code IS 12269 – 1987. The fine aggregate i.e., sand quarried from Krishna River confirming to grading zone II of IS 383 - 1970 was used. The coarse aggregate, crushed basalt stone aggregates were used.

Ordinary tap water was used for mixing of all concretes. The size of SF particles are much smaller than cement grains so these can fill the space between cement particles in fresh paste and restrain the movement and re-agglomeration of carbon Nano fibers [24]. The mineral admixtures SF & MK are very reactive pozzolana materials, to be used in concrete due to its fine particles, large surface area and the high silicon dioxide content. By addition of these pozzolana materials to concrete, the various mechanical properties such as workability, durability, strength, resistance to cracks and permeability can be improved. NBD materials are non-corrosive, resistant to chemical attack, light in weight, easy to handle. The empty plastic cement bags are collected from construction sites to make HDPE fibers with aspect ratio between 50 to 150. The physical properties of HDPE material were evaluated by CIPET tests as per IS: 9755:1999 / IS: 1969. The physical properties of HDPE are tensile breaking load at yield, tensile breaking load at break, tensile elongation break are 61.79Kg, 2.78Kg and 26.91% respectively. The following eight types of concretes are selected to carry a detailed investigation and they are

- Control concrete (CC),
- Fiber reinforced concrete (FRC),
- High performance concrete containing SF (HPC-S),
- High performance concrete containing MK (HPC-M),
- High performance concrete containing SF & MK (HPC-SM),
- Fiber reinforced high performance concrete containing SF (FRHPC-S),
- Fiber reinforced high performance concrete containing MK (FRHPC-M) and
- Fiber reinforced high performance concrete containing SF & MK (FRHPC-SM).

The IS Code method followed for mix design. The mix proportions for the above selected concretes are presented in Table 1. Finally, the authors concluded that the use of 15% SF and 15% rice husk ash as optimum dosage for a partial replacement of cement by weight led to increase in concrete densification and all properties which were proved by microstructure investigations [25]. After a robust collection of literature and from it the optimum percentage of mineral admixtures for replacement of cement was considered as 15% by weight of cement. The preliminary experimental work was completed to select the optimum percentage of HDPE fibers by volume of concrete and fixed as 3% based on the 28 days cube

ISBN: 978-93-8830-599-0

compressive strengths for this study. The agglomerations are seen by naked eye at the time of mixing admixture and these are not found after thorough mixing of ingredients.

Table 1 Mix Proportions for M40 Grade Concretes

Type of Concrete	Materials in Kg/m^3							
	Water binder ratio	Cement	Fine aggregate	Coarse aggregate	Water	HDPE fibers	Silica fume	Metakaolin
CC	0.4	450	635	1156	165	---	---	---
FRC	0.4	450	635	1156	165	2.878	---	---
HPC-S	0.4	450	635	1156	165	---	67.5	---
HPC-M	0.4	450	635	1156	165	---	---	67.5
HPC-SM	0.4	450	635	1156	165	---	33.75	33.75
FRHPC-S	0.4	450	635	1156	165	2.878	67.5	---
FRHPC-M	0.4	450	635	1156	165	2.878	---	67.5
FRHPC-SM	0.4	450	635	1156	165	2.878	33.75	33.75

2.2. Municipal Solid Waste

The collection, seperation and safe disposal of solid waste is big tasks for the municipalities. The total MSW generated from twin cities of Hyderabad and Secunderabad was 5030 MT/day in the year 2014 as per report of Greater Hyderabad Municipal Corporation (GHMC), Telangana State, India. The MSW was generated per day per head in GHMC was approximately 0.599 grams. In this the rate of accumulation of plastic waste is high. Using new technology and integrated approach is a major component to ensure that solid waste problems are addressed in a manner which provides for the greatest common benefit [26]. The major task of the present study is to prove that the reuse of plastic waste as building material in fiber form in construction industry.

2.3. Sample Preparation and Tests

The BIS samples are tested for compressive strength after 28 days curing. The powder form specimens are used to examine the mineralogical tests and cubes of side 10mm are used for morphological tests. The workability tests on fresh concrete samples was done and found as high for all selected concretes. The values are presented in Table 2. The pozzolana glass powder blended concrete can enhance the fresh concrete properties such as handling and placing [27].

Table 2 Workability Test Results of M40 Grade Concrete

Type of concrete	Slump in mm	Compaction factor
CC	120	0.865
FRC	100	0.861
HPC-S	110	0.874
HPC-M	115	0.882
HPC-SM	125	0.885
FRHPC-S	105	0.867
FRHPC-M	110	0.876
FRHPC-SM	125	0.882

2.4. Energy Dispersive Spectroscopy

The chemical composition of all samples was examined by EDS tests and presented in Table 3. In the long term, mineral admixtures in HPC are increasing compressive strength and transport properties of concretes [28]. The change in base materials was observed from CC to moderated concretes. The available calcium atomic percentage was decreased for moderated concretes compared to CC. The silica and alumina atomic percentages were increased for moderated concretes when compared with CC.

2.5. X-Ray Diffraction

The mineralogical analysis was completed by using XRD techniques. The diffractogram was show, where the 2θ value is on X-axis from 0^0 to 80^0 and the intensity is on Y-axis from 0 to 900. The XRD patterns of all selected concretes specimens are shown in Fig. 1. The decrease in Quartz peak of specimen cured beyond 90 days results in a nominal increase in unconfined

ISBN: 978-93-8830-599-0

compressive strength of the reference mix with dry or treated tire chips [29]. The number of high peaks for quartz was observed more in control concrete diffractogram compared to other concretes. The silica content of blended concrete with rice husk ash and MK is greater than that of unblended self-compacted concrete proved by the energy dispersive x-ray analysis (EDAX) spectrum analysis [30]. The C-S-H gel peak intensity is more in CC compared to blended concretes. The ettringite peak was not traced in HPC-SM x-ray diffractogram Fig. 1(e). The portlandite and magnetite peaks were also observed in all concretes and less in blended concretes compared to control concrete. These observations are compared with the results of EDS tests of all concretes and satisfied.

2.6. Scanning Electron Microscopy

The morphological analysis was done using SEM tests. The microstructure images of the eight hardened concretes are displayed in Fig. 2 & 3. In the amorphous silica blended concrete it made the calcium silicate hydrate microstructure denser and increased the amount of isolated nm-sized pores [31]. The images are revealing the microstructure, pore sizes and densification, cracks, homogeneity, etc. The blended concrete SEM images are presented very less capillary pores and its dispersion when compared to CC.

The 28 days compressive strength results of all selected concretes are tabulated in Table 4. The improvements in compressive strength of moderated concretes are perceived from Table 4. The relative strength is the ratio of moderated concrete compressive strength to CC. The FRHPC-SM concrete shows the highest improved compressive strength when compared to all other concretes. The addition of activated South African MK up to 30% can increase the strength of concrete [32]. The additive and mineral admixtures improved concrete compressive strength due to the reaction of SF and MK with cement gel. The FRC enhances the ductility property and the mode of failure changes from catastrophic to gradual [33]. The failure load capacity of FRC specimens are little more compared to other concretes. The plastic waste is to be converted as fibers and mixed with concrete to use in mass concrete construction works which will reduce the environmental pollution. The results of materials flow analysis of solid waste serve to identify a few potential strategies for improving local solid waste management [34]. The fibers when added to concrete it changes the behaviour of concrete from brittle nature to ductile which will improve the failure load capacity of concrete. The FRC specimens revealed a pseudo ductile behaviour and observed that the fibers bridging the beam crack because of this the energy absorption during flexural failure was higher than that for plain concrete [35]. The minor peaks in the diffractogram were observed due to existence of non-crystalline products resulted from un-reacted particles in concrete samples.

3. RESULTS AND DISCUSSIONS

Table 3 Chemical Analysis by EDS Test for (a) CC (b) FRC (c) HPC-S (d) HPC-M (e) HPC-SM (f) FRHPC-S (g) FRHPC-M and (h) FRHPC-SM

(a)Element	Weight %	Atomic %	(b)Element	Weight %	Atomic %	(c)Element	Weight %	Atomic %
O K	65.97	81.40	O K	59.75	76.46	O K	64.95	79.80
Mg K	0.43	0.35	Mg K	0.64	0.54	Mg K	0.44	0.36
Al K	1.20	0.88	Al K	1.92	1.46	Al K	1.71	1.24
Si K	6.95	4.89	Si K	10.94	7.97	Si K	10.95	7.66
S K	0.51	0.32	S K	0.70	0.45	S K	0.39	0.24
Ca K	24.10	11.87	Ca K	24.77	12.65	Ca K	20.26	9.94
Fe K	0.83	0.29	Fe K	1.29	0.47	Fe K	0.70	0.25
Na K	0.00	0.00	Na K	0.00	0.00	Na K	0.60	0.51
K K	0.00	0.00	K K	0.00	0.00	K K	0.00	0.00
Total	100.00		Total	100.00		Total	100.00	
(d)Element	Weight %	Atomic %	(e)Element	Weight %	Atomic %	(f)Element	Weight %	Atomic %
O K	68.43	81.91	O K	67.30	81.51	O K	64.42	79.01
Mg K	0.34	0.27	Mg K	0.44	0.35	Mg K	0.40	0.33
Al K	2.64	1.87	Al K	1.81	1.30	Al K	3.55	2.58
Si K	10.69	7.29	Si K	10.46	7.22	Si K	12.00	8.39

Contd...

S K	0.29	0.17	S K	0.56	0.34		S K	0.44	0.27
Ca K	16.62	7.94	Ca K	18.61	9.00		Ca K	16.53	8.09
Fe K	0.57	0.20	Fe K	0.83	0.29		Fe K	0.56	0.20
Na K	0.41	0.34	Na K	0.00	0.00		Na K	0.25	0.21
K K	0.00	0.00	K K	0.00	0.00		K K	1.85	0.93
Total	100.00		Total	100.00			Total	100.00	
(g)Element	**Weight %**	**Atomic %**	**(h)Element**	**Weight %**	**Atomic %**				
O K	66.92	80.38	O K	63.12	78.92				
Mg K	0.72	0.57	Mg K	0.49	0.40				
Al K	4.04	2.88	Al K	2.30	1.71				
Si K	12.74	8.72	Si K	9.66	6.88				
S K	0.26	0.16	S K	0.51	0.32				
Ca K	13.10	6.28	Ca K	22.50	11.23				
Fe K	1.00	0.35	Fe K	1.22	0.44				
Na K	0.23	0.19	Na K	0.00	0.00				
K K	0.98	0.48	K K	0.20	0.10				
Total	100.00		Total	100.00					

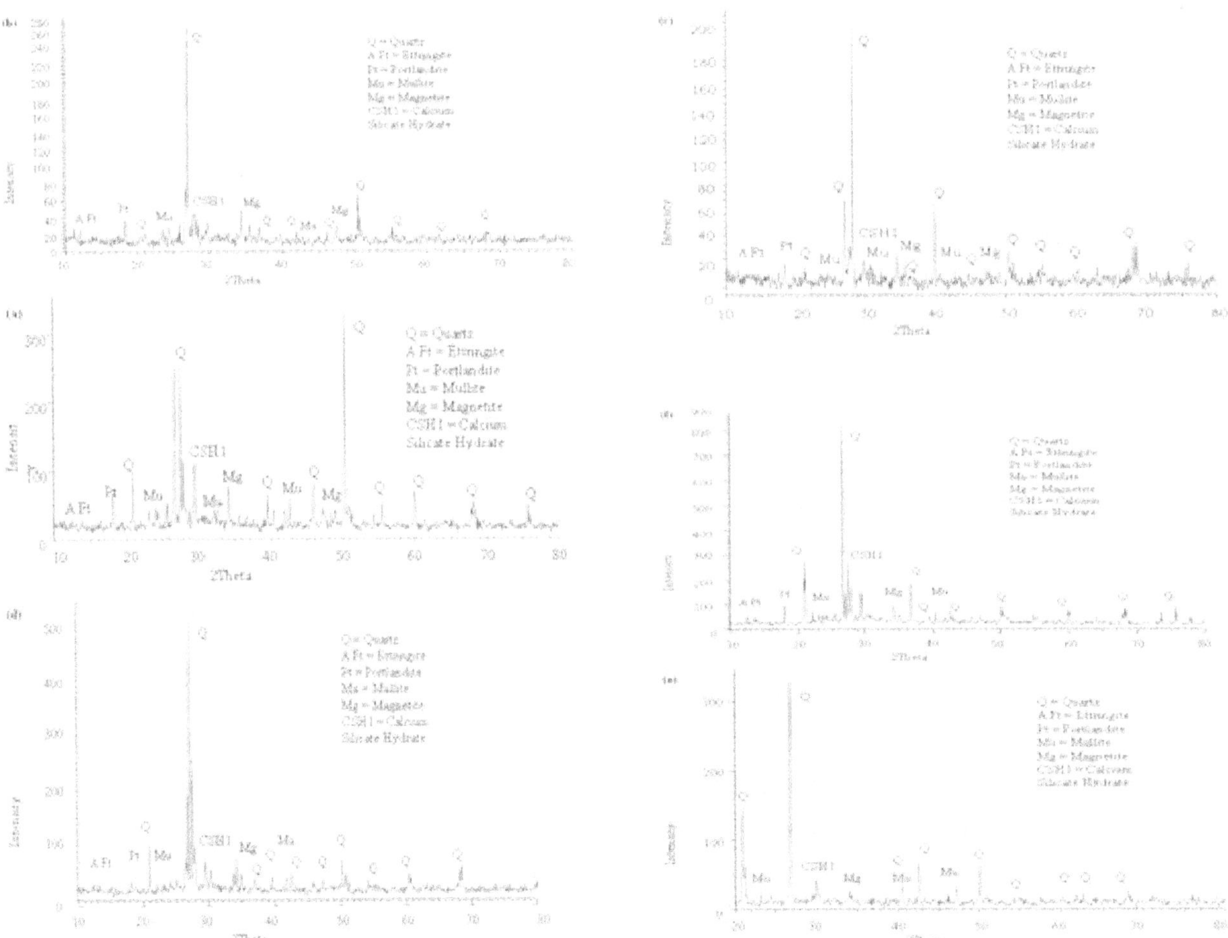

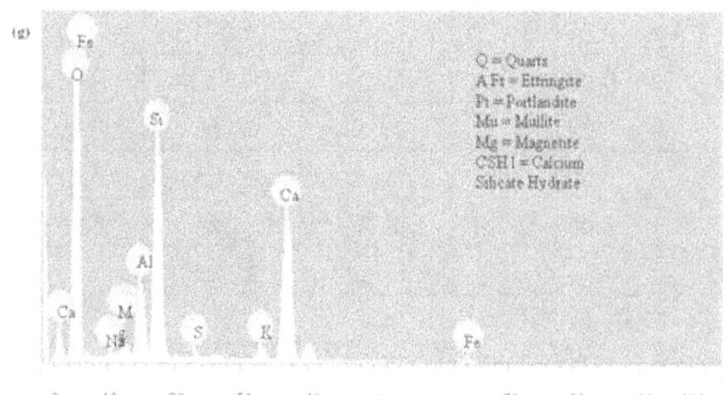

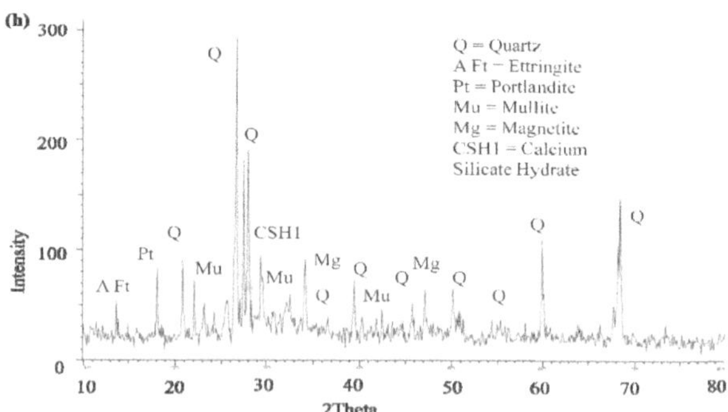

Fig. 1 X-Ray diffraction analysis of (a) CC (b) FRC (c) HPC-S (d) HPC-M (e) HPC-SM (f) FRHPC-S (g) FRHPC-M (h) FRHPC-SM

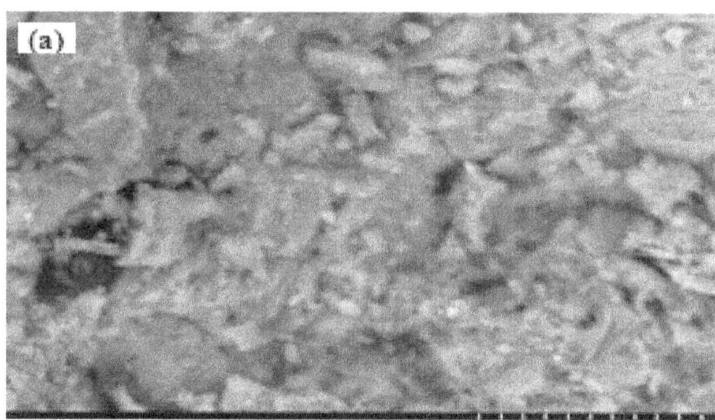

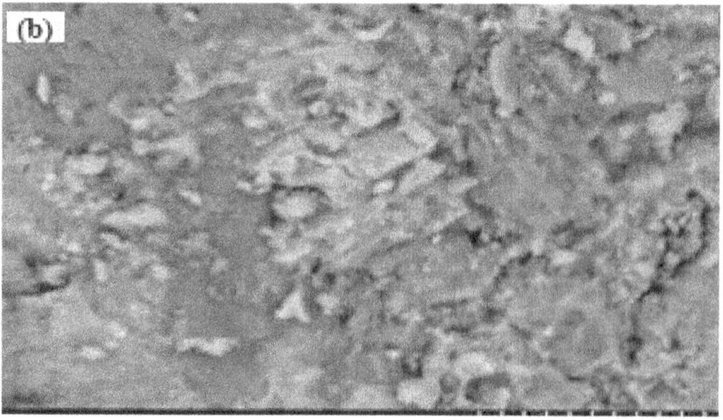

Fig. 2 Scanning electron spectroscopy of (a) HPC-S (15.0kV; x2000 magnification; 20μm) and (b) HPC-SM (15.0kV; x2000 magnification; 20μm)

The surface area of SF and MK particles are high when compared to cement, so these are pack the gap between the aggregates very closely and quickly. The experimental results reveal that there is an interaction between fly ash and micro silica particles in concrete with their age of curing [36]. The SF & MK are therefore improve the durability of concrete by reacting with the calcium hydroxide (CH) produced upon cement hydration, by reducing the porosity of the hardened concrete and by densifying and reducing the thickness of the interfacial zone. The MK blended concrete samples shows an increase in the rate of heat evolution and enhanced aluminates peaks [37]. The existence of more Al_2O_3 in SF & MK supports to sustain the hardening process. This surplus hardening process increases the early compressive strength of blended concretes. The enriched performance of UHPFRC samples was accomplished by optimizing the packing density of the cementitious matrix, using very high strength steel fibers, tailoring the geometry of the fibers and optimizing the matrix-fiber interfaces properties [38]. The physical action of the pozzolana provides a denser, more homogeneous and uniform matrix in the concrete. The succeeding modification of microstructure of cement composites improves the mechanical properties, durability and service life properties. Primarily, due to the high reactivity of the SF & MK, the CH produced by the hydrating cement will be rapidly consumed by the admixtures which will prevent ettringite production. The reduction of CH and the increase of C-S-H gel in the cement matrix of blended concretes results from the pozzolana reaction of admixtures. These interpretations were verified through XRD diffractograms and SEM images. The ettringite formed in all concretes due to the hydrate reaction are less in blended concretes compared to control concrete which was proved by EDS mapping. The size and densification of capillary pores are observed more in control concrete than blended concretes, which may be the main reason of poor strength of concrete. The microstructure of the all blended concrete exhibits a homogeneous, compact structure and relatively low porosity. During the SEM test, the presence of HDPE fibers found nil, because they burn due to high heat, but the impressions were observed on the samples.

ISBN: 978-93-8830-599-0

Table 4. Compressive Strength of M40 Grade Concretes

Type of concrete	Load in kN	Compressive strength in N/mm^2	Average compressive strength in N/mm^2	Percentage of Improvement in compressive strength	Relative strength
CC	1005	44.67	44.2	------	------
	986	43.82			
	993	44.13			
FRC	1194	53.07	52.0	17.7	1.18
	1158	51.47			
	1161	51.60			
HPC-S	1084	48.18	46.5	5.3	1.05
	1040	46.22			
	1018	45.24			
HPC-M	1189	52.84	54.3	22.8	1.23
	1263	56.13			
	1213	53.91			
HPC-SM	1201	53.38	52.3	18.2	1.18
	1187	52.76			
	1139	50.26			
FRHPC-S	1210	53.78	53.9	21.8	1.22
	1235	54.89			
	1190	52.89			
FRHPC-M	1290	57.33	56.4	27.5	1.28
	1270	56.44			
	1245	55.33			
FRHPC-SM	1285	57.11	57.2	29.3	1.29
	1278	56.80			
	1296	57.60			

4. CONCLUSIONS

Based on the experimental results and images of eight concretes the following conclusions drawn for microstructure study of NBDFRHPC:

1. In the EDS spectrum analysis, the silica content of moderated concretes with SF & MK is clearly greater than that of CC. The high silica content observed in the all moderated concretes indicates that a good pozzolana reaction and a synergic effect have occurred in moderated concretes. When the CC moderated with admixtures SF & MK the atomic percentage of silica is enhanced by nearly 50%.

2. The maximum consumption of calcium in moderated concretes which are revealed by the atomic percentage as 6.28% and in CC as 11.87%. The consumption of calcium increased nearly 25% by adding of admixtures to CC.

3. The clear crystal formation in control concrete established by the number of peaks of Quartz displayed in x-ray diffractogram Fig. 1(a) and

ISBN: 978-93-8830-599-0

this number of peaks is 10 where it is highest among all the selected concretes.

4. The presence of C-S-H gel in control concrete detected through x-ray diffractogram and its intensity as 110 from Fig. 1(a) which recorded highest among all the concretes reveal that this gel flow resisting the reaction in concrete.

5. The portlandite and magnetite peaks identified in all selected concretes X-ray diffractograms which improves the reactions in concretes.

6. The formation of cluster of needles in SEM images of concretes indicates the presence of ettringite that makes the specimen denser and contributes to increases in the compressive strength of selected concretes.

7. Incorporating of admixtures SF & MK as replacement of cement at 15% by weight of cement and additive HDPE fibers at 3% by volume of concrete significantly increased the compressive strength up to $57.2 N/mm^2$, which increased up to 29.3% when compared with the CC.

ACKNOWLEDGEMENT

The author's gratefully thanks to the IST, Nano Dept., JNTU Hyderabad, for providing resources to carry out XRD tests. The authors also thankful to NGRI, Hyderabad for supporting to conduct the EDS and SEM laboratory works. The results and interpretation of the data analyses completely the views of the authors.

REFERENCES

1. Micelli F, Mazzotta R, Leone M and Aiello M A, (2015), "Review Study on the Durability of FRP–Confined Concrete," ASCE, Journal of Composites for Constructions, ISSN 1090-0268-04014056(16)

2. Ahmed A Abouhussien and Assem A A Hassan, (2014), "Application of Statistical Analysis for Mixture Design of High-Strength Self-Consolidating Concrete Containing Metakaolin", ASCE, Journal of Materials in Civil Engineering, ISSN 0899-1561/04014016(9)

3. Medine Lspir, (2014), "Monotonic and Cyclic Compression Tests on Concrete Confined with PET–FRP," ASCE, Journal of Composites for Construction, ISSN 1090-0268/04014034(10)

4. Natalie Williams Portal, Karin Lundgren, Holger Wallbaum and Katarina Malaga, (2014), "Sustainable Potential of Textile-Reinforced Concrete", ASCE, Journal of Materials in Civil Engineering, ISSN 0899-1561/04014207(12)

5. Navid H Jafari, Timothy D Starkand Korry Rowe R, (2014), "Service Life of HDPE Geo-membranes Subjected to Elevated Temperatures", ASCE, Journal of Hazards and Toxic Radioact Waste, Vol. 18, No. 1, pp 16-26

6. Erich D Rodriguez, Susan A Bernal, John L Provis and Jordi Paya, (2012), "Structure of Portland Cement Pastes Blended with Sonicated Silica Fume", ASCE, Journal of Materials in Civil Engineering, Vol. 24, No. 10, pp 1295-1304.

7. Khadiranaikar R B and Mahesh M Awati, (2012), "Concrete Stress Distribution Factors for High-Performance Concrete", ASCE, Journal of Structural Engineering, Vol. 138, No. 3, pp 402-415

8. Pedram Zohrevand and Amir Mirmiran, (2012), "Cyclic Behavior of Hybrid Columns Made of Ultra High Performance Concrete and Fiber Reinforced Polymers", ASCE, Journal of Composites for Construction, Vol. 16, No. 1, pp 91-99

9. Khater H M, (2011), "Influence of Metakaolin on Resistivity of Cement Mortar to Magnesium Chloride Solution," ASCE, Journal of Materials in Civil Engineering, Vol. 23, No. 9, pp 1295-1301

10. Erhan Guneyisi, Mehmet Gesoglu and Kasim Mermerdas, (2010), "Strength Deterioration of Plain and Metakaolin Concretes in Aggressive Sulphate Environments", ASCE, Journal of Materials in Civil Engineering, Vol. 22, No. 4, pp 403-407

11. Stephanie J Barnett, Jean Francois Lataste, Tony Parry, Steve G Millard and Marios N Soutsos, (2010), "Assessment of Fibre Orientation in Ultra High Performance Fibre Reinforced Concrete and Its Effect on Flexural Strength", Materials and Structures, Vol. 43, pp 1009-1023

12. Hampel T, Speck K, Scheerer S, Ritter R and Curbach M, (2009), "High-Performance Concrete under Biaxial and Triaxial Loads," ASCE, Journal of Engineering Mechanics, Vol. 135, No. 11, pp 1274-1280

13. Khatib J M, Kayali O and Siddique R, (2009), "Dimensional Change and Strength of Mortars Containing Fly Ash and Metakaolin," ASCE, Journal of Materials in Civil Engineering, Vol. 21, No.9, pp 523-528

14. Khatib J M, (2009), "Low Temperature Curing of Metakaolin Concrete," ASCE, Journal of Materials in Civil Engineering, Vol. 21, No. 8, pp 362-367

15. Yazdani N, Filsaime M and Islam S,(2008), "Accelerated Curing of Silica-Fume Concrete ", ASCE, Journal of Materials in Civil Engineering, Vol. 20, No. 8, pp 521-529

16. Katrin Habel, Emmanuel Denarie and Eugen Bruhwiler, (2006), "Structural Response of Elements Combining Ultrahigh-Performance Fiber-Reinforced Concretes and Reinforced Concrete", ASCE, Journal of Structural Engineering, Vol. 132, No.11, pp 1793-1800

17. Schleser M, Walk–Lauffer B, Raupach M and Dilthey U, (2006), "Application of Polymers to Textile–Reinforced Concrete," ASCE, Journal of Materials in Civil Engineering, Vol. 18, No. 5, pp 670-676

18. Christopher J Waldron, Thomas E Cousins, Adil J Nassar and Jose P Gomez, (2005), "Demonstration of Use of High Performance Light Weight Concrete in Bridge Superstructure in Virginia", Journal of Performance in Constructions & Facilities, Vol. 19, No. 2, pp 146-154

19. Khaled Sobhan and Mehedy Mashnad, (2003), "Fatigue Behaviour of a Pavement Foundation with Recycled Aggregate and Waste HDPE Strips", ASCE, Journal of Geotechnical and Geo-environmental Engineering, Vol. 129, No.7, pp 630-638

20. Meda A and Rosati G, (2003), "Design and Construction of a Bridge in Very High Performance Fiber-Reinforced Concrete", ASCE, Journal of Bridge Engineering, Vol. 8, No. 5, pp 281-287

21. Xianyu Jin and Zongjin Li, (2003), "Effects of Mineral Admixture on Properties of Young Concrete", ASCE, Journal of Materials in Civil Engineering, Vol.15, No. 5, pp 435-442

22. Kypros Pilakoutas, Kyriacos Neocleous and Maurizio Guadagnini, (2002), "Design Philosophy Issues of Fibre Reinforced Polymer Reinforced Concrete Structures", ASCE, Journal of Composites for Construction, Vol. 6, No. 3, pp 154-161

23. Joo-Hwa Tay, Sze-Yunn Hong and Kuan-Yeow Show, (2000), "Reuse of Industrial Sludge as Pelletized Aggregate for Concrete", ASCE, Journal of Environmental Engineering, Vol. 126, No. 3, pp 279-287

24. Ardavan Yazdanbakhsh and Zachary Grasley, (2014), "Utilization of Silica Fume to Stabilize the Dispersion of Carbon Nano Filaments in Cement Paste", ASCE, Journal of Materials in Civil Engineering, ISSN 0899-1561/06014010(5)

25. Sakr K, (2006), "Effects of Silica Fume and Rice Hush Ash on the Properties of Heavy Weight Concrete", ASCE, Journal of Materials in Civil Engineering, Vol. 18, No. 3, pp 367-376

26. Kuruva Syamala Devi, Arza V V S Swamy and Ravuri Hema Krishna, (2014), "Studies on the Solid Waste Collection by Rag Pickers at Greater Hyderabad Municipal Corporation, India", International Research Journal of Environment Sciences, ISSN 2319-1414 Vol. 3(1), pp 13-22

27. Bashar Taha and Ghassan Nounu, (2009), "Utilizing Waste Recycled Glass as Sand/Cement Replacement in Concrete", ASCE, Journal of Materials in Civil Engineering, Vol. 21, No. 12, pp 709-721

28. Hassan K E, Cabrera J G and Maliehe R S, (2000), "The Effect of Mineral Admixtures on the Properties of High–Performance Concrete", Cement and Concrete Composites, Vol. 22, No.4, pp 267-271

29. Guleria S P and Dutta R K, (2011), "Unconfined Compressive Strength of Fly Ash-Lime-Gypsum Composite Mixed with Treated Tire Chips", ASCE, Journal of Materials in Civil Engineering, Vol. 23, No. 8, pp 1255-1263

30. Kannan V and Ganesan K, (2016), "Effect of Tricalcium Aluminate on Durability Properties of Self-Compacting Concrete Incorporating Rice Husk Ash and Metakaolin," ASCE, Journal of Materials in Civil Engineering, ISSN 0899-1561/04015063(10)

31. Amitava Roy, Nicholas Moelders, Paul JSchilling and Roger K Seals, (2006), "Role of an Amorphous Silica in Portland Cement Concrete", ASCE, Journal of Materials in Civil Engineering, Vol. 18, No. 6, pp 747-753

32. Potgieter-Vermaak S S and Potgieter J H, (2006), "Metakaolin as an Extender in South African Cement," ASCE, Journal of Materials in Civil Engineering, Vol. 18, No. 4, pp 619-623

33. Appa Rao G and Raghu Prasad B K, (2005), "Fracture energy of fibre reinforced high strength concrete", ASCE, Journal of Structural Engineering, pp 249

34. Emily L Owens, Qiong Zhang and James R Mihelcic, (2011), "Material Flow Analysis Applied to Household Solid Waste and Marine Litter on a Small Island Developing State", ASCE, Journal of Environmental Engineering, Vol. 137, No. 10, pp 937-944

35. Youjiang Wang, HC Wu and Victor C Li, (2000), "Concrete Reinforcement with Recycled Fibers", ASCE, Journal of Materials in Civil Engineering, Vol. 12, No. 4, pp 314-319

36. Khan M I, (2003), "Permeation of High Performance Concrete", ASCE, Journal of Materials in Civil Engineering, Vol. 15, No. 1, pp 84-92

37. Justice J M and Kurtis K E, (2007), "Influence of Metakaolin Surface Area on Properties of Cement-Based Materials", ASCE, Journal of Materials in Civil Engineering, Vol. 19, No. 9, pp 762-771

38. Kay Wille, Antoine E Naaman, Sherif El-Tawil and Gustavo J Parra-Montesinos, (2011), "Ultra-High Performance Concrete and Fiber Reinforced Concrete: Achieving Strength and Ductility Without Heat Curing", Materials and Structures, 10.1617/s11527-011-9767-0

39. Goa Tourism Department, (2014), "Tourist Arrivals (Year Wise)", Department of Tourism, Government of Goa, India

ISBN: 978-93-8830-599-0

Effect of Recycled Aggregate Processing Techniques on the Performance of Concrete

B. Rajesh[1], V. Murali Sharma[1], K. Sai Kiran[1] and V. Srinivasa Reddy[2]
[1]B.Tech Students, [2]Professor , Department of Civil Engineering, GRIET Hyderabad.
vempada@gmail.com

Abstract

Due to depletion of natural aggregates the need for the usage of recycled aggregate in concrete has gained significance. In this regard, the present study is an attempt to evaluate the performance of M20 grade of concrete made with 100% recycled aggregate processed using various techniques. Handpicked aggregate from concrete rubble is also used to prepare concrete. The use of chemical admixture is mandatory to compensate the extra water (3 to 6%) required by the RCA (Recycled Aggregates).Results show that surface modification of RCA through microbial carbonate precipitation (MCP) is feasible evidenced by increasing weight and reducing water absorption of treated RCA. The results also show that the use of different acid molarities to remove loose mortar that attaches to RCA particles can significantly improves surface contact between the new cement paste and the aggregate, which subsequently resulted in a significant improvement in the strength of concrete. A novel microwave-assisted technique also increases the quality of RCA by temperature distribution and stresses developed in RCA. In thermal –mechanical methodthe recycled aggregates have been heated to 300°C to remove weaker mortar and cement particles from the aggregate and placed in a rotating drum containing iron balls. In chemical–mechanical method the aggregate is exposed to sodium sulfate solution and mechanical stresses are created by subjecting RCA to freeze-and-thaw action to separate mortar from RCA. In acid soaking beneficiation method the mortar around RCA is removed by pre-soaking RCA in 0.1 M acidic solutions of HCl and $H_2SO_4$for 24 hours. However, these methods are not practiced yet in full scale. However, the effectiveness of these treatment methods remains dependent on several factors that require further consideration. Compressive strength and water absorption capacities of various concrete samples made with recycled aggregate prepared using above discussed processing techniques are evaluated.

Keywords: Recycled aggregate concrete, processing techniques, recycled aggregate, CWD.

I. INTRODUCTION

The management of construction and demolition waste is a major concern due to increase in quantity of demolition rubble, continuing shortage of dumping sites, increase in cost of disposal and transportation and above all, the concern about environment degradation. According to a survey conducted by Central Pollution Control Board, the estimated quantity of solid waste generated in India in 2007 was around 48 million tons per annum of which 25 percent was the waste from construction. The Energy and Resources Institute (TERI) has estimated that by 2047, waste generation in Indian cities will increase five-fold to touch 260 milliontons per year, implying that the current solid waste generation is over 50 milliontons per year (Kala and Kumar, 2013). They estimated the annual increase in the quantity of solid waste in Indian cities to be at the rate of 5 per cent per annum. Presently in India this waste is disposed off in the landfill or used as an infill material. The poor management of solid waste has led to contamination of ground water and surface water through leachate. Unscientific practices in processing and disposal in reclaimed areas or river banks compound the environmental hazards posed by solid waste. With landfill spaces decrease and environment being destroyed, this inert waste needs a better strategy to manage. Thus with huge demand seen in construction industry and strategies present to fulfill the demand, an integrated and holistic approach involving design and construction engineering is required which respects the construction and economic environment of the country. Rapid strides are being made towards advancement of research in the field of construction material and technology. Recycle options have been tried to fulfill the growing demand which has led to the reuse of demolished waste in countries outside India. Research work on recycling of aggregates has also been carried out at Central Building Research Institute (CBRI), Roorkee, and Central Road Research Institute (CRRI), New Delhi, but as per the study commissioned by (TIFAC), 70 percent of the construction industry is not aware of recycling techniques. Hence creating awareness as well as promoting the use of recycled product is the need of the day to achieve the necessary goal. Thus it is seen that the problems in India are also alarming as in the west, considering the quantum of

ISBN: 978-93-8830-599-0

construction and demolishing waste generated. It is not far off when India may also have to seriously think of reusing demolished rubble and concrete for production of recycled construction material. Work on recycled concrete has been carried out at few places in India but waste and quality of raw material produced being site specific, tremendous inputs are necessary if India has to use the material in construction for producing concrete.

Concrete is one of the most important building materials [Shetty (2005)]. Concrete is mouldable, adoptable, relatively fire resistant, generally available and affordable. The properly designed and produced concrete has an excellent workability, mechanical and durability properties. Globally, 10 billion tons of concrete is required for construction industry in every year [Meyer (2009)]. Coarse aggregates comprises (approximately) 70-80% of the volume of concrete and exert significant influence on the properties of concrete [Dhir and Paine (2010)]. Hence, there is a need to search for an alternative coarse aggregate material to protect the natural aggregates. In this context, the utilization of recycled aggregate (RA) in concrete has been very popular method and encouraging by many countries in recent years. So far, most of the researchers utilized recycled aggregate (RA) for lower-grade concrete applications and for un-important works because of variations in quality of RA from source to source and the presence of adherent cement mortar on its periphery [Marios et al. (2011)]. The utilization of RA in high-strength concrete has been reported very rarely because of the problems associated with recycled aggregates. Some of the eminent researchers utilized RA for high-strength concrete also. Majority of the past researchers have utilized un-processed recycled aggregate widely in high-strength concrete production. Because of this reason, the replacement level of natural aggregates with RA has been limited to 20% only [Limbachiya et.al (2000)]. In literature, researchers have already done enormous quantity of research in the area of recycled aggregate concrete (RAC) by using many processing techniques and mixing techniques to improve the final performance. They are i) pre-soaking of RA in acid [Tam et al. (2007)b, Sallehan and Mahyuddin (2013)], ii) pre-soaking of RA in water [Nealen and Schenk (1988), WSDOT Research report (2014)], iii) micro-wave oven heating method of RA [Ong et al. (2010), Akbarnezhad (2011)], iv) ultra-sonic cleaning methods of RA [Katz (2004)], v) P.V.C. coating of RA method [Wan et al. (2005), Kou and Poon (2010)], vi) Micro carbonate precipitation treatment of RA [Anna et al. (2012), Jishen Qiu et al. (2014)], vii) pozzolanic coating methods of RA [Li et al. (2009), Deyu Kong et al. (2010)], viii) Surface treatment of RA with oil type agents [Tsujino et al. (2007)]. Similarly, some more processing techniques have suggested in literature to improve the performance of RAC. They are i) straight-forward mechanical grinding [Dhir and Paine (2010)], ii) heating and rubbing [Tateyashiki et al. (2002), Muller and Linss (2004)] and iii) eccentric–shaft rotor method etc. But these techniques are not yet practiced in full-scale experimentation of RAC [Dhir and Paine (2010)]. Majority of researchers utilized un-processed recycled aggregate. There are many factors contributing to this, from the availability of new material and the damage caused by the quarrying of NA and the increased disposal costs of waste materials. Recently, these aggregates started to be used for intermediate utility applications such as foundations for building and roads. However, more care is to be taken on physical and chemical characteristics of the RA and hence it can be used to produce recycled aggregate concrete (RAC) by minimizing environmental problems and optimizing the cost without compromising the quality. The advantages of recycling C&DW are as follows: It reduces the amount of construction and demolition waste entering landfill sites, reduces the use of natural resources in construction, contributes to the environment, provides a renewable source of construction material and used in situ, reduces haulage costs.

II. OBJECTIVES

The main aim of the present project study is to realise the compressive strength and water absorption properties of 100%recycled aggregate concrete of M20 grade. To achieve the above mentioned objectives, the experimental investigations are planned as shown below—

1. Determine the mix proportions for M20 gradeconcrete made with 100% recycled aggregate (RA)
2. Evaluation of workability in terms of slump for M20 concrete made with 100% recycled aggregate (RA)
3. Assessment of compressive strength and water absorption capacities of natural and 100% recycled aggregate M20 grade concrete at 28 days age of curing.

ISBN: 978-93-8830-599-0

III. MATERIALS

A. Cement

Ordinary Portland cement (OPC) of 53 grade [IS: 12269-1987, Specifications for 53 Grade Ordinary Portland cement] has been used in the study.

B. Fine Aggregate

The fine aggregate used was locally available river sand without any organic impurities and conforming to IS: 383.

C. Coarse Aggregate

The physical properties of coarse aggregate were investigated in accordance with IS: 2386 – 1963 Parts I to VIII. Crushed stone coarse aggregate of angular in shape obtained from local crushing plant was used in the present study.

D. Recycled aggregates

Recycled aggregates are extracted from demolished and construction waste. The C&D waste collected from locally available, poured water on C&D to remove silt and clay then removed other foreign materials such as broken brick bats, reinforcement, mortar etc. About 100mm size concrete samples separated and fed in the feeding unit of the laboratory jaw crusher of capacity 250 kg/hr. Fig 1 shows the jaw crusher. The jaw of the crusher is adjusted to obtain aggregates of desired size. The material passing through the feeder unit of the laboratory jaw crusher is broken to appropriate sizes of 20 mm (Fig 2 (b)). The crushed aggregates received from the feeder unit are then sieved through a series of IS sieves as per IS: 2386 (Part I)-1997 to segregate them in various sizes by discarding the material smaller than 20 mm size. The material in size larger is subjected to re-crushing so as to obtain the aggregates of appropriate size. The sieving process also help for the removal of adhered mortar traces loosened during the crushing process. It was observed that the jaw crushers provide the best grain-size distribution of recycled aggregate to be used in making concrete.

E. Super Plasticizer

In the present work, water-reducing admixture Conplast SP 430 conforming to IS 9103: 1999 [Specification for admixtures for concrete], ASTM C – 494 [Standard Specification for Chemical Admixtures for Concrete]

types F, G and BS 5075 part.3 [British Standards Institution] was used. Conplast SP 430 is a Sulphonated Naphthalene based Formaldehyde (SNF), super plasticizer and it was manufactured by Fosroc. Dosage range of 0.5 to 1.5% by weight of cement is normally recommended.

Fig. 1 Jaw Crusher

Fig. 2 (a) Construction Demolition Waste (b) Recycled Aggregate after crushing

V. RECYCLED AGGREGATE PROCESSING TECHNIQUES

The table 1 below gives various recycled aggregate processing techniques used in the present study to treat recycled aggregate obtained from construction demolition waste.

V. CONCRETE MIX DESIGN

The grade of the concrete used in the present investigation isM20 made with 100% recycled aggregate (RCA). The quantity of materials per one cubic meter of concrete are designed using IS: 10262-2009 (Table -2).

Table 1 Various recycled aggregate processing techniques adopted in the present study

Type of specimen	Designation	Recycled aggregate processing technique	Description
M20 grade concrete made with Natural aggregate	Method 1	-	Naturally available crushed angular coarse aggregate
M20 grade concrete made with 100% Recycled aggregate	Method 2	Hand-picked	Hand-picked coarse aggregate from Construction and demolition waste
	Method 3	Microbial carbonate precipitation (MCP)	Hand-picked coarse aggregate from Construction and demolition waste are pre-soaked in S. pasteurii bacterial suspension of 10^5 cells/ ml concentration
	Method 4	Acid soaking beneficiation method	Hand-picked coarse aggregate from Construction and demolition waste are pre-soaked in 0.1 M HCl for 24 hours
	Method 5	Acid soaking beneficiation method	Hand-picked coarse aggregate from Construction and demolition waste are pre-soaked in 0.1 M H_2SO_4 for 24 hours
	Method 6	Thermal – mechanical method	Hand-picked coarse aggregate from Construction and demolition waste are heated to 300°C on microwave and placed in a rotating drum containing iron balls
	Method 7	Chemical–mechanical method	Hand-picked coarse aggregate from Construction and demolition waste areexposed to 1% sodium sulfate solution and subjected to repeated freeze-and-thaw action (7 cycles)
	Method 8	Jaw crusher	Construction and Demolition waste is crushed in Jaw crusher into recycled aggregate

Table 2 Quantity of materials required for one cubic meter of concrete

Grade	Cement kg	Fine aggregate kg	Coarse aggregate kg		Water L	W/C ratio
			20 mm	10mm		
M20	333	739.11	464.34	703.28	160.02	0.50

VI. WORKABILITY

Slump test is the most commonly used method of measuring consistency of concrete which can be employed either in laboratory or at site of work. It does not measure all factors contributing to workability. However, it is used conveniently as a quality control test and gives an indication of the uniformity of concrete from batch to batch. Slump values for are evaluated.

VII. COMPRESSIVE STRENGTH

This investigation is carried out to study the compressive strength of M20grade concrete mix made with 100% recycled aggegate (RCA) at 28 days.For this concrete cubes of 150mm are cast and tested to study the compressive strength under axial M20 grade concrete mixes made with 100% recycled aggregate (RCA) and natural aggregates compression on completion of 28 days as per IS: 516-1999.

VIII. WATER ABSORPTION CAPACITY STUDIES

The aim of this study is to determine the total water absorption capacity ofM20 grade concrete mixes made with 100% recycled aggregate (RCA) and natural aggregates as per ASTM C642-13.

IX. TEST RESULTS AND DISCUSSIONS

A. Workability

In the present study, workability of concrete is assessed using slump cone test. Slump values for M20gradeconcrete made with 100% recycled aggregate and naturalaggregate are presented in table 3.

ISBN: 978-93-8830-599-0

Table 3 Slump values of recycled aggregate concrete

Grade of concrete	Type of Concrete	Slump mm
M20	Natural Aggregate	90
	Recycled aggregate	45

To enhance workability, mineral admixtures such as GGBS can be added in suitable quantity so that the desired strength and workability can be achieved.

For 100% recycled aggregate concrete, workability in terms of slump values decreases when compared to concrete made with natural aggregate. The decrease in workability in recycled aggregate concrete is attributed to the fact that the recycle aggregate has some amount of mortar adhered to the surface which increases the water absorbtion capacity of the aggregate. So some quantity of mixing water, calculated during mix design trials intended for desired workability and strength, may get absorbed by the recycled aggregate reducing the workability of the concrete mix and prolongs the mixing time. This limitation can be overcome by admixing the recycled aggregate concrete with optimum amount of ground granulated blast furnace slag (GGBS).

B. Compressive Strength and Water absorption Studies

Compressive strengths of M20 grade concrete mixes made with 100% recycled aggregate (RCA) and natural aggregate are tabulated in table 4. For this concrete cubes of 150x150x150 mm are cast and tested as per IS: 516-1999 to study the compressive strength under axial compression on completion of 28 days. Similarly the water absorption capacity of M20 grade concrete mixes made with 100% recycled aggregate (RCA) and natural aggregates as per ASTM C642-13 are presented in table 4.

Table 4 Compressive strength and water absorption capacity of M20 grade concrete made with 100% recycled aggregate prepared using various processing techniques

Designation	Recycled aggregate processing technique	Compressive Strength (MPa) @ 28 days Age of Curing	% increase / decrease of Compressive strength	Water absorption capacity %
Method 1	-	27.86	-	5.62
Method 2	Hand-picked	17.86	-35.89	11.79
Method 3	Microbial carbonate precipitation (MCP)	35.91	28.89	2.79
Method 4	Acid soaking beneficiation method	26.13	-6.21	6.43
Method 5	Acid soaking beneficiation method	27.19	-2.40	6.55
Method 6	Thermal – mechanical method	22.17	-20.42	8.56
Method 7	Chemical– mechanical method	24.89	-10.66	7.75
Method 8	Jaw crusher	22.19	-20.35	7.87

Contd...

The following observations were made-

1. In concrete made with 100% handpicked recycled aggregate, compressive strength decreases by about 36% at all ages of curing when compared to concrete made with natural aggregate. If the handpicked recycled aggregate is treated with bacterial solution then the compressive strength increases by 29% and water absorbtion is reduced by 11.8%.

2. The decrease in compressive strength in recycled aggregate concrete is attributed to the fact that the recycle aggregate has some amount of mortar adhered to the surface which increases the water absorbtion capacity of the aggregate. This behaviour is observed for all the grades of concrete made with recycled aggregate. Interfacial Transition Zone (ITZ) in concrete made with 100% recycled aggregate is weak than in concrete make with natural aggregate.

3. Concrete on a macroscopic scale, is a mixture of cement paste and fine and coarse aggregates, with a range of sizes and shapes. With regard to its mechanical behavior, concrete is often considered to be a three-phase composite structure, consisting of aggregate particles, the cement paste matrix in which they are dispersed, and the interfacial transition zone (ITZ) around the aggregate particles and cement paste. Strength of ITZ depends on the bond between aggregate and cement paste.

4. In 100% recycled aggregate concrete, bond between recycled aggregate and cement paste surrounding is weak so development of CSH

ISBN: 978-93-8830-599-0

crystals in this zone are structured weakly which leads to failure in compression. So compressive strength of concrete made with 100% recycled aggregate reduces drastically. It can also reported that due to higher porosity and absorption capacity of recycled aggregate can contribute to the porous ITZ microstructure of concrete.

5. Increase in the compressive strength of recycled aggregate concrete by increasing the proportions of cement content in recycled concrete mixes is possible, since the w/c ratio of mix will be reduced by increasing cement content to achieve 28-day compressive strength equivalent to corresponding natural aggregate concrete. This may not be a sustainable justified solution so the use of optimum dosage of mineral admixtures in recycled aggregate concrete is encouraged to achieve equivalent compressive strength corresponding to natural aggregate concrete.

6. Of all the recycled aggregate processing techniques, recycled treated with mineral precipitating bacterial solution yields maximum compressive strength and has less water absorption capacity.

X. CONCLUSIONS

Based on the results reported in this research work and key findings during the experimental investigations on M20 grade concrete mixes made with 100% recycled aggregate (RCA) and natural aggregate, the following conclusions are drawn:

1. Specific gravity of recycled aggregate in saturated surface dry condition (SSD) is less than that of natural aggregates and particle shape of recycled aggregate is angular with more rough surface texture.

2. It is observed that for all the grades of the concrete mixes made with 100% recycled aggregate, workability decreases due to loss of water in the form of absorbtion by recycled aggregates.

3. The recycled aggregates being porous in their characteristics may have a tendency to absorb more water from the concrete mix leading to a porous structure in concrete, which when subjected to drying may lead to higher shrinkage strains. Thus the optimum amount of water/cement ratio for a recycled mix is to be decided not only by taking into account workability of the mix but also the durability criteria. Recycled aggregate concrete also contain a higher volume of void space on account of the additional water added to compensate for the water absorption for providing a workable mix

and may thus lead to higher creep strains on loading.

4. There is a significant increase in the compressive strength of concrete made with bacteria treatedrecycled aggregate due to formation of calcite mineral precipitation modifying the concrete pore structure within the cement-sand matrix.

5. Untreated handpicked recycled aggregate yield very less compressive strength and has more water absorption capacity.

REFERENCES

1. Ajdukiewicz, A. and Kliszczewicz, A. (2002), "Influence of Recycled Aggregates on Mechanical Properties of High strength/High performance concrete", *Cement and Concrete Composites,* Vol.24, pp.269-279.

2. Akbarnezhad, A., Ong, K.C.G., Zhang, M.H., Tam, C.T. and Foo, T.W.J. (2011), "Microwave assisted beneficiation of recycled concrete aggregates", *Construction and Building Materials,* Vol.25, No.8, pp.3469-3479.

3. Anna, G.M., Justyana, K., Daniel, Z. and Krupa, D. (2012), "Modification of recycled concrete aggregate by calcium carbonate biodeposition", *Construction and Building Materials,* Vol.34, pp.145-150.

4. Barra, M. and Vázquez, E. (1998), "Properties of concrete with recycled aggregates: influence of properties of the aggregates and their interpretation, Use of Recycled Concrete Aggregate", Proceeding of the *International Symposium on sustainable Construction*, London, UK, pp.19–30.

5. Chakradhara, R.M., Bhattacharyya, S. and Barai, S. (2011), "Influence of field recycled coarse aggregate on properties of concrete", *Materials and Structures*, Netherlands, pp. 205–220.

6. Etxeberria, M., Vázquez, E. and Mari, A.R. (2006), "Microstructure analysis of hardened recycled aggregate concrete", *Magazine of Concrete Research*, Vol.58, pp.683–690.

7. Katz, A. (2004), "Treatments for the improvement of recycled aggregate", *Journal of Materials in Civil Engineering, ASCE*, Vol.16, No.6, pp.597–603.

8. Ong, G.C., Akbarnezhad, A., Zhang, M.H., Tam, C.T., Hao, J. and Foo, T.W.J. (2010), "Mechanical properties of concrete incorporating Microwave-treated recycled concrete aggregates, 35 th conference on our world in concrete structures", *Singapore,* pp.25-27.

9. Sallehan, I. and Mahyuddin, R. (2013), "Engineering properties of treated recycled concrete aggregate (RCA) for structural applications", *Construction and Building Materials*, Vol.44, pp.464–476.

10. Sanchez, D.M. and Gutierrez, P.A. (2009), "Study on

ISBN: 978-93-8830-599-0

the influence of attached mortar content on the properties of recycled concrete aggregate", *Construction and Building Materials,* Vol.23, No.2, pp. 872–877.

11. Tam,V.W.Y., Tam,C.M. and Le, K.N. (2007)b, "Removal of cement mortar remains from recycled aggregate using pre-soaking approaches", *Resources Conservation and Recycling*, Vol.50, pp.82-101.

12. Zaharieva, H., Wirquin, E., and Buyle-Bodin, F. (2000), "Use of water absorption by concrete as a criterion of the durability of concrete-application to recycled aggregate concrete", *Materials and Structures*, Vol.33, pp.403–408.

ISBN: 978-93-8830-599-0

Optimally Locating Rain Harvesting Pits Using GIS as A Tool

M. Madhuri[1] and P. Sai Shraddha[2]

[1,2]Assistant Professor, Department of Civil Engineering in Vignan Institute of Technology and science, Deshmukhi, Hyderabad.

Abstract

The conventional means to record hydrological parameters of urban flood often fail to record the operation of flood mapping and flood risk assessment. GIS plays a major role in the management of multi-dimensional natural hazard with an inherent spatial component. GIS generates a visualization of flooding and also creates possibilities to analyze probable damage estimate of the urban flood. Hence, there's got to review the present literature with a holistic read of managing varied prospects and constraints of using the technology of remote sensing and GIS of flood management. This study focuses on identifying the least elevation point for locating rain harvesting pit optimally using GIS.

Keywords: Urban Flood Control, Rain Harvesting Pits, GIS.

I. INTRODUCTION

Urban flooding is the most common problem arising in an urban area during the monsoon season due to high-intensity rain falls. Improper drainage system, defective flood water disposal techniques do not have the necessary capacity to drain the amounts of rainwater that is falling. Sometimes the water disposed of in sewage system leaks on the roads through the drainage manholes results in foul smell, growth of bacteria and other sewage troubles.

Urban flooding creates immense troubles like traffic congestion due to water stagnation on roads, roads accidents, economic damages, etc. During heavy rains, the water slowly rises on roads and flows towards slopes resulting in water stagnation at lowest points. Quick rainwater disposal from the road surface is essential to maintain the quality and lifespan of the road. The installation of a suitable drainage system is an essential part of urban road design and construction.

Disposing of rainwater is an important feature in the determination of pavement ability to withstand the traffic and environmental effects. Other than the rainwater pavements experience poor conditions due to many other reasons out of which poor drainage is one of the reason. The strength, ability, and performance of pavement decrease with increase in moisture content. Defective drainage system causes premature failure of the pavement. Due to the lack of supportive infrastructure, the rainwater is not quickly getting disposed of resulting in damage of road structure.

Greater Hyderabad Metropolitan City with 10 million population is the most urban flood-affected areas is specific in the fact that has a lack of drainage issues and rainwater disposal problems. To overcome these problems a keen observation is needed hence this study focuses on identifying the optimal location for rain harvesting pits for Mansoorabad area by analyzing the three-dimensional data using Arc GIS software as a tool.

II. GEOGRAPHICAL INFORMATION SYSTEM

Geographical information system is an application of acquisition satellite data remotely which contains earth information. Through this information one can study, access and analyze various topographical parameters of earth. Recent advancements in this technology helping the researchers to utilize the application for studies related to transportation planning, flood monitoring, land use planning etc.

Various GIS software's are available in the market applied for wide varieties of data. For carrying out this study Arc GIS 10.1 Software is utilized to study the DEM (Digital Elevation Model) of the study area.

III. DATA COLLECTION

Greater Hyderabad is recognized as the most populated urban area in our country experiencing heavy urban flooding during monsoon. Heavy traffic congestion, delays, and economic losses. Mansoorabad area comes under the zone where the rainwater stagnation percentage is high. The 3 Dimensional data of the chosen study area is gathered using Arc GIS 10.1 in the following steps:

Fishnet

Create a fishnet tool generates a feature class of a rectangular grid cell creating three sets of basic information. a fishnet is created of grid size 200m X 200m covering an area of 9 Sqkms for the chosen study area as shown in figure 1.

ISBN: 978-93-8830-599-0

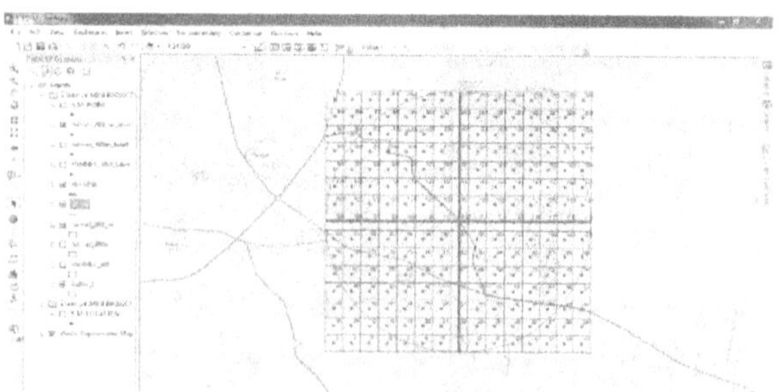

Fig. 1 Fishnet Created of grid size 200m X 200m

From the generated fishnet a total of 200 fishnet label points are produced along with it for which the elevation values and X, Y coordinate values have to be extracted. Usually, the X,Y coordinates are taken as latitudes and longitudinal values, here in this study the X, Y values are taken are the distances of the fishnet label point from the axis to identify the path the water flows during the flood.

Slope calculation

The slope can be calculated from elevations extracted from fishnet grid cells. The slope is defined as an angle. Slope measure the steepness of the surface at any particular location is often measured in percent rise of ground from MSL. In this study, the slope of the ground is estimated from an online web tool called GPS Visualizer.

From the generated fishnet label points, label point layer is convertedtoKMLformat for the extraction of three-dimensional data covering the latitude, longitude and elevation data using GPS Visualizer. The extracted elevation points are labeled on map and bifurcated into 4 quadrants anti-clockwise direction for easy identification of rain harvesting pit location as shown in figure 2.

Fig. 2 Labelled elevation points in the map

IV.   METHODOLOGY

The optimal location is found out by adopting a statistical equation called the method of moments. In this methodology, the feasible location for a Rain harvesting pit is identified by using the formula, elevation values of the nodes E_1, E_2, E_3… multiplied by horizontal and vertical distances i.e. X1, X2, X3… and Y1, Y2, Y3…. taken from the reference line (0,0). The optimal location points can be calculated from the below-presented formula.

$$X = \frac{E_1 x_1 + E_2 x_2 + \cdots..}{E1 + E2 + E3 + \cdots}$$

$$Y = \frac{E_1 y_1 + E_2 y_2 + \cdots..}{E1 + E2 + E3 + \cdots}$$

The x and y coordinate values obtained in four quadrants for each point are a substitute in the formula mentioned. The outputs received are mentioned in table 1

Table 1 Output Values after substituting in the method of moments formula.

QUADRANT	X⁻	Y⁻
1	798.290306	806.220242
2	-799.848895	796.111672
3	-798.2903	-806.2202
4	799.0301	-818.519

From the result obtained the value of X⁻, Y⁻ are known from 4 node points. Hence it will be opted as low-level land and used as the optimal location for facility location. Comparing with the graphical real-time data the values X⁻, Y⁻ gives the exact locality for placing a pit rather than any other methods. Hence, it can be taken as a feasible location for locating a Rain harvesting pit. Below figure 3 shows the map plotted with values of identified X and Y values

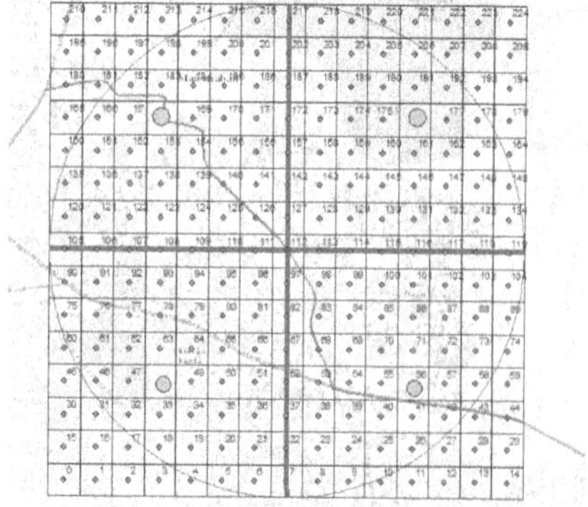

Fig. 3 Optimal locations of rain harvesting pits

ISBN: 978-93-8830-599-0

V. RESULT

The obtained X^-, Y^- satisfy as the lowest elevation points which represent the flow direction of water during the flood. Hence these location points serve as the optimal places to place the rain harvesting pits for disposing of the flood water in various methods and helps in avoiding traffic congestion, delays and pavement surface failures.

VI.   CONCLUSION

This study concludes that through GIS-based analysis of land slope analysis and positioning of rainwater harvesting pits is very much useful in future planning of the decision support system of rainwater harvesting. This kind of GIS-based analysis helps in identifying the water flow path to plan accordingly the supportive infrastructure for its easy disposal and storage.

REFERENCES

1.   Swathi Vemula,   K. Srinivasa Raju,   S. Sai Veena, A. Santosh Kumar "Urban floods in Hyderabad, India, under present and future rainfall scenarios: a case study"

2.   Ahmed Z, Rao DRM, Reddy KRM, Raj YE (2013) Urban flooding-case study of Hyderabad global journal of engineering. Des Technol 2:63–66

3.   GHMC   Disaster   Management   cell (2018) http://www.ghmc.gov.in/Disaster.aspx. Accessed in July 2018

4.   Chatterjee M (2010) Slum dwellers response to flooding events in the megacities of India.

5.   GPS   Visualizer   an   online   map   utility http://www.gpsvisualizer.com/

6.   Zameer Ahmed , D. Ram Mohan Rao , Dr.K.Ram Mohan Reddy, &Dr. Y. Ellam Raj " urban flooding – case study of Hyderabad"

7.   Wan B, James W (2002) SWMM calibration using genetic algorithms. In: International conference on Global Solutions for Urban Drainage. American Society of Civil Engineers.

8.   T.I. Eldho, ... A.T. Kulkarni, in Integrating Disaster Science and Management, 2018 " Urban Flood Management in Coastal Regions Using Numerical Simulation and Geographic Information System ".

9.   Jeffrey scott broke June 2015 " locating optimal water quality monitoring locations using demand coverage index method ".

10.   Guide on Artificial Recharge to Groundwater, Ministry of Water Resources, New Delhi,2000

11.   Sayed Joinal Hossain Abedin and Haroon Stephen " GIS Framework for Spatiotemporal Mapping of Urban Flooding"

12.   Sunmin Lee, ID , Saro Lee , ID , Moung-Jin Lee,ID and Hyung-Sup Jung. " Spatial Assessment of Urban Flood Susceptibility Using Data Mining and Geographic Information System (GIS) Tools".

13.   Sarup, Jyoti, and Vimal Shukla."Web-Based solution for Mapping Application using Open- Source Software Server." International

14.   HuiminLiua,b, , Qingming Zhana,b, Meng Zhan " the uncertainties on the gis based land suitability assessment for urban and rural planning"

15.   Bhaskar, N.R., James W.P., and Devulapalli, R.S (1992)."Hydrologic Parameter Estimation Using Geographic Information System (GIS)", *Journal of Water Resources Planning and Management*, ASCE,118(5),492-512.

ISBN: 978-93-8830-599-0

Automatic Ship Detection Method Using Spaceborne SAR Images

V. Madhavi Supriya[1*], B. Asha Rani[2] and K. Manjula Vani[3]

1-JRF, NRSC, ISRO, 2-Sci/Eng.-SE, ADRIN, DOS,ISRO, 3- Professor, Dept. of Civil Engineering, JNTUH-CEH

venigallasupriya@gmail.com

Abstract

The microwave remote sensing provides a powerful surveillance capability allowing the observation of broad expanses, independently from weather effects and from the day and night cycle. Since the development of SAR imaging technology many models for ship detection have been developed such as the K-distribution Constant False Alarm Rate (CFAR) method and two-parameter CFAR method, alpha-stable distribution method, etc. The proposed method takes the advantage of K-CFAR method and two-parameter CFAR method and a new improved method has been proposed which has high detection rate and improves the processing speed.

Keywords: SAR, ship detection, CFAR, K-distribution, Two-parameter distribution

1. INTRODUCTION

SAR imagery of the ocean is typically inhomogeneous and a large number of false alarms can arise because of the existence of the speckle noise. Historically, the ship detection is related to the invention of radar by Taylor and Youngman in Wisconsin, USA in 1922. Many attempts have been made for automated detection of ships using optical and microwave data, namely Gamma distribution, G_o distribution, Inverse gamma distribution, Rician inverse Gaussian, Log-normal model, Weibull model including K-distribution Constant False Alarm Rate (CFAR) method developed by Vachon (1997), and Qingshan et al.(1999) and two-parameter CFAR method of Eldhuset (1996); David(l999); Wakerman et al.(2001) are noteworthy. The K-distribution CFAR method is efficient when the SAR imagery background is homogenous as it doesn't need *a priori* knowledge of ship target characters but only needs the estimation of a global threshold of back scattering coefficient (sigma knot) obtained from the statistics of the whole SAR imagery. Two-parameter method requires the fine-tuning and analysis of the look parameter (L) and shape parameter (v).

2. OBJECTIVES AND SATELLITE DATA

The objectives of the study are 1. To detect the ships automatically in the image

2. To reduce the false alarms in the image.

ENVISAT ASAR (C- Band) of Wide Swath Mode (WSM) with the spatial resolution of 150m×150min VV polarizationnear coast of Galicia, Spain and SENTINEL-1A (C-Band) image of Interferometric Wide Swath mode (IWSM) with the spatial resolution of 5m×20m in VV polarization in North Sea near to Dunkirk, France is taken.

3. METHODOLOGY

The steps involved in the methodology development within the scheme are as follows:

- Reading the SAR imagery.
- Preprocessing of the SAR imagery involves radiometric calibration, speckle filtering
- A moving mean of the SAR imagery is computed by using the window size and Two-parameter such as shape parameter (v) and number of statistically independent looks (L).
- K-distribution CFAR probability density function formula is implemented with Two-parameter such as the v and L values.
- A global threshold values should be applied to the entire image to obtain a binary image. The pixels in the imagery, whose grey values lying above the threshold, are marked as possible ship targets and their values are set as 1, while the other pixels are marked as background sea area and their values are set as 0.

Thus, as the result, a binary image is given.
- Ship detection from the image.

ISBN: 978-93-8830-599-0

The K-distribution of probability density function of sea clutter in SAR images is

$$f(x) = \frac{2}{x\Gamma(v)\Gamma L}\left(\frac{Lvx}{\mu}\right)^{\frac{L+v}{2}} K_{v-L}\left(2\left(\frac{Lvx}{\mu}\right)^{\frac{1}{2}}\right)$$

Where µ-mean,

L- Number of statistically independent looks,

v- Shape parameter,

$K_{v\text{-}L}$ (.)-Modified Bessel function of (-L) th order

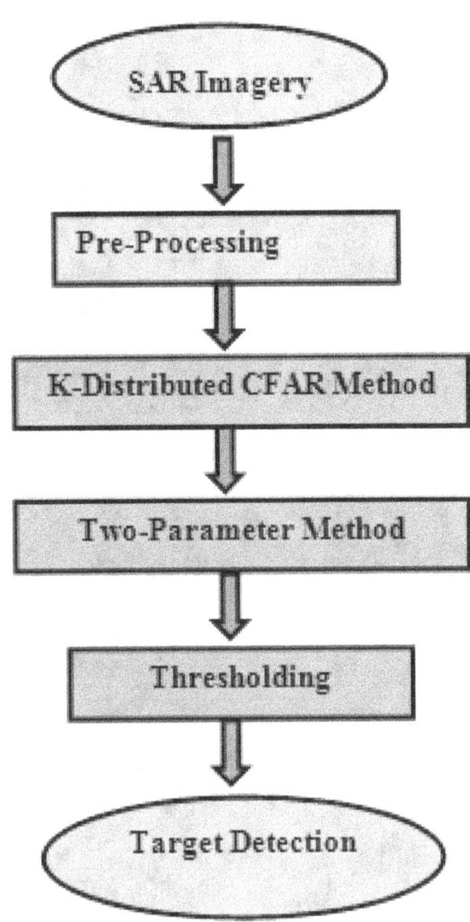

Fig. 1 Methodology

4. RESULTS AND CONCLUSION

The proposed detector has been applied to test ENVISAT ASAR and SENTINEL-1 SAR real SAR images, considering crucial situations, for evaluating its performance. Performance under heterogeneous background condition is a very important evaluation criterion to any target detector. Generally, ships are brighter than background clutters in marine SAR images, since the scattering of ship targets can last longer than sea clutters in azimuth. This method combines the advantages of K-distribution CFAR method and two-parameter CFAR method. The method also provides a great benefit to the faint target detection under the higher sea state conditions. This method has given robust and reliable results when applied to the SENTINEL-1 SAR and ENVISAT images, compared with CFAR method.

ISBN: 978-93-8830-599-0

4.1: Sentinel-1 (C-Band) SAR Image in VV polarisation

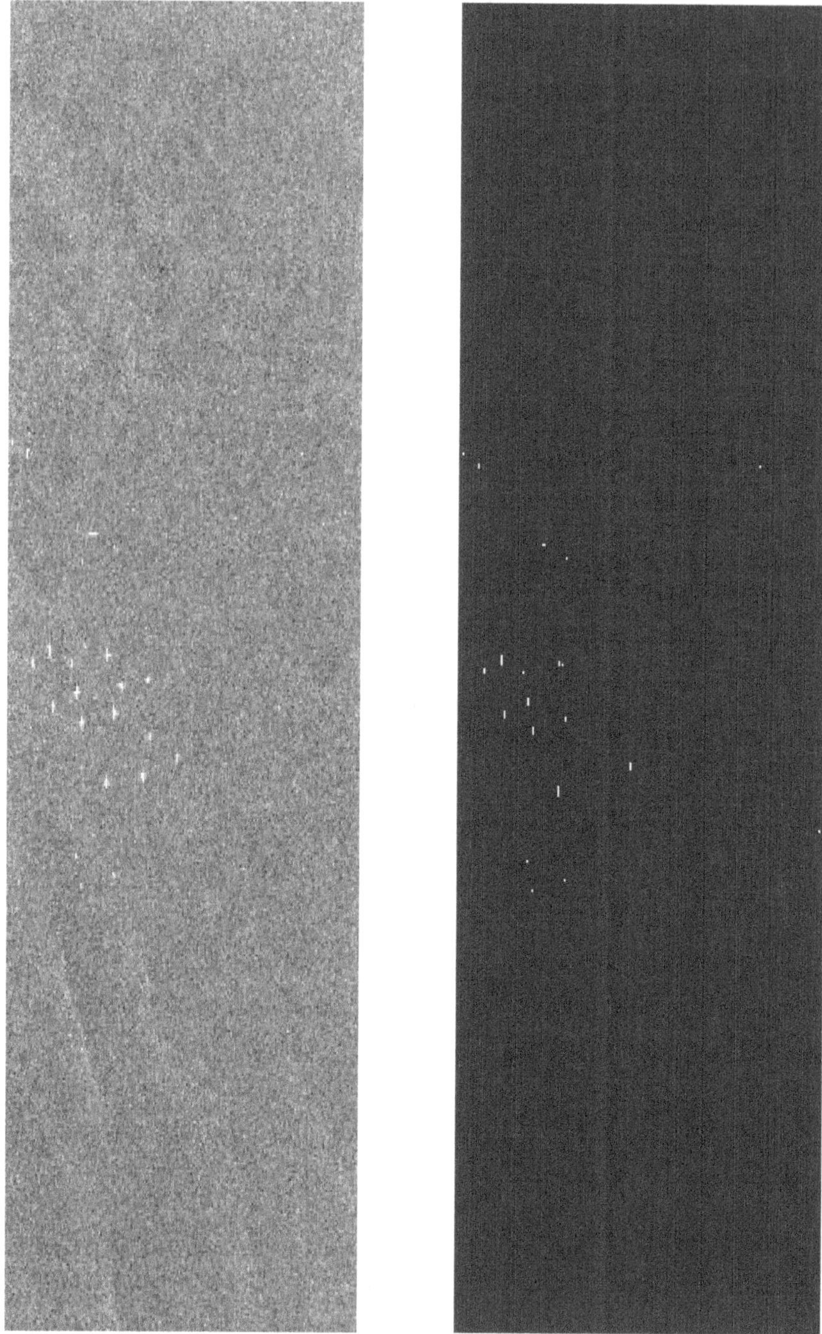

Fig. 2 a) Sentinel-1 SAR image, b) Detected output

978-93-8830-599-0

4.2: Envisat ASAR (C-Band) SAR Image

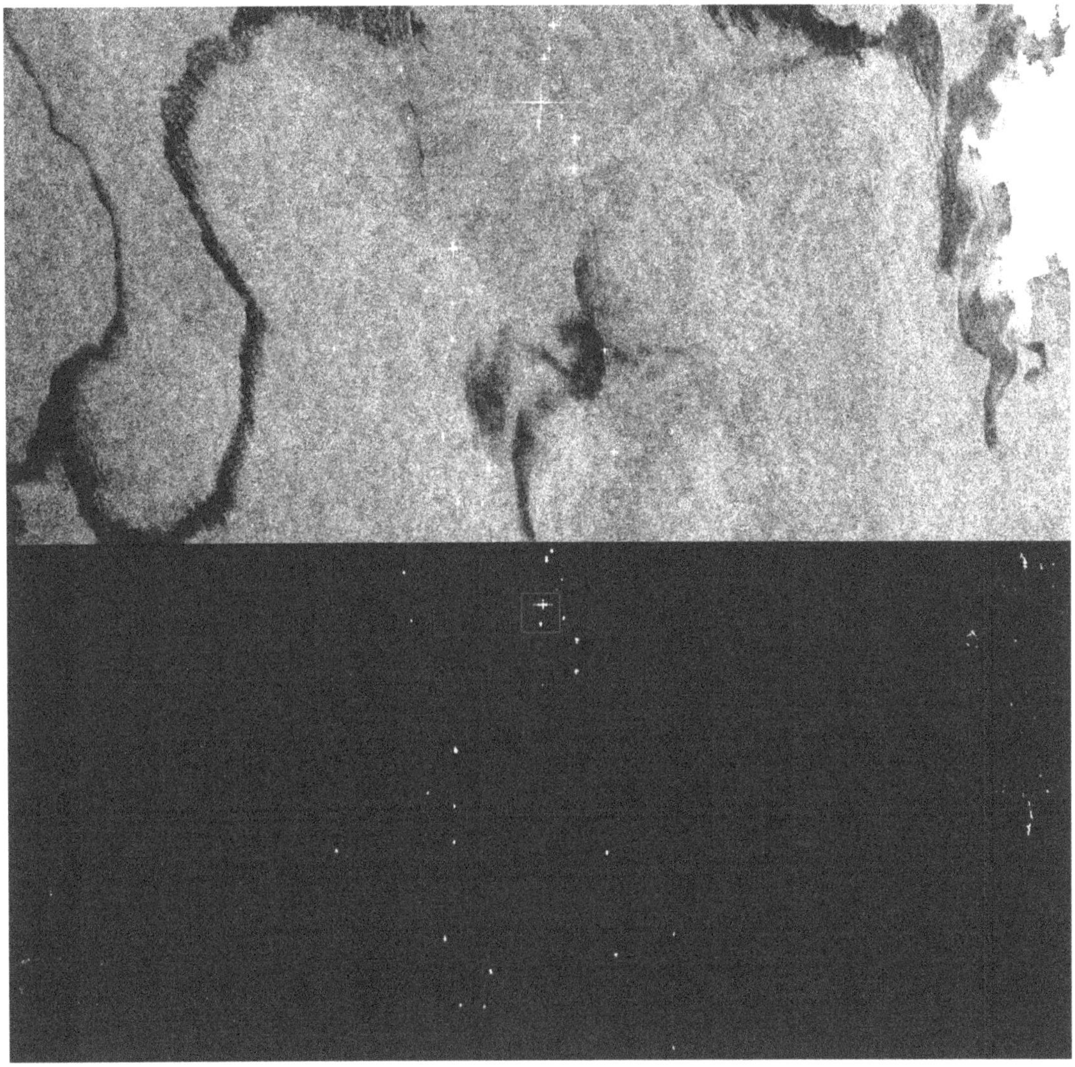

REFERENCES

1. Biao Hou, Xingzhong Chen and Licheng Jiao, "Multilayer CFAR detection of ship targets in very high resolution SAR images", IEEE geoscience and remote sensing letters, vol. 1, no. 4, pp. 811- 815, April 2015.

2. MariviTello, Carlos Lopez- Martinez and Jordi J. Mallorqui, " A novel algorithm for ship detection in SAR imagery based on the wavelet transform", IEEE geoscience and remote sensing letters, vol. 2, no. 2, pp. 201-205, April 2005.

3. C.C. Wackerman, K.S. Friedman, W.G. Pichel, P. Clemente-Colon & X. Li, "Automatic Detection of ships in RADARSAT-1 SAR imagery", Canadian Journal of Remote Sensing, vol. 27, issue-5,pp. 568-577, October 2001.

4. F. Zhang & B. Wu , "A scheme for ship detection in inhomogeneous regions based on segmentation of SAR images", International Journal of Remote Sensing, Vol. 29, No. 19, pp. 5733–5747, 10 October 2008.

5. Andreas Arnold-Bos, Ali Khenchaf, Arnaud Martin, "An evaluation of current ship wake detection algorithms in SAR images", IEEE transactions on geoscience and remote sensing, vol. 42, October 2006.

Detection of Spatio-Temporal changes in Singareni Coal Fields: Effect of Open Cast mining

G. Sri Priya[1], C. Sarala[2], B. Harish[3] and L. Ravi[3]
[1]Mtech Student [2]Head of the Department [3]Assitant Professor [3]Reasearch Scholar
CSIT,IST,JNTUH

Abstract

The aim of the study is to, show that change detection of vegetative cover and land surface temperature in the Singareni coal field areas, Telangana due to open cast mining, based on the analysis of United states Geological Survey(USGS) landsat data . To do comparative study on the LST and Normalized difference Vegetative index(NDVI).Thermal infrared band(TIR) data have been used to retrieve Land Surface Temperature(LST).LST is an important parameter in the studies of urban thermal environment and dynamics. The change detection in Singareni coal field area were analysed for a period of 29years i.e, from the year 1989 to 2018.The changes were detected on a certain time interval using Landsat-5TM,Landsat-8 OLI and TIRS satellite images and human impact on the environment are discussed. The study reveals that the Other Classes like waterbodies, sand, mines and ash dumpyards increases from 11.60% in 1989 to 16.86% in 2018.Barren land, shrubs and grassland increases from 55.62% in 1989 to 71.52% in 2018.On the other hand, medium vegetation and high vegetation (dense forest)decreases from 31.03% to 10.735% and 1.72% to 0.878% in 1989 and 2018 respectively. Land surface temperature increases from 39.92°c in 1988 to 48.74°c in 2018.

Keywords: NDVI, ,LST, Change detection,open cast mining and Landsat.

I. INTRODUCTION

In this case study , there is a extensive and rapid underground and open cast mining is going on continuously. Due to ,this results in environmental impacts of mining can occurs at local, regional and global levels through direct or indirect mining practices. Impacts can be results like soil erosion, loss of biodiversity, the contamination of soil, surface water and groundwater by the chemicals emitted by the mining processes. These mining processes also have an impact on the atmosphere by the emission of carbon which effect on the quality of human health and biodiversity. As day by day ,the energy requirement is increasing in India ,the coal minng companies are gradually increasing their production to meet the energy production demand through thermal power plants, where coal is used for electricity generation(Provisional   Coal   Statistics,2015-16).Coal production by the Singareni collieries company limited, has produced 64.40 million tonnes of coal during 2018-2019 an has dispatched 67.67 million tonnes of coal to various categories   of   consumers during 2018-2019(Provisional

Production and Dispatches performance of SCCL ,2019).For proper Land Management and Decision making it is necessary or essential to identify the effect of mining on land use and land cover to minimize its impact on environment(Laskar 2003,Bocco et al.,2001,Turner et al.,2007).

To analyse the change detection Remote Sensing and GIS techniques were used for periodical NDVI (Normalised difference   vegetative   index)   and   Land   surface temperature(LST) changes. Vegetative cover plays an important role within maintenance of ecosystem and land condition. Therefore, NDVI is an powerful tool to get information on degree of forest degradation including forest fragmentation it is need to be defined to formulate the strategy for habitat and ecosystem management(4).   NDVI also gives information related to primary production of vegetation(6).Vegetation index is the most useful index to quickly identify the vegetated areas by using multispectral remote sensing data(5). Spatio temporal Landsat satellite images are used to determine NDVI. Land surface temperature (LST) is a important parameter in the surface temperature monitoring between the Earth's surface and its atmosphere(Niclos et al.,2009)and   often assimilated into land surface models(Brunsell & Gillies,2003;Rodell et at.,2004;Jiang.2006;Anderson   et   al.,2008;Zang   et al.,2008;Kustas & Anderson.,2009;Karnieli et al.2010).LST plays a major role in land surface processes instead of not only, because of climatic importance ,but also due to its sensible and latent heat lux exchange(Aires,2001;Sun.,2003).

LST is applicable in many fields like evapotranspiration, climate change, hydrological cycle, vegetation monitoring urban climate and environmental studies (Bastiaansenetal., 1998; Kogan., 2001; Su., 2002; Arnfield., 2003; Voogt and Oke., 2003; Weng et al., 2004; Klama et al., 2008; Weng., 2009; Hansen et AL.,2010). LST plays a key role in land surface processes, not only, because of its climatic importance, but also due to its control on sensible and latent heat flux exchange (Aires, 2001; Sun, 2003). LST has also wide application in many fields viz; evapotranspiration, climate change, hydrological cycle, vegetation monitoring, urban climate and environmental studies (Bastiaanssen et al.

1998; Kogan, 2001; Su, 2002; Arnfield, 2003; Voogt and Oke, 2003; Weng et al. 2004; Kalma et al. 2008; Weng, 2009; Hansen et al. 2010). Vegetative cover can be effectively influenced by the LST because o selectively absorbing and reflecting solar radiation energy and regulating latent and sensible heat exchange.Due to mining   ,the

ISBN: 978-93-8830-599-0

intensity of human activities is enchanced and surface cover is rapidly changed.Therefore, the relationship between NDVI and LST should be investigated for further analyse the ecological effects of LST and identify the regional environmental problems.

STUDY AREA

The Singareni Coal field area is located in Telangana State it covers newly formed 6 Districts namely Kummarambheem Asifabad, Mancherial, Peddapalli, Jayashankar Bhoopalpally, Khammam and Bhadradri Kothagudem. The Singareni collieries company limited was established on 23rd December 1920.Sccl is a government coal mining company on 51:49 equity basis between government of telangana and government of India.The singareni coal fields are stretch across 350km of Pranahitha and Godavari valley of Telangana. Currently Sccl operating 29 underground mines and 18 Open cast mines within the 6 districts. Entire Singareni Coal field area cover geographical area of 29,238.4sq kms.

The study area lies between 16° 45' 25" to 19° 53' 56" N latitude and 78° 18' 36 "to 80° 33' 43 "E longitude .

Fig. 1 Map of the study area showing Singareni coal field area

METHODOLOGY

1. Image Acquisition and Source

To carry out change detection analysis ,the images of Singareni Coalfield area were acquired for 4years namely 1989,1997,2007 and 2018.Considering Atmosphere criteria satellite images of all 4years are acquired for same month that is the December.

The images are obtained from the USGS website(United States Geological Survey)(https://earthexplorer.usgs.gov/). USGS is America's largest water, earth, and biological science and civilian mapping 8 agency, the U.S. Geological Survey (USGS) collects, monitors, analyzes, and provides

scientific understanding about natural resource conditions, issues, and problems.

The images of 1988,1997 and 2007 were collected from Landsat 4-5 Thematic Mapper (TM) images having resolution of 30 meters. The 2018 images are collected from Landsat 8 Operational Land Imager(OLI) and Thermal Infrared Sensors(TIRS) having spatial resolution of 30 meters.

The study area is covered by six tiles – 142*47,142*48,143*47,143*48,144*46 and 144*47. Topohseets of the years were used, and collected from Survey of India, Hyderabad.

Toposheet were georeferenced because all the satellite images were obtained are georeferenced. By using the ArcGis software or Erdas Imagine software georeferencing of toposheet is done.

2. NDVI calculation

Normalized difference vegetative index was calculated for all the 4 images .NDVI values ranges from -1 to +1.NDVI calculation using following formula

$$NDVI = \frac{(NIR - Red)}{(NIR + Red)}$$

Table 1 Shows the NDVI ranges of the study area.

YEAR	IMAGE TYPE	NDVI FUNCTION	NDVI RANGE
1988	LANDSAT-5TM	(band4-band3)/(band4+band3)	-1 to +1
1997	LANDSAT-5TM	(band4-band3)/(band4+band3)	-0.76to 0.832
2007	LANDSAT-5TM	(band4-band3)/(band4+band3)	-0.724 to 0.972
2018	LANDSAT-8OLI	(band5-band4)/(band5+band4)	-0.397 to 0.571

3. LST calculation:

Landsat-5(TM)

As per Landsat-5 Science Data Users Handbook Land Suface Temperature provided by NASA(National Aeronautics and Space Adimistration) Land Suface Temperature is calculated.

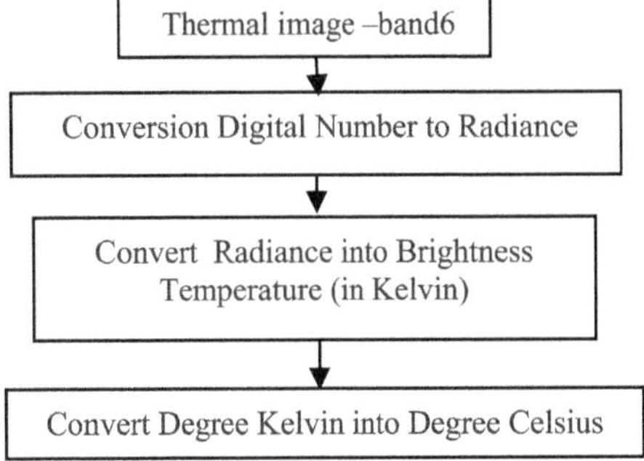

ISBN: 978-93-8830-599-0

LANDSAT-8OLI and TIRS

Landsat-8 TIRS land surface temperature is calculated based on Landsat-8 science Data Users Handbook provided by NASA.

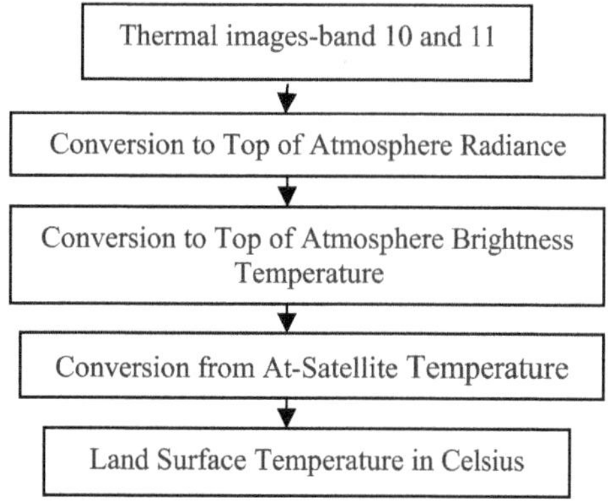

RESULTS

Its is very interesting fact that LST distribution is very closely related to the distribution of NDVI and NDWI. In general,Land Surface Temperature(LST) presents an inverse relationship with Normalized difference Vegetative Index(NDVI).From Fig.2, Fig.3, Fig.4 and Fig.5 it is observed that as NDVI values increases LST values decreases.

Table 2 Shows the NDVI range and LST max temperature.

Year	NDVI		LST
	min	max	Max(°c)
1989	-1	+1	39.92
1997	-0.768	0.832	42.33
2007	-0.724	0.972	47.39
2018	-0.397	0.571	48.74

Relation between LST and NDVI

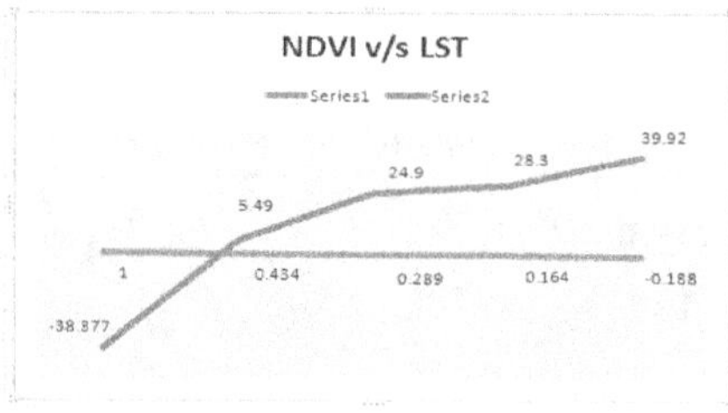

Fig. 2 The above graph shows the relation between NDVI and LST for the year 1989.

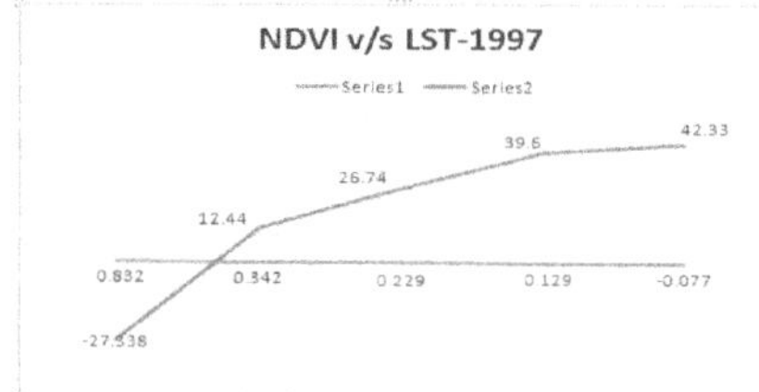

Fig. 3 The above graph shows the relation between NDVI and LST for the year 1997.

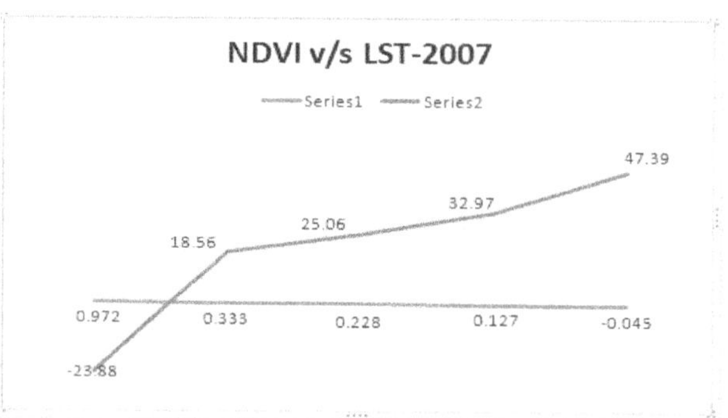

Fig. 4 The above graph shows the relation between NDVI and LST for the year 2007.

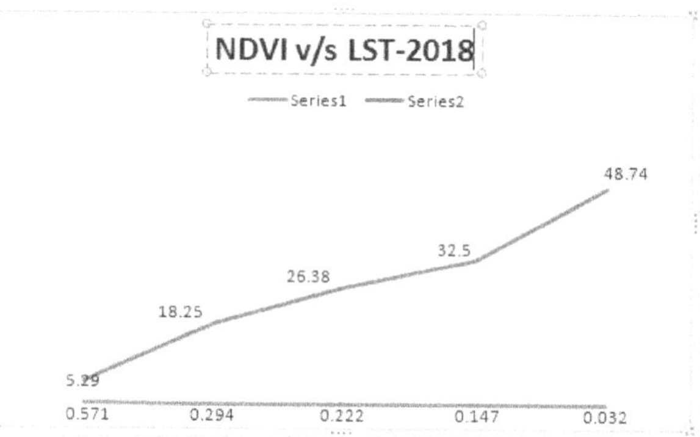

Fig. 5 The above graph shows the relation between NDVI and LST for the year 2018.

Table 3 Areal distribution of vegetative classes of the study area.

Classes	1989	1997	2007	2018
Other Classes	11.60%	13.35%	13.66%	16.86%
Low Vegetation	55.62%	61.07%	64.329%	71.52%
Medium Vegetation	31.03%	24.68%	19.31%	10.735%
High Vegetation	1.72%	0.89%	0.889%	0.878%

From table.3.The other classes of the study area shows increasing trend from 1989 to 2018 covering an area of 3394.77km²(11.60%) in the year 1989 to 4929.76(16.86%) in the year 2018.The open cast mining area that had spread over the study area ,mines increases from 0.0239%, 0.0581%, 0.1744% to 0.424% in the year 1989,1997,2007 and 2018 respectively.

Low vegetation increases from 55.621%, 61.07%, 64.329% and 71.52% in the 1989, 1997, 2007 and 2018 respectively. Medium Vegetation decreases gradually from 31.03%,24.68%,19.31% and 10.735% in the year 1989, 1997, 2007 and 2018 respectively. High Vegetation is drastically decreased from 1.72% in the year1989 to 0.89% in the year 1997,then gradually decreased from 0.889% to 0.878% in the year 2007 and 2018.

ISBN: 978-93-8830-599-0

CONCLUSION

From the study it was found that as the LST increases with decrease in NDVI and it reveals that coal mine areas increased and high and medium vegetation is decreased from table.3.Hence it is necessary that use of proper mine closure plans and Land reclamation practices must be adopted. So, it is necessary to adopt ecofriendly mining and environmental management plans to reduce the deforestration and extinct of flora and fauna.

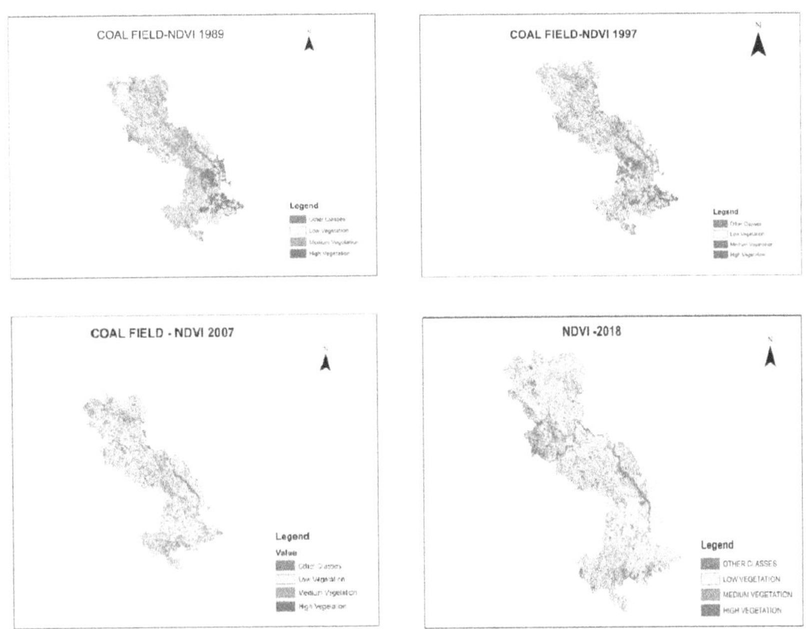

Fig. 6 The maps depicting the normalized difference vegetative index(NDVI) stretched over Singareni coal field area from 1989 to 2018.

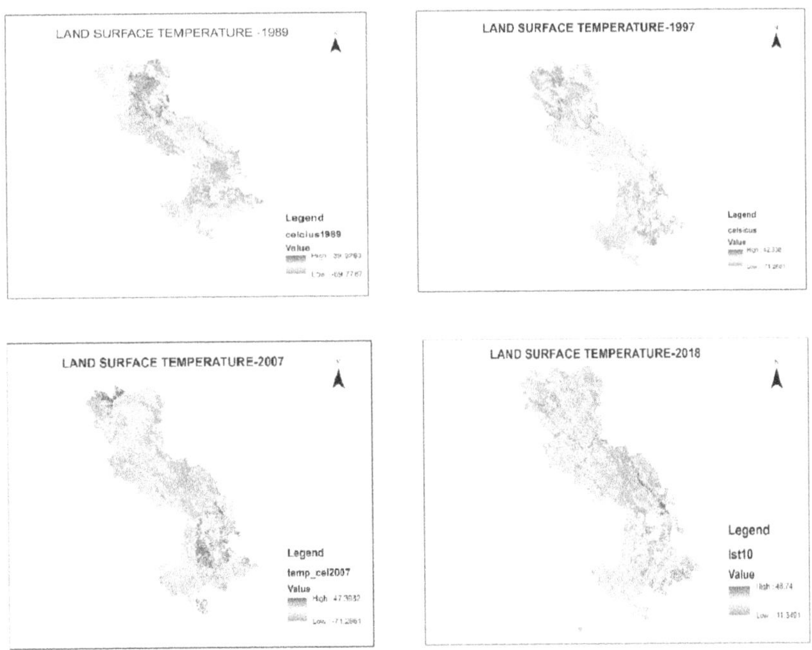

Fig. 7 The maps depicting the Land Surface temperature (LST) stretched over Singareni coal field area from 1989 to 2018.

ISBN: 978-93-8830-599-0

References

1. Crop Drought monitoring using serial NDVI&NDWI in Northern China. Chenglin liu, Bingfang Wu, Yichen Tian, Wenbo Xu, Jianxi Huang, institute of Remote sensing applications, Chinese academy of sciences.
2. Relationship among land surface temperature and LULC,NDVI in typical Karst area. Yuanhong deng, Shijie wang, Xiaoyong bai, Yichao tian, Luhua wu, Jianyong Xiao, Fei Chen & Qinghuan Qian.
3. Normalized difference vegetation index (ndvi) analysis for land cover types using landsat 8 oli in besitang watershed, Indonesia A Zaitunah1, Samsuri1, A G Ahmad1 and R A Safitri2 1 Forestry Management Department, Forestry Faculty, Universitas of Sumatera Utara.
4. D Priatna ,Y Santosa,L B Prasetyo and A P Kartono 2012 Habitat selection and activity pattern of gps collared Sumateran   Tigers Journal Manajemen Hutan Tropika 18(3) pp 155-163.
5. Pirottia F, Parragab M A, Stuarob E, Dubbinib M, Masieroa A and Ramanzin M 2014 NDVI from Landsat 8 Vegetation Indices to Study Movement Dynamics of Capra Ibex in Mountain Areas. The International Archives of the Photogrammetry, Remote Sensing and Spatial Information Sciences, ISPRS Technical Commission VII Symposium (Istanbul, Turkey) (XL-7) pp 147-155.
6. Pettorelli N, Ryan S, Mueller T, Bunnefeld N, Jedrzejewska B, Lima M and Kausrud K 2011 The Normalized Difference Vegetation Index (NDVI): unforeseen successes in animal ecology Climate Research 46 (1) pp 15-27
7. Retrieving of Land Surface Temperature Using Thermal Remote Sensing and GIS Techniques in Kandaihimmat Watershed, Hoshangabad, Madhya Pradesh Mohammad Subzar Malik* and J. P. Shukla
8. Normalized difference vegetation index (ndvi) analysis for land cover types using landsat 8 oli in besitang watershed, Indonesia To cite this article: A Zaitunah et al 2018 IOP Conf. Ser.: Earth Environ. Sci. 126 012112
9. Retrieving of Land Surface Temperature Using Thermal Remote Sensing and GIS Techniques in Kandaihimmat Watershed, Hoshangabad, Madhya Pradesh,AUTHOR Mohammad Malik n Journal of the Geological Society of India · September 2018.
10. A remote sensing & hydrogeomorphological study. Environ.monitor.assess.142,309-318. Akiwumi, F.A. Butler, D.R, 2008.mining & environmental changing sierra leone.
11. Land use & land cover changes in the mining area of godhavari coal field of southern India. Debashri garai, A.C.Narayana, centre for earth &space sciences, university of Hyderabad, India.

ISBN: 978-93-8830-599-0

Surface Water Change Detection using Temporal Series of Landsat Data
A case Study of Hyderabad, Telangana

L. Ravi*[1], K. Manjula Vani[2] and M. Anji Reddy[3]
*Research Scholar, CSIT Department, [2]Professor of Civil Engineering,
[3]Professor of Environmental Science & Technology
ravi.rams.77@gmail.com, manjulavani@jntuh.ac.in and mareddyanjireddi@gmail.com

Urbanization with its extend growth in built up affords favourable to human in technological expansion, substitute it leads to ecological hazards the major concern is surface water encroachments. Surface water is the water contained in our lakes, rivers, creeks, wetlands and streams. Remote sensing with its emerging technology designed as earth observatory and providing information to do research for prior data. Spatial and Spectral water index including the Normalized difference water index (NDWI), Modified NDWI (MNDWI) were used for extraction of surface water from satellite data.

Surface water for the past four decades of study area Hyderabad from temporal data of Landsat 1977, 1995, 2004, 2011 & 2018, apart of its massive growth in urbanisation study area has 183 lakes and 30km stretch of urban river. The results shows the spread of surface waterbodies in time series interval.

INTRODUCTION

The normalized difference water index (NDWI),is proposed for remote sensing of vegetation liquid water from space. NDWI is defined as (p(0.86μm) - p(1.24 μm))/ (p(0.86 μm) + p(1.24μm)), where p represents the radiance in reflectance units. Both the 0.86-1 μm and the1.24-1 tm channels are located in the high reflectance plateau of vegetation canopies (**GAO, 1996**). The NDWI is expressed as follows (**McFeeters 1996**):

$$NDWI = \frac{Green-NIR}{Green+NIR}$$

where Green is a green band such as TM band 2, and NIR is a near infrared band such as TM band 4.Thematic band differs for different sensors. This index is designed to

1. maximize reflectance of water by using green wavelengths;

2. minimize the low reflectance of NIR by water features; and

3. take advantage of the high reflectance of NIR by vegetation and soil features.

As a result, water features have positive values and thus are enhanced, while vegetation and soil usually have zero or negative values and therefore are suppressed.

The modification of the NDWI using a MIR band instead of a NIR band can considerably improve the enhancement of open water features. It can quickly and accurately discriminate water from non-water features. The MNDWI is more suitable for enhancement of water with many built-up land areas in the background than the NDWI because it can efficiently reduce and even remove built-up land noise. The threshold values for the MNDWI to achieve best water extraction result are usually much less than those of the NDWI, suggesting using zero as a default threshold value can produce better water extraction accuracy for the MNDWI than for the NDWI. This would be very useful for the MNDWI to be automated.(HANQIU XU,2006)

$$MNDWI = \frac{Green-MIR}{Green+MIR}$$

Water body extraction is an important task in differ-ent disciplines, such as lake coastal zone management, coastline change and erosion monitoring, flood pre-diction and evaluation of water resources.(Gulcan Sarp 2016)

Study Area Hyderabad is an historical city constructed on banks of Musiriver in 1591 at present city is the capital of southern Indian state of Telangana and de jure capital of Andhrapradesh.Hyderabad is in 6th place in urban agglomeration in India, Hyderabad municipal administration study area Greater Hyderabad municipal administration (GHMC) consists of 650 sq.km(GHMC-website)

ISBN: 978-93-8830-599-0

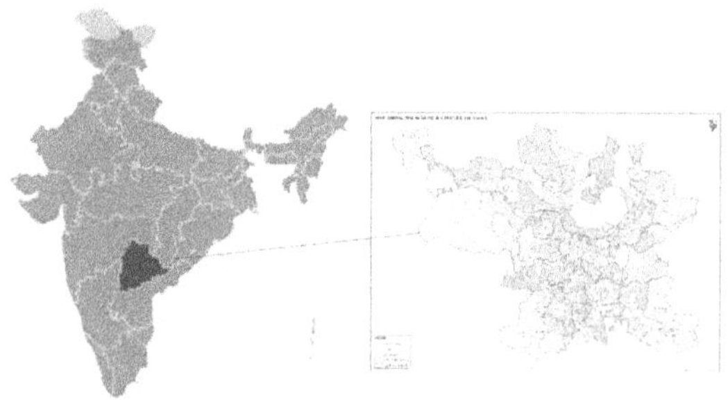

Fig. 1 Map showing study area GHMC

Methodology

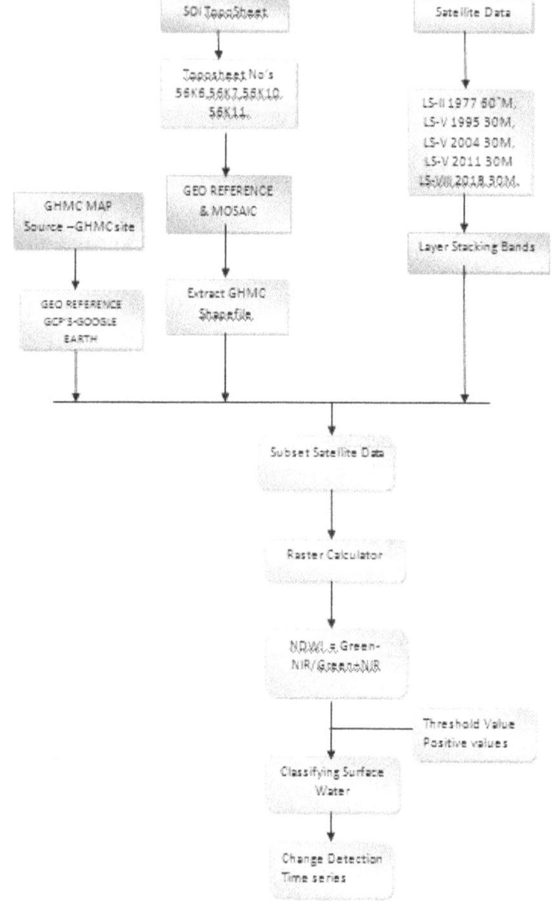

The above flowchart decribes the process followed,Toposheets from Survey of India to identify the boundaries of Greater Hyderabad Municipal Corporation(GHMC) after following the geometric correction of images in JPG format,it is compared with GHMC (Geo referenced Image with Google earth ground control points) map provided by website of http://www.ghmc.gov.in/.

GHMC is the local governing authority covering an area of 650sq.km with perimeter of 240km

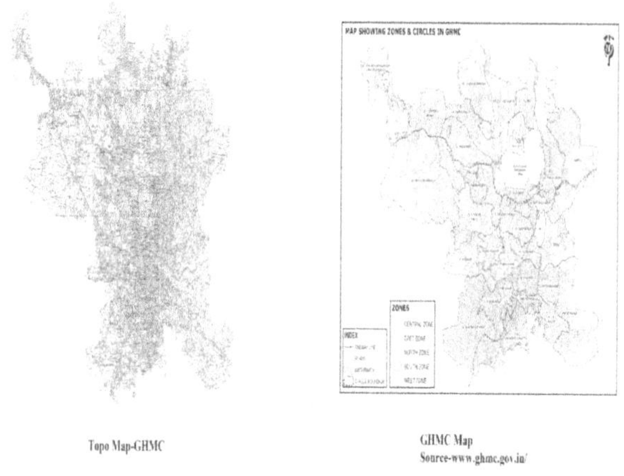

Fig. 2 Topomaps of GHMC(Hyderabad city)

USGS earth explorer is providing users the ability to query,search and order satellite images, It started with the first Landsat satellite's launch in 1972 and is continuing with Landsat 8, still operational. For almost 40 years, the Landsat program has continuously collected spectral information from Earth's surface; this unparalleled data archive gives scientist the ability to assess changes in Earth's landscape.

Table 1 Satellite Info

Satellite_Name	Path/Row	Spatial Resolution	Date of Acquisition
Landsat-II	154/048	60*m	30th Nov 1977
Landsat-V	144/048	30m	22nd Feb 1989
Landsat-V	144/048	30m	16th Dec 2004
Landsat-V	144/048	30m	18th Jan 2011
Landsat-VIII	144/048	30m	05th Jan 2018

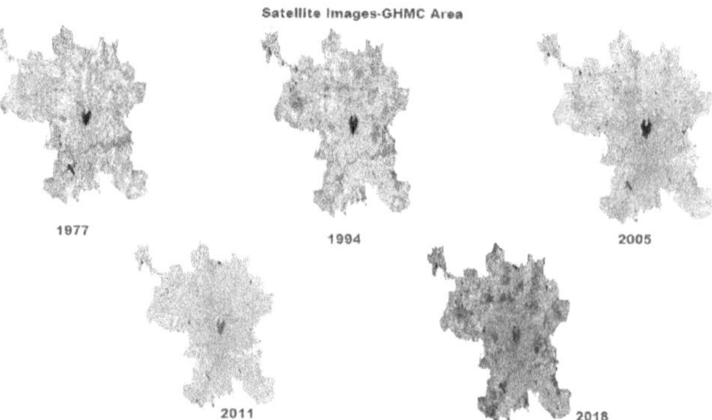

Fig. 3 Satellite images obtained from Landsat Satellite imageries.

RESULTS & DISCUSSION

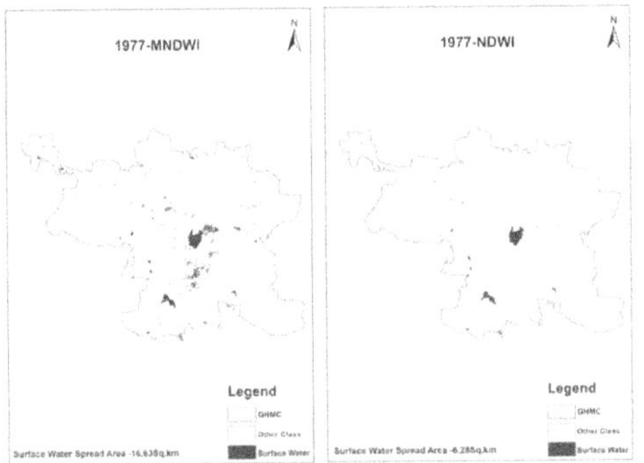

The above result shows Modified Normalized difference water index(MNDWI) and Normalized difference water index (NDWI) with a threshold value of positive value o.oo1 for landsat-02 Satellite imagery. The MNDWI value range from -0.65 to 0.93 with raster calculating of band number 4 & 7,for NDWI value range from -0.64 to 0.79 band number 4&6 .

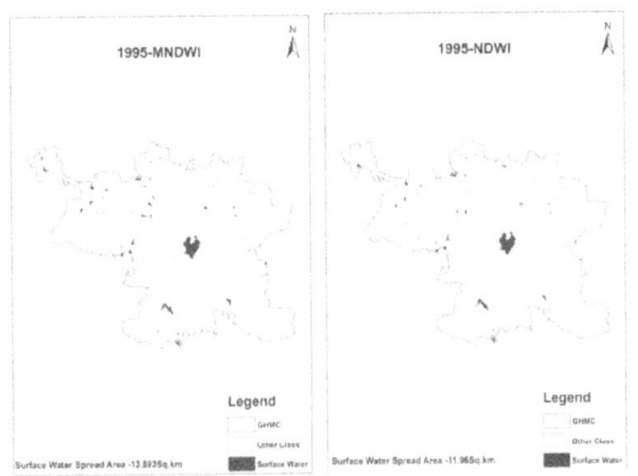

The above result shows Modified Normalized difference water index(MNDWI) and Normalized difference water index (NDWI) with a threshold value of positive value o.oo1 for landsat-05 Satellite imagery. The MNDWI value range from -0.67 to 0.68 with raster calculating of band number 2& 5, for NDWI value range from -0.62 to 0.44 with raster calculating of band number 2&4 .

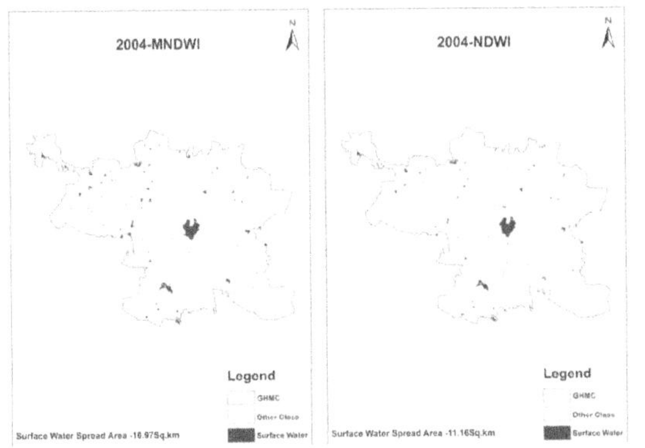

The above result shows Modified Normalized difference water index(MNDWI) and Normalized difference water index (NDWI) with a threshold value of positive value o.oo1 for landsat-05 Satellite imagery. The MNDWI value range from -0.65 to 0.38 with raster calculating of band number 2 & 5, for NDWI value range from -0.63 to 0.38 with raster calculating of band number 2&4.

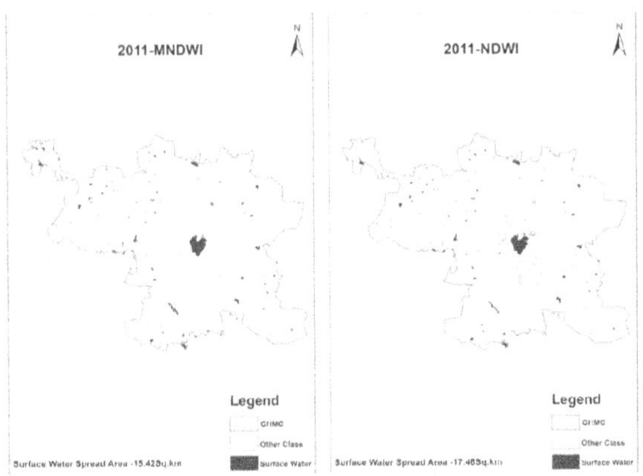

The above result shows Modified Normalized difference water index(MNDWI) and Normalized difference water index (NDWI) with a threshold value of positive value o.oo1 for landsat-05 Satellite imagery. The MNDWI value range from -0.63to 0.78 with raster calculating of band number 2 & 5, for NDWI value range from -0.63 to 0.52 with raster calculating of band number 2&4 .

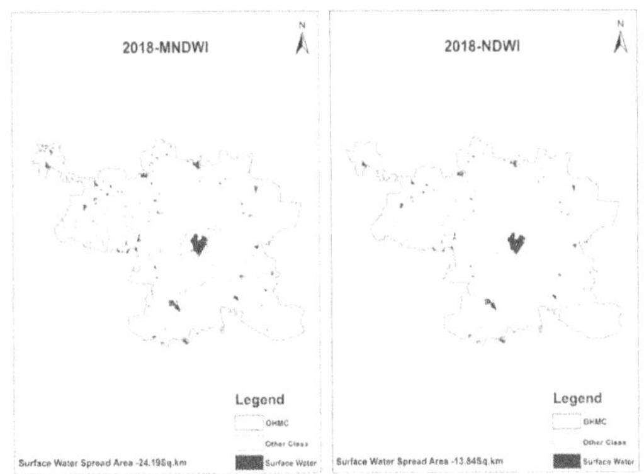

The above result shows Modified Normalized difference water index(MNDWI) and Normalized difference water index (NDWI) with a threshold value of positive value o.oo1 for landsat-05 Satellite imagery. The MNDWI value range from -0.38 to 0.30 with raster calculating of band number 3& 6, for NDWI value range from -0.45 to 0.14 with raster calculating of band number 3&5.

ISBN: 978-93-8830-599-0

Table 2 Comparison of MNDWI & NDWI Output in Sq.Km

Year	MNDWI		NDWI	
	Other class (Sq.Km)	Water Bodies (Sq.Km)	Other class (Sq.Km)	Water Bodies (Sq.Km)
1977	633.61	16.63	642.6	7.28
1995	635.96	13.89	637.89	11.96
2004	638.88	10.97	638.68	11.16
2011	634.52	13.48	631.58	16.42
2018	625.67	24.18	636.14	13.84

GHMC administration of Hyderabad covers an area of 650 Sq.Km above table represents the variation of NDWI & MNDWI values in square kilometers from 1977 to 2018.

CONCLUSION

1. The total water spread area of GHMC jurisdiction is 24.18sq.km MNDWI & 13.84sq.km NDWI in 2018.
2. The water spread has been increased in the study area by 1.22% (MNDWI) &1.00% (NDWI) for a period of 30 years.
3. Water bodies in MNDWI method than the NDWI method is noticed that the reflectance shows roads have been merged with streams.

REFERENCES

1. GAO, B.C., 1996, (NDWI—a normalized difference water index for remote sensing of
2. Vegetation liquid water from space. Remote Sensing of Environment, 58, pp. 257–266.
3. Gulcan Sarp(2016), Water body extraction and change detection using time series:A case study of Lake Burdur, Turkey, Journal of Taibah University for Science 11 (2017) 381–391.
4. HANQIU XU(2006), Modification of normalised difference water index (NDWI) to enhance open water features in remotely sensed imagery, International Journal of Remote Sensing.
5. Vol. 27, No. 14, 20 July 2006, 3025–3033.
6. Ravi. L, "Change detection of Vegetation Spectrum for Urban Area through NDVI analysis". National conference on Remote sensing applications, JNTUH-CEH, April 2018.
7. S. K. McFEETERS (1996), The use of the Normalized Difference Water Index (NDWI) in the delineation of open water features, International Journal of Remote Sensing, 17:7, 1425-1432, DOI: 10.1080/01431169608948714.

ISBN: 978-93-8830-599-0

A Study on Urban Sustainbility and Resilience of Visakhapatnam City

B. Priyanka[1] and J. Swaraj[2]
[1]Andhra University, Vishakhapatnam
[2]Scientist (SG), EPTRI Hyderabad.
priya.bdd.0806@gmail.com, swarajeptri@gmail.com

Abstract

Rapid growth of urban population has unregulated physical expansion of a city's built-up area has serious social and environmental consequences, including the segregation of low-income groups in the worst located and often the most dangerous areas. Urban poverty is another challenge facing city governments. Carbon dioxide emissions (the main greenhouse gas) rise with economic growth and are concentrated in cities. This study is conducted to go in depth details of management of Visakhapatnam city by its city administration, then it is tried to propose strategies for Sustainable Urban Environmental Management. The study will help in better understanding of the present status of urbanization and its management by city administration. This study is focused on identification of options to develop city as a sustainable, resilient and livable habitat.

2.0 INTRODUCTION

2.2.1 Key Features of Urban Planning

- Town planning
- Regulation of land use for residential and commercial purposes
- Planning for economic and social development
- Construction of buildings, roads, bridges
- Water supply for domestic use, industrial and commercial purposes
- Public health care, sewerage, sanitation and solid waste management
- Proper fire services
- Urban forestation, protection of environment through sustainable development
- Promotion of ecological balance and maintenance
- Safeguarding the interests of weaker sections of society
- Organized slum improvement and phased removal or alleviation of urban poverty
- Increased provision of basic urban facilities Increased public amenities including street lighting, parking lots, bus-stop etc. Continual promotion of cultural, educational and aesthetic aspects of the environment
- Increased number of burial/cremation grounds and electric crematoria
- Proper regulation of slaughter houses and tanneries
- Proper maintenance of population statistics

2.2.2 STUDY AREA

Visakhapatnam city is situated at east cost of India and adjacent to Bay of Bengal with port on East, industrial belt on the west and green space and hill in between them. In 2001 the urban area of Visakhapatnam is 111 sq.km and present it is spread over 534 sq.km including 32 Peripheral areas and Gajuwaka Municipality with 8 zones and 83 wards. After merging Anakapalli & Bhimili the area of GVMC is 621 sq.km.

3.0 METHODOLOGY

- Collection of secondary data: Data collected from city administration, technical organizations reports on the city or on urban planning.
- Field visits: Visited various areas in the city covering hazard risk areas, vulnerable areas, critical infrastructures, old towns, satellite towns, developing areas, etc.
- Interviews: Conducted interviews with city administration officials, urban development officials, administrators, planners, engineers, NGOs, academicians, scientists and other stakeholders.
- Brainstorming: Conducted group discussions and brainstorming sessions to analyze the data and to identify best options for the city developments.

4.0 STATUS OF URBAN ENVIRONMENT OF VISAKHAPATNAM GROWTH OF SLUMS AND SLUM POPULATION

Greater Visakhapatnam Municipal Corporation characterized by a very significant presence of the urban poor, with a growing poverty profile. At present there are ~700 slums and the slum population is over 6 lakhs.

ISBN: 978-93-8830-599-0

Table 1 Greater Visakhapatnam Municipal Corporation At A Glance

GENERAL			PHYSICAL INFARASTRUCTURE	
Area		621Sq.km	**Water Supply**	
Population	2011	1883000	Protected Water Supply Reservoirs:	11 Nos
			Total installed capacity of WS:	382 MLD
			No. of House Service Connections	154010
			No. of Public Stand Posts	9800
Total no. of HHs		435863	Length of distribution pipe line:	1435 Km
No. of municipal wards		72	Unserved area	15%
Total no. of slums		786	Water Supply through Borewells:	
			No. of Power Bores	285 Nos.
Slum Population		8,69,495	No. of Hand Bores	6700
			Roads	
Annual Expenditure (2014-15)		826.26 Cr	CC. Roads	447 km
Community Amenities:			BT.Roads	1199 Km
Health & Education			W.B.M	1470 Km
No. of Government Hospitals		95	Gravel roads	360 Km
No. of Government Schools		24	**Drains**	
High Schools		223	Pucca Drains	854 km
Upper Primary Schools		679	Kutcha Drains	1570 km
Primary Schools		3367	Storm Water Drains	25 km
Anganwadi Centers		215	Unserved area	
No. of Burial Grounds		254	**Street Lighting**	
Vegetable Market		250	Total no. of street lights	
Play Grounds		86	High Mast Lights	1323
Parks		158	Central Lighting, LED	92392
Religious Structures			SV. Lamps	--
Temple		350	MV. Lamps	---
Mosque		35	Tube Lights	---
Church		258	Unserved area	1.5%
Community Halls		154	**Sanitation**	
E-Seva Centers		72	Garbage Generation/day	350 MT/day
Auditoriums		25	Garbage Lifted/day	350 MT/day
Stadiums		2	Service Deficiency	

Municipality	Area (Sq.Km)	Population in Lakhs			Density (2011) (Persons/Sq.Km)
		Year 1991	Year 2001	Year 2011	
GVMC	621	9.87	13.45	18.83	2761

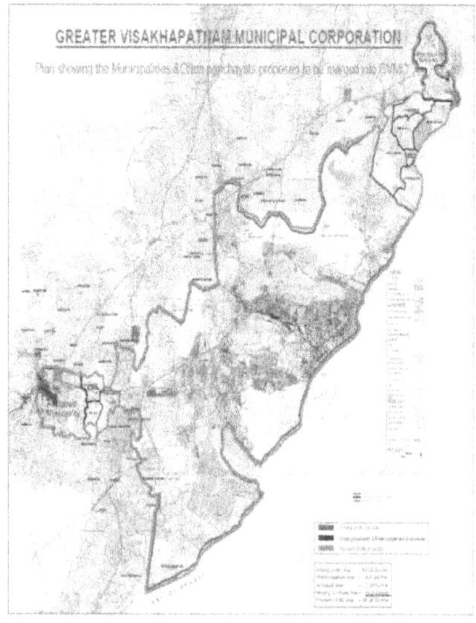
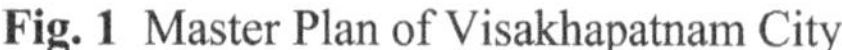

Fig. 1 Master Plan of Visakhapatnam City

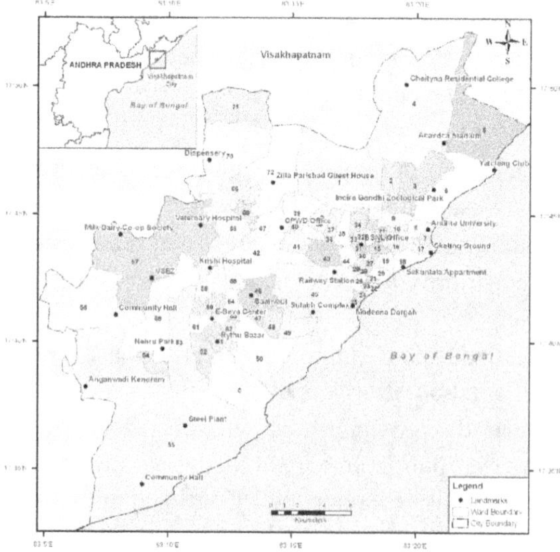

Fig. 2 Ward boundary map of Visakhapatnam city

ISBN: 978-93-8830-599-0

5.0 HAZARD RISK AND VULNERABILITY PROFILE OF   VISAKHAPATNAM CITY

Visakhapatnam, as it is a coastal industrial city, the peoples living in its urban and fringe areas having various levels of vulnerability and is at risk to expose natural and anthropogenic hazards. The city has its unique geo climatic conditions. The vulnerability varies from location to location, within the city.

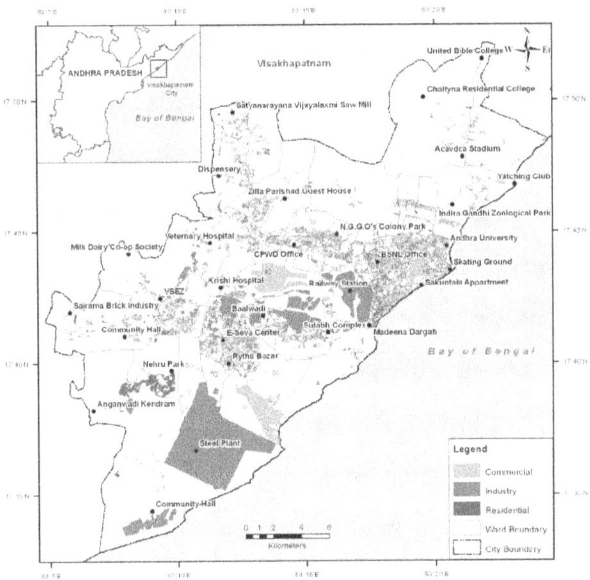

Fig 3 Spatial distribution of residential, commercial and industrial built-up

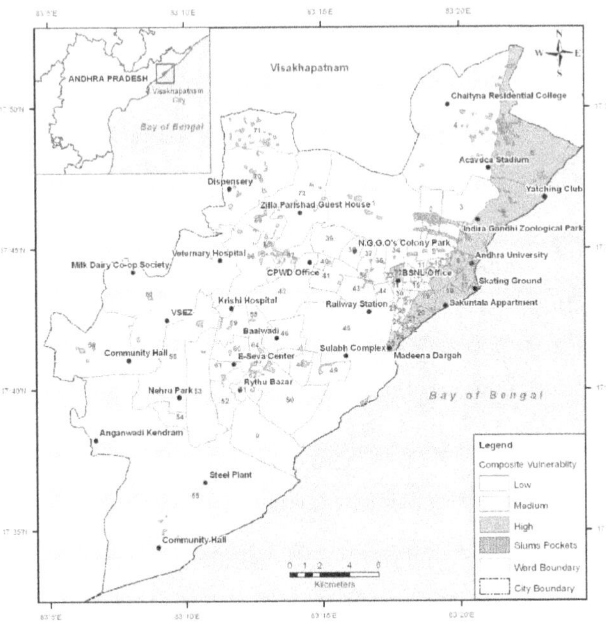

Fig 4 Composite vulnerability map of Visakhapatnam city

6.0   CONCLUSION

The urban environment in the Visakhapatnam is multi factorial and mostly influenced by unplanned urbanization, intensive transportation, irresponsible industrialization, location specific hazard risk with vulnerability conditions i.e., loss in vegetation cover, increase in built-up area, ill planned developmental activities and carbon intensive lifestyles.

As Visakhapatnam is a coastal industrial city, there is a threat of natural and man-made disasters. The temperatures in the city were in increasing trend and the effects of climatic change are evident. The natural resources and environmental quality are in deteriorating condition. In this situation it is very important to mainstreaming disaster risk reduction, climate change adaptation principles and environmental considerations in urban planning, to develop Visakhapatnam city as a sustainable, resilient and livable habitat.

6.0   REFERENCES

1. Allen, P. M., & Sanglier, M. (1979). A dynamic model of urban growth: II. Journal of Social and Biological Structure, 2, 269–278.

2. Bowers, J., 1997. Sustainability and Environmental Economics - An Alternative Text.

3. Longman, England.

4. Burgess, E. W. (1925). The growth of the city. An introduction to a research project. In R. E. Park, E. W. Burgess, & R. McKenzie (Eds.), The city (pp. 47–62). Chicago: University of Chicago Press.

5. Central Pollution Control Board, 2003, Water Quality in India Status and Trend(1990-2001), New Delhi, MINARS/20/2001-2002, pp.

6. Census of India, 2001, Ministry of Home Affairs, Office of the Registrar General, India, 2A, Mansingh Road, New Delhi-110011,Website http://www.censusindia.net

7. Guidelines for air quality, World health organization publication, Geneva

8. Haimes, Y. Y., (1991). Total risk management. Risk Analysis, 11:1-8.

9. Hoyt, H. (1939). The structure and growth of residential neighborhoods in American cities. Washington, DC, USA: Federal Housing Administration.

10. IDNDR, (1996). Disaster reduction in urban systems. Volume 28 of Stop disasters. IDNDR, Geneva.

11. Sudhira, H.S. et al (2002), Urban sprawl pattern recognition and modeling using GIS.

ISBN: 978-93-8830-599-0

12. Taubenböck, H., Esch, T., Thiel, M., Wurm, M., Ullmann, T., Roth, A., et al. (2008a). Urban structure analysis of mega city Mexico City using multi-sensoral remote sensing data. In Proceedings of SPIE-Europe (international society for optical engineering) conference, Cardiff, Wales.

13. Zhang, L., Wu, J., Zhen, Y., & Shu, J. (2004). A GIS-based gradient analysis of urban landscape pattern of Shanghai metropolitan area, China. Landscape Urban Plan, 69, 1–16.

ISBN: 978-93-8830-599-0

Estimation of Land use Impacts on Shallow Ground Water Quality
– A Case Study from Krishna Basin

A. Vamshi Krishna Reddy[1], A. Umashankar Kumar[2] and S. Jyothi[3]

[1]Assistant Professor(c), Institute of Science & Technology, JNT University Hyderabad, India.
[2]Assistant Professor, Swamy Vivekananda institute of Technology, Hyderabad, India.
[3]Assistant Professor, Department of Chemistry, Kakatiya University, Warangal, India.
vamshiavkreddy@gmail.com

ABSTRACT

Water is essential to people and the largest available source of fresh water lies underground. Increased demands for water have stimulated exploration of underground water resources. Water resources get polluted due to rapid industrialization, advancement in agricultural techniques, increasing population and other adverse impacts of environments. All these factors may result in changing the hydrological cycle. As the urban & rural environmental quality depends on the land usage. The quality of the environment is determined by studying the land use features and their impact are analyzed.

In present study an attempt is made to evaluate the impact of landuse / landcover on ground water quality of Krishna Basin (Krishna & Godavari districts) area. Various thematic maps are prepared from the toposheet on 1:50000 scale using ArcGIS Software. The landuse / landcover map of the study area is prepared from the linearly enhanced fused data of IRS-1D PAN and LISS-III imagery.

Ground water samples were collected at pre-determined sampling locations based on Satellite Imagery of the study area. All the samples were analyzed for various physico-chemical parameters adopting standard protocols for the generation of attribute data. Based on the results obtained maps showing spatial distribution of selected water quality parameters are prepared for the study area.

The variations in the concentrations of water quality parameters indicated high concentrations of TDS, Fluoride, Hardness, Nitrates, Sulphates, Chlorides which may be attributed to seepage of domestic wastes and agriculture waste through open nallahs and industrial wastes.

Keywords: Water pollution, Land use/ Land cover, Remote sensing, GIS, Spatial maps.

STUDY AREA DESCRIPTION

Krishna

Krishna District is located on the east coast of India covering an area of about 8,727 Sq.Km. It lies between 15°43'N latitude and 17°10'N Latitude and between 80° E longitude and 81°33'E Longitude. The district is surrounded in the Eastern & Southern side by the Bay of Bengal, Guntur and Nalagonda on the western and Khammam & West Godavari district on the northern side. The only major river in the district is Krishna, Other riverlets include Muneru, Tammileru and Budameru. Besides there are some hill streams like Jayanthi, Kattaleru, Ippalavagu, Upputeru, Paleru, Ballaleru, Nidimiyeru etc. This district shares with the West Godavari district one of the large fresh waterlakes on the east coast, namely, Kolleru.

Guntur

Guntur district is one of the Central coastal districts of Andhra Pradesh. It has a geographical area of 11,328 sq. kms. It lies between North latitudes 15°18' & 16°50' and East longitudes 79°10'00" & 80°55'00". The annual normal rainfall of the district is 889.1 mm. South west and northeast monsoon contributes59% and 26% respectively. Krishna, Nagulleru, Chandravanka and Gundlakamma rivers drains thedistrict. The district has been gifted with the vast surface and ground water resources. About 3.01 lakh ha area is irrigated by canals and it has ground water recourses of 1.72 lakh ham.

Objectives of the study

The main objectives of the study are:

1. Assessment of existing environmental problems of Krishna Basin (Krishna & Godavari districts) with a special reference to ground water pollution.

2. Preparation of spatial digital database consisting of land use/land cover, drainage, road network maps with the help of IRS-ID PAN & LISS-III merged data, SOI toposheets and ground data on Arc GIS platform.

3. To create attribute digital database of Krishna Basin (Krishna & Godavari districts) for water quality parameters using standard methods of analysis for environmental studies.

4. Preparation of Spatial distribution maps of Ground water quality in the study area for easy understanding of the water quality phenomenon.

MATERIALS & METHODOLOGY

Source of Data Products: 1) Satellite data from National Remote Sensing Centre (NRSC) (IRS-ID LISS-III + PAN), Toposheets of 1:50,000 scale from

ISBN: 978-93-8830-599-0

Survey of India (SoI), 2) Water samples are collected from the study area, 3) Ground control points were collected with Global positioning System.

The methodology considered for the analysis of environmental impacts in Krishna Basin ground water (study area) is basically a Four steps process. They are as following.

(i)   Spatial digital database and Attribute digital database creation.

(ii) Integration of spatial database and Attribute database.

(iii) Estimation of Ground Water Quality.

(iv) Preparation of Spatial distribution maps for the study area using water quality data.

(v)   Estimation of Land Use/Land Cover categories.

1. Land Use / Land Cover Map: Comprehensive information on land use / land cover is the basic prerequisite for land resources evaluation, assessment, utilization and management. Today, with increasing population pressure on land and the resulting changes in the land use pattern and processes, a considerable degree of land transformation and environmental deterioration is being witnessed. Therefore it is important to understand the cause and effect of the changes through scientific studies.

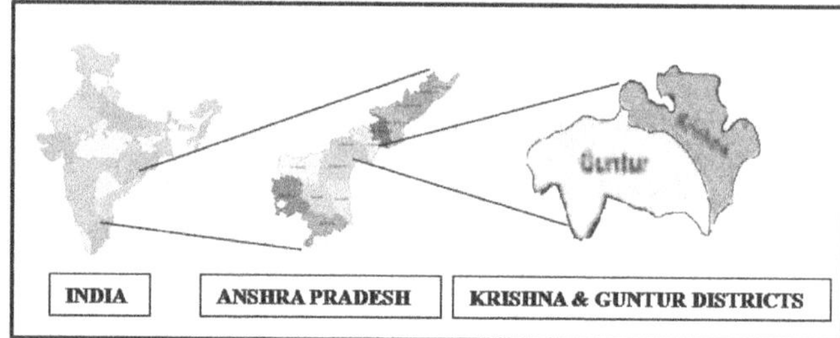

Fig.1 Location Map of the study area

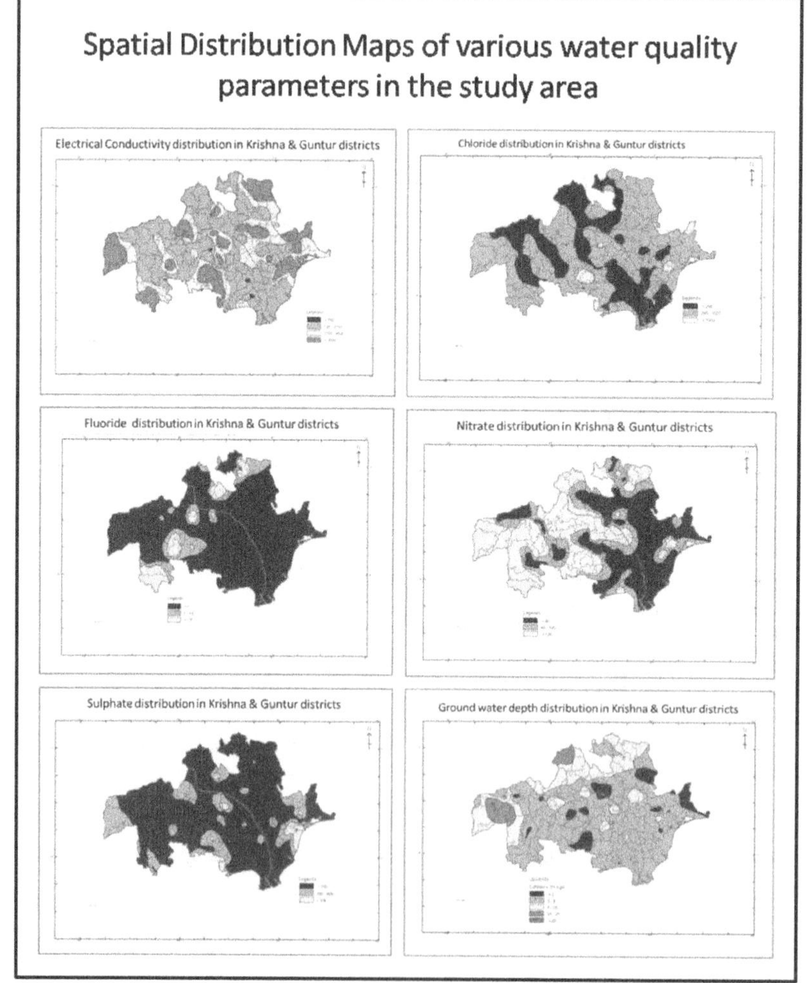

ISBN: 978-93-8830-599-0

Table 1 Existing Landuse Pattern of the study area

S.No	LAND USE class	Area in lakh Hect
1	Geographical area	8.73
2	Area under Forests	0.76
3	Barren and uncultivable Land	0.38
4	Land put to non Agricultural use	1.45
5	Culturable waste	0.27
6	Permanent pasteures and other grazing lands	0.10
7	Land under Misc. tree crops & groves	0.09
8	Area under Fish & Prawn culture	0.328
9	Current Fallow lands	0.28
10	Other Fallow lands	0.27
11	Net area sown	4.76
12	Area sown more than once	2.61
13	Total cropped area	7.37
14	Net area irrigated	3.44
15	Area irrigated more than once	2.02
16	Gross area irrigated	4.38
Percentage of net area sown to Geographical area		54.52

2. Estimation of Water Quality Index (WQI) for the study area: A total number of nine water quality parameters are estimated from 43 water samples randomly collected at different locations in the study area. These water quality parameters are estimated based on the standard methods of water quality analysis. Spatial distribution of each one of the water quality parameter is generated on ArcGIS software. These maps are generated based on the integration of spatial and attribute database. The spatial distribution maps may gives a clear assumption of the water quality in the study area. The nine water quality parameters pH, alkalinity, chlorides, sulfates, nitrates, sodium, total hardness, TDS and fluorides are considered for the estimation of water quality index for this zone.

RESULT AND DISCUSSION

1. Ground water quality in Krishna District:

➤ The pH of water indicates the form in which the carbon dioxide is present. A change of pH from 6 to 7 indicates that there is a tenfold decrease in the hydrogen ion concentration. In a similar fashion, a change of pH 7 to 8 indicates a tenfold increase in hydroxyl ions. In all the sampling locations of the study area, the pH values of ground water vary from 7.74 to 9.23.

➤ The EC values for the samples are in the range from 712 to 13600µS/cm. The electrical conductivity value is directly proportional to the total dissolved matter. The results showed that the electrical values of all the samples are exceeded as per drinking water standards.

➤ In the present study total dissolved solids ranged from 377 to 8166 mg/lit. The results showed that the total dissolved solids values of all the samples are exceeded. The higher values in Upparigudem, Vissannapet and kaikaluru1.

➤ The hardness due to dissolution of alkaline earth metal salts from geological matter. The total hardness ranges between 100 to 1601 mg/lit. The results showed that the total hardness values of all the samples are exceeded to desirable limit. Some villages' results are showed higher values.

➤ The alkalinity values ranges between 177 to1159mg/lit. The results showed that the total alkalinity values of the samples are exceeded to desirable limit.

➤ In the study area, the concentrations of the cations, calcium, magnesium, sodium and potassium are varied from 4.0 to 140, 5 to 340, 35 to 2484 and 0 to 390 mg/lit respectively. Calcium values are not exceeded in the permissible limit. The magnesium, sodium and potassium values are exceeded to desirable limit.

➤ In the present study the amounts of ions carbonates, chlorides, sulphates, nitrates and fluorides are varied from 0 to 36, 71 to 3687, 1 to 1016, 1 to 170 and 0.02 to 2.20 mg/lit respectively. The results showed that the chloride, Nitrate, Sulphate & Fluoride content of all the samples is exceeded the desirable limit.

2. Ground water quality in Guntur District:

➤ The pH of water indicates the form in which the carbon dioxide is present. A change of pH from 6 to 7 indicates that there is a tenfold decrease in the hydrogen ion concentration. In a similar fashion, a change of pH 7 to 8 indicates a tenfold increase in hydroxyl ions. In all the sampling locations of the study area, the pH values of ground water vary from 7.34 to 8.56.

➤ The EC values for the samples are in the range from 522 to 8670µS/cm. The results showed that the electrical values of all the samples are exceeded to desirable limit and permissible limit as per drinking water standards.

➤ The level of total dissolved solids is one of the characteristics, which decides the quality of drinking water. In the present study total dissolved solids ranged from 312 to 5688 mg/lit. The results showed that the total dissolved solids values of all the samples are exceeded to desirable limit and permissible limit prescribed by Indian Standards.

➤ The hardness due to dissolution of alkaline earth metal salts from geological matter. The total hardness ranges between 70 to 2082 mg/lit. The results showed that the total hardness values of all the samples are exceeded to desirable limit.

➤ The alkalinity values ranges between 122 to 1159 mg/lit. The results showed that the total alkalinity values of the samples are exceeded to desirable limit.

➤ In the study area, the concentrations of the cations, calcium, magnesium, sodium and potassium are varied from 8.0 to 421, 9 to 468, 49 to 1086 and 1 to 100 mg/lit respectively Calcium values are not exceeded in the permissible limit. The magnesium, sodium and potassium values are exceeded to desirable limit.

➤ In the present study the amounts of ions carbonates, chlorides, sulphates, nitrates and fluorides are varied from 0 to 24, 57 to 2099, 24 to 643, 0 to 2410 and 0.02 to 3.10 mg/lit respectively and the observations are presented in the Table 4.2. The results showed that the chloride, Nitrate, Sulphate & Fluoride content of all the samples is exceeded the desirable limit.

CONCLUSIONS

➤ It is seen from the exploration data that most of the potential zones were encountered within the depth range of 20-150 m and beyond this depth, potential fractures through occur, but rare.

➤ Environmental awareness among the people regarding environmental pollutions, particularly different ways of ground water pollution aspects to be created.

➤ The provision of underground drainage system has to be adopted, to check the percolation of sewage water into the sub basins.

➤ Providing adequate drainage system with proper treatment before disposal of waste and also removal of faulty construction lined or unlined septic tanks and cesspools, which may restrict the deterioration of ground water quality.

➤ Conjunctive use practices have to be adopted in the command area by utilizing both surface and ground water resources. Ground water potential zones in the command area are to be identified and developed. Ground water development through bore wells can be restricted to 40-120 m.

➤ Exploring the possibilities of diversion of surface water through canals/pipes for filling up of existing dried up tanks in over-exploited mandals.

➤ Rainwater harvesting structures like contour bunding, check dams, percolation tanks, farm ponds are already in vogue. The construction of the artificial recharge structures should be taken up scientifically for 50% of non-committed run-off so as to not to deprive the downstream watersheds. A technical team consisting of Scientists, Engineers, and Bureaucrats should be constituted for monitor the structures on regular basis.

➤ In 'Safe' mandals, the artificial recharge to ground water should go hand-in-hand with ground water development further development of ground water should be restricted up to a depth of 150 m to avoid failures of bore wells.

➤ Since the district is water scarce, land use system should place emphasis on cultivation of high value and low water requiring crops such as pulses, oilseeds. The suggestions of Agriculture Department have to be followed, according to seasons.

➤ Modern irrigation systems using drip and sprinkler irrigation equipment have to be used for reducing the stress on ground water system and help in enhancing the availability of resource.

REFERENCES

1. BIS, (1994),"Indian Standards Specifications for Drinking Water", Bureau of Indian Standards, IS:10500 -2012).

2. Standard methods for the Examination of Water and Waste Water – APHA American Public Health Association 2012.

3. NEERI manual on water and waste water.

4. WHO (1996), "Guidelines for drinking-water quality, health criteria and other supporting information", V.2, 2nd edn, World Health Organisation, Geneva, 940-949 pp.

5. Goel P.K.,(2000), " Water Pollution-causes, effects and control", New Delhi, New Age Int. (P) Ltd.

6. Pandey Sandeep K. and Tiwari S.,(2009), "Physico-chemical analysis of groundwater of selected area of Ghazipur city-A case study, Nature and Science, 7(1).

7. Sharma, V.V.J., (1982) "Ground Water Resources of Northern Eastern Ghats" Proceedings of the Seminar on Resources Development and Environment in the Eastern Ghats, Visakhapatnam, pp.69-75.

8. Subba Rao, N., Krishna Rao. G., (1991) "Intensity of Pollution of Groundwater in Visakhapatnam Area, A.P., India, Journal of Geological Society India, Vol. 36, pp. 670-673. [6] Todd D.K., (2001), "Groundwater Hydrology". John Wiley and Sons Publication, Canada, 280–281p.

9. A. TOR Desalination 2006

10. Ground water quality around Krishna and guntur.

11. Central Ground water Board standards

12. J Sirajuddin et al: Archives of applied science research (2013), 5(3)

13. Davies, S.N., DeWiest, R.J.M: Journal of Hydrogeology, (1966) Vol. 463, New York: wiley.

14. N. Janardhana Raju and T.V. Krishna Reddy, Urban development and the looming water crisis – a case study, The Geographical Society of London, 2006.

15. V. Srinivasa Rao, S. Prasanthi, J. V. Shanmukha and K.R.S.Prasad, Physicochemical analysis of water samples of Nujendla area in Guntur district, Andhra Pradesh, India, International Journal of ChemTech Research, 2(4), 2012, 691-699.

16. U.S. Salinity laboratory (1954). Diagnosis and improvement of saline and alkali spoils US department of agriculture.

17. USSL (1954). Diagnosis and improvement of salinity and alkaline soil. USDA Hand Book no. 60, Washington.

ISBN: 978-93-8830-599-0

Selection of Suitable Site for Rain Water Harvesting Structure by using Rs & GIS for Knigunta Watershed, Sangareddy TS

Rathod Ravinder[1], Suruboyina Datta Sai Yadav[2], M. Sai Kumar Goud[3],
Amisetty Avinash[4] and Saicharan Dubey[5]

[1]Assistant Professor GRIET, [2,3,4,5] Students of GRIET, Bachupally, Hyderabad.
rathod506ravinder@gmail.com.

ABSTRACT

Flood is one of the most devastating natural hazards which lead to the loss of lives, properties and resources. Floods resulting from excessive rainfall within a short duration of time and consequent high river discharge damage crops and infrastructures. They also result in siltation of the reservoirs and hence limit the capacity of existing dams to control floods. The purpose of flood risk assessment is to identify the areas within a plan that are at risk of flooding base on factors that are relevant to flood risks. It has therefore become important to create easily read, rapidly accessible flood map. Maps give a more direct and stronger impression of the spatial distribution of the flood risk than other forms of presentation (verbal description, diagrams).

Remote sensing is a reliable way of providing required data over a wide area in a very cost-effective manner. It also overcomes the limitation of the ground stations to register data in an extreme condition. This paper is aimed at assessing flood risk in the Anigunta region, Sangareddy district, Telangana state, India. Remote sensing technology along with geographic information system (GIS) is the key tool for flood monitoring. The map will be made using Geographic Information System (GIS). A GIS database of indicators for the evaluation of hazard will be created. Each indicator will be analysed and weighted, after which, the weights of the indicators will be combined to obtain the final map. The results obtained can provide useful information to suggest artificial recharge structures for decision making.

1. INTRODUCTION

In India, more than 75% of the population depends on agriculture for their livelihood, which plays a important role in our country's economy. More than 70% of the total geographical area is under distress due to the frequent occurrence of droughts and agriculture is seriously affected by uncertain monsoon.

In order to meet the growing demand for food, fuel, and fodder of ever-increasing population, land and water resources need to be optimally utilized. It requires timely and reliable information on available land and water resources, which could be derived from spaceborne multispectral data. Remote Sensing and Geographic information system together provide an information base for efficient management of land and water resources.

A watershed is a hydrological unit, draining runoff water at a common point and is demarcated based on the ridge and gully lines. The programmes under the watershed approach fall into soil and water conservation, dry land and rainfed farming, ravine reclamation, control of shifting cultivation and improvement in the vegetative cover.

1.2. STATEMENT OF THE PROBLEM

As India is one of the developing countries, due to increased urbanization and deforestation the vegetative cover is being decreased as a result of which the hydrological cycle is being affected and this is resulting in reducing the levels of the groundwater and increased risk of flood. All these consequences finally result in desertification of that piece of land.

Anigunta in Zaheerabad Mandal of Medak District, Telangana state, is selected because of its low levels of groundwater and frequent floods. The open wells have become dry forever and even bore wells are becoming dry after certain period. The urban areas in the study area which are densely populated are overexploiting the water resources for various uses. So there is a need to raise the water table to a reasonable level and controlling flood by various means such as artificial recharge and constructing recharge structures in the present study area.

1.3. OBJECTIVES OF THE PRESENT STUDY

The following are the objectives for the present study

1. To generate the thematic maps like a base map, drainage map, contour map, slope and aspect maps, land use/land cover, geomorphology, soils, groundwater prospectus maps.

2. To develop an action plan map by overlaying all the thematic maps.

3. To propose artificial recharge structures.

2. LITERATURE REVIEW

Venkataramana (2001) in an evaluation study at Chevella watershed in the southwest part of the Ranga Reddy district of Andhra Pradesh using remote sensing techniques observed raised water levels in the

ISBN: 978-93-8830-599-0

watershed. This was attributed to the fact that the runoff and water retained by various soils and water conservation structures within the watershed boundaries in the long run helped for the recharge of groundwater in the watershed.

Jyoti syrup et al., (2002) integrated remote sensing and GIS for identification of sites for artificial recharge for Kalwantaluka of the northern part of Nasik district in the state of Maharashtra. They used geocoded Indian remote sensing satellite (IRS) linear imaging self-scanning (LISS-III) data and IRS 1D PAN imagery and digital data (with a spatial resolution of 5.8m) along path number 94 and row number 58 have been used in their study. Based on the cumulative weighted index, the composite map classes were divided into seven groundwater potential zones. These identified sites were used for the construction of water harvesting structures like check dams, percolation tanks, and rockfill dams.

Khan et al., (2002) have delineated groundwater prospect zones in Neem-ka-thana tehsil in Sikar district of western Rajasthan by using IRS-ID LISS III satellite geo-coded data on 1:50,000 scale. The information on lithology, structure, geomorphology, and hydrology was generated to prepare the groundwater prospect map. The information on nature and type of aquifers, type of wells, depth range, yield range, success rate, and sustainability were supplemented to form a good database for identification of favorable zones. The study area has complex geomorphology and geology is dominated by rocks of Delhi Intrusive and Quaternary sediments. On the basis of hydrogeology and geomorphic characteristics, four categories of groundwater prospect zones like high, moderate, low and very low were delineated. The high prospect zones are alluvial plains and valley fills mainly influenced by quaternary formations with yield expectations between 100 - 200 lpm. The moderate zone has pediment surfaces covered by shallow soil cover in addition to weathered and fractured aquifer material with an expected yield of water between 50-100 lpm. The very low prospect zones act as run-off generating zones.

Shanwad (2006) has carried out a project work titled Action Plan Preparation (MedakNala) and Impact Assessment (Katanga Nala) of Watersheds in Gulbarga district using Remote Sensing and GIS technologies. LISS 3 and PAN merged data were used for preparing the thematic maps, land use/land cover map, soil map and lineament map, groundwater potential maps. Suitable weights are assigned to each thematic feature for proposing suitable water harvesting structures Medak and Katanga Nala.

AninditaLahiri (2011) have carried out an Integrated approach using Remote Sensing and GIS techniques for delineating groundwater potential zones in Dwarakeswar watershed, Bankura district, West Bengal. In the study area, the geological formation consisting of quartz, schist, slate show a fine texture of drainage pattern which indicates a high potential for groundwater occurrence. The final map prepared in the form of a prospect map would provide first-hand information to local authorities and planners about the areas suitable for searching groundwater followed by its suitable exploration.

3. METHODOLOGY AND STUDY

Based on the literature review and the objectives identified in the study the area chosen for the study and its various details is also presented in this chapter.

Fig. 1 Location Map of the Study Area

3.1. METHODOLOGY ADOPTED

The following is the methodology adopted for the present study

1. All the thematic maps such as road network and settlements map, drainage map, contour map, triangulated irregular network, digital elevation model, slope, aspect, flow accumulation map, flow direction map, soil map, geomorphology map, and groundwater maps are prepared in ArcGIS 10.3 software.

2. Using the satellite image obtained from National Remote Sensing Agency and Survey of India toposheet number 56 G/10/SW of scale 1:25,000

ISBN: 978-93-8830-599-0

land use/land cover map is prepared in ArcGIS 10.3 software.

3. By overlaying all the above-said maps and by performing raster calculations action plan map is created.

3.2.   PREPARATION OF THEMATIC MAPS

Athematic map is a type of map or chart especially designed to show a particular theme connected with a specific geographic area. The various thematic maps include road network and settlements map, drainage map, contour map, slope map aspect map, triangulated irregular network map, digital elevation model map, flow direction map, flow accumulation map, land use/land cover map, soil map, geomorphology map, and groundwater prospectus map are prepared.

3.3. ROAD NETWORK AND SETTLEMENTS MAP

Road network and settlements map are essential to locating all the human inhabitations like rural and urban settlements with important economic facilities such as health care, educational institutions, drinking water facilities hospitals, communication facilities, etc. Road network is used to locate and create facilities which would strengthen the economy of the watershed.

This road network map helps to know the connectivity to different human inhabitants within a watershed, which is very important to take up developing programmes that are to be carried. It is also used for knowing the various transport facilities within that region. Road layer would be useful to locate new development and its relationship to existing development.

Road network and settlements map are prepared by using Survey of India toposheet number 56 G/10/SW for a scale of 1:25,000 and updated with the help of satellite imagery in ArcGIS 10.3 software.

In the road network and settlements map, all the settlements and their connectivity within the study area are digitized from Survey of India toposheet and each settlement is labelled.

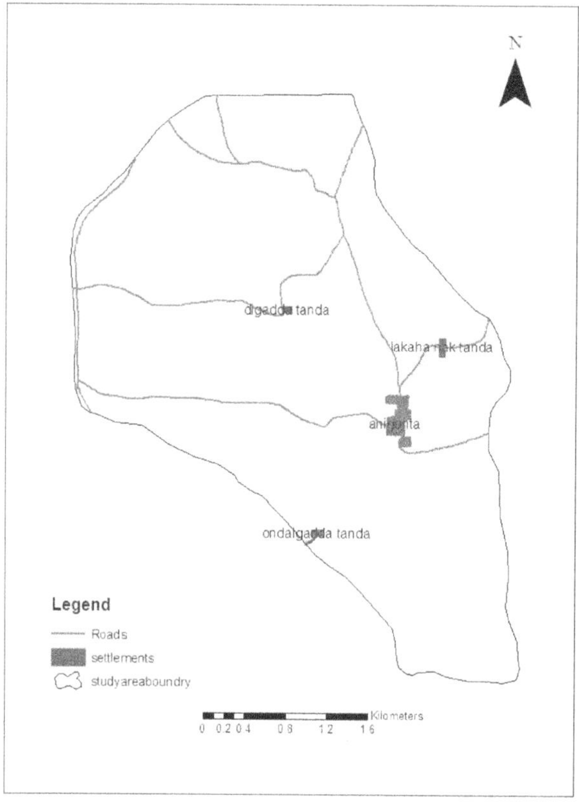

Fig. 2 Road network and settlements of the study area

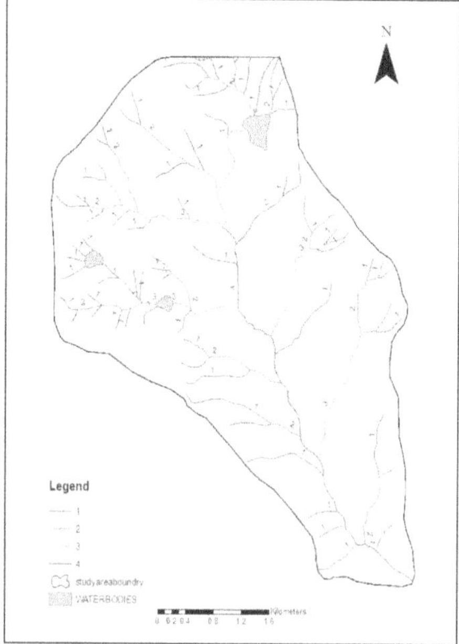

Fig. 3 Drainage Map of the Study Area

ISBN: 978-93-8830-599-0

Table 1 Number of Streams in the Study Area

S. No.	Stream Order	No. of Streams
1	First order	68
2	Second order	16
3	Third order	5
4	Fourth order	1

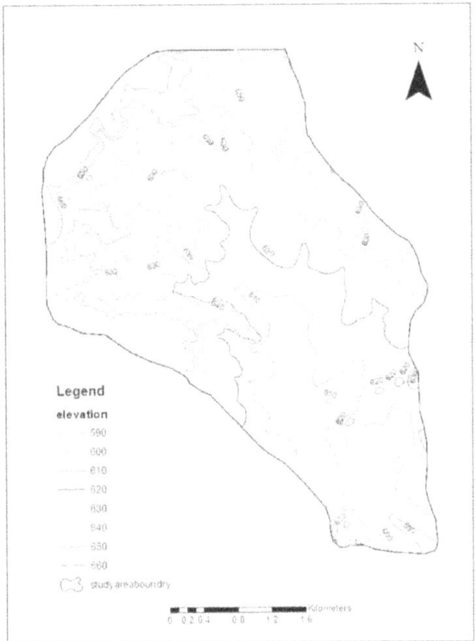

Fig. 4 Contour Map of the Study Area

Table 2 Slope Categories in Anigunta Micro-Watershed

Class	Percentage	Slope Category
1	0-1	Nearly level
2	1-3	Very gently sloping
3	3-5	Gently sloping
4	5-10	Moderately sloping
5	10-15	Strongly sloping
6	15-35	Moderately steep to steep sloping
7	>35	Very steep sloping

4. SUMMARY, CONCLUSIONS, AND RECOMMENDATIONS

4.1 Summary

For the present study, various thematic maps such as road network and settlements map, drainage map, and contour map are prepared from a survey of India toposheet. GIS analysis is done using a contour map prepared for the study area. Soil map geomorphology map and groundwater prospectus map are prepared using satellite data available and thematic maps of the study area. These maps are digitized using ArcGIS 10.3 software.

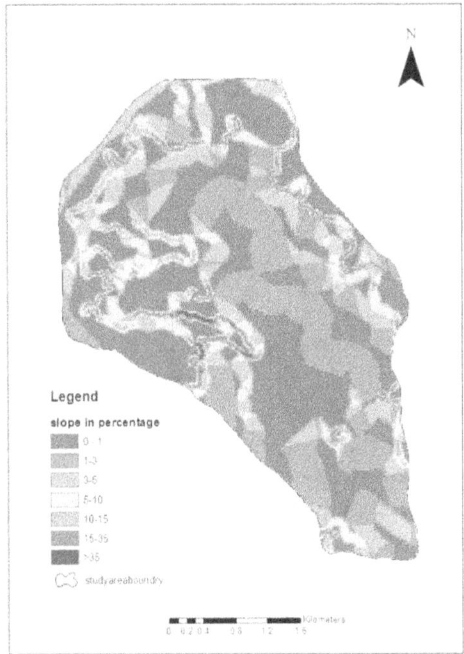

Fig. 5 Slope Map of the Study Area

4.2 CONCLUSIONS

The following are the few conclusions drawn from the present study

(1) The drainage pattern of the study area is a dendritic type of network which provides information for understanding both structural controls of surface flow as well as recharge zones.

(2) The highest elevation contour is 660 meters and the lowest elevation contour identified is of 590 meters above the mean sea level.

(3) The hydrological analysis is carried with the help of the prepared raster. The very steep slope is present towards the northern direction in the western region and most the study area has a nearly level slope in the study area. Flow direction is mainly seen in east and south direction and maximum flow accumulation is in the central region of the study area.

(4) From the Landuse/ Landcover map prepared, it is concluded that land with scrub and crop covered most part of the study area and the main source for irrigation is groundwater.

(5) Most of the study area is covered with lateritic plateau but there are few weathered zones where the ground is moderate to good. The

study area is mainly characterized by loamy soils.

(6) In the study area ten check dams of which seven are proposed on first-order streams. Two percolation tanks are proposed on fourth order streams and those areas have less than two percent slope. Contour bunding is proposed on steep slope areas of the study area. All these structures are proposed at appropriate locations based on the raster theme weights.

4.3   RECOMMENDATIONS

(1) Strip cropping, Mixed cropping is a practice of growing more than one crop in the same field so there will be one main crop and one or two subsidiary crops. These two methods can be recommended in the existing fields which can help in the prevention of soil erosion, land capability, improves soil.

(2) Wastelands can be regenerated by sowing seeds of grasses, and leguminous plants can be grown to increase the green canopy and for improving soil moisture, raising levels of groundwater levels and also serves as a fodder for animals.

(3) Artificial recharge methods and various types of irrigational practices can be adapted to recharge groundwater in the vast unutilized area like land without scrub.

(4) Proper Planning and care have to be taken in the selection of sites for bore wells in the future with the help of Remote Sensing maps, geophysical and hydro-geological studies.

(5) Awareness among farmers has to be created at village / mandal level by organizing campaigns.

5. REFERENCES

1. Burrough (1986).Principles of geographical information systems for land resources assessment. Oxford, University Press, pp 191.

2. Dixit (1976) Drainage basins of Konkan forms and characteristics. National Geographical Journal India, 22, pp 79-105.

3. Khan & Moharana (2002) Use of remote sensing and geographic information system in the delineation and characterization of groundwater prospect zones. Indian Society of Remote Sensing 30(3), pp 131-141.

4. Krishnamurthy &Srinivas (1995) Role of geological and geomorphologic factors in groundwater exploration: a study using IRS LISS data. International Journal of Remote Sensing 16(14), pp 2595–2618.

5. Krishnamurthy, Venkatesesa Kumar, Jayaraman and Manivel, 1996, An approach to demarcate groundwater potential zones through remote sensing and a geographical information system, International Journal of Remote Sensing, pp 7 (12): 1867-1884.

6. MadhavaRao, Herman, and KesavaRao, Generation of action plans for watershed development using satellite imagery (2010), NIRD, Hyderabad. pp 56-84.

7. Prasad, Integrated land and water resources conservation and management development plan using remote sensing and geographic information system of Chevelle sub-watershed Rangareddy district, Andhra Pradesh, pp 55-68.

8. Rajesh Rojora, Integrated watershed management (1989), Rawat Publications, Jaipur and New Delhi, pp 339-343.

9. Dwarakanath, Groundwater information on Medak district, Andhra Pradesh (July 2007), a pdf manual, pp 27-29.

ISBN: 978-93-8830-599-0

Assessment of Groundwater Quality Parameters in and Around Jawaharnagar, Hyderabad

Mudavath Mothilal and Kalakuntla Mounika

Structural Engineering, JNTU College of Engineering, INDIA

Abstract

The water and environment has become an emotive issue with the people and policy makers. The chief causes for the pollution of water and environment are anthropogenic activities of human beings. The primary objective of this paper is to study the groundwater quality parameters in the surrounding wells of Jawaharnagar, in upper Musi catchment area of Ranga Reddy district in Andhra Pradesh. The bore wells data is collected from the study area. The groundwater contour analysis is done by using Arc GIS software. The study reveals that the concentration of major constituents are well within the permissible limits of IS (10500-1994), except in few cases where total hardness and fluoride concentrations are high. From the analysis it has been observed that the groundwater is polluted in the entire study area. Due to this reason during the monsoon seasons the rainwater drains into the solid waste polluting the land leachate existing in the surrounding areas and in the low lying areas. During last few years, the utilization of surface and groundwater for drinking, industrial and agricultural purposes has increased manifolds but consequently it is observed that the water is polluted and affecting the human health, soil nutrients, livestock, biomass and environment in certain areas. Hence a study has been carried out for the quality of the available groundwater.

Keywords: Groundwater, pollution, leachate, soil nutrients, human health.

1. INTRODUCTION

Groundwater may be considered as one of the most precious and one of the basic requirements for human existence and the survival of mankind providing him the luxuries and comforts in addition to fulfilling his basic necessities of life and also for industrial and agricultural development thus being a very important constituent of our eco-sysem. Telengana region is classified as hard rock area where surface water resources are limited, with the result; groundwater has become a major source of supply to the village population. The problem connected with supply of safe water to rural communities from individual wells is often neglected. Added to this with rapid solid waste leachate pollution of groundwater due to rapid increasing of population, urbanization and industrialization, the available groundwater is rapidly getting polluted. Unfortunately the surrounding villages and catchment area aquifers affecting from this pollution has resulted mostly from urban human activities.

2. RESEARCH SIGNIFICANCE

The primary objective of this research paper is to study the groundwater quality parameters in the surrounding wells of Jawaharnagar, in upper Musi catchment area of Ranga Reddy district in Telengana. The bore wells data is collected from the study area for seasons i.e., pre monsoon January to May 2016. The groundwater contour analysis is done by using Arc GIS software. The study reveals that the concentration of major constituents are well within the permissible limits of IS (10500-1994), except in few cases where total hardness and fluoride concentrations are high. From the analysis it has been observed that the groundwater is polluted in the entire study area. Due to this reason during the monsoon seasons the rainwater drains into the solid waste polluting the land leachate existing in the surrounding areas and in the low lying areas. During last few years, the utilization of surface and groundwater for drinking, industrial and agricultural purposes has increased manifolds but consequently it is observed that the water is polluted and affecting the human health, soil nutrients, livestock, biomass and environment in certain areas. Hence a study has been carried out for the quality of the available groundwater.

STUDY AREA

1 GENERAL

Location North Latitude1

17030'01" N to 17032'03" N

ISBN: 978-93-8830-599-0"

East Longitude

Geographical area (sq.km) : 13.25

Major rivers : Musi, Musa, Kagna, Manjeera

Soils: Red soils, Medium Black soils, Mixed soils

2 RAINFALL

Normal annual rainfall 833 mm

3 Geology

Granites, Gneisses, Limestones, Shales,

Deccan Trap, Laterite, granites, basalts

Major aquifers - Weathered and fractured

78034'13" E to 78037'47

2. METHODS & MATERIAL

I Delineating a Watershed

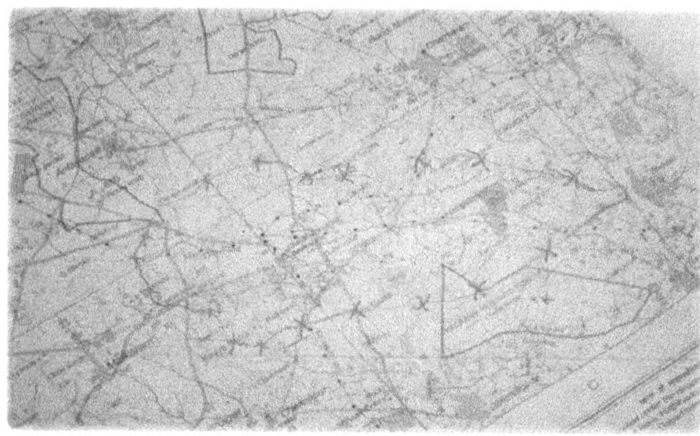

1. Draw a circle at the outlet or downstream point of the wetland in question (the wetland is the hatched area shown in Figure 3.1.2.1 to the right)
2. Put small "X's" at the high points along both sides of the watercourse, working your way upstream towards the headwaters of the watershed.
3. Starting at the circle that was made in step one, draw a line connecting the "X's" along one side of the watercourse (Figure 3.5, below left). This line should always cross the contours at right angles (i.e. it should be perpendicular to each contour line it crosses).
4. Continue the line until it passes around the head of the watershed and down the opposite side of the watercourse. Eventually it will connect with the circle from which you started.

I.   ESTABLISH OF GROUND WATER MONETRING STATION

The groundwater samples are collected during the post monsoon period i.e., December 2015 and pre monsoon period i.e., June 2016 from the seven bore wells located in the study area. The well locations in the study area are represented in figure. The quality analysis has been carried out for the parameters like pH, total alkalinity, electrical conductivity, total dissolved solids, total hardness, calcium hardness, magnesium hardness, nitrites, nitrates, sulphates, chlorides fluorides by following the standard methods prescribed as per IS: 10500-1994 codes.

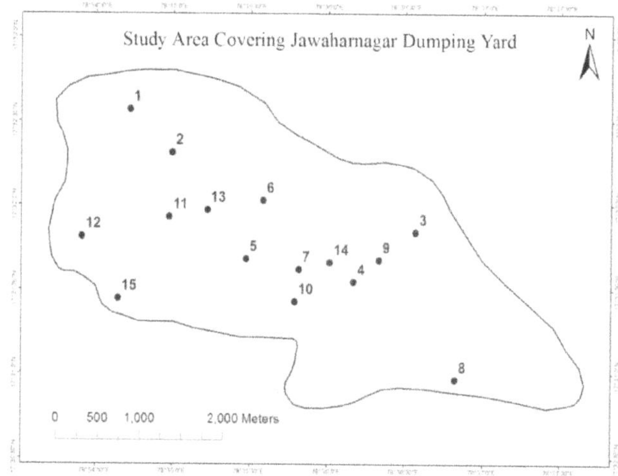

Fig. 1 Ground water monitoring stations

In the above basin area represent the monetring stations for the collection of the samples from the ground monetring points as we are located in the map. The ground monitoring points are nothing but the bore wells are dug wells present at the site of our located map above in the figure From this point samples are taken it is the best way to collected samples from the site in this method easily we can find outing the bore wells and longitude &latitude of the ground monetaring points and we can easily collects the water samples from the site

II. SAMPLE COLLECTION AT SITE

The groundwater samples are collected during the pre-monsoon i.e., January to May 2016 fromfifteen bore wells located in the study area

Table Collected data from study area

SampleNo	Longitude	Latitude	DTW_ftbgl	Total Depth	TypeofWell	name of location
1	78.545692	17.483539	150	400ft	BW	new ambedker nager
2	78.4867	17.385	22	200ft	BW	darga(jawaharnager)
3	78.609464	17.53071	500	600ft	BW	Goshala,malkaram
4	78.602811	17.525789	400	500ft	BW	hari das palli
5	78.591197	17.528046	260	400ft	BW	Malkaram
6	78.593055	17.533846	240	450ft	BW	hari das palli
7	78.59689	17.527032	150	30ft	BW	rajev gruha kalpa ,hari das palli
8	78.613896	17.513762	50		open well	chirayala village
9	78.60554	17.527941	60	150ft	BW	Mallakaram
10	78.596453	17.523851	20		surface pond	dumping yard,jawaharnager
11	78.5829	17.5322	130	500ft	BW	Malkaram
12	78.573482	17.530256	60	300 ft	BW	nursury,mallkaram
13	78.277513	17.251365	150		open well	Malkaram
14	78.29752	17.313659	235	300ft	BW	giriprasad nager
15	78.577383	17.52415	650	750ft	HP	devender nager

3. RESULTS & DISCUSSION

Table Seasonal Wise Concentrations of Water Quality Parameters in Groundwater Samples

Sno	pH	Cond	TDS Mic sim./ Cm	Co3 Mg/l	HCo3 mg/l.	CL mg/l	F mg/l	NO3_N mg/l.	So4 mg/l	Na mg/l	K mg/l	Ca mg/l	Mg mg/l	TH mg/l
1	7.85	1073	687	0	225	160	2.29	1.7	72	92	1.9	48	53	340
2	7.2	7090	4538	0	129	2260	0.58	11.0	148	401	6.5	592	297	2700
3	7.61	3630	2323	0	137	1060	0.86	9.7	125	204	7.1	192	219	1380
4	7.92	785	502	0	140	110	1.46	2.9	80	105	6.3	56	5	160
5	7.74	944	604	0	156	130	1.28	8.4	94	53	1.9	72	44	360
6	7.56	4710	3014	0	135	1470	1.21	5.0	102	401	4.8	240	219	1500
7	7.46	10540	6746	0	Highly saline		0.42	17.9						
8	7.84	1829	1171	0	174	400	1.75	5.4	142	101	2.6	96	112	700
9	7.11	5200	3328	0	179	1620	0.55	4.2	92	314	2.4	472	185	1940
11	7.91	1131	724	0	98	180	1.44	37.8	70	77	3.9	80	49	400
12	7.77	2200	1408	0	315	300	2.25	62.4	123	223	4.3	104	88	620
13	8.00	774	495	0	146	90	1.80	15.1	54	49	3.5	56	34	280

ISBN: 978-93-8830-599-0

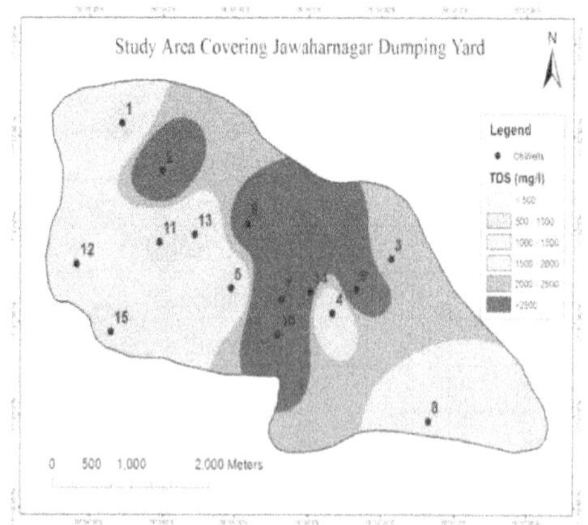

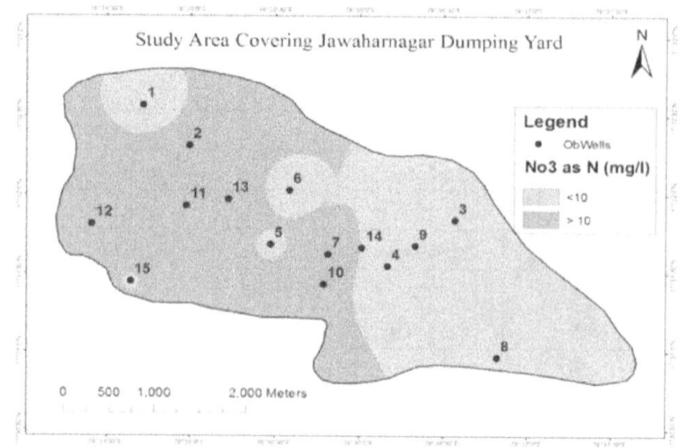

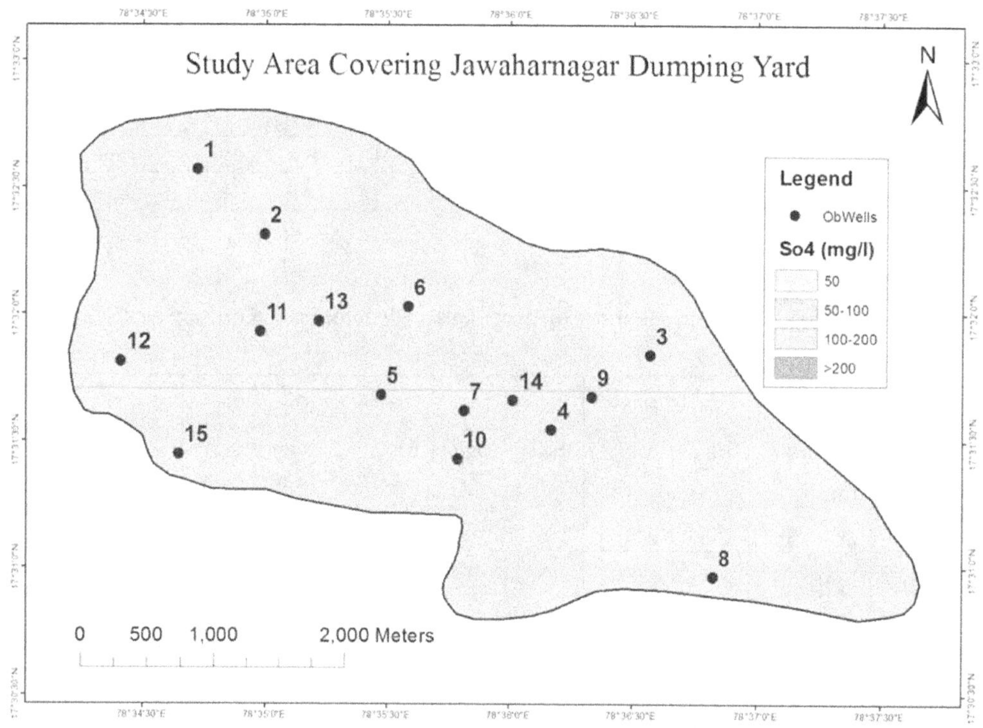

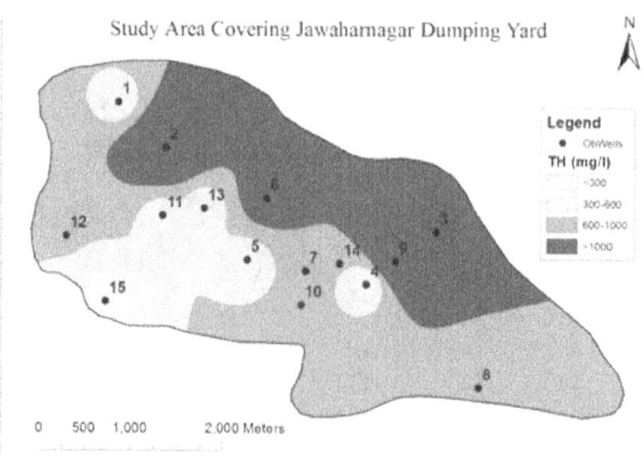

ISBN: 978-93-8830-599-0

4. CONCLUSIONS

Groundwater quality in and around Jawaharnagar, Hyderabad has been analysed in the present work. The groundwater is acidic in nature and total hardness observed in all samples fall under hard to very hard category. The total dissolved solids falls under fresh water to saline categories. The fluoride concentration in the northern and southern region exceeded the permissible limit. The concentration of physiochemical constituents in the water samples were compared with the Bureau of Indian Standards to know the suitability of water for drinking. Based on the analysis, most of the area at many locations near the solid waste dumping site falls in moderately polluted to severely polluted category indicating that the water is unsuitable for drinking purpose. The influences of solid waste dumping site, aquifer material mineralogy together with semiarid climate, other anthropogenic activities and increased human interventions have adversely affected the groundwater quality in the study area.

5. RECOMMENDATIONS

- The overall quality of groundwater is low and this situation needs urgent and strategic solutions since domestic waste and agricultural activities has factor effect in the hydrochemistry, it is an important issue to: Complete and improve the sanitation services in the governorate and implement Monitoring programs to ensure the proper use in of fertilizers and pesticides.

6. REFERENCE

1. BIS, (1994), "Indian Standards Specifications for Drinking Water", Bureau of Indian Standards, IS: 10500 -1994).
2. Goel P.K., (2000), "Water Pollution-causes, effects and control", New Delhi, New Age Int. (P) Ltd.
3. Pandey Sandeep K. and Tiwari S.,(2009), "Physico-chemical analysis of groundwater of selected area of Ghazipur city-A case study, Nature and Science, 7(1).
4. Sharma, V.V.J., (1982) "Ground Water Resources of Northern Eastern Ghats" Proceedings of the Seminar on Resources Development and Environment in the Eastern Ghats, Visakhapatnam, pp.69-75.
5. Subba Rao, N., Krishna Rao., G., (1991) "Intensity of Pollution of Groundwater in Visakhapatnam Area, A.P., India, Journal of Geological Society India., Vol. 36., pp. 670-673.
6. Todd D.K., (2001), "Groundwater Hydrology". John Wiley and Sons Publication, Canada, 280–281p.
7. WHO., (1996), "Guidelines for drinking-water quality, health criteria and other supporting information", V.2, 2nd edn, World Health Organisation, Geneva, 940-949 pp.
8. Chenini I. , Khemiri S. , (2009), "Evaluation of ground water quality using multiple linear regression and structural equation modeling". Int. J. Environ. Sci. Tech., 6 (3), 509-519.

ISBN: 978-93-8830-599-0

Prediction of Small Pelagic Fish Resources in North West Bay of Bengal using Artificial Neural Network

M. Pandu[1], Ballu Harish[2] and J.Venkatesh[3]
[1]Student, Centre for Spatial Information Technology-JNTUH
[2,3]Assistant Professor, Centre for spatial Information Technology-JNTUH

Abstract

Global marine fisheries are under pressure from increasing demands for protein, driven by rapidly growing human populations. Many commercial fish stocks remain either fully or over-exploited and show continued declines. This situation can be regularized by forecasting species-specific occurrence and abundance for sustainable fishing to conserve and manage. For this, five environmental variables such as sea surface temperature, dissolved oxygen, sea surface salinity, chlorophyll-a data from ROMS model, rainfall satellite data against Hilsa fishery data for a period 2009-2016 were utilized to develop species-specific habitat suitability model by advanced data mining methods and neural network backpropagation. Based on the present environmental data analyses for the region 20°N-22°N and 86.5°E-90°E, the suitable ranges of environmental variables for Hilsa was 27–32°C for SST, 24 to 31ppm for SSS, 0–0.14 mg/m^3, for Chl-a concentration, 187-206 ppm for dissolved oxygen. The independent variable neural network analysis with Levenberg-Marquardt backpropagation algorithm (trainlm) showed that SST and Chl-a have a high impact on CPUE prediction compared to salinity, rainfall and Dissolved Oxygen. A neural network model was developed with backpropagation learning method and used to predict the spatial distribution of Hilsa with 732 monthly samples for the year 2016. However, we need to work further to authenticate the model accuracy through validation experiment as the ecological preferences of Hilsa shad in the ocean is very less known fact.

INTRODUCTION

India is a seventh largest country with second highest populated nation in the world, with the coast line more than 8100km. In this peninsular country over about 7 million people live along the coast line of India with fishing as their major occupation. It is a difficult task for fishermen to locate and catch fish, without prior knowledge on fish habitats which leads to high investment and low profits. Especially in the monsoon season due to non-availability of satellite information due to over casting, there is a need of advisories on cloudy days for fisher men to locate fish hotspots. Also due to global warming, illegal fishing times and over exploitation fish production dropped to alarming stage which is leading to threatening of species. This situation can be regularized by sustainable fishing. Sustainable fishing ensures a guarantee that there will be a marine population in ocean and fresh water for the future by hunting for species specific at different times of the year leads to health stocks known as sustainable fishing.

To address above such problems, using artificial intelligence which has gained an increasing importance in studies of ocean systems as well as for extracting and monitoring the dynamics of the oceanic environment. Analysis of oceanic variables data and fisheries data in machine learning and geographic information system (GIS) environment have facilitated to understand fundamental relationships between fishery resources and their oceanic environment. The marine ecosystem is complex with respect to its species composition as well as the processes occurring within it. The prediction of marine ecosystem structure and function depends on a thorough understanding of the physical and biological processes which govern the abundance (Dhyey Bhatpuria & Prakash Chauhan at el), distribution and productivity of the species on a wide range of time and space scales.

OBJECTIVES

The following are the objectives of this paper.

- To find correlation between catch per unit effort and individual oceanographic variable.
- To map suitable habitat of Hilsa.
- To find suitable ranges of each ocean variables that have direct or indirect impact on the distribution pattern.
- To convert near real-time fishery advisory services into forecasting system for Hilsa shad.
- To forecast during overcasting days and satellite data interruption problems.

STUDY AREA

Bay of Bengal, occupying an area of about 839,000 square miles (2,173,000 square km). It lies roughly between latitudes 5° and 22° N and longitudes 80° and 90° E. It is bordered by Sri Lanka and India to the west, Bangladesh to the north, and Myanmar (Burma) and the northern part of the Malay Peninsula to the east. The maximum depth is 15,400 feet (4,694 meters). A number of large rivers the Mahanadi, Godavari, Krishna, and Kaveri (Cauvery) on the west and the Ganges (Ganga) and Brahmaputra on the north—

ISBN: 978-93-8830-599-0

flow into the Bay of Bengal. For the current project work study area is northwest Bay of Bengal for the region Latitude: 20°N-22°N and Longitude: 87°E-90°E which are surrounded by Odisha and west Bengal states.

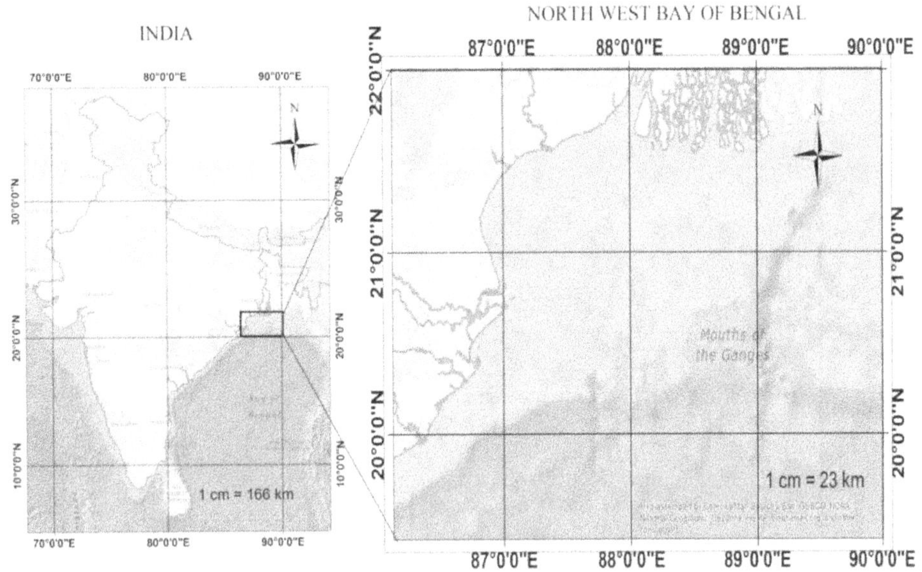

Fig. 1 Study area

METHODOLOGY

The methodology of this project towards establishment of artificial neural network model using fishery data and oceanographic variable data. In this model, a feed-forward back-propagation with scaled conjugate gradient algorithm, was used to train the data to predict fishing ground levels, and Levenberg-Marquardt algorithm for regression analysis to find impact of (correlation coefficient) of individual variables on CPUE. Workflow of the above methodology is explained in fig2.

DATA EXTRACTION

Data extraction for the environmental variable is done using ferret from .netcdf file e.g. code is shown below to extract sea surface temperature value at a particular *Hilsa* caught location for the day, on surface of Bay of Bengal ocean. Similarly for all variables data extracted to create a data base of 5055 samples of all fish. Out of which the Hilsa species samples are filtered for the current project.

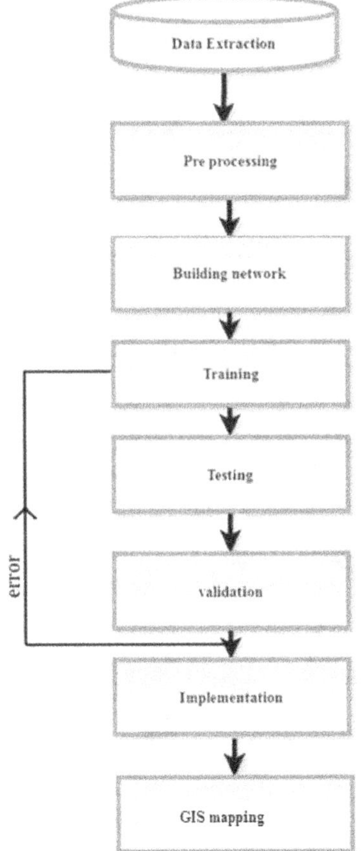

Fig. 2 Work flow Chart

ISBN: 978-93-8830-599-0

RESULTS AND DISCUSSION

FISHERY DATA ANALYSIS

The monthly average CPUE of Hilsa fishery for years 2009-2016 from fishing ground fluctuated in between 1.48-2.86 (Figure 5-1, Table 5-1). The high mean of CPUE was 2.8 in the September and the lowest 1.48 in the January.

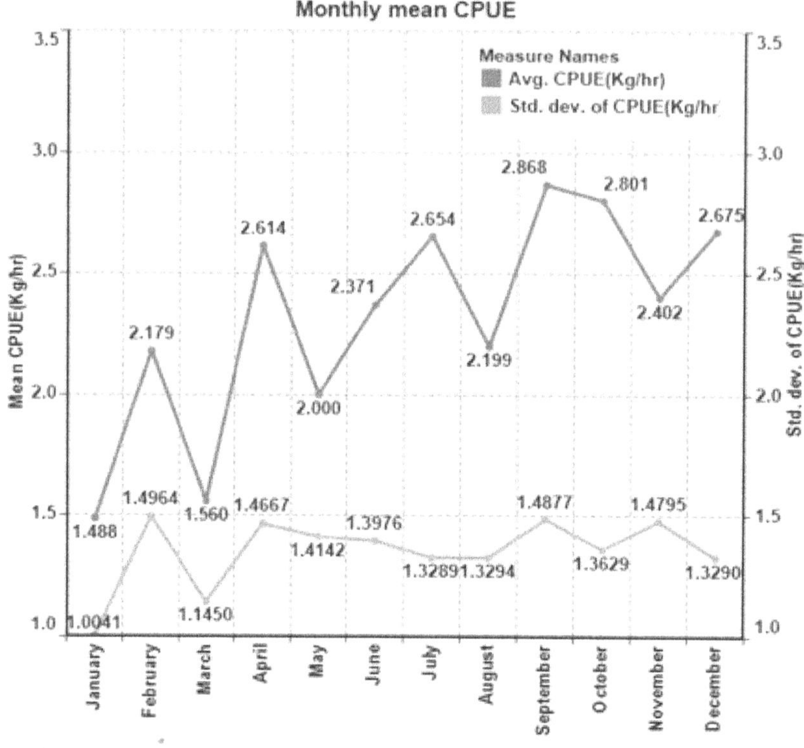

Fig. 3 Temporal variation of mean and std dev of CPUE of Hilsa

Table 1 Temporal variation of Hilsa CPUE

Month	Mean CPUE	Std. dev-CPUE
January	1.5	1.0
February	2.2	1.5
March	1.6	1.1
April	2.6	1.5
May	2.0	1.4
June	2.4	1.4
July	2.7	1.3
August	2.2	1.3
September	2.9	1.5
October	2.8	1.4
November	2.4	1.5
December	2.7	1.3

From the mean graph (Figure 5-1)CPUE linearly increasing from January to December ,in the month September and October has high mean CPUE .It is also observed that in pre monsoon has low abundance and monsoon and post monsoon has significant Hilsa abundance in study area.

From the above histogram (Figure 0-1) June to December has increase in frequency i.e. monsoon has high catch and in the post monsoon October to January CPUE frequency decreasing. In the pre monsoon it is observed that it has very low frequency.

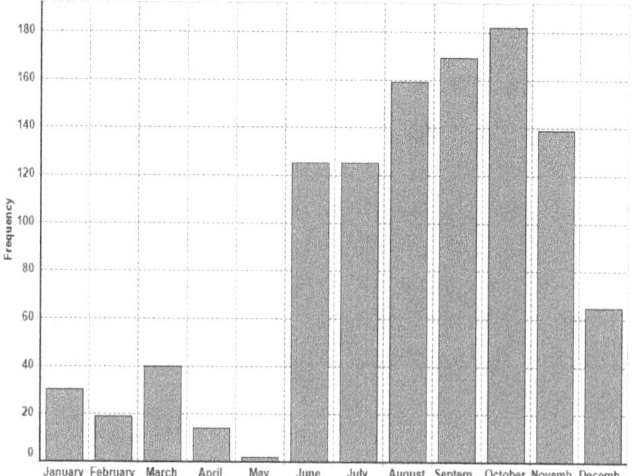

Fig. 4 Histogram of monthly CPUE

ISBN: 978-93-8830-599-0

DISCUSSION

The impacts of environmental variations on fish abundance and fishing ground distribution are well recognized by understanding how fish species react to climate change and variations in the regional environments. Predicting the dynamics of the fish population are essential for the effective management of marine resources. This project presents a neural network approach for correlation analysis between CPUE and environmental variables for the region north west Bay of Bengal .In this study, the neural network backpropagation suggests that the process for exploring PFZ was complicated. The relationship between the independent variables and classification levels of fishing grounds was non-linear to salinity Chl-a, dissolved oxygen and linear to sea surface temperature and rain fall. The backpropagation forecasting model not only predicted PFZ, but also provided a critical evaluation of suitable ranges of environmental variables for Hilsa through neural network regression and classification analyses. This shift in the distribution of fishing ground for Hilsa was closely related to SST and Chl-a. In this project, according to neural interpretation and the correlation analysis it's found that the SST and Chl-a are the most important environmental factor in the formation of potential fishing grounds and it had the greatest influence on the prediction model, suggesting that SST and chl-a could be used as an major indicators to explore PFZ. The favorable ranges of Hilsa 27–32°C for SST, 24 to 31ppm for salinity,0–0.14 mg/m^3 for Chl-a concentration,187-206 ppm for dissolved Oxygen and finally most fish catch records observed on non rainy days.

REFERENCES

1. Anton Emmanuel Selvakumar Patrick, "Species composition and abundance of fishes with seasonal fluctuations of rainfall and water level in Vavuniya reservoir, Sri Lanka "2017.
2. Aziz Ahmad1, Tim Anderson1 "Global Solar Radiation Prediction using Artificial Neural Network", 2014.
3. Atesmachew, Girma, Yasin." Application of GIS for Natural Resource Management".
4. Brey, T, A. Jarre-Teichmann, and O. Borlich. 1996. "Artificial Neural Network versus Multiple   Linear Regression: Predicting P/B Ratios from Empirical Data." Marine Ecology Progress Series 140: 251–256. doi:10.3354/meps140251.
5. Beale, M. H, M. T. Hagan, and H. B. Demuth. 2010. "Neural Network Toolbox User's Guide."
6. Cao, J, X. J. Chen, and Y. Chen. 2009. "Influence of Surface Oceanographic Variability on Abundance of the Western Winter-Spring Cohort of Neon Flying Squid Ommastrephes bartramii in the NW Pacific Ocean." Marine Ecology Progress Series 381: 119–127. doi:10.3354/ meps07969.

Comparison of Object-based and Pixel based Classification of LULC Changes in Gadchiroli District

Mahbooba Asra*[1], C. S. Jha*[1], R. Suraj Reddy*[1], Praveen MSS*[1], G. Rajashekar[1], T. Mayamanikandan[1]
V. Krishna Sandeep Kumar[2] and Natalia Grace Bird[2]
[1]Forestry and Ecology Group, National Remote Sensing Centre, Hyderabad
[2]Jawaharlal Nehru Technological University, Hyderabad.
asramajeed1223, chandra.s.jha, suraj.nrsc,mutyala45,grajashekar,@gmail.com
vkskreddy,nataliagracebird@gmail.com

Abstract

Land Use/Land Cover (LULC) classifications for mapping purposes using remote-sensing imagery has attracted significant attention in recent years. Classification of imagery have proven to be valuable assets for natural resource managers interested in landscape characteristics and the changes that occur over time. This study compares the results of an object-oriented classification with supervised pixel-based classification for mapping land use/land cover in the Gadchiroli District, Maharastra, India. The object-oriented approach involved the segmentation of imagery into objects at multiple scale levels. The supervised pixel-based classification involved the selection of training areas and a classification using maximum likelihood algorithm. These approaches were evaluate the relative importance of multi-temporal and multi-spatial imagery to classification accuracy. A comparison of the results shows better accuracy of the object-oriented classification over the pixel-based classification. This object-oriented analysis has great potential for extracting land use /land cover information from satellite imagery captured over study area. Results indicate that the overall accuracy of the object-based classification was 86%, while the pixel-based classification was 70.2%. The result shows a significant difference in the overall accuracies of the classifications.

Keywords: LULC, remote sensing, segmentation, supervised classification.

INTRODUCTION

Remote sensing land use/land cover terminology relates to presentation of information about human activity which connected to specific land and the type of feature on the surface of the earth (Lille sand et al., 2004). LULC monitoring can provide a baseline reference to help delineate the current limits of land cover types, can become standards with which to compare future land cover changes, can provide a basis for judging what constitutes ecological threats or impairments, and can help identify the need for corrective management actions (DeBacker et al. 2005). Land use policies must consider the importance of LULC change to human welfare (e.g. increased food production) while minimizing environmental effects (Foley et al. 2005). Multi-resolution segmentation of object-based image analysis, on the other hand, results in objects with different sizes and shapes, which are meaningful and better represent the real size and shape of land cover types (Salehi et al.,2011). The segmentation process separates the image into segments that are arranged according to the classification of spectral, geometric, textural and other characteristics of the objects (Veljanovski et al., 2011). Similar research are also done by (Oruc et al., 2004), (Xiaoxia et al., 2005), (Qian et al., 2007), (Bruce, 2008), (Gholoobi et al., 2010), (Weih and Riggan, 2010), (Avci et al., 2011), (Aguirre-Gutiérrez et al., 2012), which make compared pixel-based and object-based method. Shridhar D et al., 2015 a review on Pixel Oriented and Object Oriented Methods for Information Extraction from Remotely Sensed Satellite Images with on Cryospheric Applications. Results may obtained from remotely sensed imagery with medium spatial resolution, traditionally change detection methods base on pixel provide a low accuracy and a poor automation to improve the accuracy, the object-oriented approach will be introduced to land-used and land-cover change detection for the period of 2013-2017 years.

STUDY AREA

Gadchiroli district is the adobe of diversity. the erstwhile forted hillocks (gad=fort; chiroli = hillocks).Gadchiroli district in the state of Maharashtra, part of central India. it is located at eastern side of Maharashtra and it has total geographical area of 14412.0 km^2 out of this area 11694.0 km^2 i.e., 78.40% of the land is included in the reserve or protected forest

ISBN: 978-93-8830-599-0

category. and remaining area is under cultivation and waste lands which constitutes 21.60%. It lies between 79°1'38.873"E   23°21'26.61"N and   84°20'32.836"E 15°23'52.509"N geographically with an average elevation of 217m. As it is located at centre of Indian peninsula far from Bay of Bengal and the Arabian Sea, Gadchiroli has a tropical wet and dry climate with dry conditions prevailing for most of the year. it receives rainfall of about 1000mm from monsoon rains during June to September.

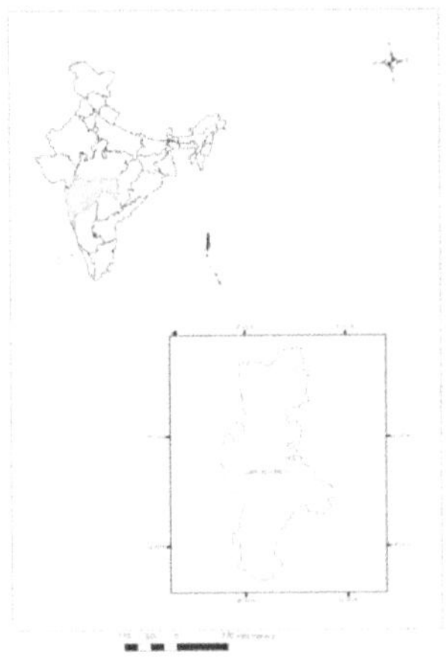

Fig. 1 Location Map of the Study area

Data used

Satellite imagery were procured from Landsat-8 (OLI & TIRS) sensors in the path 143 row 46 for 31st October 2013 and 11th December 2017 and the path 143 row 47 for 31st October 2013 and 27th November 2017 in World Reference System 2 (WRS 2) for the study area. Fig-2 shows data were acquired from Global Land Cover Facility (GLCF) with a spatial resolution of 30 meters. Satellite data acquired from landsat-8 (OLI) for two different periods 2013 and 2017 have been used to show the change detection in Gadchiroli district using Classification techniques.

METHODOLOGY

As land cover classes, the study area is covered by agricultural land, built-up areas, water/rivers, open or dense forests and scrub lands. Therefore we focused our classification over five land use/land cover classes: forests, settlements, water, scrub land, agricultural land etc. After determination of assessing the quality of data preprocessing steps has been carried out. The main objective of this paper is to evaluate the performance of pixel-based and object-based approaches through two classification methods, Nearest Neighbor and Maximum Likelihood Classification. Figure-3 shows the flowchart for data preprocessing and classification using both methods. The land use/land cover classification map has been classified into five categories using Object based and pixel based techniques. The change detection in the LULC of the district (Dhanora, Etapali, Aheri and Sironcha talukas in the eastern part of the district are covered by forest) was also analyzed for the period of five years. After the classification of LULC classes the ground truth verification were carried out and the errors in it have been corrected to prepare the accurate LULC maps.

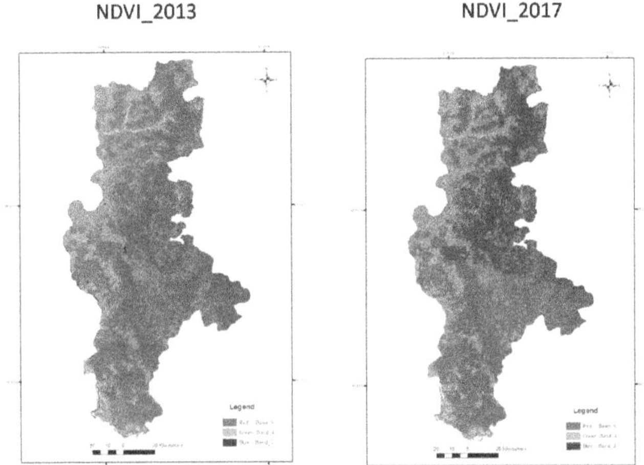

Fig. 2 Landsat-8(OLI) imagery for 2013 & 2017 with FCC band Combination Red-5; Green-4;Blue-3

ISBN: 978-93-8830-599-0

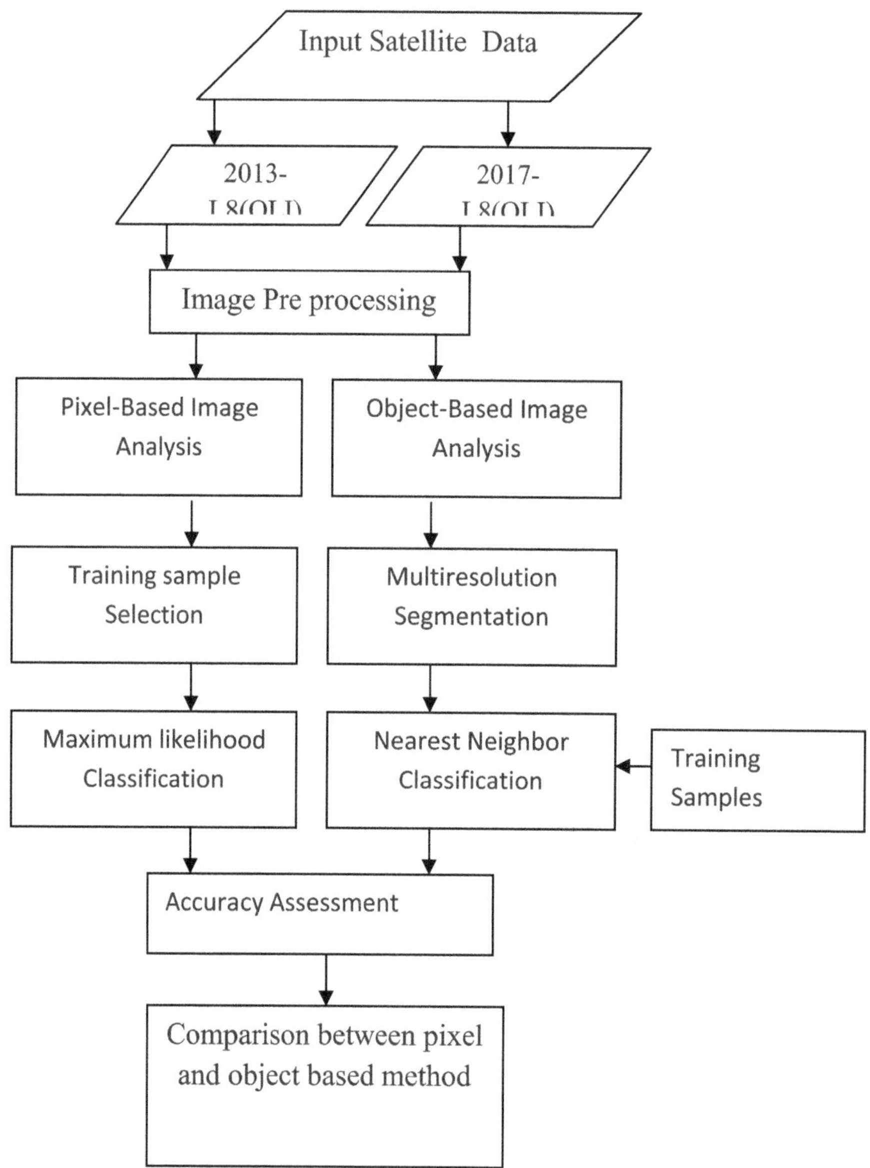

Fig. 3 Flowchart for Satellite data processing and classification approach using object-based and pixel-based approaches

Object-Based classification

Object-based classification not only use spectral features but also spatial measurement that characterized the shape and compactness of the region. The object-oriented classification conducted using eCognition v4 software [Baatzet al., 2004] has previously been described [Whiteside and Ahmad, 2004]. The object-based classification implemented using eCognition software, which applied segmentation with multi resolution segmentation and supervised classification with nearest neighbour method. (Jonsson, 2015), multi resolution segmentation (MRS) is a bottom up method, where segmentation starts with one pixel and, by loops, merges pixels into pairs and into larger objects. Parameter setting in segmentation step for the data as follows:

scale:. a higher value creates larger objects, sets the spatial resolution of MRS

shape: a higher shape value , during segmentation less value will be placed on color.

Compactness: a higher value, after segmentation more bound objects created.

Table 1 Segmentation parameters

Year	Scale level	Scale parameter	Shape factor	Compactness
2013	1	30	0.5	0.7
2017	2	20	0.4	0.6

ISBN: 978-93-8830-599-0

Nearest Neighbor classification in eCognition is based on a fuzzy classification algorithm, where each class contains can consist sets of fuzzy expressions based on parameters. Image classification using nearest neighbor, start with select samples for each land use land cover class which has been defined the criteria for

classification. The pixel based classification cannot statisy image classification with precision and it produce large data redundancy. Object based uses geometry and structure information. it provide truly revolutionary and accessible approach.

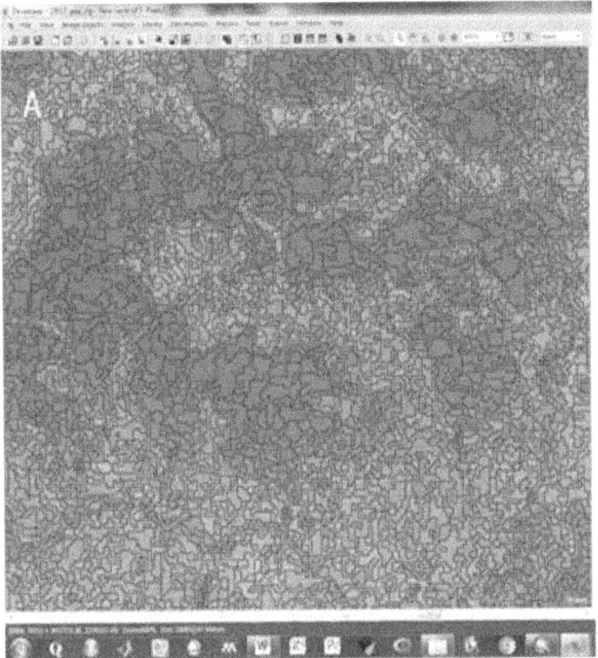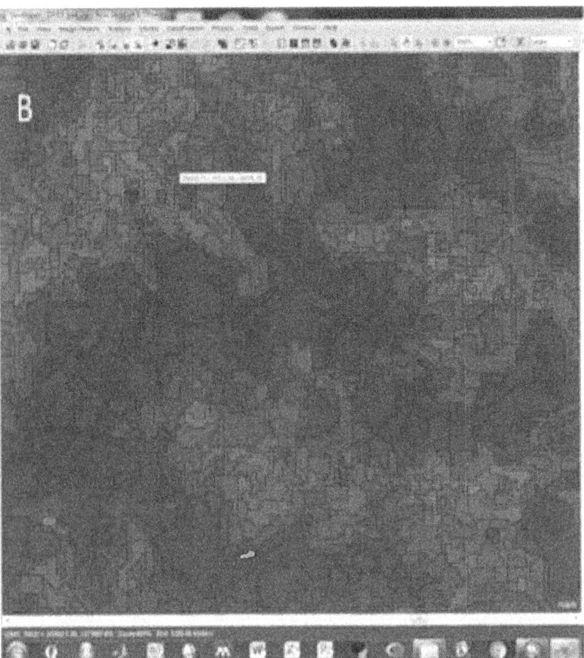

Fig. 4 A section of the study area showing Multi resolution segmentation at scale level 1 (a) 2013 and level 2 (b)2017

PIXEL-BASED CLASSIFICATION

Pixel-based methods aim to identifying land use/ land cover classes. This approach is referred to pixel-based methods given that each pixel is compared from one image to another independently from its' neighbours. The maximum likelihood technique was selected for supervised classification and for comparison with the object-based classification. in this study MLC were used and MLC classifier using parametric approach which the spectral statistic data should be distributed normally, and also the decision rule in MLC to estimate from training data determined by Bayes□ theorem (Richards and Xiuping, 2006). Maximum likelihood classification use „mean value and covariances to compute the probability of individual pixels belonging to specific class for the variation in spectral data□ (Ghosh and Joshi, 2014). The pixel-based classification was undertaken using ERDAS Imagine software. It was a standard supervised classification using the maximum likelihood algorithm [Jensen, 1996; Lillesand and

Kiefer, 2000]. This involved the selection of training areas representative of the five land use/land cover classes. A number of training 1228 areas were selected to represent each class. The signature (or spectral mean) of the training area was then used to determine to which class the pixels were assigned. Five classes i.e, scrub land, agricultre, settlment, forest and water bodies.

ACCURACY ASSESSMENT

Accuracy Assesment adopted for generated thematic maps. Accuracy assessment has three basic components, i.e. sampling design, response design for each training unit, estimation and analysis protocols (Stehman and Czaplewski, 1998. Error matrix or confusion matrix is common manners to explicit percentage of the map area which has been classified and compared with reference data (Story and Congalton, 1986). In this study overall accuracy and kappa value which generated from statistical computation training area to data coverage will be compared, and also tested

ISBN: 978-93-8830-599-0

RESULTS

The detailed areal extent of each class is shown in Figure5(a) and Table-1. First the individual year assessment have been calculated, from which it has been found that, total 9510.5 km^2 area were having densed /open vegetation in study area.

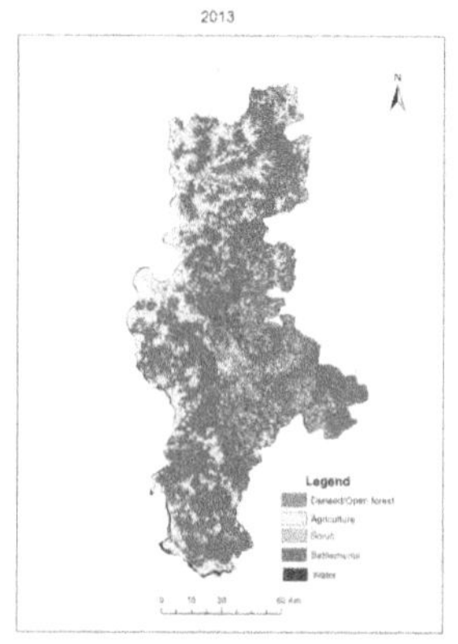
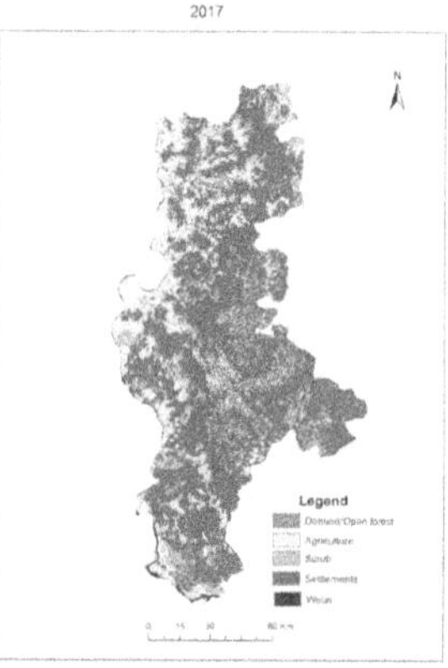

Fig. 5(a) Object-based classification result

Table 2(a) Number of features and area by LULC

Classes	2013(sq.km.)	Area(%)	2017(sq.km.)	Area(%)
Open/Densed Forest	9484.5	70.0	9350.1	69.0
Scrub	145.1	1.1	281.5	2.1
Agriculture	3521.1	26.0	3518.9	26.0
Water	303.2	2.2	303.4	2.2
Settlements	90.3	0.7	90.3	0.7

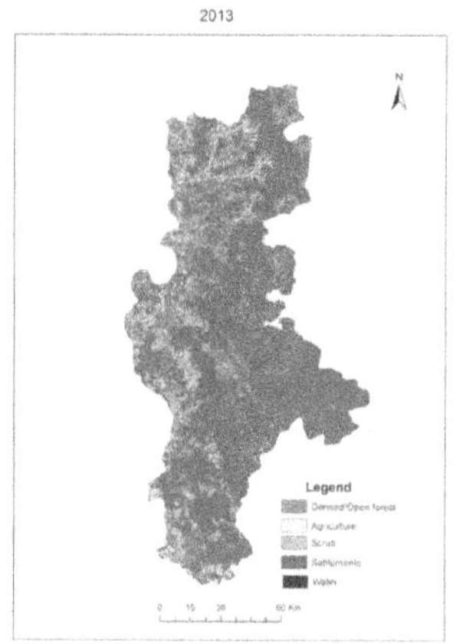
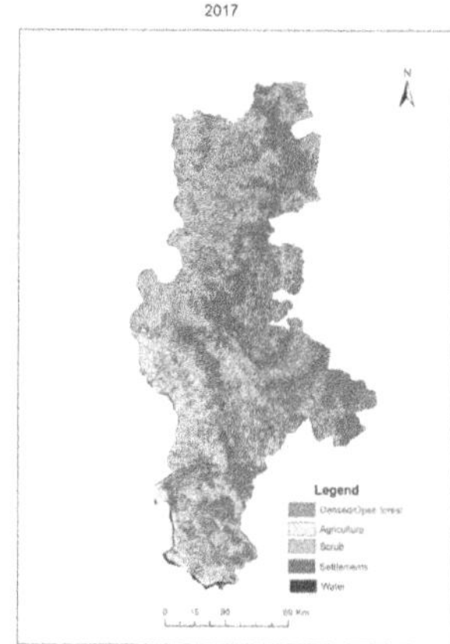

Fig. 5(b) Pixel-based classification result

Table 2(b) Number of features and area by LULC

Classes	2013(sq.km.)	Area(%)	2017(sq.km.)	Area(%)
Open/Densed Forest	9300.5	68.7	9158.2	67.6
Scrub	485.5	3.6	419.2	24.8
Agriculture	3350.4	0.0	3357.5	24.8
Water	314.7	2.3	323.5	2.4
Settlements	93.1	0.7	285.8	2.1

Table 3 Accuracy Assessment for 2017 imagery in comparision with pixel and object based

Classes	NN		MLC	
	Overall Accuracy	Kappa	Overall Accuracy	Kappa
Open/Densed Forest	95.61%	0.8335	85.43%	0.609
Scrub	91.63%	0.8488	81.18%	0.7084
Agriculture	80.11%	0.7316	78.59%	0.7191
Water	84.79%	0.7439	80.03%	0.669
Settlements	71.80%	0.5617	74.90%	0.6395

This study indicates that an object-based methodology produce an accurate LULC classification when applied to medium-spatial, multi-spectral satellite imagery. When compared with the overall accuracy of a pixel based classification of the same imagery, the object-based method was significantly more accurate. MLC run on single pixels, so the result provided by this method is influenced strongly by "salt and pepper" effect, although the filters were applied. In terms of object based, when segmentation image is operated, the structure and shape of objects have been taken into account, so this approach shows its more accurate than pixel based.

REFERENCES

1. Blaschke, T., 2010. Object based image analysis for remote sensing. ISPRS Journal of photogrammetry and remote sensing 65, 2-16.
2. Clinton, N., Holt, A., Yan, L., and Gong, P., 2008. "An accuracy assessment measure for object based image segmentation". In: XXIth ISPRS Congress, July 2008, Beijing, China, 2008, IAPRS, vol. XXXVII.
3. Dean, A. M., and Smith, G. M., 2003. "An evaluation of Perparcel Land Cover Mapping using Maximum Likelihood Class Probabilities". International Journal of Remote Sensing, 24(14), 2905–2920

ISBN: 978-93-8830-599-0

4. Definiens 2007 Definiens Imaging Developer 7. eCognition User Guide(München) Definiens AG. (Germany)

5. E. Juniati, E. N. Arrofiqoh-Comparison Of Pixel-Based And Object-Based Classification Using Parameters and Non-Parameters Approach for the Pattern Consistency Of Multi Scale Landcover, The International Archives of the Photogrammetry, Remote Sensing and Spatial Information Sciences, Volume XLII-2/W7, 2017 ISPRS Geospatial Week 2017, 18–22 September 2017, Wuhan, China

6. Huth J Kuenzer C Wehrmann T Gebhardt S Tuan V and Dech S 2012 Land cover and land use classification with TWOPAC: towards automated processing for pixel- and object-based image classification. J. Remote Sens. 4 pp 2530-2553.

7. Journal of the Arkansas Academy of Science, Vol. 63 [2009], Art. 18

8. Lillesand, T., Kiefer, R.W., Chipman, J., 2004. Remote sensing and image interpretation, Fifth ed. John Wiley & Sons.

9. N.D. Riggan, Jr. and R.C. Weih, Jr- A Comparison of Pixel-based versus Object-based Land Use/ Land Cover Classification Methodologies

10. Shridhar D. Jawak, PraptiDevliyal and Alvarinho J. Luis, —A Comprehensive Review on Pixel Oriented and Object Oriented Methods for Information Extraction from Remotely Sensed Satellite Images with a Special Emphasis on Cryospheric Applications, Advances in Remote Sensing, Vol. 4, pp. 177-195, 2015

11. Salehi, Bahram, Yun Zhang, Ming Zhong, and Vivek Dey., 2012. "Object-Based Classification of Urban Areas Using VHR Imagery and Height Points Ancillary Data." Remote Sensing, 2256-276.

ISBN: 978-93-8830-599-0

Detection of Forest Cover Changes using Geospatial Techniques

G. Ramesh[1], A. V. Subba Rao[2], L. Ravi[3] and Ballu Harish[3]
[1]Student, [3]Research Scholar, [3]Assistant Professor
Centre for Spatial Information Technology-IST-JNTUH
[2]Senior Scientific Officer TRAC, Planning Department, Hyderabad.

ABSTARCT

Land cover (LC) refers to the physical and biological cover and the surface of the land including, water, bare soil and artificial structure. Land use (LU) is a more complicated term involving human activities such as building construction that alter land surface processes including biogeochemistry, hydrology and biodiversity. Land use land cover (LULC) is an important component in understanding the interaction of the human activities with the environment and thus it is necessary understand changes, in the context of nature resource management.

The present study focuses on the role of remote sensing and geographic information system (GIS) in assessment of changes in forest cover based on Land use/Land Cover using remote sensing Techniques, between 2006and 2014, in the khammam district, telangana. The trend of forest cover changes over the time span of 8 years was precisely analysed using high resolution Satellite data. The study analyses the forest cover change in the tropical deciduous forest region of the Eastern Ghats of India. It is envisaged that the study would prove the usefulness of Remote Sensing and GIS in forest restoration planning.

This study aims at generation of land use land cover information using remote sensing data belonging to different time period and integration of remote sensing derived LULC information with socio-economic data to predict future scenarios in LULC pattern in the study area, using GIS capabilities.

INTRODUCTION

Land Use" refers to human activities that take place on the earth's surface (how the land is being used) such as residential housing or agriculture cropping. "Land cover" refers to natural or man-made physical properties of the land surface. Both systems utilize multiple and varying classification schema with distinct overlaps and important differences.

There are few landscapes remaining on the Earth's surface that have not been significantly altered or are not being altered by humans in some manner. Mankind's presence o the Earth and his modification of the landscape have had a profound effect upon the natural environment. These anthropogenic influences on shifting patterns of land use are a primary component of many current environmental concerns as land use and land cover change is gaining recognition as a key driver of environmental change. Changes in land use land cover are pervasive, increasingly rapid, and can have adverse impacts and implication at local, regional and global scales.

By utilizing remote sensing technology and implementing GIS mapping techniques, land use land cover change of designated areas can be monitored and mapped for specific research and analysis. The principal objective is to detect, delineate, and map areas that have experienced land use and land cover change, specifically, the conversion of forests to agriculture pastures, in Khammam District as over the 8 year time period of 2006-2014 utilizing multi-spectral satellite imagery.

The image processing techniques employed in this study were conducting using ERDAS imagine software package with advance vector capabilities. The classified images were exported into Arcinfo and Arcview Gis software and converted from a raster to a vector format for additional GIS analysis.

Once the areas of land use change had been delineated, an additional land assessment of these areas was conducted to determine the wisdom of such land use practices and land productivity classifications. The final hardcopy land use maps were produced in Arcview and portray the areal extent of land use land cover change in khammam District over a eight year span from 2006 to 2014. All the computer software and hardware required in this study was utilized at Telangana remote sensing application centre (TRAC),

R. Sakthivel et al (2002) focused on the role of remote sensing and GIS in the assessment of changes in forest cover in the Kalarayani hills Tamil Nadu. For the purpose, he used 70 years of forest cover images. ArcGIS and ERDAS Imagine software was used for the study. The study found that the forest cover area decreased 30%.

Sarma et al (2005) have worked on coal mining impact on land use/land cover in Jaintia hills district of Meghalaya, India using remote sensing and GIS technique. The studies were used LANDSAT data of 1975, 1987, 1999 and 2000. Visual interpretation technique was used for land use/land cover mapping for

ISBN: 978-93-8830-599-0

the different data of four years. The study concludes that there was fourfold increase in mining area from 1975 to 2007 accompanied by threefold decrease in forest area.

Yuan et al (2005) have argued that land cover classification and change analysis of twin cities area by multi-temporal landsat remote sensing. They have developed a methodology to map and monitor land cover change using multi temporal landsat thematic mapper (TM) data. The results quantify the land cover change patterns in the metropolitan area and demonstrate the potential of landsat data to provide an accurate economic means to map. The study analyzed changes in land cover over time that can be used as inputs to land management and policy decisions. Zubair et al (2006) have used land use / land cover map to detect changes in Ilorin capital of Kwara state, Nigeria. The study used Landsat imagery of 1972, 1986and 2001. The result of the study shows rapid growth in the built up land between 1972 and 1986 while the periods between 1986 and 2001 witnessed a reduction in the class. It was also observed that change by 2015 may likely follow the trend.A detailed study has been done on the work of Edward et al (2009)

Land use change due to mining employed over a twenty year period was analyzed within the Golden Star Resources Bogoso Prestea Limited (GSRBPL) concession. The study revealed that mining in the area increased by 12.1 % in land coverage from 1986 to 2006 with decrease in agricultural land use from 97.8% in 1986 to 82.7% in 2006. Settlements increased from 0.45 % in 1986 to 4.95 % in 2006 due to a rural – urban migration.

Anil et al (2010) carried out a study that discusses the land use land cover changes in valley of Ballarpur region, In this present study IRS-P5, LANDSAT -5 TM images are used. The results showed that almost all dense vegetation have been converted into mine land and overburden dump which causes pollution level to increase enormously in the surrounding area; recently it has attained the critical level.Manonmani et al(2010) have worked on the change detection analysis of Villivakkam block of Thiruvallur district west of Chennai. In this study Landsat 7 etm+, Liss III data are used. The study shows that built up area has increased by 15.83%and area with irrigated land farms have been decreased by 2.48% and scrub land decreased to 5.19%.

Sourav choudary et al (2011) carried out a study on change detection and land use analysis at Talcher region, Odisha. The land cover changes in Talcher area was analyzed for the three different years. Landsat mss and landsat TM images were used.

NDVI maps of the areas have been created by using ERDAS Imagine software.

The results show a tremendous increase in mining area is about 34km2 and total loss of forest area is 43km2 in 1973 and 2009.

2. STUDY AREA

Khammam district occupies an area of approximately 16,029 square kilometers (6,189 sq mi). Khammam District is a district in Telangana, India. It had a population of 2,797,370 of which 19.81% were urban as of 2001 census. Khammam district is located between 16 45` and 18 35` of Northern Latitude and 79 47` and 80 47` of the Eastern Longitude

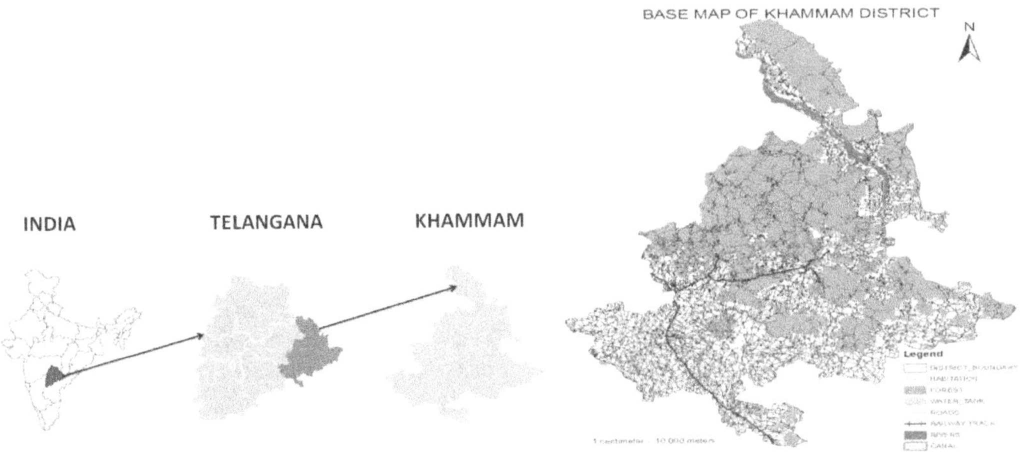

Fig. 1 Study area

ISBN: 978-93-8830-599-0

The base map is prepared from SOI toposheet. The base details like road network, railway track, settlements, drainage, boundary of the study the area forest boundary etc. using base map, various thematic maps are prepared.

Objectives & Methodology

1. To Generation of land use land cover information using remote sensing data belonging to different time periods 2006 to 2014

2. To determine the trend, rate location magnitude of land use land cover change
3. Integration of remote sensing derived LULC information with socio economic data to predict future scenarios in LULC patterns in the study area
4. To study and analyze the probable areas of LULC

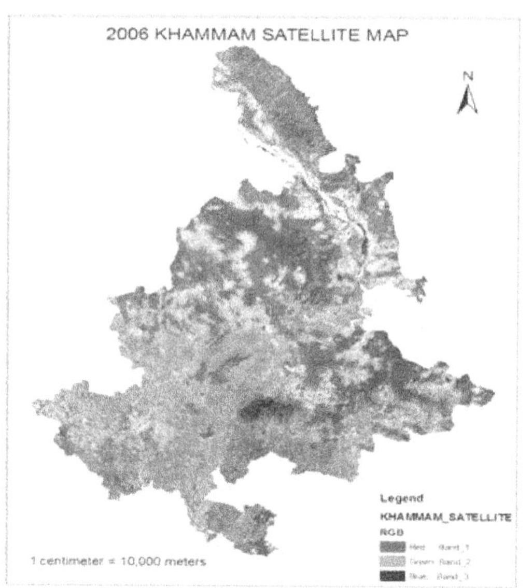

Fig. 2 2006 Khammam Satellite Map

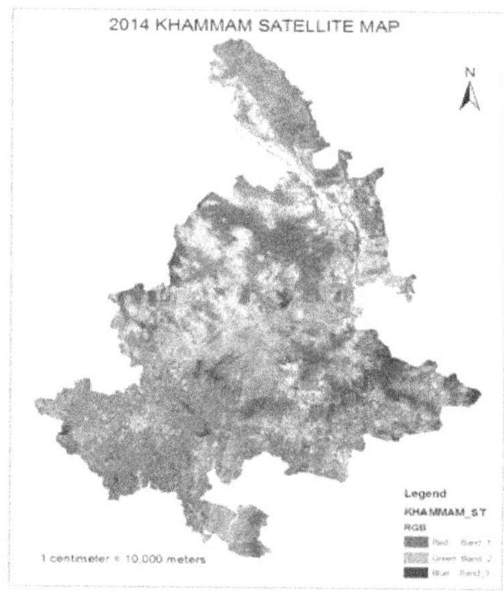

Fig. 3 Khammam Satellite Map

Methodology of the study involved

1. Creation of base layers like district boundary, road network, drainage network, contours, mapping of water bodies, etc. from the SOI toposheets of scale 1:250000 and 1:50000.
2. Extraction of bands (Landsat TM with 30 m resolution of 1986 and LISSIII with resolution 23.5 m of 2003) from the data procured from NRSA.
3. Identification of ground control points (GCP's) and geo-correction of bands through resembling.
4. Cropping and mosaiming of data corresponding to the study area.
5. Computation and analysis of various vegetation indices.
6. Generation of FCC (False Colour Composite) and identification of training sites on FCC.
7. Collection of attribute information from field corresponding to the chosen training sites using GPS.
8. Classification of remote sensing data (1986 and 2003): Land cover and land use analyses.

9. Change detection analysis using different techniques (Image differencing, Image rationing, etc.).
10. Detection, visualisation and assessment of change analysis.
11. Statistical analysis and report generation.

RESULTS AND DISCUSSIONS

Land use and Land Cover maps have to be prepared adopting visual or digital interpretation techniques in conjunction with collateral data such as topo maps and census data. If the digital method is followed then digital output is converted into line maps with corresponds to other theme maps. The total number of classes are would vary from area to area. The imagery has to be interpreted and ground checked for corrections.

Land use /land cover are classified into classes. They are

1. Built-up land
2. Agriculture land
3. Forest

ISBN: 978-93-8830-599-0

4. Water bodies
5. Waste land
6. Gullied / ravenous land
7. Land with or without scrub

8. Barren rocky/stony waste /sheetrock area
9. salt affected land
10. waterlogged land
11. marshy/ swampy land

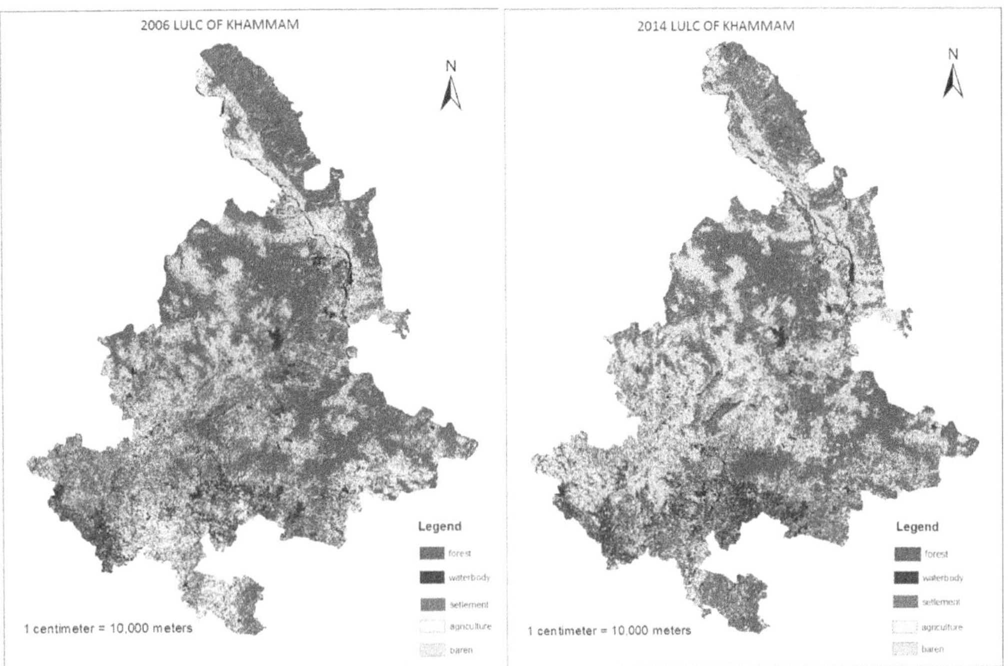

Fig. 4 Land use /land cover classification

Table 1 Areal Extent of Land Use/Land Cover 2006 and 2014

Description	Area (Km^2)2006	Area (Km^2)2014
Forest Plantation	302.94	205.94
Forest	5841.20	4533.75
Built Up(Urban)	19.45	28.75
Built Up(Rural)	241	296.55
Mining/Industrial	73.5	63.49
Agriculture Plantation	450.6	540.72
Aquaculture/Pisciculture	1.85	2.04
Agriculture Crop land	4715	5959.48
Transportation	5.45	7.81
Barren	10.21	12.88
Scrub Land-Dense	16.11	16.11
Scrub Land-Open	745.58	605.42
Gullied/Ravenous	2.87	2.85
Salt Affected Land	6.57	6.31
Sandy Area	1.52	0.59
River/Stream/Dam	415	335.18
Reservoir/Tanks	272.52	288.08
Lakes/Ponds	5.74	2.83
Canal	16.32	16.32

PieChart of LULC 2006&2014

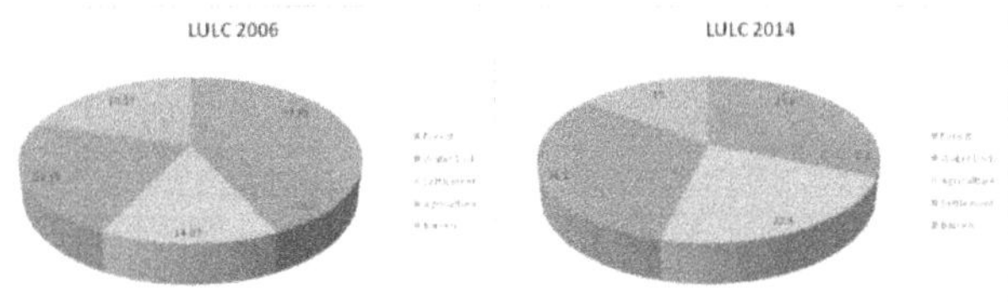

Fig. 5 PieChart of LULC

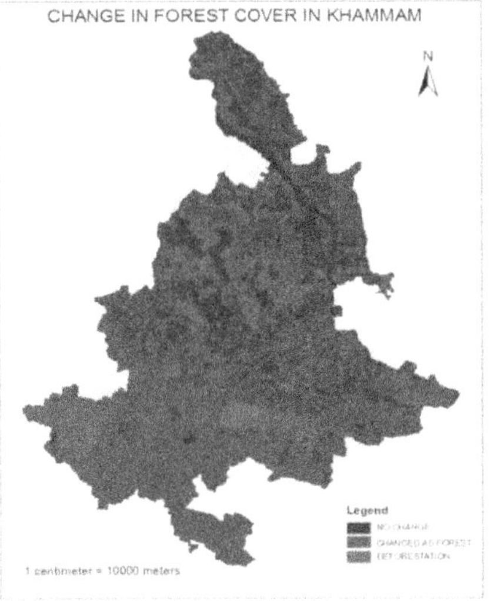

Fig. 6 Change detection of LULC

Normalised Difference Vegetation Index map:

NDVI = ((NIR − RED) /(NIR + RED))

Where

NIR is the near-infrared band

RED is the red band

The difference between the vegetation indices computed separately for each image gives a measure of the vegetation change in the area

NIR is the near-infrared band

RED is the red band

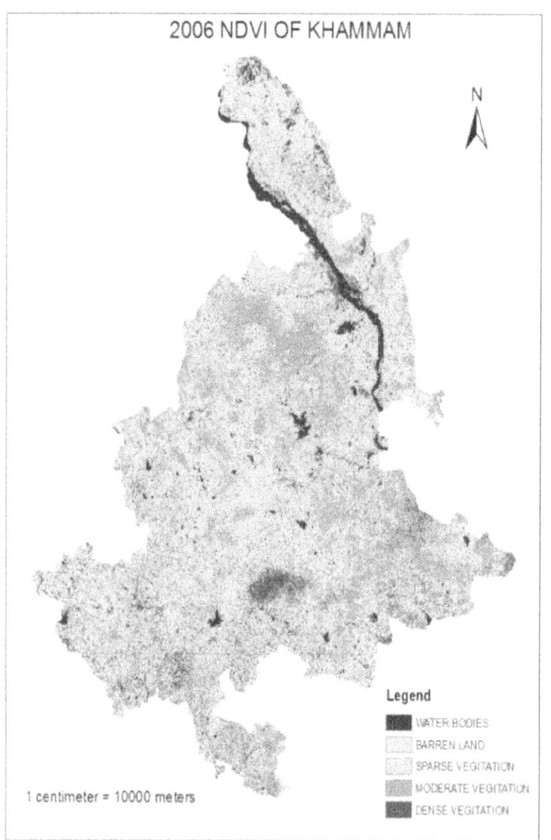

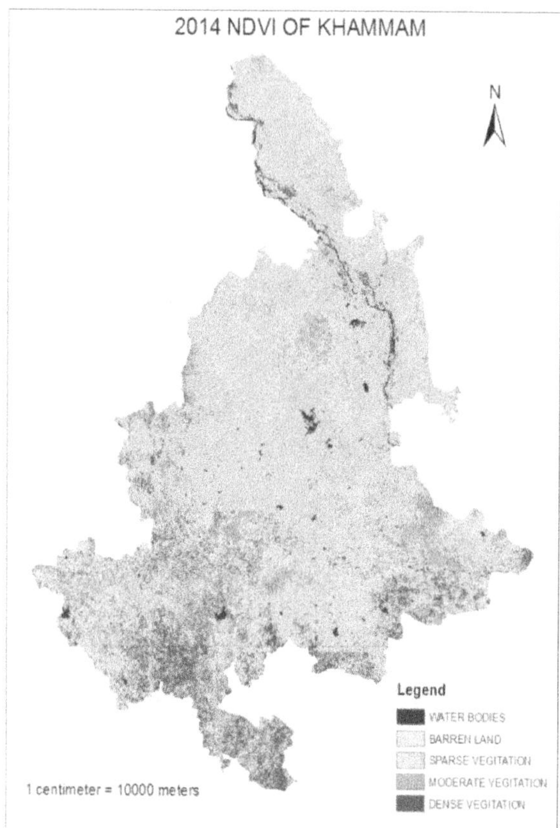

Fig. 7 Normalised Difference Vegetation Index map

Change Detection of NDVI 2006-2014

Table 2 Change Detection of NDVI 2006-2014

S.NO	Vegetation	2006 (Sq. Km)	2014 (Sq.km)	CHANGE
1	Sparse vegetation	3008	3560	-552
2	Moderate vegetation	5010	4785	225
3	Dense vegetation	8011	7684	327
		16029	16029	0

Conversion of water lands to agriculture lands

Out of the 100% of the wastelands, 50% of the wasteland cannot have the availability of water resources and ground water potential such as area under bhadrachalam, vajedu, yellandu, ect other 50% of the wasteland is covered by the scrub land , depending upon the availability of the natural resources , the lands can be brought under the cultivation lands Socio-economic parameters like agriculture labour and SC-ST population density are taken into account for the development of the agriculture most of agriculture labours are based on the rain fed irrigation so , most of lands are fallow if the rain fall is not sufficient . Government should take initiative in providing water focalises in those lands such that fallow can also be brought into the agricultural land. Drip irrigation mush better than normal irrigation and gives the double output then the normal production. Govt should come out with the development schemes such as providing loans developing the nurseries and ect. SC-ST population density is also and important parameter for developing the agricultural land.

The Sc St population and total population villages are identified, total irrigation, un-irrigation and

ISBN: 978-93-8830-599-0

cultivated wasteland etc are generated by agriculture is brought into development

By following the recommended suggestions, the wasteland can be reclaimed into useful lands (cultivable land). The model developed for stimulating the spatial dynamic process can be used as a planning tool to taste the affect of different land use change scenarios. Cellar automata are scene not only as a pradigisim for thinking about complex special temporal phenomena and experimental laboratory for testing ideas.

CONCLUSION

Though information on land use land cover is very important and is in fact central to many natural resource planning programs but due consideration should be given to the existing wastelands .proper management and development of these wastelands should be proposed ,so that it is reclaimed to meet the needs of the rapidly growing populations .The study thus shows that remote sensing is the best tool in today's rapidly changing environments and it can be very effectively dm resources mapping and management. The basic of this research compares the multi temporal classification of Landsat thematic map per satellite imaginary to detect and map the conversation of forest to pastures in Khammam district, Telangana 2006 to 2014.Thus the aim of this project was to produce both current and past land use land cover maps of Khammam district from recent and past land cover changes in the period of 8 years.

In the study area the analysis of LULC changes during 2006-2014 revealed the following.

- **Built Up**: Built up land in study area considerable increased by 27.13 km^2 from 2006 to 2014 (14.07 Km^2 to 31.2 Km^2)
- **Agriculture**: Agriculture land has considerable increased by 0.25 Km^2 from 2006 to 2014 (22.15 Km^2 to 22.40 Km^2)
- **Forest**: Forest area considerable decreased by 9.81 Km^2 because forest area not degraded most of the contently
- **Wasteland**: Wasteland decreased by 5.57 Km^2 in 2014 because increasing land cover changes due to rapid population growth and cultivation of agricultural land
- **Water bodies:** Water bodies has considerable decreased 2 Km2 in 2014

REFERENCES

1. Balak and Kolarkar, 1993. Ram Balak, A.S. Kolarkar Remote sensing application in monitoring land use changes in arid Rajasthan Int. J. Remote Sens., 14 (17), pp. 3191–3200.
2. Baulies, X and Szejwach, G. 1997. LUCC Data Requirements Workshop Survey of Needs, Gaps and Priorities on Data for Land-Use/Land-Cover Change Research Organized by IGBP/IHDP-LUCC and IGBP-DIS Barcelona, Spai.
3. E.F. Lambin, H.J. Geist, E. Lepers 2003. Dynamics of land use/cover change in tropical regions Annu. Rev. Environ. Resour., 28, pp. 205–241.FAO (1997a), FAO Yearbook Vol. 80 - Fishery statistics, Catches and landings 1995. Rome, FAO.
4. National Remote Sensing Center 2006, "Manual of National and Use/Land Cover Mapping using Multi Temporal satellite Data", Dept. of Space, Govt. of India, Hyderabad.
5. National Remote Sensing Center 2007. "Manual of Rajiv Gandhi National Drinking water Mission Project", Dept. of Space, Govt. of India, Hyderabad.
6. Raghavswamy, V. 1994. Remote sensing for urban planning and management. In: Space Technology and Geography, Eds. Gautam, N.C., Raghavswamy, V. Nagaraja, R.NRSA Publishers, and Hyderabad, India, pp.34-36.Robert Pontius, 2013. Land use and Land cover changes in climate changes, The encyclopedia of Earth., Article 154143.
7. Selcuk, R., Nisanci, R., Uzun, B., Yalcin, A., Inan, H.,Yomralioglu, T., 2003. Monitoring land-use changes by GIS and remote sensing techniques: case study of Trabzon,
8. Wilkie, D.S and Finn, J.T. 1996. Remote Sensing Imagery for Natural Resources Monitoring. A Guide for First-TimeUsers. Methods and Cases in Conservation ScienceSeries.xxi+295 pp. New York, Chichester: Columbia University Press. ISBN 0231079281.
9. Zubair A.O., 2006. Change detection in land use and Landcover using remote sensing data and GIS: a case study of Ilorin and its environs in Kwara State, Msc Thesis,
10. Sourav Choudhury (2012) change detection analysis of talcher coalfield using remote sensing and GIS Reoffered from
11. Zubair and Ayodeji opeyemi (2006) change detection in land use and land cover using remote sensing data and GIS
12. Biswajit Majumder (2011) land use and land cover change detection study at sukinda valley using remote sensing and GIS.

The Use of Generalized Additive Model (GAM) To Assess Fish Abundance and Spatial Occupancy in North-West Bay of Bengal

Bandanadam Swathi[1], Swarnalatha V[2] and Venkatesh Jogu[3]
Master of Technology in Geoinformatics and Surveying Technology,
Jawaharlal Nehru Technological University Hyderabad, Telangana, India

Abstract

The remote sensing data, such as sea surface temperature & chlorophyll concentration obtained from various satellites are utilized by Indian National Centre for Ocean Information Services (INCOIS) to provide Potential Fishing Zone (PFZ) advisories to the Indian fishing community which plays a vital role in national GDP. The data on Sea Surface Temperature (SST) is retrieved regularly from thermal-infrared channels of NOAA-AVHRR and chlorophyll concentration (CC) from optical bands of Oceansat-II and MODIS Aqua satellites for the identification of Potential Fishing Zones (PFZ) in Indian water. PFZ information has certain limitations, such as it can't predict the type of fish available in the notified fishing zone. In this dissertation, I have worked towards the development of short-term Hilsa shad predictive capabilities in a sustainable way. An effort has been taken to categorize all essential biological, environmental and climatic signals that have a direct or indirect impact on the Hilsa shad distribution. Remote sensing, ocean biogeochemical modelling, and statistical modelling approach have gained an increasing importance to study the marine ecosystems as-well-as for understanding the dynamics of the oceanic environment. Shad habitat has been studied from the geo-tagged fish catch data and oceanic/ecological indicators as predictor variables. For short-term prediction, the variables have been derived from a biophysical model, configured at INCOIS, using Regional Ocean Model System (ROMS) and remote sensing data. Using generalized additive model (GAM) Catch per Unit Effort (kg h−1) has been calculated as a response variable. Probability maps of predicted habitat with no fishing zone information have been generated using geographic information system.

Keywords: Remote Sensing, Potential Fishing Zones, Chlorophyll-A Concentration, Sea Surface Temperature.

I. INTRODUCTION

The abilities to develop satellite remote detecting innovation, joined with traditional information gathering strategies, give a capable instrument to the proficient and savvy administration of living assets. Varieties in maritime condition assume a key part in deciding the common dissemination of fish stocks. Laurs et al. (1984) utilized NOAA-AVHRR and Coastal Zone Color Scanner (CZCS) information to comprehend the circulation of Albacore tuna in the waters of California. They found that Albacore fish had a tendency to be moved in the region of temperature fronts on the hotter side the water. Later Arone (1987) utilized bunching method to group distinctive water masses and to describe the connection between the satellite-inferred Sea Surface Temperature (SST) and fish distribution. Solanki et al. (1988) utilized Indian Remote Sensing Satellite ModulerOpto-electronic Scanner (IRS-MOS) and NOAA-AVHRR information to portray the connection between the physical and biological factors. They presumed that the chlorophyll (Chl) and sea surface temperature has conversely corresponded. Synergistic investigation of satellite-determined CHL and SST highlights were completed by Solanki et al. (2001). They found that CHL and SST highlights match at a few areas demonstrating a nearby coupling amongst biological and physical parameters. Potential Fishing Zones (PFZ) are delineated from SST and CHL images and provided operationally as livelihood information to the marginal fishing community of India.

OBJECTIVES

As with the existing method of PFZ advisory generation can't provide information on the type of fish available in the Potential Fishing Zones, this study aims to illustrate how remotely sensed and modeled oceanic variables and fishing operations data can be used to predict suitable habitat of small pelagic fishery resources, such as Hilsa shad using GIS and Generalised Additive Model (GAM). Also, how the habitat suitability modelling approaches can be utilized in forecasting the fishing grounds of Hilsa shad.

II. LITERATURE REVIEW

PFZ: Solanki et al (2005) built up a satellite-based potential Fishing Zone (PFZs) advisory framework created utilizing incorporation of Ocean Color Monitor (OCM) inferred Chl and NOAA-AVHRR determined SST. Co-enrolling these two information sorts, an ace advisory was generated utilizing ocean features, such as eddies, fronts, and gyres situated on both the images. Validation of the PFZ estimates was done amid the years 1999-2002 utilizing the fish Catch Per Unit Effort (CPUE) data collected from Fishery Survey of India's vessels. Base trawl gets and its percent commitment by

ISBN: 978-93-8830-599-0

species was resolved from 30m to 100m profundity extend.

Choudhury et al (2007) have advanced an 'integrated Potential Fishing Zone (IPFZ)' map where the facts on capabilities which includes fronts, eddies, gyres, currents and upwelling derived from satellites were included from geo-coded co-registered CHL and SST pics. Fishery Survey of India (FSI) fishing vessels of Visakhapatnam base became used to validate the concerned IPFZs. The capture consistent with the unit attempt CPUE becomes computed taking total trap divided through actual fishing hours. The validation shows that the CPUE is high in the IPFZ regions compared with an environment. In the maximum of the cases, greater than an order growth in fish catch is observed within the IPFZ notified regions.

Zainuddin et al (2011) described the short-term relationship between skipjack tuna and oceanographic conditions and to visualize the predicted high catch areas, remotely sensed satellite based-oceanographic SST and CHL together (Aqua / MODIS) from fisheries data were used. Results indicated that the highest skipjack CPUEs were mainly found in the SST region of 29.0-31.5°C. The preferred ranges provide a good indicator for initially detecting potential skipjack fishing grounds.

Lan et al (2012) used cloud-free Advanced Microwave Scanning Radiometer for the Earth Observing System (AMSRE) SST and daily set longline fishery data, to study the relationship between albacore (ALB) fishing grounds and thermal conditions in the southern Indian Ocean. SST and Jensen–Shannon divergence (JSD) maps with a daily spatiotemporal resolution were related to sites with high catches per unit effort (CPUE) (>11 fish/103 hooks). He also concluded that a high JSD is considered to be an index of an SST front. In winter, high CPUE occurred in the vicinity of the North Subtropical Front, where SST was 15–19° C. Monthly maps of joint probability density (JPD) with the optimum ranges of SST and JSD revealed that high CPUEs are located in the narrow bands with high JPD (>50%).

GAM: Venable et al (2004) provided a summary of the modelling method of Generalized Linear Models (GLMs), Generalized Additive Models (GAMs) and Generalized Linear Mixed Models (GLMMs), particularly with relation to fisheries analysis. The author additionally delineated the essential facet of model interpretation and construction thus on succeed its correct application. The analysis starts with the best models and shows the progression from GLMs to either GAMs or GLMMs. though this is often not a comprehensive review, a stress was set on topics relevant to fisheries science like transformation choices, link functions, adding model flexibility through splines, and victimization random and glued effects square measure reviewed during this study. Finally, the authors have mentioned the assorted aspects of those models and their variants and provided a read on their relative edges to fisheries analysis.

Estrada et al (2009) have analyzed the performances of computational neural networks (CNN), multiple linear regressions (MLRs) and generalized additive models (GAMs) to predict Pacific sardine landings and to analyze their relationships with environmental factors in the north area off Chile. For this purpose, several local and global environment variables and indexes (sea surface temperature, sea level, and Ekman transport index in the Chilean coast and the southern oscillation index) were considered as inputs or independent variables. Additionally, several CNN's were calibrated and validated adding the anchovy landings in the same area as model inputs. In conclusion, the results of this paper support that the quantitative prediction of sardine landings can be carried out from environmental variables, although the predictive capacity of the CNN has significantly improved if the anchovy landings are added as the input variable.

Zainuddin et al (2008) derived SST, sea surface CHL concentration (SSC), and sea surface height anomaly (SSHA) and catch data based potential habitats for albacore in the western North Pacific Ocean using the relationship between albacore fishing ground and oceanographic conditions. Empirical cumulative distribution function and high catch data analyses were used to calculate preferred ranges of the three oceanographic conditions. Results indicate that highest CPUE corresponded with areas having SST ranges 18.5°–21.5° C, SSC ranges 0.2–0.4 mg/m3, and SSHA ranges 5.0 to 32.2 cm during the winter in the period 1998– 2000. The author used these ranges to generate a simple prediction map for detecting potential fishing grounds. Statistically, to predict spatial patterns of potential albacore habitats, a combined Generalized Additive Model (GAM) Generalized Linear Model (GLM) were used. To build the model, first, a GAM was used as an exploratory tool used to identify the functional relationships between the environmental variables and CPUE. Then made parameters were derived out of these relationships using the GLM to generate a robust prediction tool. The results of this model predict the spatial pattern of the potential albacore habitat in good agreement with the simple prediction map and observation data.

ISBN: 978-93-8830-599-0

Chang et al (2012) developed a habitat suitability index (HSI) \model to examine the relationships between their spatiotemporal distribution and environmental factors and to identify potential fishing grounds for the swordfish in the South Atlantic Ocean using the Taiwanese distant-water longline fishery data and remote-sensing oceanographic data for 1998–2007. All the environmental factors considered such as SST, mixed layer depth (MLD), sea surface height anomaly (SSHA), CHL and ocean bathymetry (BAH) – were highly significant with most of the CPUE variation explained by SST. The most optimum habitat (i.e. hotspot) was found in the areas with SST of 27°–28° C, SSHA of -0.05 to 0.05 m, CHLs of 0.1-0.2 mg/m3 and BAHs of 4000 to 4500 m. The arithmetic mean model with five environmental variables was found to be the most appropriate according to the information theory based on the evaluation of different empirical HSI models in combination with different environmental factors. The bimonthly Geographic Information System maps of the predicted HSI values were cross-validated by the observed CPUE, suggesting that the model can be used as a tool for reliable prediction of potential fishing grounds.

DISTRIBUTION & ECOLOGICAL PREFERENCES OF HILSA SHAD

Indian shad (Tenualosa ilisha) ordinarily known as Hilsa is a very prized fish in India. It is a critical transitory species in the Indo-Pacific districts. It has an extensive variety of circulation and happens in marine, estuarine and riverine conditions. The fish is found in the Persian Gulf, Red Sea, Arabian Sea, Bay of Bengal, Vietnam Sea and China Sea. The riverine living space covers the Satil Arab and the Tigris and Euphrates of Iran and Iraq, The Indus and Pakistan, the Irrawaddy of Myanmar, the streams of eastern and western India in particular the Ganga, Bhagirathi, Hooghly, Rup Narayan, Brahmaputra, Godavari, Narmada, Tapti and other beachfront streams and the Padma, Yamuna, Meghna, Karnafuly and other beachfront streams of Bangladesh. Truth be told, the Hilsa fishery in India and Bangladesh is subject to the single species viz. Tenualosa ilisha having a place with the living spaces of Indo-Gangetic and Brahmaputra stream bowls The Ganga and Brahmaputra River System is the biggest seepage where Hilsa is winning. It is broadly disseminated species under clupeid, occupying the seaside waters mostly from the Bay of Bengal and using the vast majority of the estuaries, waterways, and backwaters to breed. The present record depicts an outline of Hilsa fisheries in the Hooghly-Bhagirathi stream framework. Numerous specialists have taken a shot at Hilsa amid the most recent couple of decades, yet exhibit situation in regard of the hydrology, natural surroundings, environmental change and negative human mediations has changed to a substantial degree. Unpredictable catch of brood fish and adolescents and over-misuse have been distinguished as the most overwhelming attributes influencing the Hilsa fishery of stream framework. It is trusted that we need to attempt essential preservation measures for guaranteeing maintainable Hilsa fishery.

Hilsa sustained a lucrative fishery in the middle stretch of Ganga which declined after the commissioning of Farrakka barrage in 1972. Since 1972, the fishery has severely declined in the upstream areas and is now mainly in downstream rivers, estuaries, coastal areas and the sea. Being an anadromous species, during the southwest monsoon, Hilsa starts spawning migration to upstream and flooded regions of all rivers. Eggs are laid in freshwater and hatching takes place within ~24 hours at an average temperature of 23°C; in about 6-10 weeks the fry grow to about 12-20 cm, the stage is popularly known as 'Jatka' and juveniles make their way downstream to the sea after a freshwater (Figure 05) period of 5-6 months (Mome 2007). The sea phase is about 1 year; makes Hilsa mature and undertake spawning migration towards freshwater and repeats the cycle (Haroon 1998). But some controversies are there in typical anadromous nature of Hilsa shad.

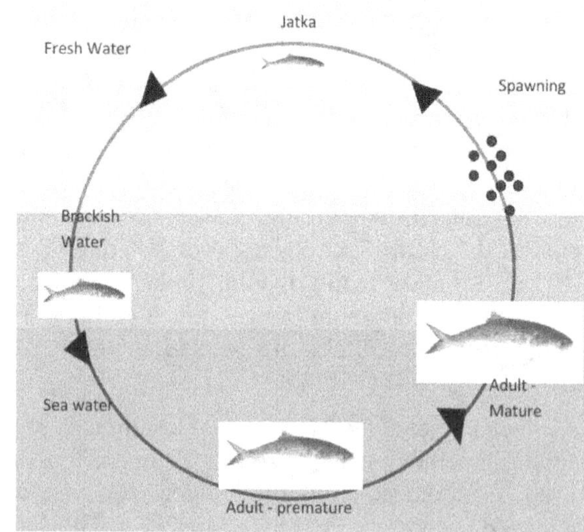

Fig. 1 Movement pattern of Tenualosa ilisha into different habitats (Mome 2007)

It is still not very clear, why Hilsa migrates towards river during breeding and which are the limiting factors

ISBN: 978-93-8830-599-0

influence the appearance and disappearance of Hilsa shoal along the coastal waters? Generally post-breeding in the estuary while returning to the open ocean spent fishes are open for harvest. Low water discharge from the river Ganges and associated heavy siltation, indiscriminate exploitation of juveniles, disruption of their migration routes, loss of spawning, feeding and nursery grounds and increased fishing pressure have all contributed to a decline in the catch per unit effort (CPUE) in both the marine and river Hilsa fishery.

III. METHODS AND MATERIAL
FISH CATCH INFORMATION

Fishery information utilized as a part of this investigation was gathered from fishing logbooks of different fishing boats operating in the study area. The logbook is intended to gather institutionalized data for the fish data set, the species, a number of snares and the quantity of individual fish recovered in each Fishing operation. The information from 2009–2016 are put away in a MySQL database where an inquiry is sent to pick the required datasets for examinations. Catch Per Unit Effort (CPUE) is measured dividing the total Hilsa fish catch per haul by the fishing hours of each haul.

REGIONAL OCEAN MODELING SYSTEM (ROMS) DATA

Major oceanic/ecological indicators, such as sea surface temperature, salinity, and chlorophyll are derived from a biophysical model configured at INCOIS using Regional Ocean Model System (ROMS). The coupled biophysical model has been validated against a wide range of observational data and has demonstrated considerable capability in reproducing marine ecosystem dynamics at synoptic and seasonal time-scales in the Indian Ocean.It is a three- dimensional, free-surface, terrain-following numerical model. It simplifies equation of motion. The motivation to build a free-surface oceanic model is two-fold.

The ROMS model was set-up for the entire Indian Ocean basin from 30 °E to 120°E in the east-west direction and 30 °S to 30 °N in the north-south direction. The model has a horizontal resolution of 1/12 (9.5 km) and has 40 sigma levels in the vertical plane. The vertical profile has a resolution of 0.7 m near shore. Off the shelf to a bottom depth of 5000 m, the vertical resolution ranges from 15 m at the surface to 500 m at

the bottom. The bathymetry is supplied by Sindhu et al. (2007).

As the assimilation of the data available on real-time is a key component of operational ocean forecast systems, INCOIS is in the process of developing an

ensemble Kalman Filter (EnKF) based data assimilation scheme in the ROMS setup. It is proposed to assimilate temperature and salinity profiles from the moored buoys, XBTs and the ARGO floats. By using the above ARGO floats and moored buoys, these salinity and SST data are modelled in ROMS covering most of the Indian seas and depths up to hundreds of meters.

The ROMS model was spun up for a period of 10 years from the initial state. Using Estimating the Circulation and Climate of the Ocean (ECCO) data, monthly climatological values for January are used to derive the tracer fields (temperature and salinity), Chakraborty et al. 2016

SEA SURFACE TEMPERATURE

This data output is in the form of NetCDF form for array analysis of locations and their corresponding SST values in various mathematical software as well as Displaying in Raster Images in GIS-based software's available.

The SST data is provided and processed in INCOIS in NetCDF format which has been dated from "01-01-2009 to 31-12-2015" in the region of Bay of Bengal is been collected, these figures are shown below

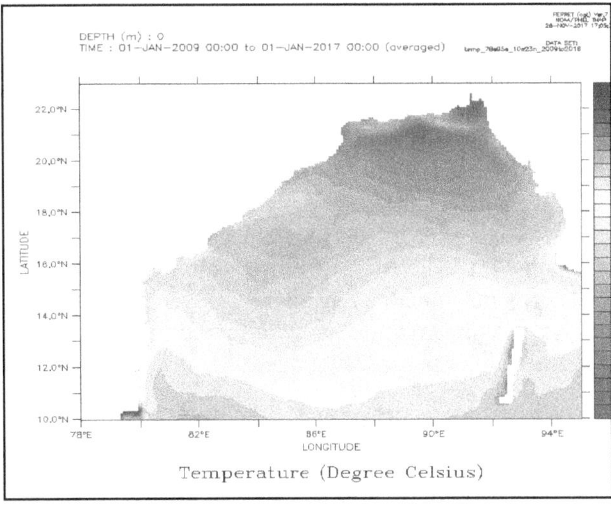

Fig. 2 SST

ISBN: 978-93-8830-599-0

SEA SURFACE SALINITY:

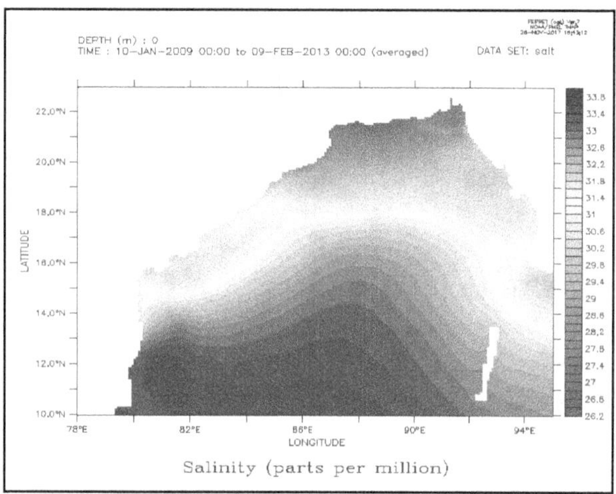

Fig. 3 SALINITY

The ocean Surface Salinity data is provided and processed in INCOIS in NetCDF format that has been dated from "01-01-2009 to 31-12-2015" within the region of Bay of the geographical area is been collected, these figures area unit shown above.

GENERALIZED ADDITIVE MODEL

Using generalized additive model (GAM) catch per unit effort (kg h−1) is calculated as a response variable for the pre-fixed geo-locations in the study area using following equation.

G(CPUE)=a+s(SST)+s(Chl.)+s(Salinity)+s(Rainfall)

Where g is the link function, a is a constant, s (.) is a spline smoothing function of the predictors

ROMS simulated sea surface temperature (SST), chlorophyll (Chl.), salinity and remotely sensed rainfall over the sea were used as predictor variables. Best-fit model was selected based on the significance of model terms, reduction in Akaike's Information Criterion and increase in cumulative deviance explained. The selected model has been used to predict catch per unit effort (CPUE) of Hilsa shad using predictor variables and the model was validated using a linear model. Based on geo-tagged predicted CPUE, probability maps (beta version) of suitable habitat is generated using geographic information system for further validation and fine-tuning purpose.

IV. RESULTS AND DISCUSSION

This shows the development of monthly Hilsa shad predictive capabilities in a sustainable way. An effort has been taken to categorize essential physical, biological, and environmental signals that have a direct or indirect impact on the Hilsa shad distribution. Shad habitat has been studied from the geo-tagged fish catch data and oceanic/ecological indicators as predictor variables. For prediction, the variables are derived from a biophysical model, configured at INCOIS, using Regional Ocean Model System (ROMS) and remote sensing data. The investigation indicates the utilization of models for deciding and appreciate the spacious living space conveyance and accessibility of Hilsa in NWBOB. It is the primary Indian exertion in the Bay of Bengal wherein fishers get information is connected with modelled and satellite-inferred factors to build up an expectation framework utilizing an added substance demonstrate. Formerly, Mugo et al 2010 have attributed the habitat of skipjack tuna in the western North Pacific Sea and Rajapaksha et al. 2010 concentrated to find out the relation between oceanographic variables (from satellite) and yellowfin tuna habitat in the northeast Indian ocean using fish catch data from Sri Lankan long-line fleets using GAM.

The Fishing operations information alongside maritime covariates was isolated into a database which clarifies the maritime ecological conditions symmetric to Fishing operations in time and space. Information from this Master Sheet was utilized to create GAM models that give more data on the natural surroundings reasonableness of the Hilsa shad species.

4.1 GAM PLOTS

Below figures are generated using GAM and the plots indicate distribution range with 95% confidence level. The narrow confidence level indicates high relevance among each other and wider confidence level indicates less relevance.

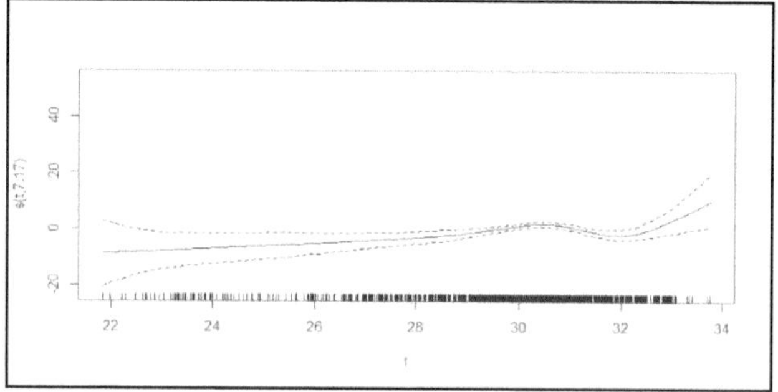

Fig. 4 Plot showing CPUE variation with Temperature in "best-fit model"

Based on the GAM plots & AIC values, the favorable oceanographic parameters for Hilsa shad aggregation off North West Bay of Bengal were characterized by; SST of 29–32 °C in the above **Plot** Solid dashed line in with lines converging gives 'best

ISBN: 978-93-8830-599-0

fit' region plots indicates the 95% confidence intervals and most influential region on Response Predicted CPUE.

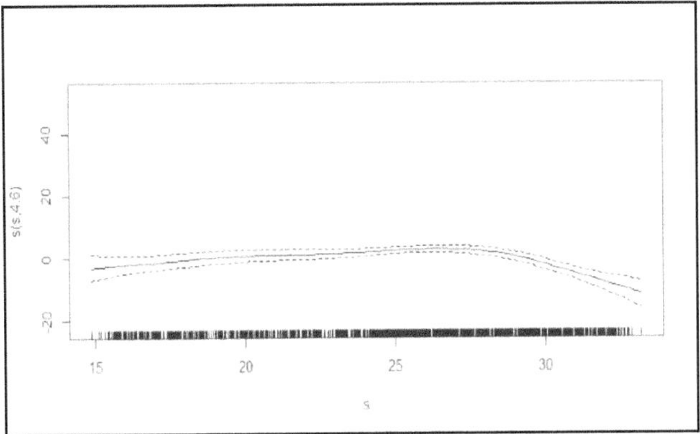

Fig. 5 Plot Showing CPUE variation with Sea Surface Salinity in Best-Fit Model

Based on the GAM plots & AIC values, the favorable oceanographic parameters for Hilsa shad aggregation off North West Bay of Bengal were characterized by;

Sea Surface Salinity is 25–30 in the above **Plot** Solid dashed line in with lines converging gives 'best fit' region plots indicates the 95% confidence intervals and most influential region on Response Predicted CPUE along with other three Predictor terms salinity has third best influence on Response CPUE.

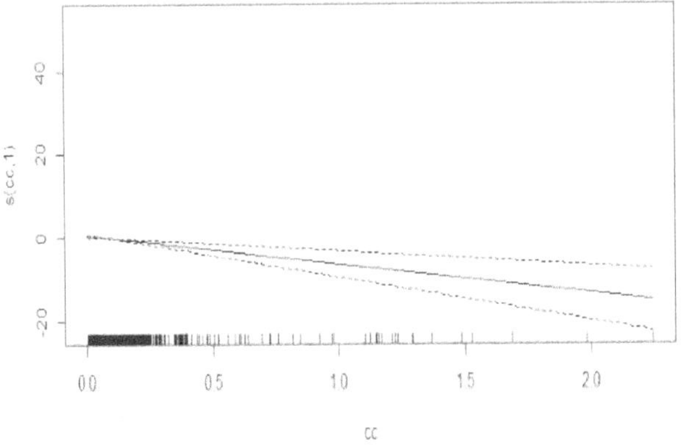

Fig. 6 Plot showing CPUE variation with chlorophyll in "best-fit model"

The favorable oceanographic parameters for Hilsa shad aggregation off North West Bay of Bengal were characterized by; Chl-a of range 0.05-0.5 mg/m³ in the above **Plot is** greatly affecting the GAM-derived Response of Predicted Solid dashed line in with lines converging gives 'best fit' region plots indicates the 95% confidence intervals and most influential region on Response Predicted CPUE

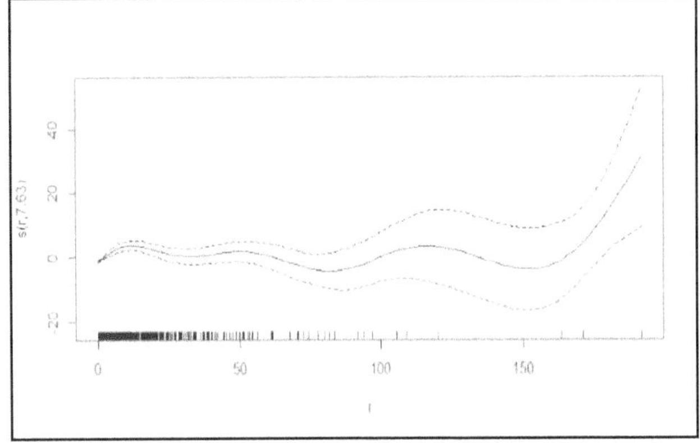

Fig. 7 Plot showing cpue variation with TRMM Rainfall in "best-fit model"

From Table 2 as mentioned below containing GAM-Derived AIC values & % Deviance Explained the parameter which is Satellite-derived TRMM-Rainfall(mm/hr) is having highest weightage and great infulence on Hilsa Shad migration and availabilty in Monsoon Season and time period having good rainfall, from above **Plot** we can clearly see the influence of TRMM on Response Predicted CPUE as in the range of 0-50 mm/hr rainfall has best influence of 95% confidence intervals as lines are converging and 50-100 mm/hr has medium and above 100mm /hr has minimal influence, hence the most important predictor in this modelling is TRMM- Rainfall

Table 1 GAM-derived deviances and AIC values

Hilsa	% deviance_explained	AIC
ONE VARIABLE		
Rainfall	4.85	14729.60
SST	4.86	15550.00
Salinity	1.69	15610.20
Chlorophyll	1.69	15617.88
TWO VARIABLES		
SST+Rainfall	7.86	14527.40
Salinity+Rainfall	6.00	14556.86
SST+Salinity	6.37	15530.00
SST+Chlorophyll	6.17	15536.17
Salinity+Chlorophyll	3.18	15593.10

Contd….

ISBN: 978-93-8830-599-0

THREE VARIABLES		
SST+Salinity+Rainfall	9.37	14505.00
Rainfall+Chlorophyll+ SST	8.95	14510.00
SST+Salinity+Chlorop hyll	7.63	15512.00

ALL VARIABLES		
SST+Salinity+Chlorophyll+ Rainfall	10.4	14486.44

Above GAM derived variables along with those AIC values of all possible combinations in increasing order of parameters with Best Fitted models have been generated on the yearly scale and placed in the above Table. The model explained the influence of each parameter on CPUE through the AIC values. The individual parameter with least AIC value will be in the top priority which has a huge impact on the CPUE. In case of this study, rainfall is on the top with AIC value of 14729.60. The second one is SST that had AIC value of 15550. The third one will be salinity with AIC value of 15610.20. And, the least prior parameter is chlorophyll with AIC value of 15617.88. GAM model result for the Hilsa species is depicted in the last part of the table considering all parameters, SST, salinity, chlorophyll, and rainfall. The model explained 10.4% deviance considering all variables. The selected model had the lowest AIC value of 14486.44. Using the same model monthly CPUE for September to December 2016 has been calculated in different predetermined points falling in the study area. The excel data, containing CPUE of predetermined points are incorporated in the GIS platform using ArcGIS software and exported into point layer form. Using point to raster conversion tool the point data converted into a raster layer. The catch information brings to a range of 0 to 1 by taking log10(CPUE) and stored as an additional column in the attribute table. Raster layer is generated on the basis of logarithmic information of catch data. The best way to represent data with High, Medium and Low catch data information using GIS Software is thematic map representation. The value between 1 to 0.6 is considered as High Catch and the value between 0.6 to 0.4 is considered as medium catch and the value below 0.4 is considered as a low catch.

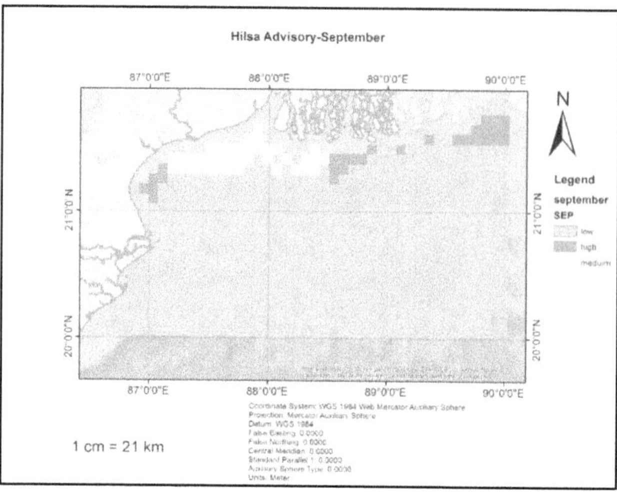

Fig. 8 Habitat Suitability map (September)

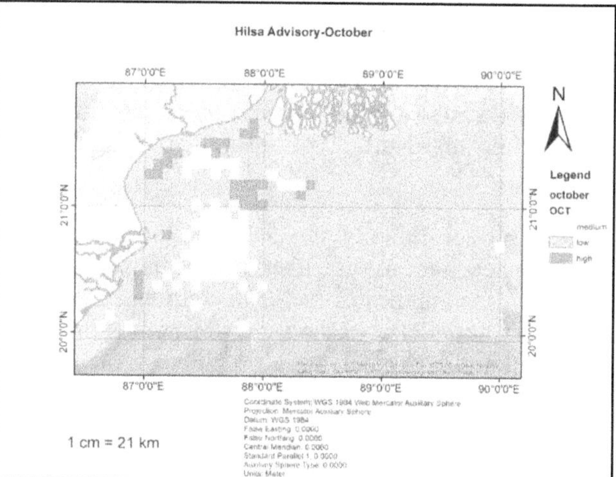

Fig. 9 : Habitat Suitability map (October)

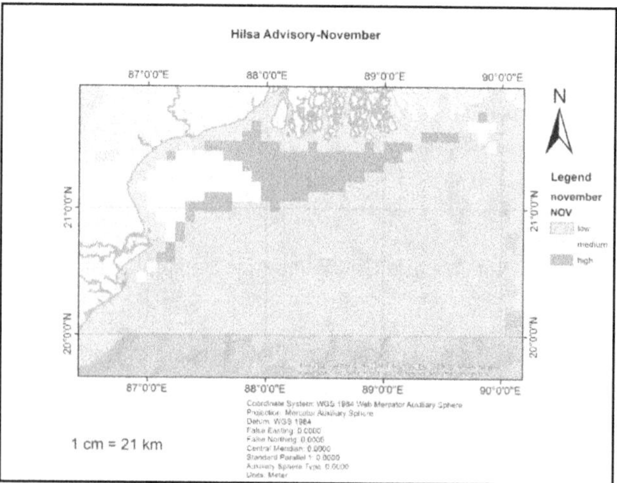

Fig. 10 : Habitat Suitability map (November)

ISBN: 978-93-8830-599-0

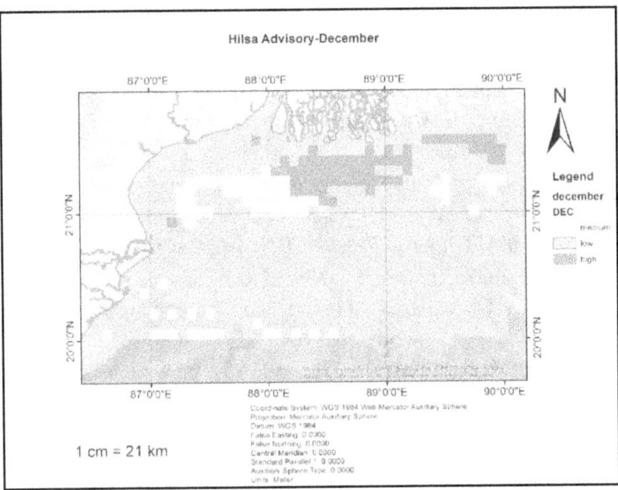

Fig. 11 Habitat Suitability map (December)

V. CONCLUSION

This study characterized the suitable forecast regions for HILSA distribution in along the Coast of NORTHWEST BAY OF BENGAL using fishing operations data and satellite-derived biophysical variables. Habitat and food indicators were linked with satellite-derived biochemical and physical parameters. GAM model was applied to predict HILSA distribution in the North West Bay Of Bengal. Model outputs are in good coherence with the species response to the environmental condition. A Good Percentage deviance was explained by the BEST FIT model. Based on AIC and deviance metrics GAM was trained on several combinations of input variables to generate the best fit model. But it needs a minimum number of adequate points to validate the model and to forecast HILSA CPUE. Classification results from the GAM model used to forecast the levels of CPUE as High, Medium & Low. Model results were validated through in-situ data.

The model generated from GAM analysis indicated a good match with higher CPUE points in the potential predicted areas in the validation exercise. Factors affecting marine fishery species can be multiple and hence, take into account as many as possible can provide a better understanding of their interactions with the environment. Although the present study provides some good preliminary results, implementation of a near real-time dynamic model with the input of available oceanographic variables may further improve the accuracy of predicting fish abundance areas.

REFERENCES

1. Anon, 2012, Report of Department of Fisheries, Government of West Bengal.

2. R.A. Arnone, Satellite-derived colour– temperature relationship within the Alboran ocean, Remote Sensing of setting twenty-three (1987), pp. 417–437

3. Chang Y, Sun C, Chen Y, Yeh S, Dinardo G. 2012. Habitat suitability analysis and identification of potential fishing grounds for swordfish, Xiphias gladius, in the South Atlantic Ocean. Int J Remote Sens. 33:7523– 7541.

4. Yen K, Lu H, Chang Y, Lee M. 2012. Using remote-sensing data to detect habitat suitability for yellowfin tuna in the western and Central Pacific Ocean.Int J Remote Sens. 33:7507–7522.

5. S. B. Choudhury, B. Jena, M. V. Rao, K. H. Rao, V. S. Somvanshi, D. K. Gulati & S. K. Sahu International Journal of Remote Sensing Vol. 28, Iss. 12,2007 Validation of integrated Potential fishing Zone (IPFZ) forecast using satellite-based chlorophyll and sea surface temperature along the east coast of India.

6. Pacific sardine (Sardinops sagax, Jenyns 1842) landings prediction. A neural network ecosystemic approach Juan Carlos Gutiérrez- Estradaa EleuterioYáñezbInmaculadaPulido-CalvoaClaudioSilvabFranciscoPlazabCinthyaBó rquezb 2009

7. Hosoda, Kohtaro & Kawamura, Hiroshi & Lan, Kuo-Wei & Shimada, Teruhisa & Sakaida, Futoki. (2012). Temporal Scale of Sea Surface Temperature Fronts Revealed by Microwave Observations. IEEE Geosci. Remote Sensing Lett.. 9. 3-7. 10.1109/LGRS.2011.2158512.

8. Hastie T, Tibshirani R. 1986. Generalized additive models. Stat Sci. 1:297–310

9. Haroon, Y, 1998. Hilsa shad: Fish for the teeming millions, new management alternatives needed for the hilsa young. Shad Journal, 3:7

10. Laurs R.M., Fielder, P. C. and Montgomery, D.R. (1984). Albacore tuna catch distribution relative to environmental features observed from satellite. Deep Sea Research, 31:1085-1099

11. Mugo, R., S. Saitoh, A. Nihira, and T. Kuroyama. 2010. "Habitat Characteristics of Skipjack Tuna (Katsuwonus Pelamis) in the Western North Pacific: A Remote Sensing Perspective." Fisheries Oceanography 19 (5): 382–396.doi:10.1111/j.1365 2419.2010.00552.x.[Crossref], [Web of Science®], [Google Scholar]

12. Mome M.A., 2007. The potential of the artisanal hilsa fishery in Bangladesh: an economically efficient fisheries Policy. The Fisheries Training Programme, The United Nations University, Iceland.

13. Rajapaksha JK, Nishida T, Samarakoon L. 2010. Environmental preferences of yellowfin tuna (Thunnus albacores) in the Northeast Indian Ocean: an application of remote sensing data to longline catches. Report of the Sixth Session of the IOTC. Victoria: Working Party on Ecosystems and bycatch. (Working

ISBN: 978-93-8830-599-0

Party on Ecosystems and Bycatch; IOTC-2010-WPTT- 43, 27–30 October 2010).

14. R Core Team. 2014. R: A language and environment for statistical computing. R foundation for statistical computing, Vienna, Austria; [2014 Dec 11]. Available from: http://www.R-project.org/

15. Solanki HU, Mankodi PC, Nayak SR, Somvanshi VS. 2005. Evaluation of remote- sensing-based potential fishing zones (PFZs) forecast methodology. Cont Shelf Res. 25:2163– 2173

16. Sanchez P, Demestre M, Recasens L, Maynou F, Martin P. 2008. Combining GIS and GAMs to identify potential habitats of squid Loligo vulgaris in the Northwestern Mediterranean.Hydrobiologia. 612:91–98.

17. Applications of a generalized additive model (GAM) to satellite-derived variables and fishery data for prediction of   fishery resources distributions in the Arabian Sea. H. U. Solanki, Dhyey Bhatpuria & PrakashChauhanGeocarto International Published Online: 13 Jan 2016

18. Predicting potential fishing zones of Japanese common squid (Todarodespacificus) using remotely sensed images in coastal waters of south-western Hokkaido, Japan Xun Zhang, Sei-Ichi Saitoh& Toru Hirawake International Journal of Remote SensingXun Zhang, Sei-Ichi Saitoh& Toru Hirawake Pages 6129-6146 | Published online: 11 Dec 2016

19. Venables, W.N., Dichmont, C.M., 2004. A generalized linear model for catch allocation: an example of Australia's Northern Prawn Fishery. Fish. Res. 70, 405–422.

20. Wood S. 2006. Generalized additive models: an introduction to R. Chapman and Hall/CRC Boca Raton (FL): CRC Press. ISBN: 978-1- 58488-474-3.

21. Yamanaka, I., Ito, S., Niwa, K., Tanabe, R., Yabuta, Y., &Chikuni, S. (1988).The fisheries forecasting system in Japan for coastal pelagic fish. FAO Fisheries Technical Paper, 301 (72 pp.) stem in Japan for coastal pelagic fish. FAO Fisheries Technical Paper, 301 (72 pp.)

22. Zainuddin M, Saitoh K, Saitoh S. 2008. Albacore (Thunnusalalunga) fishing ground in relation to oceanographic conditions in the western North Pacific Ocean using remotely sensed satellite data. Fish Oceanography. 17:61–73.

23. Zainuddin, Mukti. (2011). Skipjack tuna in relation to sea surface temperature and chlorophyll-a concentration of bone bay using remotely sensed satellite data.

24. Jurnal Ilmu dan Teknologi Kelautan Tropics. 3.. 10.28930/jitkt.v3i1.7837.

ISBN: 978-93-8830-599-0

Estimation and Design Concepts of MEP in Ecofriendly Commercial Structures – A case study

A. Vamshi Krishna Reddy[1], MA. Rouf Khan[2*] and A. Umashankar Kumar [3]

[1]Assistant Professor(c), Institute of Science & Technology, JNT University Hyderabad, India.
[2*]M.Tech Student, Centre for Environment, IST, JNT University Hyderabad, India.
[3]Assistant Professor, Swamy Vivekananda Institute of Technology, Hyderabad, India
vamshiavkreddy@gmail.com, maraufkhan@gmail.com

Abstract

The construction industry is one of the biggest producers of demolition waste and greenhouse gases. As we are aware of various environmental issues like global warming, water pollution, air pollution etc. To minimize all these effects on environment we have to take up the eco-friendly structure. The construction industry is one of the biggest producer of emissions during the construction stage and later when it is occupied. The major consumption of energy in buildings are HVAC, Water Heaters and Lighting. In order to reduce the energy consumption of the above amenities, we have to plan and design the MEP systems which are more energy saving and eco-friendly. As we are aware that fresh water is less than 2% of total water available on earth and in that 87 % of water is ice. As we don't have enough fresh water we can have the water treatment plants to recycle the water and reuse it.

Ecofriendly commercial structures need to be designed, constructed, and operated to have a minimum impact on the environment, both indoor and outdoor. The Estimation and Design concepts which we are going to implement should be a zero pollution to environment. In the present study we are not only going for green building design concepts but also making a building structure as environmental friendly. The methods will be planned as 1) maximize the recycling and reuse of water, including rainwater harvesting, storm water and gray water, 2) Use of renewable energy resources like sun. We can use solar energy to convert into electricity with the help of solar panels 3) Ground water recharge pits 4) Solid waste management practices etc. The Construction of eco friendly buildings are not about a increasing a little efficiency, it is about constructing buildings that optimize the uses of local materials and ecology. Most importantly they are build to reduce the energy consumption, water resources and natural resources.

Keywords: MEP, Eco-friendly, Renewable sources, Energy, HVAC

OBJECTIVES

- To study the implementation of Optimum usage of building materials, Air ,water and energy efficiency, Conserving the natural environment &Waste management
- To promote and advise improvement in energy efficient sustainable Buildings which leads to the reduction in Carbon foot print and fuel dependency.

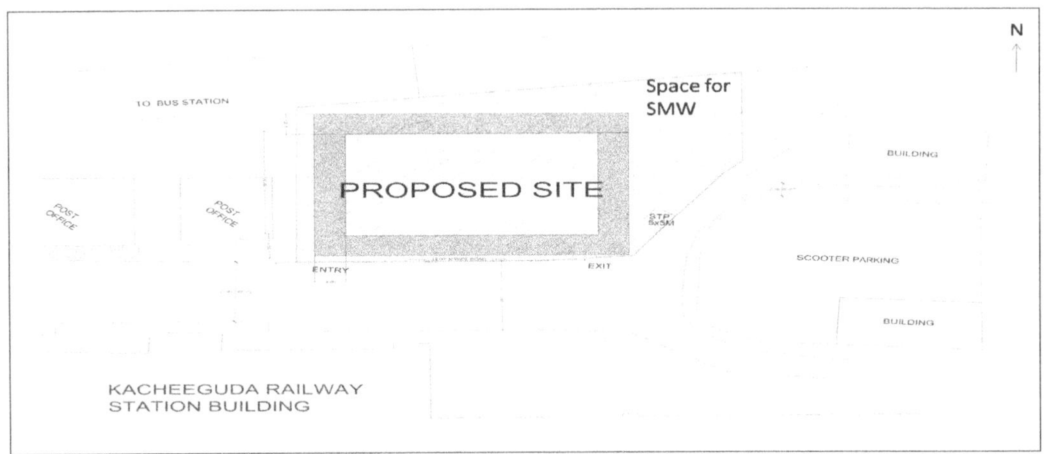

Fig. 1 Location map of the Study arae

ISBN: 978-93-8830-599-0

- Explore the Opportunities for initial savings, cost reduction during running operation and to improve energy efficiency of buildings by identifying major Energy and Environment issues related significant with MEP Systems.

- Assessment of suitable sustainable methods, that help to avoid the harmful gases, by maintaining a healthy climate and surroundings. (STP,SWM).

- To promote green technologies and designs, to secure the future generations and fuels exhaustion, increasing energy conservation and reducing carbon emissions.

Study Area Description:

- The Multi-functional commercial building is located at kachiguda near railway station in Hyderbad, Telangana state.

- The building has 2 basements + Ground floor + 5 floors

- The total land area is about 4,800 sq.m

- The build up area is 3,200 sq.m

- Total area of the floors combined are 20,600 sq.m

- Occupancy capacity is around 1500 people

METHODOLOGY

MEP (Mechanical, Electrical, Plumbing) services of building structures such as apartment, malls, hotels, industries, skyscrapers, hospitals, plays an important role in the contribution of pollution to environment if not properly designed.

In this project we are going to stress more on the MEP systems of the building which we are going to propose for this building.

The MEP systems which we are proposing will be Energy Efficient and cost saving.

The following systems are proposed for this building 1) VRF system for HVAC; 2) STP ; 3) Solar Panels

1. VRF systems for HVAC

Working of VRF system: In a VRF system, multiple indoor fan coil units may be connected to one outdoor unit. The outdoor unit has one or more compressors that are inverter driven, so their speed can be varied by changing the frequency of the power supply to the compressor. As the compressor speed changes, so does the amount of refrigerant delivered by the compressor. Each indoor fan coil unit has its own metering device that is controlled by the indoor unit itself, or by the outdoor unit. As each indoor unit sends a demand to the outdoor unit, the outdoor unit delivers the amount of refrigerant needed to meet the individual requirements of each indoor unit (Fig. 2). These features make the VRF system ideally suited for all applications that have part load requirements based on usage or building orientation, as well as applications that require zoning. [Carrier Corporation Syracuse, New York January 2013]

The major power consumption of the building is by HVAC systems: In order to reduce the power consumption and reduce the impact on environment we are proposing here VRF system instead conventional split ac system.

Here are the few advantages of VRF system explained

1. Energy Efficiency: The VRF system provides exactly the amount of cooling needed in current space or conditions, which means it uses less amount of energy to generate cooling for that space. Example: 1) Normal Split ac system- For 5 rooms 10 TR is required at peak time, we keep 2 TR ac in each room. The ac no matter what cooling is required they run at their full capacity and runs on full energy.

2. VRF System: For same 5 rooms 10 TR is requires. If we go for the VRF System the VRF system automatically detects the amount of cooling required and runs accordingly to it saving the energy consumption by the condenser.

3. Quiet Operation: In VRF system, the noisier condensing unit is kept on the roof and where as in split ac the outdoor units are just behind the indoor unit.

4. HEAT and COOL simultaneously: In VRF system we can Heat and cool the rooms simultaneously using the same outdoor unit. Example: The VRF system captures residual amount of heat from the air during the cooling process and redirects that same heat to other parts where heating is required in that same building.

5. Consistent Comfort. The VRF system's compressor can detect the precise requirements of each zone, and send the precise amount of refrigerant needed to do the job. As a result, each area of our space is consistently comfortable with well-controlled humidity and no hot or cold spots.

6. Less Downtime. Since the VRF-HVAC system is designed to run only when needed and under partial-load conditions, there is less wear and tear on the parts. That means fewer breakdowns. Also, if something goes wrong with one air

handler, often the others are unaffected. That means your whole space won't be without air conditioning all at once.

7. Requires Less Space. Since the air handlers are smaller and VRF-HVAC systems don't usually require ducts, they don't require as much wall and ceiling space for the equipment. That means you get to keep those gorgeous high ceilings in your apartment.

8. Modern Controls. For residences, you can take advantage of mobile control technology that lets you adjust temperature settings for each zone from your mobile device. For commercial settings, the VRF system's built-in controls may allow you to skip purchasing expensive building management software.

1) STP (Sewage treatment plants):

Sources of waste water: Wastewater can be defined as the flow of used water discharged from homes, businesses, industries, commercial activities and institutions which is directed to treatment plants by a carefully designed and engineered network of pipes. This wastewater is further categorized and defined according to its sources of origin. The term "domestic wastewater" refers to flows discharged principally from residential sources generated by such activities as food preparation, laundry, cleaning and personal hygiene.

Typically 200 to 500 liters of wastewater are generated for every person connected to the system each day. The amount of flow handled by a treatment plant varies with the time of day and with the season of the year.

Need of wastewater treatment?

We need to remove the wastewater pollutants to protect the environment and protect public health. When water is used by our society, the water becomes contaminated with pollutants. If left untreated, these pollutants would negatively affect our water environment. For example, organic matter can cause oxygen depletion in lakes, rivers, and streams. This biological decomposition of organics could result in fish kills and/or foul odors. Waterborne diseases are also eliminated through proper wastewater treatment. Additionally, there are many pollutants that could exhibit toxic effects on aquatic life and the public [Handbook on Wastewater Management]. The process of sewage treatment shown in Fig 3.

2) SOLAR PANELS

Introduction to solar panels: Now days we are seeing a high amount of increase in consumption electric power which generated by using non-renewable resources such as fossil fuels which are also very harmful for the environment. In order to avoid this pollution and minimize the usage of non- renewable resources solar panels are used. The solar panels convert the solar energy (sun light) into electric energy using the PV cells. Solar energy is cheaper when and eco friendly when compared to conventional electricity.

Working principle of solar Panel: PV cells Convert Sunlight to Direct Current (DC) electricity. Charge Controller work as control the power from solar panel which reverses back to solar panel get cause of panel damage. Battery System act as storage of electric power is used when sunlight not available (i.e. night).From this system connected to inverter for convert Direct Current (DC) into Alternating Current (AC).

From the figure we can observe that solar enrgy is observed by the panels which turn solar energy to electric current and pass on to charge controller then the enrgy is stored in batteries and then the batteries are connected to inverter which is then directly connected to AC power. The process of Solar panel working was shown in Fig 4.

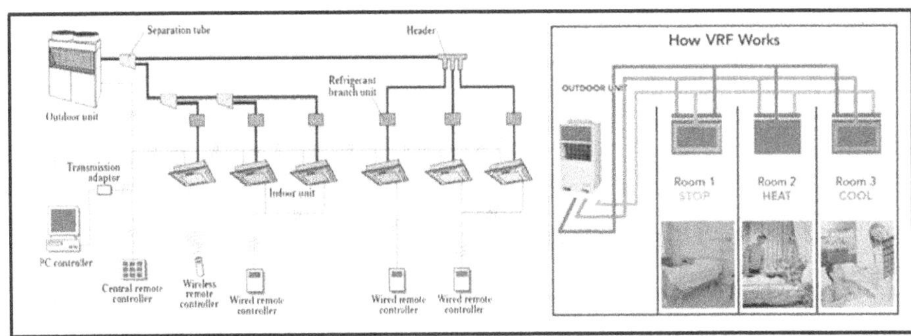

Fig. 2 Flow chart of VRF-HVAC Working principle

ISBN: 978-93-8830-599-0

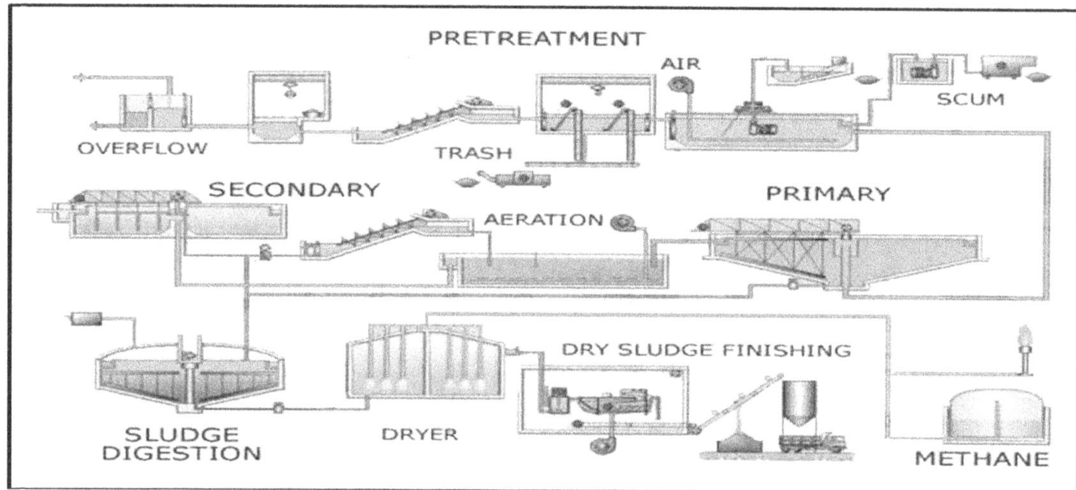

Fig. 3 Flow chart of Sewage treatment plant

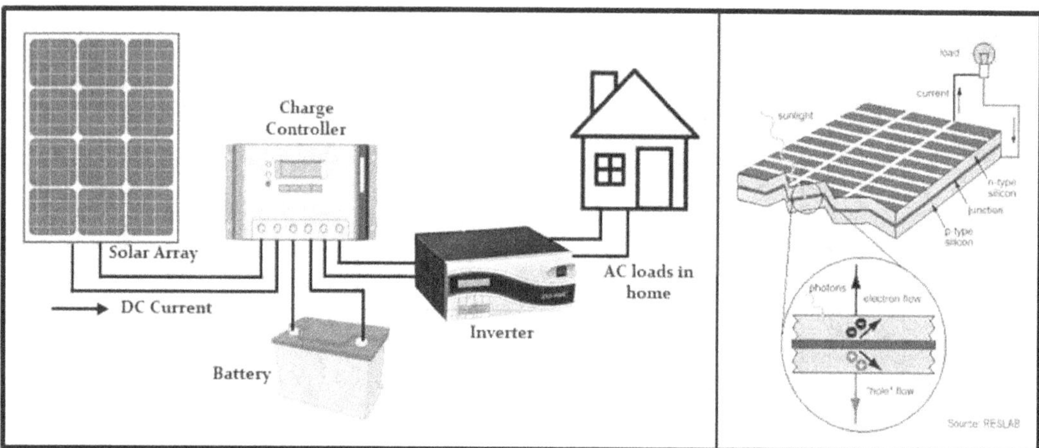

Fig. 4 Renewable energy generation through solar panels

RESULTS

1. **HVAC (Heating, Ventilation & Air Conditioning) System:** In this building we are going to propose the VRF system for HVAC rather than split air conditioner system which takes lot of power consumption. By choosing the VRF system we can reduce the consumption of power. In this building we are getting approx 320 TR of load. We are planning for 16 VRF outdoor of capacities each 20 TR each.   The Table:1 shows the heat load in the proposed project area.

2. **STP (Sewage treatment plants):** In this building we are going to install the sewage treatment plant and reuse the filtered water for flushing of toilets. We are considering the 70 KLD water treatment plant with 172 KLD membranes. Excess treated water can be discharged into municipal line. The Table:2 shows the sewage treatment plant calculations in the proposed project area.

3. **Solar Panels:** In this building we are planning to install solar panels on the roof. The energy generated by these solar panels can be used for the common floor lightings, Lobby, Building façade lights. The total power consumption by these common floor lightings, Lobby, Building façade lights is 15.5 KW.

According to the space we have on the roof of this building we are planning for 70 solar panels. One solar panel produces approx. 250 watts/day(size 2x1m). Total energy produce by solar panels will be 17.5 KW/day. 200 batteries are required to store 15.5 KW of power with 12hrs of backup. We are also going to install solar powered water heaters and solar powered street lights in garden area.

ISBN: 978-93-8830-599-0

Table 1 Heat load calculations

HEAT LOAD CALCULATIONS			
VRV/VRF SYSTEM - INDOOR UNITS			
SNO	TYPE	CAPACITY (TR)	QUANTITY
1	CASSETTE	1.5	39
2	CASSETTE	2	78
3	CASSETTE	3	15
VRV/VRF SYSTEM OUTDOOR UNITS			
SNO	FLOOR (AREA SERVED)	HP	
1	SECOND FLOOR	104	
2	THIRD FLOOR	45	
3	FOURTH FLOOR	54	
4	FIFTH FLOOR	80	
SPLIT SYSTEM UNITS - HI WALL			
SNO	TYPE	CAPACITY (TR)	QUANTITY
1	HI-WALL	1.5	77
2	HI-WALL	2	110
3	CASSETTE	1.5	5
4	CASSETTE	2	5
5	CASSETTE	3	24

Table 2 The STP calculations of the proposed building.

S.No	STP CALCULATIONS — Description	Units	No. of Persons	Remarks		
1	No. of Hotel Rooms	Nos.	80			
2	No. of Persons in Hotel Room (A)	Nos.	160			
	Note : Considering 2 persons in each room					
3	No. of Dormitory Beds	Nos.	140			
4	No. of Persons in Dormitory (B)	Nos.	140			
5	No . Of Staff (C)	Nos.	100			
6	No. of Visitors (D)	Nos.	100			
7	No. of Persons in Restaurant & Food Court (E)	Nos.	100			
8	No. of Persons in Conference Room (F)	Nos.	500			
	Total No. of Persons (A+B+C+D+E+F)	**Nos.**	**1100**			
Part A	**Waste Water Calculation**					
S.No	Description	Unit	Quantity	Unit	Quantity	Remarks
1	Estimated Quantity of Water Supply	Ltrs	172000	KL	172	As per Water Requirement Calculations
2	Estimated Quantity of Waste Water Generated	Ltrs	172000	KL	172	
3	Qauntity of Sludge Generated Per Person	gm	100	kg	0.1	Assumption
4	Total Quantity of Sludge	gm	110000	kg	110	
5	Total Quantity of Waste Water Per Day Excluding Sludge	Ltrs	171890	KL	171.89	
Part B	**Water Requirement in Toilet Flush**					
S.No	Description	Unit	Quantity	Unit	Quantity	Remarks
1	Total No. of Toilets	Nos.	120			
a	No. of Toilets in Hotel Room	Nos	80			
b	No. of Times Used Per Day	Nos	480			
	(Assuming 6 Time use per day)					
c	**Quantity of Water Required for Flushing Per Day in Hotel Room**	**Ltrs**	**4800**	**KL**	**4.8**	10 Litres per Fush
d	No. of Toilets in Dormitory, Food Court etc	Nos	80			
e	No. of Times Used Per Day	Nos	4800			(Assuming 60 Times Use per day)
f	**Quantity of Water Required for Flushing Per Day in Dormitory**	**Ltrs**	**48000**	**KL**	**48**	10 Litres per Fush
2	**Estimated Quantity of Water Required for Flush in Toilets Per Day (c+f)**	**Ltrs**	**52800**	**KL**	**52.8**	
Note: 1. At Full Capacity Water Requirent is 172 KLD.						
2. Considering 70 KLD STP with 172 KLD Membrane , Excess treated Water Can be sent as municipal waste.						

ISBN: 978-93-8830-599-0

Acknowledgement

The author wishes to express their sincere gratitude to the Mr. Abdul Qhuddus Khan, Director of AASNAA ENGINEERS PRIVATE LIMITED for their generous support in helping me prepares and publish this paper.

References

1. http://efc.syr.edu/wpcontent/uploads/2015/03/Chapter1-web.pdf

2. http://www.utcccscdn.com/hvac/docs/1001/Public/0B/04-581067-01.pdf

3. https://guelph.ca/wpcontent/uploads/IntroductionTo Wastewater.pdf

4. https://aristair.com/blog/7-advantages-of-a-vrf-system-for-offices/

Microbial Desalination Cell: A Recent Advancement in Wastewater Treatment and Desalination

Harapriya Pradhan[1]*, Modem Varsha[2] and Shaik Sofia[2]
[1]Assistant Professor, [2]UG Student
Department of Civil Engineering, Guru Nanak Institute of Technology, Ibrahimpatnam, Hyderabad India
harapriya.pradhan@gmail.com

Abstract

A new bioelectrochemical system called microbial desalination cell (MDC) is an emerging technology for simultaneous wastewater treatment, saline water desalination and energy production. This novel technology was first invented in 2009 and able to convert the chemical energy stored in the organic matter directly into electricity through electrogenic bacterial activity. Hence energy is generated by this process and utilized for saline water desalination. In this way, MDC offers a great technology towards energy efficient desalination and can be used either as a stand-alone process or as a pre-treatment option before reverse osmosis. In MDC, the microorganisms take the lead role to produce electricity and develop a potential gradient between the electrodes. The potential gradient acts as a driving force to desalinate the liquid present in middle desalination chamber through a pair of ion-exchange membranes. Therefore, MDC integrates the microbial fuel cell (MFC) process with electrodialysis to achieve desalination and generates bioenergy.

Recently, several modifications to the structural design of MDC have been applied to test its effect on desalination performance. This review article confers the working principle of conventional MDC system with recent advances in MDC configuration, their advantages and limitations. The performance of different MDCs in terms of wastewater treatment, desalination, and electricity generation are discussed. It is essential to review MDCs to provide comprehensive and quantitative information regarding its performance and the possibility of its practical field application.

Keywords: Ion exchange membrane, Microbial desalination cell, Organic matter, Total dissolved solids removal, Wastewater treatment

I. INTRODUCTION

The fresh water unavailability and concurrent energy depletion demand for an efficient and low energy consuming technology for desalination. In addition to this, water bodies are getting polluted by periodic discharge of wastewater effluents into them. The effluents discharged from domestic, industrial, and agricultural activities create a pollutant to the environment, if not treated adequately. The primary pollutant such as organic and inorganic matter present in wastewater needs proper treatment before discharging into the water bodies. Organic contaminants can be stabilized by the widely used biological process during secondary treatment of wastewater; while, dissolved inorganic salts mostly remains untreated. For removal of inorganic salts, several physicochemical processes, such as distillation, reverse osmosis (RO) and electrodialysis (ED) have been made for reliable water desalination. However, intensive energy requirement (3 to 4 kWh/m^3) of these technologies for desalination is not sustainable [1]. Alternately, renewable energy sources such as solar, tide, the wind, and hydropower can be explored to drive the desalination process. All these renewable energy technologies need high capital cost to install and operate the desalination process [1]. Thus, there is a demand for a cost effective and low energy consuming technology to drive the desalination process to make it sustainable.

Bioenergy can be produced from wastewater either in the form of biogas or bioelectricity. In recent years, the bioelectrochemical system (BES) such as microbial fuel cells (MFCs) has demonstrated capability of harvesting direct electricity through microbial metabolism of organic matter present in wastewater [2]. The MFC technology has shown promise towards the conversion of organic waste into electricity from both domestic and industrial wastewaters [3]. This electrical energy generated by bacterial action can be utilized for removal of inorganic ions from saline water present in the desalination chamber through ion exchange membranes (IEMs) based systems. Recently, BES has been used by researchers for desalination of saline water or wastewater using IEMs [4-10]. This new BES, called microbial desalination cell (MDC), integrates the MFC technology and ED process for organic matter removal from wastewater, saline water desalination and energy production.

The concept of MDC was first developed by Cao et al. [4] by using three chamber configurations. A typical MDC consists of three compartments, namely, anaerobic anodic, aerobic cathodic and middle desalination chamber separated by a pair of IEMs (Fig. 1).

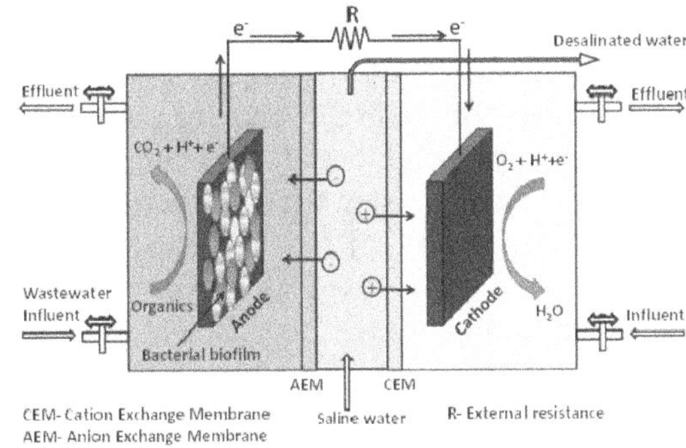

Fig. 1 Schematic diagram of a three chamber microbial desalination cell

ISBN: 978-93-8830-599-0

An anion exchange membrane (AEM) was placed near anode and cation exchange membrane (CEM) near cathode to drive the anions and cations through them from the saline water present in desalination chamber towards the anodic and cathodic chamber, respectively. Electrogenic bacteria metabolize the organic matter from wastewater in the anodic chamber and release electrons to the anode and protons into the anolyte. The AEM prevents protons and other positively charged ions from leaving the anodic chamber, so the charge is balanced by anions (Cl^-) moving from the middle desalination chamber into the anodic chamber. The electrons from anodic chamber travel through external resistance towards cathodic chamber and consumed by either chemical cathodic electron acceptor or oxygen, which combines with the proton to form water. Hence, a potential gradient is created between the electrodes. Thus, the charge is balanced by cations (Na^+) passing from the desalination chamber through the CEM into the cathodic chamber. As a consequence of these two charge-transfer processes, NaCl in the middle desalination chamber is removed, and the saline water is being desalinated.

The concept of saline water desalination in MDC is similar to water electrodialysis by the provision of an applied voltage between the chambers to drive the ions out of the desalination chamber. However, in MDC, the desalination is achieved with the current and potential generated by electrogenic bacteria without any external source of electrical energy [6]. In this way, MDC achieves three goals while treating wastewater: organic matter removal, desalination, and energy production [11]. The wastewater containing oxidizable organic matter is used as the fuel in MDC. The energy available in domestic wastewater typically ranges from 1.8 to 2.1 kWh/m^3 [12, 13], which is comparable to the minimum energy needed for practical desalination of seawater (1.8 – 2.2 kWh/m^3) [11] Thus, the use of MDCs represents a new approach for desalination, but the operational conditions and reactor designs have varied widely. Wastewater can be a good source for energy to desalinate salt water; however acetate has been used as the fuel for most of the MDC studies in order to create uniform operating conditions for desalination process. Recently, domestic wastewater [8], dewatered sludge [14], waste engine oil [15] and dye house effluent [16] have been successfully used as the substrate for the practical application of MDC.

The performance of MDC is evaluated by using chemical catholytes such as ferricyanide, phosphate buffer solution and acidified water to improve electrode potentials in MDCs [4, 7, 17, 18]. These catholytes generally produce higher cathode potentials and attain accelerated reduction kinetics which improves the overall extent of desalination. However, they are not sustainable for large scale applications due to cost, energy and environmental toxicity issues [4]. Air cathode could be a good alternative to these chemicals. However, the use of a costly cathode catalyst such as platinum makes the desalination process too expensive for large scale application. In recent past, bacterial catalyst, and microalgae have been used in MDC for efficient desalination and power recovery [19, 20].

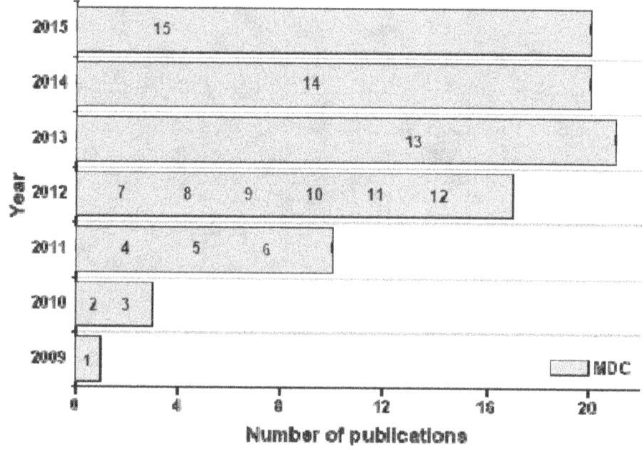

Fig. 2 Number of journal publications on MDCs. Source: Scopus on 22/03/2016. 1: Invention of three chamber MDC; 2: Air cathode MDC; 3: Microbial electrodialysis cell (MEDC); 4: Upflow MDC; 5: Stacked MDC; 6: Litre-scale MDC; 7: Electrolyte recirculation MDC; 8: Bipolar membrane MDC; 9: Osmotic MDC; 10: Resin packed MDC; 11: Capacitive MDC; 12: Biocathode MDC; 13: Photosynthetic MDC; 14: Ten-liter stacked MDC; 15: Scaling up MDC

Researches on the design of MDC are still increasing since from the commencement of its invention (Fig. 2). The recent versions of MDCs include upflow MDC (UMDC) [7], microbial electrodialysis cell (MEDC) [5], biocathode MDC [19], bipolar membrane MDC [21], resin packed MDC (R-MDC) [22], capacitive MDC (cMDC) [23], electrolyte recirculation MDC (rMDC) [24], osmotic MDC (OsMDC) [25], stacked MDC (SMDC) [26], photosynthetic MDC (PMDC) [20], and scaling up MDC [27]. All these modifications in MDC configuration suggest the popularity and promising nature of this new bioelectrochemical desalination system.

MDC has several applications excluding sea or brackish water desalination, such as water softening [18], hydrogen gas and chemical production [5, 21, 28], metal removal, etc. It was reported that MDC has higher electricity generation capability than MFC [8]. It might be due to the increased conductivity of saline

ISBN: 978-93-8830-599-0

water which reduces the system resistances. The performance of MDC is reviewed here covering organic matter removal, total dissolved solids (TDS) removal, and energy recovery with various factors such as design and operation of MDC. This review also includes various applications of MDC; current challenges persist in MDC and its future perspectives towards practical applications.

II. FACTORS AFFECTING THE PERFORMANCE OF MDC

A. Design of MDC

MDC is a new technology, which efficiently performs the desalination of saline water in lab-scale. To the authors' knowledge, very few studies have reported the large scale application of MDC namely, the scale-up of MDC having the capacity of 10 L and 105 L [27, 29]. The MDC still needs modification by changing its design for further advancement in desalination without fail. The design of MDC is a major factor which affects the system performance and includes the different configuration, inter-membrane distance and the number of cell pairs used. Hence, proper design of MDC is required for efficient desalination process.

MDC Configuration

The MDC configuration is a significant factor affecting the performance of MDC and different configurations of MDCs used till dates are described here.

Three Chambered MDC: The conventional three chamber configuration of MDC was first invented by Cao et al. [4] for simultaneous wastewater treatment, desalination and energy recovery. This three chambered MDC was able to desalinate about 90% of salt from saline water in the middle desalination chamber along with power generation of 31 W/m^3 based on total reactor volume in a single desalination cycle. The electrolyte solution was frequently replaced to avoid substrate limitation for bacterial growth and change in pH. It was observed that while desalinating the liquid from desalination chamber, the internal resistance of the system increased from 25 Ω to 970 Ω, which resulted in a reduction of overall system performance. Hence, salt removal rate decreases at the end of the batch operation due to increase in internal resistance. This configuration is suitable for two different streams of pollutant removal.

Upflow MDC: To eliminate the cathode chamber, an UMDC was constructed by Jacobson et al. [7] using a wrapped wet-cloth cathode surrounding to the CEM. The UMDC was operated in a continuous mode for 4 months with a hydraulic retention time (HRT) of 4 days. This novel UMDC removed more than 99% of

NaCl at HRT of 4 d with TDS removal rate of 7.50 g TDS/L d. A maximum power density of 30.8 W/m^3 was achieved by this UMDC. The UMDC was later advanced to a liter-scale [17] with a aim to scale-up the system. The liter-scale UMDC was capable of desalinating both salt solution (containing NaCl) and artificial sea water (containing sea salts) with 94.3 ± 2.7% and 73.8 ± 2.1% of TDS removal efficiency, respectively; at the HRT of 4 d.

Air cathode MDC: In most of the three chambered MDC, the distance between electrodes were usually high (> 1 cm). Hence, the internal resistance of MDC increases and the poor cathodic reduction kinetics decrease the electricity generation. Use of chemical catholyte also makes the system non-sustainable for practical application. Hence, it is necessary to minimize the electrode distance and avoid the use of chemicals by the introduction of air cathode to reduce the operating cost. Air cathode MDC was first developed by Mehanna et al. [6] using oxygen as an electron acceptor to replace the ferricyanide catholyte for improving sustainability of MDC. In addition, the use of catalysts (Platinum) enhanced the cathodic reduction.

Biocathode MDC: The performance of several MDCs was evaluated with abiotic cathodes and chemical catalysts, namely ferricyanide, acidified water, and platinum catalyzed air cathodes, etc. All these methods improve electrode potential in MDCs; however, they are not feasible for large scale applications due to cost, energy and environmental toxicity issues. To solve this problem, Wen et al. [19] used an aerobic biocathode consisting of carbon felt and aerobic bacterial to act as catalyst, collected from a settling tank of a wastewater treatment plant, and compared its performance with an air cathode MDC. The biocathode MDC could produce a maximum voltage of 609 mV as compared to 136 mV produced by an air cathode MDC. The biocathode MDC was able to remove 92% of salinity with a Coulombic efficiency (CE) of 96.2 ± 3.8% and a total desalination rate (TDR) of 2.83 mg/h. In this way, the biocathode MDC provides a promising approach towards efficient and sustainable desalination of saline water.

Photosynthetic MDC: Use of chemical catholyte and costly cathode catalyst will make the MDC unsustainable for future field scale application. Alternately microalgae can be used for making a sustainable biocathode by *in situ* oxygen generation in cathodic compartment, making it a photosynthetic MDC. Recently, microalgae called *Chlorella vulgaris* sp. has been used for its tolerance to high levels of CO_2 and its utilization through photosynthesis [20]. The

ISBN: 978-93-8830-599-0

passive algae biocathode MDC was able to remove the salinity of about 40% with power generation of 84 mW/m^3 (anodic volume) and 151 mW/m^3 (cathodic volume). The chemical oxygen demand (COD) removal of 65.6% was observed in PMDC and the treated wastewater was utilized as the growth medium to obtain valuable biomass for high-value bio-products.

Stacked MDC: To enhance the desalination rate in MDC, the number of desalination chamber between the electrodes should be increased. Chen et al. [26] developed a stacked MDC by providing more number of IEMs between the anodic and cathodic chamber and consisting of desalination chambers and concentrate chambers like ED process. They constructed one, two, and three desalination chambers, along with zero, one, and two concentrated chambers, respectively, between the anode and cathode chambers to form one-, two-, and three-desalination-chambered SMDCs. It was observed that more pair of ions was separated from the desalination chamber while transferring one electron from anode to cathode. The charge transfer efficiency was improved by providing more number of desalination chambers. Chen et al. [26] observed a maximum TDR of 0.0252 g/h using a two-desalination-chambered SMDC, which was 1.4 times that of single-desalination-chambered MDC, operated with an external resistance of 10 Ω.

Electrolyte Recirculation MDC: In three chambered MDC, the pH imbalance between the electrolytes was observed as a major challenge, which affects the system operating voltage. To avoid this pH imbalance, the electrolyte can be recirculated from anodic to the cathodic chamber to balance the charged ions through a recirculation microbial desalination cell [24]. The pH imbalance also inhibits the bacterial metabolism. By recirculating the electrolyte, Qu et al. [24] observed maximum power densities of 931 ± 29 mW/m^2 and 776 ± 30 mW/m^2 using 50 mM and 25 mM of phosphate buffer solution (PBS), respectively. The power densities achieved with recirculation of electrolyte were higher than that of without electrolyte recirculation, which were 698 ± 10 mW/m^2 and 508 ± 11 mW/m^2 with 50 mM and 25 mM of PBS, respectively. The salinity of salt solution was reduced by 34 ± 1% and 37 ± 2% using 50 mM and 25 mM of PBS solution, respectively, in recirculation MDC, while these removal efficiencies were 39 ± 1% (50 mM) and 25 ± 3% (25 mM) without recirculation in a MDC. Thus, rMDC provides an efficient method to balance the electrolyte pH and helps in increasing both power output and desalination efficiency.

Osmotic MDC: An OsMDC was developed by integrating forward osmosis (FO) into MDC. The OsMDC was constructed by inserting a FO membrane in place of AEM in between the anodic and desalination chamber of a conventional three chamber MDC, and the CEM was placed in between the cathodic and desalination chamber. This system can effectively treat two streams of water, namely wastewater containing organic matter in anodic chamber and saline water in desalination chamber. The OsMDC seems more convenient for treatment of high saline waters due to the dilution effect caused by the water flux. Zhang and He [25] developed the OsMDC and achieved 65.9% and 57.8% of desalination, using 5 g/L and 10 g/L of initial salt concentrations; while only 17.7% of salinity was removed with 20 g/L of initial salt concentration. In this way, the OsMDC achieves three goals of simultaneous wastewater treatment, low saline water desalination and water reuse. Zhang and He [30] further coupled hydraulically an osmotic microbial fuel cell (OsMFC) with an MDC to treat artificial wastewater and desalinate saline water. The coupled system showed 95.9% conductivity reduction along with energy production of 0.16 kWh/m^3.

Bipolar Membrane MDC: A bipolar membrane (BPM) was applied in MDC to modify the design of conventional MDC and to achieve advantages of acid and alkali recovery along with desalination. The BPM usually composed of an anion and cation selective layers held together as a single membrane. The BPM has characteristics of high perm-selectivity, long term stability, high rate of water splitting, and low passive drop in potential. However, BPM is more vulnerable to fouling caused by organics and microbial activity if placed near anode [21, 28] and subjected to scaling when placed near cathode [31]. Chen et al. [21] demonstrated a new microbial electrolysis desalination and chemical production cell (MEDCC) by placing the BPM near anode consisting of four chambers. The pH of anolyte remained stable (pH, 7) and the microbial activity greatly enhanced in this MEDCC. The desalination rate in MEDCC reached to 46% - 86% within 18 h of operation using 10 g/L NaCl solution in the desalination chamber. The maximum current density of 8.4 ± 0.5 A/m^2 was observed at applied potential of 1.0 V. With increase in applied potential from 0.3 - 1.0 V, the CE increased from 62% - 97%. The total amounts of alkali and acid production were 2.6 ± 0.9 and 2.1 ± 0.7 mmol, based on 30 mL of cathodic chamber volume, and 10 mL of the acid-production chamber volume, respectively.

Resin Packed MDC: Most of the MDCs have shown high desalination efficiency with higher salt

ISBN: 978-93-8830-599-0

concentration (sea water) in desalination chamber as compared to lower salt concentration. With low salinity water (brackish water or industrial effluent), the desalination efficiency decreases due to high ohmic resistance. To enhance the desalination rate in low salinity water, mixed ion exchange resins can be used in desalination chamber. The MDC packed with mixed ion exchange resins can be referred as resin packed MDC. Morel et al. [22] used R-MDC to desalinate low saline water (2 - 10 g/L) with anion and cation exchange resins packed in desalination chamber. The resins worked as conductor and offset the increased ohmic resistance, while treating low salinity water. Morel et al. [22] observed an increased desalination rate of 1.5 - 8 times by using resin in desalination chamber as compared to conventional MDC using only IEMs. The ohmic resistances of R-MDC were observed as 3.0 - 4.7 Ω; whereas, those were 5.5 - 12.7 Ω in MDC without resins.

Capacitive MDC: In conventional MDCs, the dissolved ions from the saline water are transferred from desalination chamber towards the adjacent anodic and cathodic chambers through respective IEMs. During this desalination process, the anolyte and catholyte become more concentrated by addition of ions, which might restrict for their reuse. To solve this issue, capacitive deionization (CDI) concept was integrated into MDC system and referred as capacitive microbial desalination cell [23, 32]. In cMDC, an adsorptive activated carbon cloth (ACC) was used as the electrodes to provide high surface area and the ions were adsorbed electrochemically by its capacitive double layers (one layer of Ni/Cu mesh, and an ACC cathode together). In this way, the removed ions from the desalination chamber were not getting further concentrated in the anodic and cathodic chamber. The cMDC was able to remove 69.4% of the dissolved salts from the desalination chamber through electrochemical adsorption by the double layer electrode during one batch cycle and adsorbed about 61 - 82.2 mg of TDS by 1 g of ACC electrode [32]. This technology provides a new platform for wastewater treatment by simultaneously treating organic matter and salt along with energy recovery.

Scaling Up MDC: Most of the desalination studies were effectively conducted in small scale MDCs (volume in milliliters) and they removed up to 99% of TDS from the saline water. To transfer this technology from lab scale to field scale for its practical application, it is necessary to scale it up. Very few researchers have attempted the scaling up of this promising technology. Jacobson et al. [17] developed a liter-scale MDC operated in an upflow continuous mode to achieve

desalination, as discussed earlier. Zuo et al. [29] developed a ten liter stacked MDC, which was packed with mixed ion exchange resins and operated in batch mode. The desalination performance of this stacked MDC was tested with 0.5 g/L of NaCl solution (1.08 mS/cm, pH 7.0), which is used as a representative of secondary treated effluent from a practical municipal wastewater treatment plant (WWTP). The stacked MDC achieved a desalination efficiency of 95.8% with initial 0.5 g/L of NaCl solution in 12 h of operation with the final effluent concentration of 0.02 g/L, which is comparable to the effluent quality of RO in terms of salinity.

Recently, Zhang and He [27] constructed a large-scale MDC having a capacity of 105 L (total liquid volume) to investigate the scaling up related issues. The scaled-up MDC was consisting of 30 tubular modules of MDCs, which were hydraulically connected by the anolyte and the saline water. Each module was constructed similar to UMDCs [7] and it was operated in continuous mode. For uniform substrate distribution in each module, multiple feeding points were provided. However, the performance observed in each module was not uniform. The scaled-up MDC was capable of generating 670 mA of current and removed salt at a rate of 3.7 kg/m^3.d. When an external potential of 1.1 V was applied to the system, both current and salt removal rates increased to 2000 mA and 9.2 kg/m^3.d.

B. Inter-membrane Distance

The inter-membrane distance is a major factor which can greatly affect the internal resistance and the desalination efficiency of the MDC [33]. In most of the MDCs, the inter-membrane distance greater than 5 mm has been reported [4-6, 8, 26, 34] except a stacked MDC, where the inter-membrane distance was minimized to 1.3 mm [35]. The internal resistance (ohmic resistance) of 7 Ω - 8 Ω was observed by using the stack having 1.3 mm inter-membrane distance. In another study, a stacked MDC with 10 mm inter-membrane distance, the internal resistance increased from 21 Ω to 312 Ω, 39 Ω to 256 Ω, and 56 Ω to 117 Ω having one-, two-, and three-desalination chambers, respectively [26]. During desalination process, the concentration of salt solution in desalination chambers decreased and results in the increase of ohmic resistance. In conventional ED process, the inter-membrane distance usually fixed as 0.2 - 3 mm to minimize the energy loss [36]. A larger inter-membrane distance can cause a higher internal resistance due to the increase in ohmic resistance and decrease both current generation and desalination efficiency [11]. Ping and He [33] observed an internal resistance of 342

ISBN: 978-93-8830-599-0

Ω, 222 Ω, and 214 Ω with the inter-membrane distance of 0.3 cm, 1.0 cm, and 1.5 cm, respectively. With lower inter-membrane distance the conductivity of the electrolyte was decreased than those with higher inter-membrane distances. Hence, inter-membrane distance greatly influences the performance of the MDC.

C. Number of Cell Pairs

In MDC, a pair of AEM and CEM generally constitutes a cell pair (CP). The desalination rate and charge transfer efficiency in MDC increases with an increase in the number of CP. Hence, it is a good practice to increase the number of CPs between the electrodes to separate more dissolved ions in MDC. Several desalination studies in MDCs have been conducted with a single CP structure. Very few studies have reported the effect of multiple CPs on the performance of MDC. Chen et al. [26] examined the performance of MDCs containing one-, two-, and three- CPs and found that the salt removal rate could be increased by (i) adding CPs and (ii) reducing the external resistance. The maximum TDR of 0.0178 g/h, 0.0252 g/h, and 0.0196 g/h were observed in one-, two-, and three- CPs SMDC, respectively. Hence, in two-CP SMDC the TDR was 1.4 and 1.3 times higher than that of one-CP and three-CP SMDC. However, the faradaic efficiencies increased by increasing the number of CPs and achieved 120%, 223%, and 283% for one-, two-, and three- CPs SMDC, respectively. Another advantage of adding more number of CPs is the need of voltage in each cell can be reduced.

D. Mode of Operation

The mode of operation is another important factor which affects the performance of MDC. In most of the studies MDCs were operated in batch mode [4-6, 23, 34]; however, few studies have reported the cyclic batch mode [8, 34, 37, 38] and continuous flow mode [7, 17, 26, 27, 35, 39]. The extent of desalination was limited by fed batch mode of operation and the performance of MDC was decreased with time due to the low conductivity of the electrolyte in desalination chamber leading to high internal resistance of the system. In batch mode of operation, the substrate distribution is also not uniform throughout the anodic chamber. Hence, batch mode operation of MDC restricts its use for the practical applications. During continuous operation of MDC, the substrate distribution occurs uniformly in the anodic chamber leading to higher power production and desalination efficiency of the system. Jacobson et al. [7] invented tubular UMDC and operated it in continuous up-flow mode with cathode exposed to air. The UMDC demonstrated more than 99% of salt removal from 30 g/L of initial TDS

concentration with high charge transfer efficiency (98.6%) and power density (3.8 W/m^3). The UMDC in liter-scale also successfully exhibited the desalination of saline water and artificial sea water with 35 g/L of initial TDS concentration in continuous flow mode [17].

A stacked MDC was operated in continuous mode and demonstrated enhanced TDR of 0.0252 g/h with high charge transfer efficiency [26]. Kim and Logan [35] connected 4 MDCs in series and operated it in continuous mode and achieved current efficiency of 86 ± 5% and salinity reduction of 97.6% during the three stage operation of the MDC.

E. Substrates used in MDC

In a bioelectrochemical system, the substrate is used as the source of energy for any biological process and considered as an important factor affecting the bioelectricity generation. The substrate is utilized by electrogenic bacteria through metabolic activity and generates current. Acetate has been used as the substrate in most of the MDCs due to its simplicity and it is used as the carbon source to motivate the electrochemically active bacteria. However, acetate fed wastewater is not sustainable for practical application of MDC. Hence, it is essential to use real wastewater (domestic or industrial) as substrate in anodic chamber of MDC for direct onsite treatment of the effluent. Luo et al. [8] used domestic wastewater as the substrate in anodic chamber having COD concentration of 2744 ± 16 mg/L. This MDC removed 52% of COD, more than 66% of salt, and generated power of 8.01 W/m^3 during batch operation. Luo et al [40] further studied the performance of MDC for 8 months of operation and fed with domestic wastewater. It was observed that after 8 months of operation the performance of MDC was decreased by 47% in current density, 46% in Columbic efficiency, and 27% in desalination efficiency. The biofouling on the AEM resulted in increased system resistance and reduced ionic transfer and energy conversion efficiency.

Dewatered sludge collected from a wastewater treatment plant was used as both inoculum and substrate in a biocathode MDC [14]. The MDC showed improved performance under long-term operation (300 d) and removed 25.71 ± 0.15% of organic matter with better anodic pH (6.6 to 7.6). The MDC removed more than 40% salt in 24 h of operation. Maximum power output of 3.178 W/m^3 with open circuit voltage (OCV) of 1.118 V was observed on 130th day of operation. Kalleary et al. [15, 16] used waste engine oil and dye house effluent as the anodic substrate in MDC and achieved biodegradation of the complex substrate with

ISBN: 978-93-8830-599-0

considerable desalination and power generation. Using an isolated strain *Bacillus subtilis moh3*, in anodic chamber MDC demonstrated waste engine oil biodegradation efficiency of 70.1 ± 0.5%, desalination efficiency of 68.3 ± 0.6%, and power production of 3.1 ± 0.3 mW/m^2. In another study, dye house effluent was used as the substrate in MDC to achieve simultaneous biodecolourisation, desalination, and power generation [16]. The MDC fed with these dye waste in anodic chamber and salt solution of 35 g/L of NaCl in desalination chamber showed desalination of 62.2 ± 0.4% and 57.6 ± 0.2% and power density of 3.01 ± 0.04 mW/m^2 and 2.86 ± 0.25 mW/m^2 using Malachite Green and Sunset Yellow dye as substrate, respectively.

F. Inoculums used in MDC

Inoculum is generally used as biocatalyst in a BES and plays predominant role affecting the power output and also it is one of the parameter deciding the overall internal resistance of the MDC. Microorganisms can be a pure culture or mixed culture. Generally, the pure culture bacteria grow slowly and having a high risk of microbiological contamination with high substrate specificity as compared to mixed culture bacteria. The pure culture bacteria are mostly exoelectrogenic bacteria, which can directly transport electrons outside the cell membrane or produce mediators, which act as electron shuttles to transfer electrons to the anode [41]. The inoculum used in MDCs was either pre-acclimated microorganisms from acetate fed MFCs [4-6, 26, 34, 35, 39] or inoculum collected from local wastewater treatment plant [7, 8, 17, 23, 27, 40]. The microbial community was analyzed for both pre-acclimated microorganism and inoculum collected from wastewater treatment plant. The community analysis for pre-acclimated microorganism revealed the abundance of bacterial species of *Pelobacter propionicus*, *Geobacter sulfurreducens*, *Rhizobium etli* CFN 42, *Acinetobacter sp.* ADP1, *Klebsiella ornithinolytica*, *Propionibacterium sp.*, and *Bacteroides rodentium* [5, 39]. Pure culture microorganisms such as *B. subtilis moh3* [15, 16], *A. hydrophila* [16], *P. aeruginosa* [42] have been used in MDCs to degrade the complex substrates of waste engine oil, dye house effluent, and phenol, respectively.

G. Total Dissolved Solids Concentration

The desalination efficiency of MDC is primarily dependent upon the initial TDS concentration of electrolyte in desalination chamber. The dissolved salt concentrations of the saline water ranging from 0.5 – 35.0 g/L have been taken in most of the studies to observe the desalination performance in MDCs [4-7, 17, 19, 20, 22, 31, 34, 43]. The current generation of

2.82 mA, 3.06 mA, and 3.17 mA were achieved with initial TDS concentration of 5 g/L, 20 g/L, and 30 g/L, respectively. Similarly, the power density increased from 158.2 mW/m^2 to 204.5 mW/m^2 with decreased internal resistance from 2432 to 2328 Ω, while increasing the initial NaCl concentrations from 5 to 30 g/L [43]. In MDC, there is an additional energy generated by the salinity gradient between anodic and desalination chamber, similar to reverse electrodialysis (RED) process [44]. It was reported that the concentration gradient across the membrane is also contributing to TDS removal in addition to current generated in MDC [17, 33, 43].

H. Catholytes used in MDC

The current generation in MDC is considerably affected by the cathodic reduction kinetics. In a BES, the overall performance losses are affected by cathodic overpotentials. The cathode activation overpotential can be reduced by employing cathodic electron acceptor having higher redox potential. Ferricyanide has been used in several MDCs to achieve higher power output [4, 8, 23, 25]. However, use of ferricyanide is expensive and toxic to the environment and its application in large scale is not sustainable. Other catholytes such as acidified water [7, 17] and phosphate buffer solution [26, 39] have also been used to complete the cathodic reduction reactions in MDCs. The use of such chemicals will increase the operational cost of the MDC. Pradhan and Ghangrekar [45] used permanganate as cathodic electron acceptor and achieved higher TDS removal with acidic pH than that of alkaline condition. In few MDCs oxygen has been used as a terminal electron acceptor with Pt catalyst to enhance the reduction kinetics and overcome the activation overpotential [5, 26]. Use of air cathode MDC with Pt catalyst will make the desalination process very expensive for its application in broad scale. Alternately, biocatalyst and microalgae can be used in cathodic chamber of MDC for improving efficiency of wastewater treatment, desalination, and to obtained biomass production as an additional value added product.

III. CURRENT CHALLENGES IN MDC AND THE WAY AHEAD

MDC is a new technology and still in developing state for practical application of organic matter removal from wastewater and saline water desalination. There are several challenges found in MDC which needs to be addressed before field application. The major challenge observed in conventional MDC is the pH imbalance in electrolyte during desalination process. The pH imbalance can affect the system performance by

ISBN: 978-93-8830-599-0

reducing the electricity generated in the MDC. It was reported that one unit of the pH gradient can drop operating voltage of the system by 0.06 V [46]. Usually, the anolyte pH decreases in anodic chamber of MDC while degrading organic constituents from the substrates. The low pH (pH < 5) of anolyte can preclude the electrogenic microbial activity for generating current. In addition to the proton accumulation in anodic chamber, the access of Cl⁻ ions from desalination chamber towards the anodic chamber through AEM decreases the pH of anolyte, while the proton consumption and formation of hydroxyl ions in cathodic chamber increases the pH of catholyte. This imbalance in pH causes the decrease in potential of MDC. To avoid pH imbalance, the electrolytes from anodic to cathodic chamber can be recirculated. However, this recirculation can decrease the CE of the system [24]. Alternately, BPM can be used to maintain the pH by passing protons into the acid production chamber [21, 28]. Addition of concentrate chambers to both side of desalination chamber can prevent the pH imbalance by access of protons and dissolved ions accumulation in the concentrate chamber without concentrating the anolyte and catholyte [37].

MDC can be applied as a pre-desalination unit to enhance the membrane life used in the membrane process (such as RO. However, slow rate of desalination in MDC, as compared to the conventional desalination process, can limit its field application. The desalination rate with low TDS concentration can be enhanced by addition of ion exchange resins into the desalination chamber. The high capital cost of MDC includes the material cost, fabrication cost, costly membrane, electrodes, and catalyst used. The capital cost can be minimized by selection of cheaper catalyst and membrane in MDC. Membrane fouling and scaling is a typical challenge observed in all MDC systems. For long term operation of MDC, membrane life should be long-lasting to reduce the operational cost. Frequent replacement of fouled membrane will make the system uneconomic. The fouling can be prevented to minimize the frequent replacement of membrane. The high internal resistance of the MDC is a significant factor, which affects the performance of the system. Desalination chamber and initial salt concentration are the two major factors which affects the internal resistance of the system. To increase the desalination rates in stack MDC, more numbers of IEM were inserted. However, by increasing the number of IEM pairs in the stack of MDC, the internal resistance of the system increases. Hence the inter-membrane distance should be kept less than 1 mm, like ED process [11]. The internal resistances can be minimized by reducing the resistance developed by electrolyte with addition of ion exchange resins into the desalination chamber or by frequent recirculation of electrolyte. [43].

Most of the studies in MDC were conducted in lab scale to observe the various factors affecting its performance in a control condition. Scaling up of this bioelectrochemical desalination technology is a great challenge to transfer this technology form lab scale to field scale. Recently MDC was scaled-up to 105 L which was divided into several modules. The scaled-up MDC exhibited 2000 mA of electricity with salt removal rate of 9.2 kg/m³/d. However, the electricity generated in each module was non-uniform. Hence, the scaling up of MDC still needs proper consideration and the present volumes of MDC chambers are too small to meet the desalination demands.

IV. Conclusions

MDC is a new bioelectrochemical desalination technology which can simultaneously remove organic matter and dissolved inorganic salts with energy recovery. MDC is assumed as the low energy consuming technology as compared to the other conventional desalination processes by producing its own electricity through the exoelectrogenic bacteria from organic matter oxidation.. MDC technology is still in its developing state and need further modifications in its design for efficient performance. Scaling up MDC is a great challenge for its practical application and includes step wise increments in reactor volume and needs more flexibility in its design for ease in operation.

References

1. Eltawil MA, Zhengming Z, Yuan L. A review of renewable energy technologies integrated with desalination systems. Renew Sust Energ Rev. 2009;13(9):2245-62.

2. Logan BE. Microbial Fuel Cells. Wiley & Sons, Inc., Hoboken, New Jersey. 2008.

3. Lovley DR. The microbe electric: conversion of organic matter to electricity. Curr Opin Biotechnol. 2008;19(6):564-71.

4. Cao X, Huang X, Liang P, Xiao K, Zhou Y, Zhang X, Logan BE. A new method for water desalination using microbial desalination cells. Environ Sci Technol. 2009;43(18):7148-52.

5. Mehanna M, Kiely PD, Call DF, Logan BE. Microbial electrodialysis cell for simultaneous water desalination and hydrogen gas production. Environ Sci Technol. 2010a;44(24):9578-83.

6. Mehanna M, Saito T, Yan J, Hickner M, Cao X, Huang X, Logan BE. Using microbial desalination

ISBN: 978-93-8830-599-0

cells to reduce water salinity prior to reverse osmosis. Energy Environ Sci. 2010b;3(8):1114-20.

7. Jacobson KS, Drew DM, He Z. Efficient salt removal in a continuously operated upflow microbial desalination cell with an air cathode. Bioresour Technol. 2011;102(1):376-80.

8. Luo H, Xu P, Roane TM, Jenkins PE, Ren Z. Microbial desalination cells for improved performance in wastewater treatment, electricity production, and desalination. Bioresour Technol. 2012;105:60-6.

9. Kim Y, Logan BE. Simultaneous removal of organic matter and salt ions from saline wastewater in bioelectrochemical systems. Desalination. 2013;308:115-21.

10. Saeed HM, Husseini GA, Yousef S, Saif J, Al-Asheh S, Fara AA, Azzam S, Khawaga R, Aidan A. Microbial desalination cell technology: a review and a case study. Desalination. 2015;359:1-13.

11. Kim Y, Logan BE. Microbial desalination cells for energy production and desalination. Desalination. 2013;308:122-30.

12. Shizas I, Bagley DM. Experimental determination of energy content of unknown organics in municipal wastewater streams. J Energy Eng. 2009;130 (2):45–53.

13. Heidrich ES, Curtis TP, Dolfing J. Determination of the internal chemical energy of wastewater. Environ Sci Technol. 2011;45(2):827–32.

14. Meng F, Jiang J, Zhao Q, Wang K, Zhang G, Fan Q, Wei L, Ding J, Zheng Z. Bioelectrochemical desalination and electricity generation in microbial desalination cell with dewatered sludge as fuel. Bioresour Technol. 2014;157:120-26.

15. Kalleary S, Abbas FM, Ganesan A, Meenatchisundaram S, Srinivasan B, Packirisamy ASB, Kesavan R krishnan, and Muthusamy S. Microbial desalination cell for enhanced biodegradation of waste engine oil using a novel bacterial strain Bacillus subtilis moh3. Environ Technol. 2014;35(17):2194-203.

16. Kalleary S, Abbas FM, Ganesan A, Meenatchisundaram S, Srinivasan B, Packirisamy ASB, Kesavan R krishnan, Muthusamy S. Biodegradation and bioelectricity generation by microbial desalination cell. Int Biodeterior Biodegrad. 2014;92:20-25.

17. Jacobson KS, Drew DM, He Z. Use of a liter-scale microbial desalination cell as a platform to study bioelectrochemical desalination with salt solution or artificial seawater. Environ Sci Technol. 2011;45(10):4652-7.

18. Brastad KS, He Z. Water softening using microbial desalination cell technology. Desalination. 2013;309:32-7.

19. Wen Q, Zhang H, Chen Z, Li Y, Nan J, Feng Y. Using bacterial catalyst in the cathode of microbial desalination cell to improve wastewater treatment and desalination. Bioresour Technol. 2012;125:108–13.

20. Kokabian B, Gude VG. Photosynthetic microbial desalination cells (PMDCs) for clean energy, water and biomass production. Environ Sci: Processes Impacts. 2013;15(12):2178-85.

21. Chen S, Liu G, Zhang R, Qin B, Luo Y. Development of the microbial electrolysis desalination and chemical-production cell for desalination as well as acid and alkali productions. Environ Sci Technol. 2012;46(4):2467–72.

22. Morel A, Zuo K, Xia X, Wei J, Luo X, Liang P, Huang X. Microbial desalination cells packed with ion-exchange resin to enhance water desalination rate. Bioresour Technol. 2012;118:43-8.

23. Forrestal C, Xu P, Ren Z. Sustainable desalination using a microbial capacitive desalination cell. Energy Environ Sci. 2012;5(5):7161-7.

24. Qu Y, Feng Y, Wang X, Liu J, Lv J, He W, Logan BE. Simultaneous water desalination and electricity generation in a microbial desalination cell with electrolyte recirculation for pH control. Bioresour Technol. 2012;106:89-94.

25. Zhang B, He Z. Integrated salinity reduction and water recovery in an osmotic microbial desalination cell. Rsc Adv. 2012;2(8):3265-9.

26. Chen X, Xia X, Liang P, Cao X, Sun H, Huang X. Stacked microbial desalination cells to enhance water desalination efficiency. Environ Sci Technol. 2011;45(6):2465–70.

27. Zhang F, He Z. Scaling up microbial desalination cell system with a post-aerobic process for simultaneous wastewater treatment and seawater desalination. Desalination. 2015;360:28-34.

28. Chen S, Liu G, Zhang R, Qin B, Luo Y, Hou Y. Improved performance of the microbial electrolysis desalination and chemical-production cell using the stack structure. Bioresour Technol. 2012;116:507-11.

29. Zuo K, Cai J, Liang S, Wu S, Zhang C, Liang P, Huang X. A ten liter stacked microbial desalination cell packed with mixed ion-exchange resins for secondary effluent desalination. Environ Sci Technol. 2014;48(16):9917-24.

30. Zhang B, He Z. Improving water desalination by hydraulically coupling an osmotic microbial fuel cell with a microbial desalination cell. J Membr Sci. 2013;441:18-24.

31. Chen M, Zhang F, Zhang Y, Zeng RJ. Alkali production from bipolar membrane electrodialysis powered by microbial fuel cell and application for biogas upgrading. Appl Energy. 2013;103:428-34.

32. Forrestal C, Xu P, Jenkins PE, Ren Z. Microbial desalination cell with capacitive adsorption for ion

ISBN: 978-93-8830-599-0

migration control. Bioresour Technol. 2012;120:332-6.

33. Ping Q, He Z. Effects of inter-membrane distance and hydraulic retention time on the desalination performance of microbial desalination cells. Desalin Water Treat. 2014;52(7-9):1324-31.

34. Luo H, Jenkins PE, Ren Z. Concurrent desalination and hydrogen generation using microbial electrolysis and desalination cells. Environ Sci Technol. 2011;45(1):340-4.

35. Kim Y, Logan BE. Series Assembly of Microbial Desalination Cells Containing Stacked Electrodialysis Cells for Partial or Complete Seawater Desalination. Desalination. 2011;45(13):5840-45.

36. Strathmann H. Ion-exchange membrane separation processes. Amsterdam: Elsevier BV; 2004(Vol. 9).

37. Pradhan H, Ghangrekar MM. Multi-chamber microbial desalination cell for improved organic matter and dissolved solids removal from wastewater. Water Sci Technol. 2014;70(12).

38. Sevda S, Yuan H, He Z, Abu-Reesh IM. Microbial desalination cells as a versatile technology: Functions, optimization and prospective. Desalination. 2015;371:9-17.

39. Qu Y, Feng Y, Liu J, He W, Shi X, Yang Q, Lv J, Logan BE. Salt removal using multiple microbial desalination cells under continuous flow conditions.Desalination. 2013;317:17-22.

40. Luo H, Xu P, Ren Z. Long-term performance and characterization of microbial desalination cells in treating domestic wastewater. Bioresour Technol. 2012b;120:187-93.

41. Rabaey K, Rodríguez J, Blackall LL, Keller J, Gross P, Batstone D, Verstraete W, Nealson KH. Microbial ecology meets electrochemistry: Electricity-driven and driving communities. ISME J. 2007;1(1):9-18.

42. Pradhan H, Jain SC, Ghangrekar MM. Simultaneous Removal of Phenol and Dissolved Solids from Wastewater Using Multichambered Microbial Desalination Cell. Appl Biochem Biotechnol 2015;177(8):1638-53

43. Yang E, Choi MJ, Kim KY, Chae KJ, Kim IS. Effect of initial salt concentrations on cell performance and distribution of internal resistance in microbial desalination cells. Environ Technol. 2015;36(7):852-60.

44. Post JW, Hamelers HV, Buisman CJ. Energy recovery from controlled mixing salt and fresh water with a reverse electrodialysis system. Environ Sci Technol. 2008;42:5785-90.

45. Pradhan H, Ghangrekar MM. Organic matter and dissolved salts removal in a microbial desalination cell with different orientation of ion exchange membranes. Desalin Water Treat. 2015;54(6):1568-76.

46. Rozendal RA, Hamelers HV, Molenkamp RJ, Buisman CJ. Performance of single chamber biocatalyzed electrolysis with different types of ion exchange membranes. Water Res. 2007;41:1984-94.

ISBN: 978-93-8830-599-0

"E- Waste Management"

Dr. Harish Kumar Agre[1], Md Jalal[2] and V. Sireesha
[1]Symbosisiskills Open University Pune
[2]Assistant Professor, Jayaprakash Narayan College of Engineering, India

Abstract

The electronic industry is the world's largest and fastest growing manufacturing industry in the world. Discarded electronic and electrical equipment with all of their peripherals at the end of life is termed e-waste. The quantity of c-waste generated in developed countries equals 1% of total solid waste on an average and is expected to grow to 2% by 2011 and is one of the fastest growing waste streams. E-waste consists of ferrous and non ferrous metals, plastic, glass, ceramics, rubber etc. E-waste is valuable source for secondary raw material but harmful if treated and discarded improperly as it contains many toxic components such as lead, cadmium, mercury, polychiorinated biphenlys etc. The presence of lead, mercury, arsenic, cadmium, selenium and hexavalent chromium and flame retardants beyond threshold quantities in e-waste classifies them as hazardous wastes.

Rather than recycle the c-waste generated, the developed countries are finding easy way out of the problem by exporting them to developing economies. Recycling c-waste in a crude manner, as is done now will lead environmental pollution. A review of the study conducted of uncontrolled dumping and crude recycling of c-waste reveals the gravity of the problem. Technologies are suggested for environmentally sound management of c-waste. Legislation is the need of the hour for enforcing environmentally sound management.

"Today's gadgets are tomorrows E-waste." 95 percent of electronic waste is recyclable. However, unregulated recycling can cause more harm to the environment than landfilling. While many companies, have safe and effective recycling programs, the majority of recycling companies export some percentage of their electronic waste to China or poor countries in Africa, where the waste is "recycled," or destroyed and stripped of its valuable metals. Though this seems like a good thing on the surface, because components are being repurposed, unregulated recycling centers burn or dissolve the plastic components to release the precious metals: a process that releases environmental contaminants into the air, land, and water that would otherwise remain trapped and inert in landfills.

The major portion of the c-waste generated domestically as well as illegally imported are recycled in crude manner leading to pollution of the environment. Lack of legislation in our country at present is aiding this hazardous form of recycling. Therefore there is urgent need to frame and implement rules for regulating this waste and to find environmentally sound, economically viable methods for recycling and disposing of this necessary evil. The necessity of environmentally sound management of c-waste is brought out with the help of a case study of uncontrolled dumping of c-waste. we have collected data of disposed materials percentage in Mahbubnagar and which is causing harm to environment and for reducing this we have explained some techniques from which it can be controlled without causing any effects to environment.

Keywords: E-waste, composition, recycle potential, control techniques.

WHY IS E-WASTE HARMFUL?

E-waste can contain hazardous materials such as lead, mercury, and hexavalent chromium, in circuit boards, batteries, and color cathode ray tubes (CRTs). Televisions and CRT monitors contain four pounds of lead on average (the exact amount depends on size and make). Electonics have been cited as a leading source of mercury in municipal waste. In addition, brominated flame retardants, now mostly banned in Europe due to their toxicity and persistence, are commonly added to plastics used in electronics. If improperly handled, these toxics can be released into the environment, potentially placing our health and subsistence resources at risk.

- **Where did this begin?** The term electronic waste is only a recent entity, as devices of higher technology only began to be produced towards the latter half of the 20th century. Before the 1970s, there was little production of these technologically complicated items, each consisting of over 1,000 different substances, several of which are toxic and create serious pollution upon disposal. (BAN, 2002) Each year, more new devices are created, and more new devices are thrown out. This began in the most developed nations such as the United States and Japan, leading the technological revolution but also leading the electronic pollution revolution as well. And with an immense population, America

MAJOR SOURCES

- **Individuals and Small Businesses:** The useful span of a computer has come down to about two years due to improved versions being launched about every 18 months. Often, new software is incompatible or insufficient with older hardware so that customers are forced to buy new computers. Large corporations, Institutions and

ISBN: 978-93-8830-599-0

Government: Large users upgrade employee computers regularly.

- **Original Equipment Manufacturers (OEMs):** OEMs generate e-waste when units coming off the production line do not meet quality standards, and must be disposed off. Some of the computer manufacturers contract with recycling companies to handle their electronic waste, which often is exported.

- Besides computers, other major e waste source is the cellular phone., refregirators, ovens, switchboards, air conditioner, batteries, tv etc.

ELECTRONIC APPLICENCES

CLASSIFICATION OF E-WASTE

COMPONENTSOF E-WASTE

E-waste has been categorized into three main categories, viz. large household appliances, IT and Telecom and consumer equipment. Refrigerator and washing machine represent large household appliances, personal computer monitor and laptop represent IT and Telecom, while television represents consumer equipment. Each of these e-waste items has been classified with respect to twenty six common components, which could be found in them. These components form the "building blocks" of each item and therefore they are readily "identifiable" and "removable". These components are

- metal,
- motor/compressor,
- cooling, plastic,
- insulation,
- glass, (Liquid Crystal Display) LCD,
- rubber, wiring/ electrical,
- transformer,
- magnetron,
- textile,
- circuit board,

- fluorescent lamp, incandescent lamp,
- heating element, thermostat,
- BFR-containing plastic, batteries,
- fluorocarbons (CFC/HCFC/HFC/HC),
- external electric cables,
- refractory ceramic fibers,

radioactive substances and electrolyte capacitors. The kinds of components, which are found in refrigerator, washing machine, personal computers (PC) and televisions, are shown in given figures.

From this figures it can be seen that the range of different items seen in e-waste is diverse. However, e-waste from these items can be dismantled into relatively smaller number of common components for further treatments.

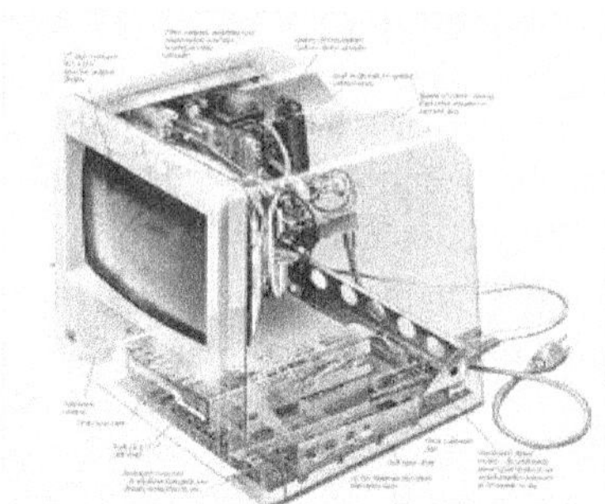

COMPUTER PARTS

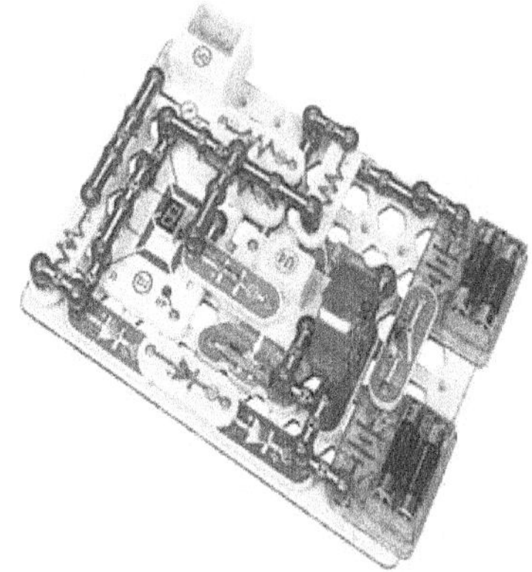

MOTHERBOARD OF CPU

CELLULAR PHONES

ISBN: 978-93-8830-599-0

BATTERIES 4.5-Volt, D, C, AA, AAA, 9-Volt,
SR41/AG3, **SR44/AG13** cell

COMPOSITION OF E-WASTE

Composition of e-waste is very diverse and differs in products across different categories. It contains more than 1000 different substances, which fall under "hazardous" and "non-hazardous" categories. Broadly, it consists of ferrous and non-ferrous metals, plastics, glass, wood & plywood, printed circuit boards, concrete and ceramics, rubber and other items. Iron and steel constitutes about50% of the e-waste followed by plastics (21%), non ferrous metals (13%) and other constituents. Non-ferrous metals consist of metals like copper, aluminium and precious metals e.g. silver, gold, platinum, palladium etc. The presence of elements like lead, mercury, arsenic, cadmium, selenium and hexavalent chromium and flame retardants beyond threshold quantities in e-waste classifies them as hazardous waste. The possible constituents of concern found in the three main categories described in given table.

Component	Possible hazardous content
Metal	
Motor/compressor	
Cooling	Ozone Depleting Substances (ODS)
Plastic	Phthalate plasticizer, brominated flame retardants (BFR)
Insulation	Insulation ODS in foam, asbestos, refractory ceramic fiber
Glass	
Cathode Ray Tube	Lead, Antimony, Mercury, Phosphor
Liquid Crystal Display	Mercury
Rubber	Phthalate plasticizer, BFR
Wiring / electrical	Phthalate plasticizer, BFR, Lead
Transformer	
Circuit Board	Lead, Beryllium, Antimony, BFR
Fluorescent lamp	Mercury, Phosphorous, Flame retardants
Incandescent lamp	
Heating element	
Thermostat	Mercury
BFR-containing plastic	BFRs
Batteries	Lead, Lithium, Cadmium, Mercury
CFC,HCFC,HFC,HC	ODS
External electric cables	BFRs, plasticizers
Electrolyte capacitors	Glycol, other unknown substances

ISBN: 978-93-8830-599-0

Possible Hazardous Substances in Components of E-waste

The substances within the above mentioned components, which cause most concern are the heavy metals such as lead, mercury, cadmium and chromium(VI), halogenated substances (e.g. CFCs), polychlorinated biphenyls, plastics and circuit boards that contain brominated flame retardants (BFRs). BFR can give rise to dioxins and furans during incineration. Other materials and substances that can be present are arsenic, asbestos, nickel and copper. These substances may act as catalysts to increase the formation of dioxins during incineration.

EFFECTS OF E- WASTE

HEALTH EFFECTS OF SOME COMMON CONSTITUENTS IN E-WASTE

The health effects of heavy metals and certain compounds found commonly in components of e-waste are described below:

- **Lead:** Lead is used in glass panels and gaskets in computer monitors and in solder in printed circuit boards and other components. Lead causes damage to the central and peripheral nervous systems, blood systems, kidney and reproductive system in humans. It also affects the endocrine system, and impedes brain development among children. Lead tends to accumulate in the environment and has high acute and chronic effects on plants, animals and micro organisms (Metcalf & Eddy,2003).
- **Cadmium:** Cadmium occurs in surface mounted device (SMD) chip resistors, infra-red detectors, and semiconductor chips. Some older cathode ray tubes contain cadmium. Toxic cadmium compounds accumulate in the human body, especially the liver, kidneys pancreas, thyroid (Metcalf & Eddy, 2003, Basel Action Network, 2002).
- **Mercury:** It is estimated that 22 % of the yearly world consumption of mercury is used in electrical and electronic equipment. Mercury is used in thermostats, sensors, relays, switches, medical equipment, lamps, mobile phones and in batteries. Mercury, used in flat panel displays, will likely increase as their use replaces cathode ray tubes.

Mercury can cause damage to central nervous system as well as the foetus. The developing foetus is highly vulnerable to mercury exposure (Metcalf & Eddy, 2003). When inorganic mercury spreads out in the water, it is transformed to methylated mercury which bio-accumulates in living organisms and concentrates through the food chain, particularly via fish (Basel Action Network, 2002).

- **Hexavalent Chromium/Chromium VI:** Chromium VI is used as corrosion protector of untreated and galvanized steel plates and as a decorative or hardener for steel housings. Chromium VI can cause damage to DNA and is extremely toxic in the environment. Long term effects are skin sensitization and kidney damage(Metcalf & Eddy, 2003).
- **Plastics (including PVC):** The largest volume of plastics (26%) used in electronics has been poly vinyl chloride (PVC). PVC elements are found in cabling and computer housings. Many computer moldings are now made with the somewhat more benign acrylonitrile butadiene (ABS) plastic. Dioxins are released when PVC is burned (Basel Action Network, 2002)..
- **Brominated Flame Retardants (BFRs):** BFRs are used in the plastic housings of electronic equipment and in circuit boards to prevent flammability. BFRs are persistent in the atmosphere and show bioaccumulation. Concerns are raised considering their potential to toxicity (Basel Action Network, 2002).
- **Barium:** Barium is a soft silvery-white metal that is usedprotect users from radiation.

Studies have shown that short-term exposure to barium causes brain swelling, muscle weakness, damage to the heart, liver, and spleen(Basel Action Network, 2002).

- **Beryllium:** Beryllium is commonly found on motherboards and finger clips. Exposure to beryllium can cause lung cancer. Beryllium also causes a skin disease that is characterised by poor wound healing and wartlike bumps. Studies have shown that people can develop beryllium disease many years following the last exposure. It is used as a copper-beryllium alloy to strengthen connectors.

Barium is a soft silvery-white metal that is used to protect users from radiation.

- **Phosphor and additives:** Phosphor is an inorganic chemical compound that is applied as a coat on the interior of the CRT faceplate. Phosphor affects the display resolution and luminance of the images that is seen in the monitor.

ISBN: 978-93-8830-599-0

The phosphor coating on cathode ray tubes contains heavy metals, such as cadmium, and other rare earth metals, for example, zinc, vanadium as additives. These metals and their compounds are very toxic. This is a serious hazard posed for those who dismantle CRTs by hand

The Environmental Impact

NEED FOR GUIDELINES FOR ENVIRONMENTALLY SOUND MANAGEMENT-

The saying waste is misplaced wealth is true in the case of e-waste. The recyclability of e-waste and the precious metals that can be extracted from the waste make recycling a lucrative business. But recycling using environmentally sound means costly business and so majority of the e-waste is recycled via the informal sector. Informal recycling involves minimal use of technology and is carried out in the poorer parts of big cities. The standard recycling drill involves physically breaking down components often without any protective gear, burning poly vinyl chloride (PVC) wires to retrieve copper, melting of lead and mercury laden parts. The extraction of gold and copper requires acid processing. The plastic parts, which contain brominated flame retardants (BFR) are also broken into small pieces prior to recycle. All these processes release toxic fumes into the atmosphere and polluted water into soil and water bodies leading to contamination. Most of those who work in the recycling sector are the urban poor with low literacy lacking awareness of the hazards of the toxic e-wastes. Children and women are routinely involved in the operations. Most of the work is done by bare hands. Waste components which do not have resale value are openly burnt or disposed off in open dumps

Rapid pace of product obsolescence resulting in short life span of computers

The recovery potential (typical values) of items of economic value from refrigerator, personal computer and television are given in tables

Recoverable Quantity of Materials in a Refrigerator

Material Type	% (by weight)
CFCs	0.20
Oil	0.32
Ferrous Metals	46.61
Non-Ferrous Metals	4.97
Plastics	13.84
Compressors	23.80
Cables/Plugs	0.55
Spent Foam	7.60
Glass	0.81
Mixed Waste	1.30
Total	100.00

ENVIRONMENTALLY SOUND TREATMENT TECHNOLOGY

Environmentally sound E-waste treatment technologies:-

Environmentally sound E-waste treatment technologies are used at three levels as described below:

1. 1st level treatment
2. 2nd level treatment
3. 3rd level treatment

Analysis

All the three levels of e-waste treatment are based on material flow. The material flows from 1st level to 3rd level treatment. Each level treatment consists of unit operations, where e-waste is treated and out put of 1st level treatment serves as input to 2nd level treatment. After the third level treatment, the residues are disposed of either in TSDF or incinerated. The efficiency of operations at first and second level determines the quantity of residues going to TSDF or incineration. The simplified version of all the three treatments is shown in figure

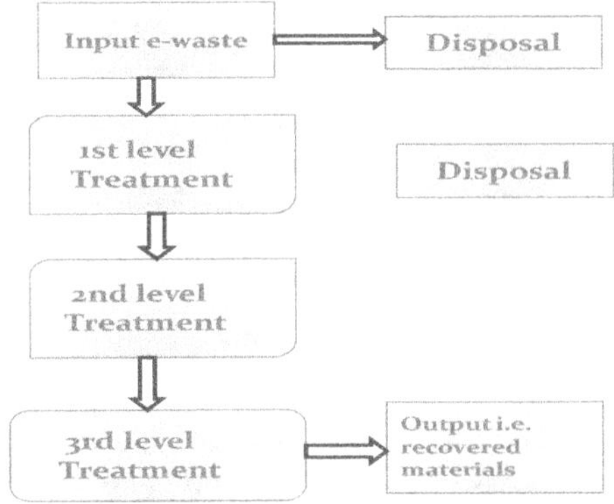

1st Level Treatment

Input: e-waste items like TV, refrigerator and Personal Computers (PC)

Unit Operations: There are three units operations at first level of e-waste treatment

1. Decontamination : Removal of all liquids and Gases
2. Dismantling -manual/mechanized breaking
3. Segregation

All the three unit operations are dry processes, which do not require usage of water.

1. **Decontamination:** The first treatment step is to decontaminate e-waste and render it nonhazardous.

   This involves removal of all types of liquids and gases (if any) under negative pressure, their recovery and storage.

2. **Dismantling:** The decontaminated e-waste or the e-waste requiring no decontamination are dismantled to remove the components from the used equipments.

ISBN: 978-93-8830-599-0

The dismantling process could be manual or mechanized requiring adequate safety measures to be followed in the operations.

3. **Segregation:** After dismantling the components are segregated into hazardous and nonhazardous components of e-waste fractions to be sent for 3rd level treatment.

Output:

1. Segregated hazardous wastes like CFC, Hg Switches, batteries and capacitors

2. Decontaminated e-waste consisting of segregated non-hazardous Ewaste like plastic, CRT, circuit board and cables

2nd Level Treatment:

Input: Decontaminated E-waste consisting segregated non hazardous e-waste like plastic, CRT, circuit board and cables.

Unit Operations: There are three unit operations at second level of E-waste treatment

1. Hammering
2. Shredding
3. Special treatment Processes comprising of
   (i) CRT treatment consisting of separation of funnels and screen glass.
   (ii) Electromagnetic separation
   (iii) Eddy current separation
   (iv) Density separation using water

Process:

The two major unit operations are hammering and shredding. The major objective of these two unit operations is size reduction. The third unit operation consists of special treatment processes. Electromagnetic and eddy current separation utilizes properties of different elements like electrical conductivity, magnetic properties and density to separate ferrous, non ferrous metal and precious metal fractions. Plastic fractions consisting of sorted plastic after 1st level treatment, plastic mixture and plastic with flame retardants after second level treatment, glass and lead are separated during this treatment. The efficiency of this treatment determines the recovery rate of metal and segregated E-waste fractions for third level treatment. The simplified version of this treatment technology showing combination of all three unit operations is given in Figure

Process flow of Non CRT based e-waste treatment:

The efficacy of the recycling system is dependent on the expected yields/ output of the recycling system. The expected yields/ output from the recycling system are dependent on the optimization of separation

E-WASTE RECYCLING & TREATMENT FACILITY

for establishment of E-waste Recycling & Treatment Facility shall be in line with the existing Guidelines/best practices/requirements in India for establishing and operating

"Recycling and Treatment and Disposal Facilities" for hazardous wastes.

Such facilities shall be set up in the organized sector. However, the activities presently operating in the informal sector need to be upgraded to provide a support system for the integrated facility. This would enable to bring the non-formal sector in the main stream of the activity and facilitate to ensure environmental compliances. The proposed mechanism for the e-waste facility is only an illustrative model and details have to beworked out to develop such facilities.

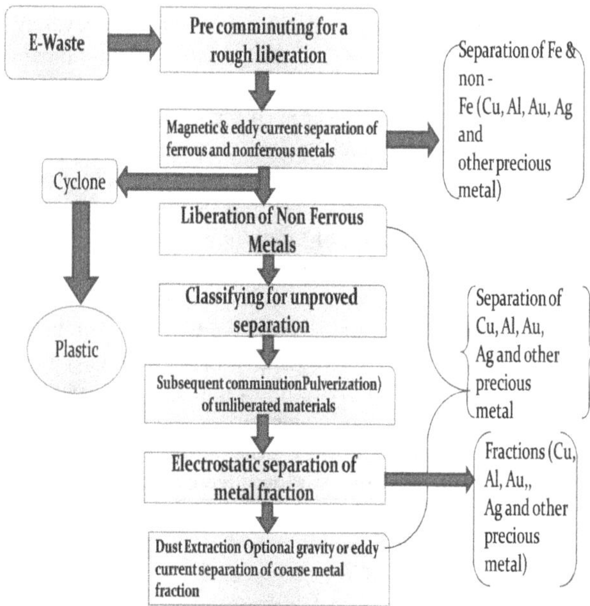

Facility Operation Requirements:
- **Collection**
- **Storage**
- **Dismantling & Segregation**
- **Recycling**
- **Treatment & Disposal**

Collection Systems for e-waste:

A producer is responsible for his products he may be involved in the establishment of the take back system for end of use electronic and electrical equipments. The producer responsibility could be either Individual or collective. Individual model requires each producer to be responsible for managing the ewaste generated by their products. The producer shall announce a take back system. The individual producers can have direct contact with dismantlers or recyclers which allows them to get back the re-usable components from their obsolete equipments. The producers can also get the data from the collector/dismantler/recycler about the specific composition and characteristics of the products. In the case of collective producer responsibility the producer would enter into a contractual agreements with a collection agency which would be responsible for collection

ISBN: 978-93-8830-599-0

of the waste from the generator. The producers through the collection agency have to pay a fixed price for their products to the generator, as in the collective responsibility model. The take back system may provide free collection or provide discount on purchase of new items. This facilitates in establishing a feasible and effective collection system to enable the channelization of the e-waste to appropriate recycling facilities and increasing reuse of certain components. The economic rationale behind is to facilitate the transfer of the benefits to the consumers enabling them to get better price on the sale of used equipments.

Collection Systems for e-waste

- **Storage areas:**

1. The storage areas for string the e-waste in a facility can be located within the facility - **on-site storage** or located at a place outside the facility – **offsite storage** including the ware houses. Such storage areas should be covered areas for storage of e-waste till such time that the waste is recycled or treated. The storages could also be the warehouses hired for this purpose.

2. Appropriate containers should be used for storing different e-waste items separately and there should be no mixing of different kinds of e-waste

3. The purpose of the weatherproof covering for storage at treatment sites is to minimize the contamination of clean surface and rain waters, to facilitate the reuse of those whole appliances and components intended for recycling and to assist in the containment of hazardous materials and fluids. The areas that are likely to require weatherproof covering will therefore include the storage areas and the treatment areas for the treating hazardous or fluid containing e-waste or whole appliances or components intended for recycling. The type of weatherproof covering required will depend of the types and quantities of waste and the storage and treatment activities undertaken. Weatherproof covering may in some circumstances simply involve a lid or cover over a container but in others it may involve the construction of a roofed building.

4. Impermeable surfaces should be provide for appropriate areas.
   "Impermeable surface" means a surface or pavement constructed and maintained to a standard sufficient to prevent the transmission of liquids beyond the pavement surface. The impermeable surface should be associated with a sealed drainage system and may be needed even where weatherproof covering is used. This means a drainage system with impermeable components which does not leak and which will ensure that no liquid will run off the pavement other than via the system and all liquids entering the system are collected in a sealed sump.

5. Appropriate spillage collection facilities should be provided. The spillage collection facilities include the impermeable pavement and sealed drainage system as the primary means of containment. However, spill kits to deal with spillages of oils, fuel and acids should be provided and used as appropriate.

6. Appropriate sites must provide appropriate storage for disassembled spare parts. Some spare parts (e.g. motors and compressors) will contain oil and/or other fluids. Such parts must be appropriately segregated and stored in containers that are secured such that oil and other fluids cannot escape from them. These containers must be stored on an area with an 49impermeable surface and a sealed drainage system.

Storage areas:

STORE ROOM

STRATEGIES FOR COMBATING E- WASTE

LEGISLATION

Separate legislation for dealing with waste electrical and electronic equipments to control aspects of production, recycle, reuse and disposal is need of the hour. Many countries have such laws in place. In India, draft e-Waste (Management and Handling) Rules have been published by the Ministry of Environment and Forests, Government of India on 14.5.2010.

EXTENDED PRODUCER RESPONSIBILITY (EPR)

Traditionally, the legislative approach toward environmental problems has been one of 'command and control', largely addressing 'end-of-pipe' pollution problems. Now, the emphasis is changing towards producer responsibility whereby those who produce good sare then responsible for the environmental impacts throughout the whole of their life cycle, from resource extraction to recycling, reuse and disposal (Nnorom et.al, 2008). Implementation of EPR in the developing countries has become necessary in the light of the present high level of trans-boundary movement of e-waste into the developing countries and the absence of basic or state-of the-art facilities for sound end-of-life material/energy recovery and disposal of e-waste.

The Organization for Economic Cooperation and Development(OECD) defined EPR as "an environmental policy approach in which a producers' responsibility for a product is extended to the post-consumer stage of a products life cycle including its final disposal"

The main goals of EPR are:
- waste prevention and reduction;
- product reuse;
- increased use of recycled materials in production;
- reduced natural resource consumption;
- internalization of environmental costs into product prices
- energy recovery when incineration is considered appropriate

Under EPR, the producer is expected to take back all electrical and electronic equipment at the end of their life.

REDUCTION IN USE OF HAZARDOUS SUBSTANCES (ROHS)

This aims at reducing the hazardous substances entering the atmosphere while dismantling the e-waste by prescribing threshold limits for use of such substances in e-waste.

HOW ELECTRONICS WASTE GET RECYCLED

EQUIPMENT STRUCTURE OF E- WASTE PLANT

FACT SHEET OF THE MAHABUBNAGAR

TOTAL E- WASTE PRODUCE FROM DIFFERENT SOURCES ARE

Social Infrastructure		PRODUCE WASTE %
Hospitals:	60	72%
Dispensaries :	90	63%
Public Health Centres :	20	15%
Education		
High Schools :	200	20%
Junior Colleges:	13	12%
Degree Colleges:	9	15%
Post Graduate Centres:	2	10%
Engineering Colleges:	3	16%
Medical Colleges:	2	41%
Polytechnical Colleges:	2	10%
Colleges of Education:	5	10%
DIET College:	1	7%

house holds

(containing PC, TV, Mobile Phone etc.)

Material	% Composition (by weight)
Mild steel	23
Stainless steel	8
Glass	27
Plastics	27
Copper	3
Aluminium	3
Other materials	8

House hold Appliances –waste %

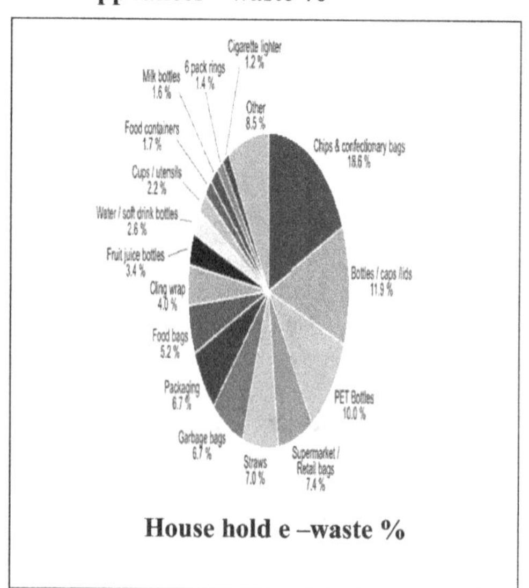

House hold e –waste %

ISBN: 978-93-8830-599-0

These are the household appliances where we use in our daily life and according to time period of the product it is workable and then it is consider as a waste and the minimum and maximum products percentage are shown in pie chart which is given above.

S.no	E waste contributors	Per day%	Per annum	Annual %
1.	**NOKIA CARE** Mobiles Chargers Battries	1kg	365	3.39
2.	**OTHER MOBILE STORES** Chips Wires Panels Keypads	2kg	730	6.78
3.	**SCRAPS SHOPS** Refregirator Televisions Washingmachines Plastics Ovens Etc.	10kg	3650	33.89
4.	**HOUSEHOLD APPLINCES** Mixers Ovens Hair dryers Vaccum cleaner Video playesrs Compact discs	5kg	1825	16.95
5.	**COMPUTER SHOPS** Chips, CDs, CRTS, Memory cards, mother boards Ram chips Pendrives Cattridges Scanners c.p.u smps etc	5kg	1825	16.95
6.	**ELECTRIC SHOPS:** Switch boards Wires Fluorescent bulbs Switches Battries Capacitors Pipes	4kg	1460	13.56
7.	**HOSPITALS:** Plastic needles Syernges Glass(injections tube) Etc.,	2.5kg	912.5	8.47

ISBN: 978-93-8830-599-0

CONCLUSION

- Electronic and electrical equipments cannot be avoided in today's world. So also is the case of waste electronic and electrical equipments.
- As long as this is a necessary evil, it has to be best managed to minimize its adverse impacts on environment.
- Through innovative changes in product design under EPR **(EXTENDED PRODUCER RESPONSIBILITY)**, use of environmentally friendly substitutes for hazardous substances, these impacts can be mitigated.
- A legal framework has to be there for enforcing EPR, RoHS for attaining this goal. Adoption of environmentally sound technologies for recycling and reuse of e-waste along with EPR and RoHS offers workable solution for environmentally sound management of e-waste.
- Manufacturers & Suppliers to set goals for reducing electronic waste. Encourage them to buy back old electronic products from consumers Disposing bulk e-waste only through authorized recyclers Send non-tradable e-waste to authorized private developers for final disposal.
- Store the electronics products in store houses which are not in use .this is one of the technique to protect environment from hazardous waste gases.
- So, we have surveyed in mahabubnagar that how much percentage of e-waste is getting disposed per annum.

- From different industries like mobile stores, electric shops. Household products, scrap shops, computers shops etc.,
- It is our duty to protect our environment from the hazardous waste to better life of further generation.
- It can also protectable to whole ecosystem.
- Todays electronics gadgets are tommrows E- waste

REFERENCES

1. Bandhopadhyay, A. (2010) "Electronic Waste Management: Indian Practices and Guidelines" *International Journal of Energy and Environment* 1(5) pp. 193-807
2. Basel Convention on the Control of Transboundary Movement of Hazardous Wastes and Their Disposal – Document accessed in 10/2010
3. Sathish Sinha (2006) E-waste Time to Act Now – Toxic Alert, accessed in 10/2010
4. Radha Gopalan, 2002, A Study on the Indian IT Sector from nautilus.org
5. Scraping the Hi-tech Myth – Computer Waste in India, 2003, Toxics Link Indian Institutes of Materials Management/Publications
6. Environmentally Sound options for E-WASTES Management. By: *Ramachandra T.V..,*Saira Varghesek. Published By: Envis Journal of Human Settlements, March 2009.
7. Global E-Waste Management & Services (GEMs), Hyderabad.
8. National Environment Agency of Singapore (NEA). March 1998. Hazardous Waste (Control of Export, Import and Transit) Regulations.

ISBN: 978-93-8830-599-0

Phytoremediation of Heavy Metals from
Pragathi Nagar Cheruvu using Aquatic Plants

Hepsi Swarupa Rani Guduri[1] and C. Sarala[2]

[1]Research Scholar and [2]Professor,

Centre for Water Resources, IST, Jawaharlal Nehru Technological University, Hyderabad, Telangana, India.

[1]hepsiguduri@gmail.com, [2]sarala2006@gmail.com

ABSTRACT

Increasing urbanization, industrialization and over population is one of the leading causes of environmental degradation and pollution. Heavy metals such as Cu,Cr,Pb, Ni, Cd etc. are one of the most toxic pollutants which show hazardous effects on all living livings. Lead is one such pollutants which disrupts the food chain and is lethal even at low concentrations. The prevailing purification technologies used for removal of contaminants from wastewater are not only very costly but causes negative impact on ecosystem subsequently. Phyto remediation, an ecofriendly technology which is both ecologically sound and economically viable is an attractive alternative to the current cleanup methods that are very expensive. This technology involves efficient use of aquatic plants to remove, detoxify or immobilize heavy metals. The purpose of this review was to assess the current state of phytoremediation as an innovative technology and to discuss its usefulness and potential in the remediation of contaminated water.

Keywords: Heavy metals, Surface water samples, Phytoremediation, Aquatic plants.

INTRODUCTION

An excess level of heavy metals are exposed into environment, for example by industrial waste and fertilizers causes serious concern in nature as they are non- biodegradable and accumulate at high levels. Heavy metal pollution is a global problem, although severity and levels of pollution differ from place to place. At least 20 metals are classified as toxic with half of them emitted into the environment that poses great risks to human health. The common heavy metals like Cd, Pb, Cu, Ni and Cr etc. are phytotoxic at both low concentration as well as very high concentration are detected in waste water. If these metals are presented in sediments then these reach the food chain through plants and aquatic animals. In small quantities, certain heavy metals are nutritionally essential for a healthy life, but large amounts of any of them may cause acute or chronic toxicity (poisoning).

So, the search for a new, simple, effective and ecofriendly technology involving the removal of toxic heavy metal from wastewater has directed attention towards phyto remediaton. The aquatic plant have been used for removal of heavy metals are Eichhorniacrassipes.

Objectives of the Study

(i) To analyze the heavy metals present in the surface water samples.

(ii) To identify the cost effective method for treating the polluted surface water.

(iii) To prepare the heavy metals concentration graphs

STUDY AREA

The present study area Pragathi Nagar is a residential colony in Kukatpally, Hyderabad, Rangareddy District in Telangana State, India. It's a major residential suburb in Kukatpally and situated at a distance of 3 kilometers from Kukatpally on National Highway 9 which leads to the city of Mumbai. It is about 2 and half kilometers from the famous university called Jawaharlal Technological University, Hyderabad. It has an average elevation of 33 metres (108 ft). Pragathinagar Study area which is lies in between 17.53'27'' North Latitude and 78.39,64'' East Longitude.

The surface water quality analysis has been carried out for the thirteen samples collected at Peddanalla Starting, Peddanalla ending, Pedda Banda, Allugu, Allugu1, Allugu2, Pragathi Nagar Bridge starting, Pragathi Nagar Bridge ending, Near fish market beginning, Near fish market ending, Sofit, Sofit1, Sofit2 Heritage Bus stop. In the present research work attempts have been made to detect the anions present in the collected samples. The location map of the study area is shown in figure1.

ISBN: 978-93-8830-599-0

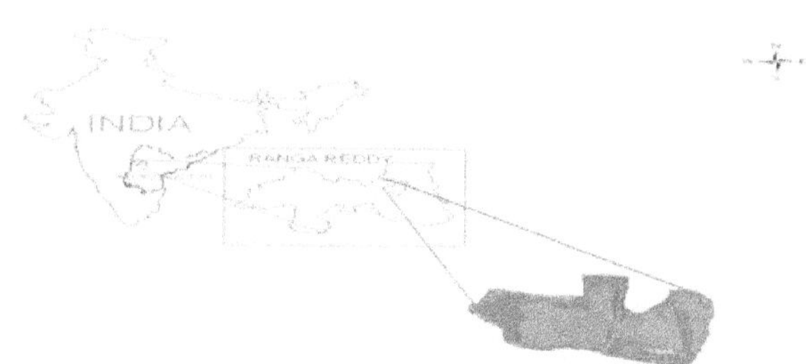

Fig. 1 Location Map of Study Area

METHODOLOGY

In this work, the data used is Survey of India Topographic maps and sample locations taken with the help of Global Positioning Systems (GPS). Thirteen surface water samples were collected from the Pragathi NagarCheruvu lake during post monsoon and pre monsoon periods of January 2015, June 2015, January 2016, June 2016 and January 2017.

The surface water samples were analyzed for heavy metals namely Copper, Nickel, Chromium, Lead and Cadmium.

The heavy metal were analyzed by using Atomic Absorption Spectrometer and the obtained values are compared with IS:10500 standards.

The surface water samples were treated for the heavy metals by adopting phyto-remediation method.

For the present study area, the phyto-remediation method is carried out by using aquaticplant

EICHHORNIA CRASSIPES.

PHYTOREMEDIATION

Phytoremediation is a cost-effective plant-based approach of remediation. It takes the advantage of the ability of plants to concentrate elements and compounds from the environment and to metabolize various molecules in their tissues.It refers to the natural ability of certain plants called hyperaccumulators to degrade, or render harmless contaminants in soils, water or air. Toxic heavy metals and organic pollutants are the major targets for phytoremediation. Knowledge of the physiological and molecular mechanisms of phytoremediation began to emerge in recent years together with biological and engineering strategies designed to optimize and improve phytoremediation. The feasibility of using plants for environmental cleanup was carried out worldwide.

RESULTS AND DISCUSSION

The surface water samples were analyzed for copper, nickel, zinc lead and iron during post monsoon and pre monsoon periods of January 2015, June 2015, January 2016, June 2016 and January 2017

Table 1 COPPER Metals Analysis for the Study Area

SAMPLE Locations	IS 10500 Standard	Copper Metal Values in mg/l				
		Jan-15	Jun-15	Jan-16	Jun-16	Jan17
PeddaNalla Starting		3.558	0.76	4.143	2.4	3.406
PeddaNalla Ending		2.592	0.84	2.616	2.3	2.973
Pedda Banda		2.39	0.25	2.514	2.13	2.643
Allugu		2.897	0.08	3.073	3.1	2.668
Allugu -I		3.965	0.08	3.835	2.2	2.796
Allugu- II		2.313	0.0508	2.6416	1.9	3.05
Pragathi Nagar Bridge Starting	0.05 - 1.5 mg/l	3.05	0.2	3.048	2.2	2.668
Pragathi Nagar Bridge Ending		3.042	0.3	3.556	2.5	2.668
Near Fish Market Beginning		2.897	0.77	3.81	2.5	2.059
Near Fish Market Ending		1.779	1.3	3.759	2.44	2.033
Sofit 1		3.101	1.14	4.419	2.3	2.491
Sofit 2		3.151	1.62	3.149	1.8	2.872
Heritage Bus Stop		3.558	0.79	4.8006	2.3	2.541

ISBN: 978-93-8830-599-0

Table 2 NICKELMetals Analysis for the Study Area

SSAMPLE Locations	IS10500 Standard	Nickel Metal Values in mg/l				
		Jan-15	Jun-15	Jan-16	Jun-16	Jan-17
PeddaNalla Starting	0.02mg/l	3.756	1.173	2.958	2.934	3.756
PeddaNalla Ending		3.59	1.244	3.052	2.582	1.995
Pedda Banda		2.653	0.634	2.629	2.371	2.817
Allugu		1.878	0.469	1.361	2.559	2.465
Allugu -I		1.737	0.351	1.638	3.052	2.277
Allugu- II		1.667	0.775	1.784	2.136	2.535
Pragathi Nagar Bridge Starting		1.76	0.916	2.911	2.582	2.629
Pragathi Nagar Bridge Ending		3.042	0.775	2.652	2.089	1.878
Near Fish Market Beginning		2.897	0.634	3.075	1.901	2.112
Near Fish Market Ending		1.779	0.705	2.793	2.629	2.347
Sofit 1		3.101	0.539	2.394	2.629	2.465
Sofit 2		3.151	0.391	2.112	2.512	2.77
Heritage Bus Stop		3.558	0.047	1.995	2.676	2.324

Table 3 CHROMIUM Metals Analysis for the Study Area

SAMPLE Locations	Chromium Metal Values in mg/l					
	IS 10500 Standard	Jan-15	Jun-15	Jan-16	Jun-16	Jan-17
PeddaNalla Starting	0.05 mg/l	2.615	2.615	2.955	2.715	3.138
PeddaNalla Ending		2.432	3.4	2.614	3.01	3.138
Pedda Banda		2.615	2.929	2.615	2.828	2.615
Allugu		2.354	3.138	2.51	3.1	3.923
Allugu -I		2.877	2.981	2.641	2.777	3.923
Allugu- II		2.589	2.781	1.15	3.871	4.211
Pragathi Nagar Bridge Starting		2.615	3.4	3.086	3.121	4.211
Pragathi Nagar Bridge Ending		2.092	2.615	3.138	3.923	4.158
Near Fish Market Beginning		2.617	3.243	4.211	4.282	4.447
Near Fish Market Ending		2.328	3.007	3.138	3.327	3.661
Sofit 1		2.798	3.138	3.191	3.289	3.426
Sofit 2		2.354	2.877	2.51	3.204	3.426
Heritage Bus Stop		2.72	2.615	2.877	3.105	3.792

ISBN: 978-93-8830-599-0

Table 4 LEAD Metals Analysis for the Study Area

SAMPLE Locations	IS 10500 Standard	Lead Metal Values in mg/l				
		Jan-15	Jun-15	Jan-16	Jun-16	Jan-17
PeddaNalla Starting	0.01mg/l	31.08	29.008	25.27	18.648	25.9
PeddaNalla Ending		30.04	29.1	22.792	19.27	29.6
Pedda Banda		33.15	28.39	21.341	4.144	30.04
Allugu		31.49	27.98	22.792	0	26.93
Allugu -I		30.04	30.46	25.071	20.93	14.71
Allugu- II		29.62	33.16	21.134	21.54	29.08
Pragathi Nagar Bridge Starting		32.53	28.81	24.864	20.51	30.04
Pragathi Nagar Bridge Ending		33.15	22.99	27.143	4.35	29.62
Near Fish Market Beginning		40.19	27.14	20.305	7.88	30.25
Near Fish Market Ending		37.29	32.53	23.413	9.32	29.42
Sofit 1		31.28	21.34	24.864	4.56	30.04
Sofit 2		38.33	31.29	20.927	1.243	30.87
Heritage Bus Stop		27.97	32.32	27.557	20.79	24.86

Table 5 CADMIUM Metals Analysis for the Study Area

SAMPLE Locations	IS 10500 Standard	Cadmium Metal Values in mg/l				
		Jan-15	Jun-15	Jan-16	Jun-16	Jan-17
PeddaNalla Starting	0.003 mg/l	0.28	0.2007	0	0.0028	0.015
PeddaNalla Ending		0.0512	0.2521	0.1822	0.001	0
Pedda Banda		0.046	0.0639	0	0	0
Allugu		0	0	0.209	0.0059	0
Allugu -I		0.212	0.064	0	0.029	0
Allugu- II		0.2004	0.2305	0	0.0032	0
Pragathi Nagar Bridge Starting		0.1928	0.1867	0.0002	0.0003	0
Pragathi Nagar Bridge Ending		0	0	0.1693	0	0
Near Fish Market Beginning		0.007	0	0.007	0.0058	0
Near Fish Market Ending		0.0053	0	0.0256	0.0034	0
Sofit 1		0.0028	0	0.0145	0.0117	0
Sofit 2		0.015	0.0021	0.0035	0.1881	0.0115
Heritage Bus Stop		0.0028	0.0845	0.0073	0.1901	0.0108

Removal of Heavy metals by phytoremediation

In the study area the aquatic plant Eichhornia Crassipes, is used to removal heavy metals concentration in the collected surface water samples.

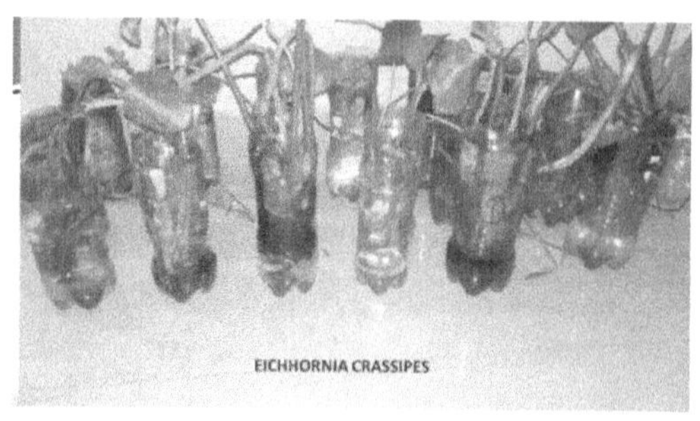

ISBN: 978-93-8830-599-0

The graphs are prepared for the percentage efficiency of heavy metal removal by each aquatic plant for all the water samples collected during the post monsoon and pre monsoon periods.

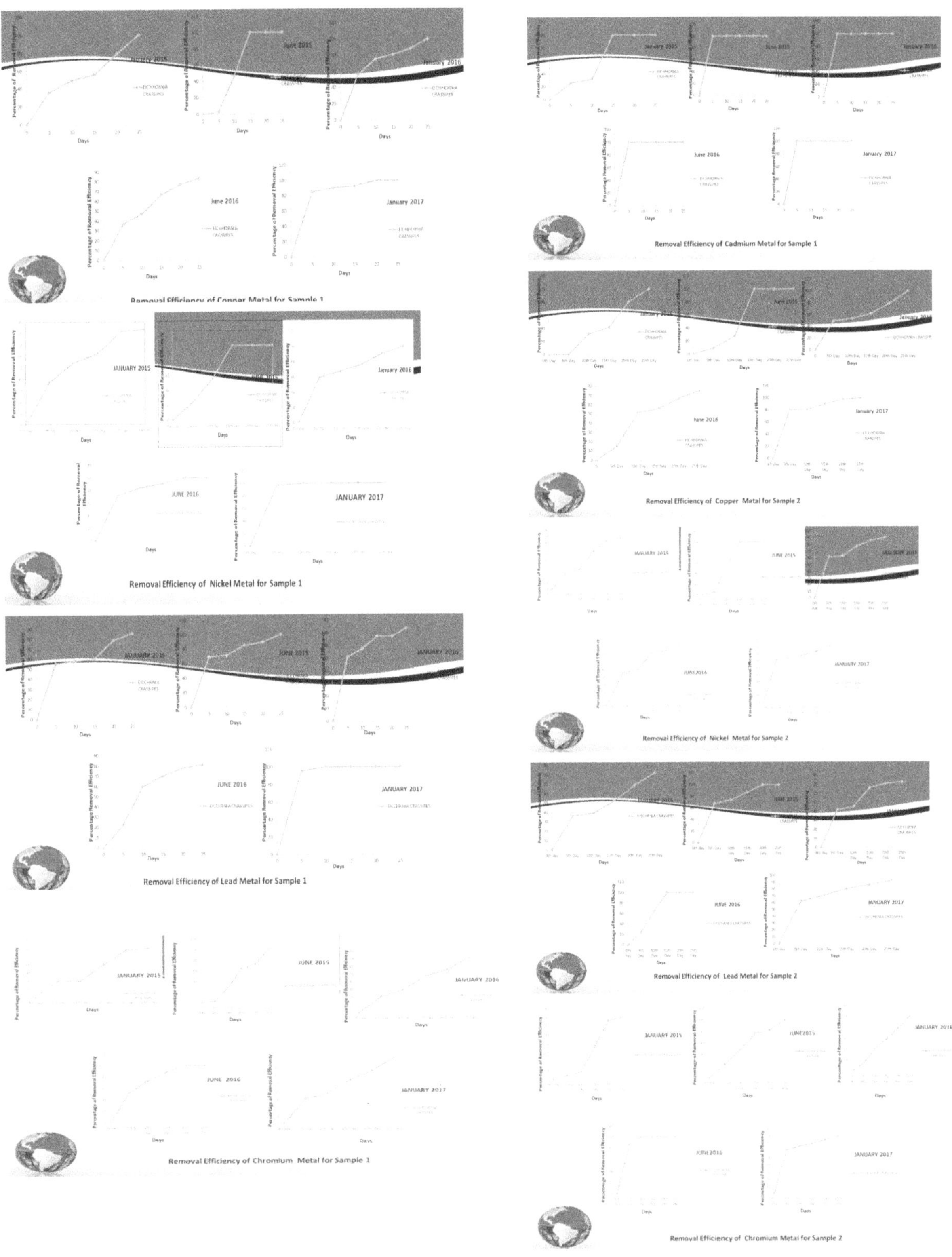

ISBN: 978-93-8830-599-0

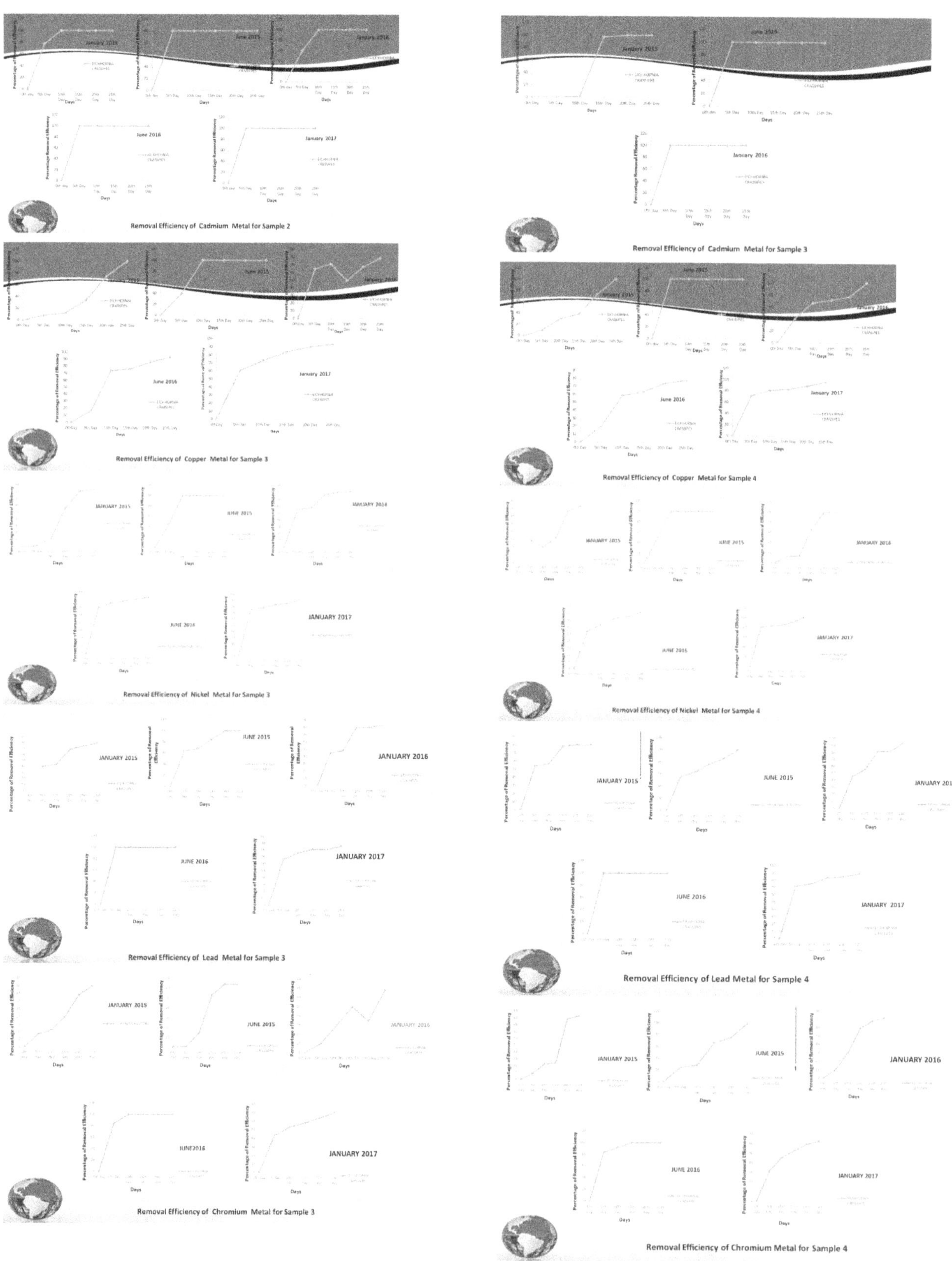

Removal Efficiency of Cadmium Metal for Sample 2

Removal Efficiency of Cadmium Metal for Sample 3

Removal Efficiency of Copper Metal for Sample 3

Removal Efficiency of Copper Metal for Sample 4

Removal Efficiency of Nickel Metal for Sample 3

Removal Efficiency of Nickel Metal for Sample 4

Removal Efficiency of Lead Metal for Sample 3

Removal Efficiency of Lead Metal for Sample 4

Removal Efficiency of Chromium Metal for Sample 3

Removal Efficiency of Chromium Metal for Sample 4

ISBN: 978-93-8830-599-0

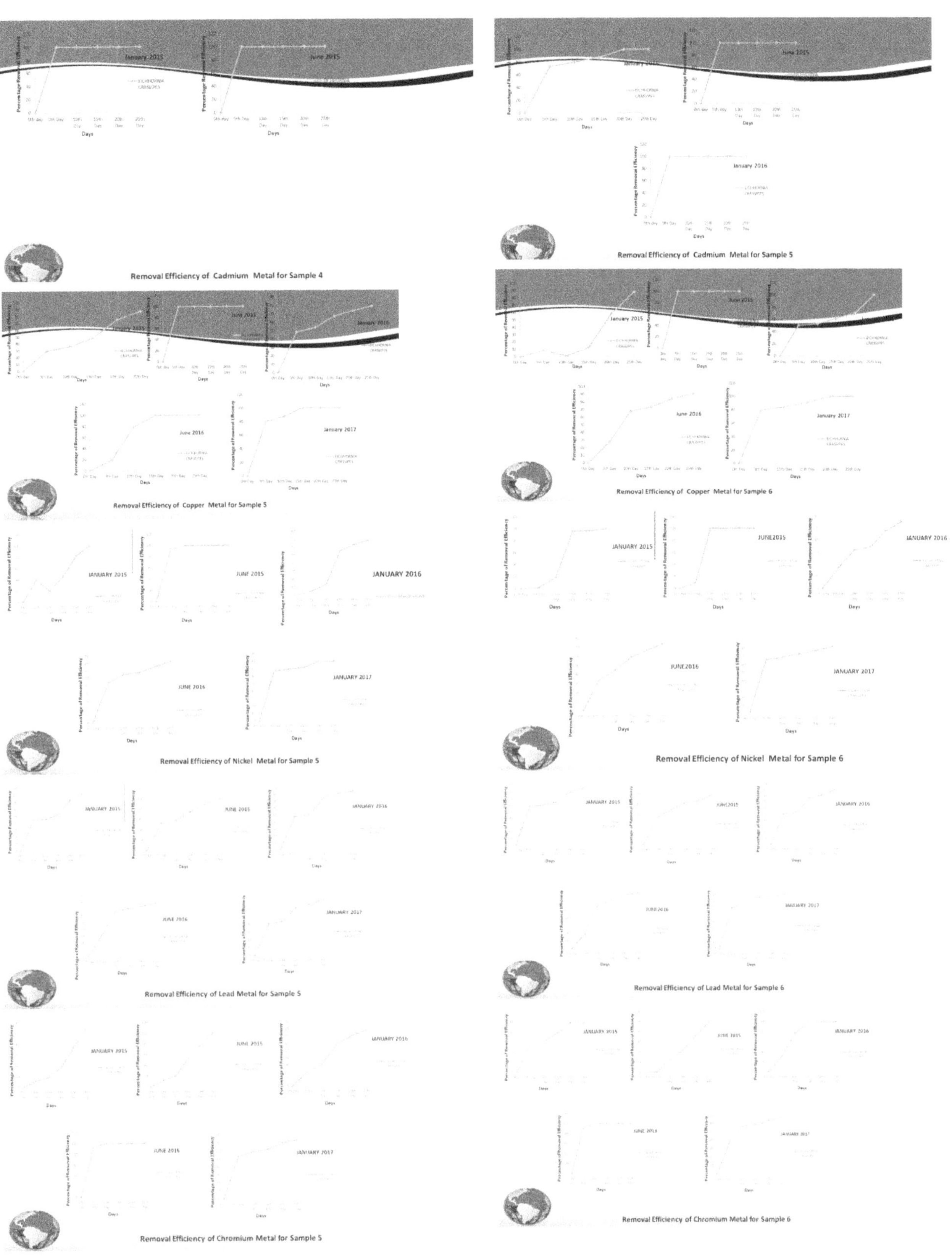

Removal Efficiency of Cadmium Metal for Sample 4

Removal Efficiency of Cadmium Metal for Sample 5

Removal Efficiency of Copper Metal for Sample 5

Removal Efficiency of Copper Metal for Sample 6

Removal Efficiency of Nickel Metal for Sample 5

Removal Efficiency of Nickel Metal for Sample 6

Removal Efficiency of Lead Metal for Sample 5

Removal Efficiency of Lead Metal for Sample 6

Removal Efficiency of Chromium Metal for Sample 5

Removal Efficiency of Chromium Metal for Sample 6

ISBN: 978-93-8830-599-0

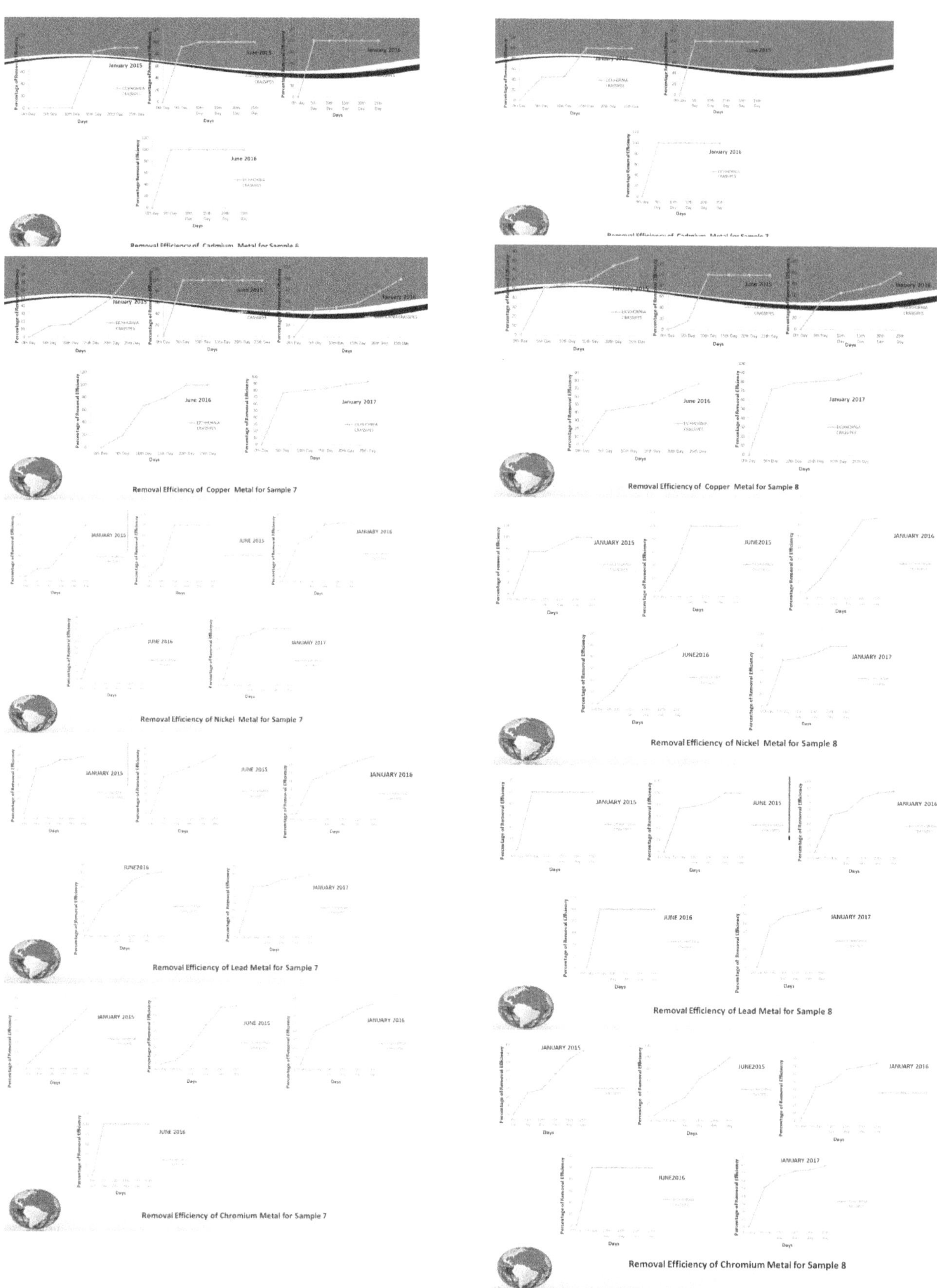

Removal Efficiency of Cadmium Metal for Sample 6

Removal Efficiency of Cadmium Metal for Sample 7

Removal Efficiency of Copper Metal for Sample 7

Removal Efficiency of Copper Metal for Sample 8

Removal Efficiency of Nickel Metal for Sample 7

Removal Efficiency of Nickel Metal for Sample 8

Removal Efficiency of Lead Metal for Sample 7

Removal Efficiency of Lead Metal for Sample 8

Removal Efficiency of Chromium Metal for Sample 7

Removal Efficiency of Chromium Metal for Sample 8

ISBN: 978-93-8830-599-0

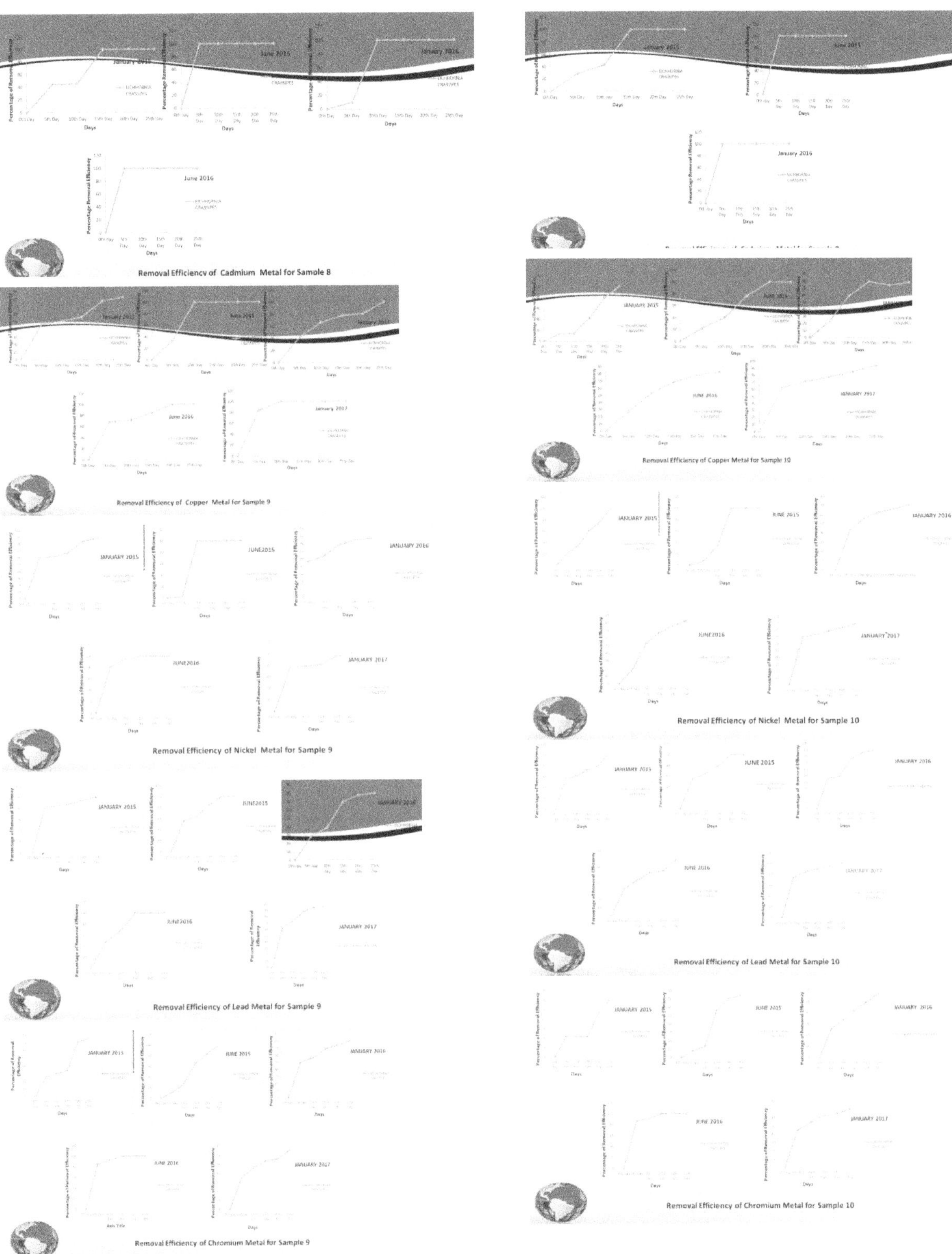

Removal Efficiency of Cadmium Metal for Sample 8

Removal Efficiency of Copper Metal for Sample 9

Removal Efficiency of Nickel Metal for Sample 9

Removal Efficiency of Lead Metal for Sample 9

Removal Efficiency of Chromium Metal for Sample 9

Removal Efficiency of Copper Metal for Sample 10

Removal Efficiency of Nickel Metal for Sample 10

Removal Efficiency of Lead Metal for Sample 10

Removal Efficiency of Chromium Metal for Sample 10

ISBN: 978-93-8830-599-0

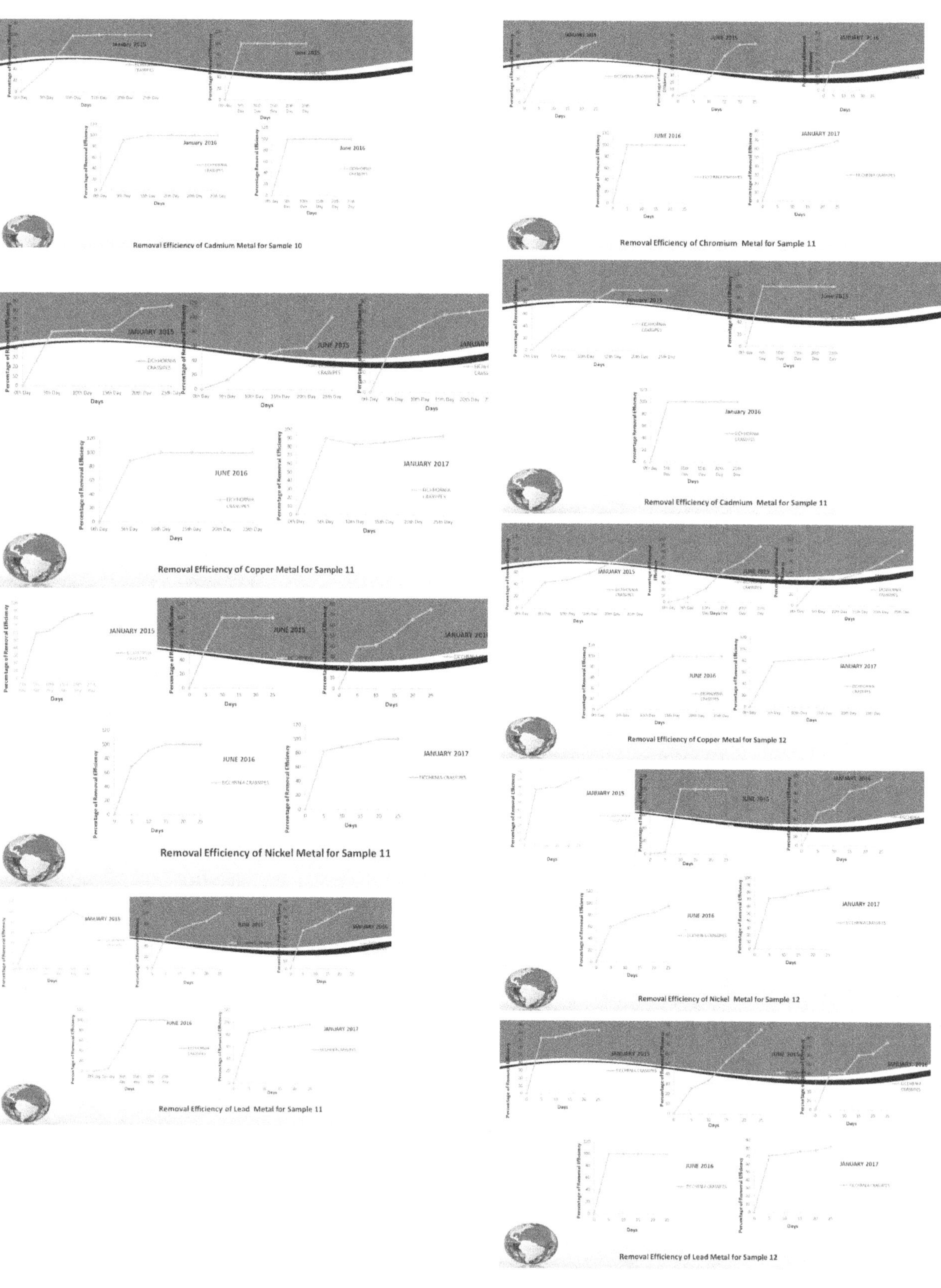

Removal Efficiency of Cadmium Metal for Sample 10

Removal Efficiency of Chromium Metal for Sample 11

Removal Efficiency of Copper Metal for Sample 11

Removal Efficiency of Cadmium Metal for Sample 11

Removal Efficiency of Nickel Metal for Sample 11

Removal Efficiency of Copper Metal for Sample 12

Removal Efficiency of Nickel Metal for Sample 12

Removal Efficiency of Lead Metal for Sample 11

Removal Efficiency of Lead Metal for Sample 12

ISBN: 978-93-8830-599-0

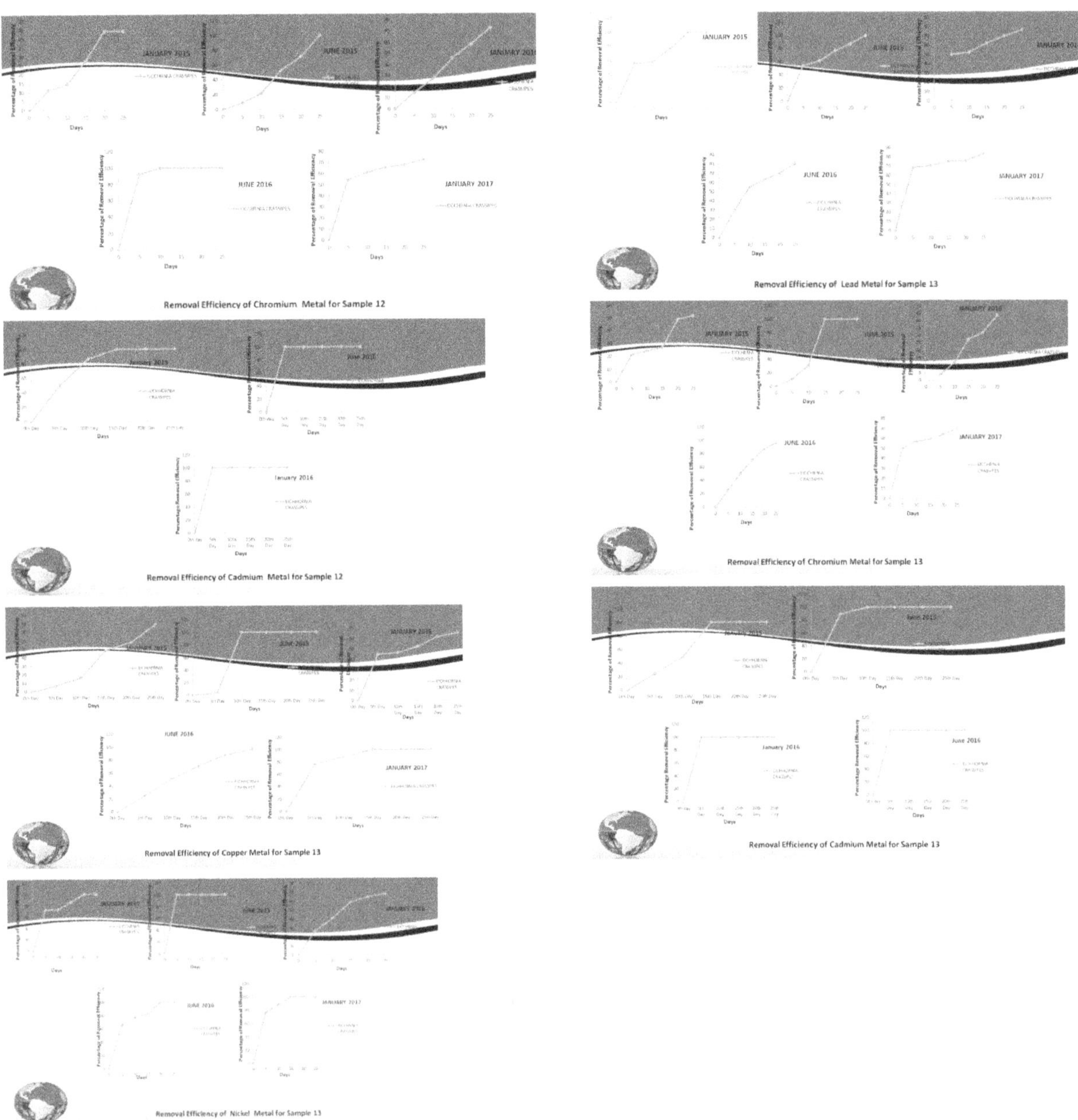

Abstract of the Heavy metals Removal in the Study area by Phyto remediation

DURATION	NAME OF HEAVY METAL	100 PERCENT REMOVAL EFFICIENCY OF AQUATIC PLANT	SAMPLE LOCATION
June-2015	COPPER	Eichhorniacrassipes 20th Day	1,2,3,4,5,6,7,8,9,10,13
Jan-2016	COPPER	Eichhorniacrassipes 25th Day	1,2,4,6,7,8,9,12
Jan-2017	COPPER	Eichhorniacrassipes 25th Day	1,2,5,6,9,12,13
Jan-2017	NICKEL	Eichhorniacrassipes 25th Day	1,4,7,8,9,10,13
Jan-2015	LEAD	Eichhorniacrassipes 25th Day	8,9,10,13
June-2015	LEAD	Eichhorniacrassipes 25th Day	1,2,3,4,5,6,8,9,10,12,13

Contd...

ISBN: 978-93-8830-599-0

DURATION	NAME OF HEAVY METAL	100 PERCENT REMOVAL EFFICIENCY OF AQUATIC PLANT	SAMPLE LOCATION
June-2016	LEAD	Eichhorniacrassipes 25th Day	2,3,4,8,9,11,12
Jan 2017	LEAD	Eichhorniacrassipes10th Day	1,10
June-2015	CHROMIUM	Eichhorniacrassipes 20th Day	1,2,4,5,6,9,10,12,13
June-2016	CHROMIUM	Eichhorniacrassipes10th Day	2,3,5,6,7,8,11,12,13
June-2015	CADMIUM	Eichhniacrassipes 5th Day	1,2,3,4,5,6,7,8,910,11,12,13
Jan-2017	CADMIUM	Eichhniacrassipes 10th Day	1,2,3,4,5,6,7,8,9,10,11,12,13
Jan-2016	CADMIUM	Eichhniacrassipes 5th Day	1,2,3,4,5,6,7,8,9,10,11,12,13
June-2016	CADMIUM	Eichhniacrassipes 20th	1,2,3,4,5,6,7,8,9,10,11,12,13

CONCLUSIONS

The Copper values in the study area varied between 0.08 – 4.8 mg/l.

- The highest value is observed at Heritage Bus Stop during June 2016.
- The Nickel values in the study area varied between 0.04 – 3.75 mg/l.
- The highest value is observed at PeddaNalla Starting during January 2017.
- The Lead values in the study area varied between 0 - 33.16 mg/l.
- The highest value is observed at Allugu-II during January 2015.
- The Chromium values in the study area varied between 1.15 – 4.44 mg/l.
- The highest value is observed at Near Fish Market Beginning during January 2017.
- The Cadmium values in the study area varied between 0 - 0.2 mg/l.
- The highest value is observed at Alluguduring January 2016.
- Whereas Copper metal can be removed completely by using Eichhorniacrassipes by 20th Day.
- In the post monsoon period samples of 2017, Cadmium metal and Chromium metal can be removed completely by using Eichhornia Crassipes plant by 5th day for all samples.
- The left over biomass of all these aquatic plants can be used for vermi composting and to increase the yield of fish in aquaculture.

REFERENCES

1. Mudgal Varsha, Madaan Nidhi and Mudgal Anurag Heavy metals in plants: phtyremediation : Plants used to remediate heavy metal pollution. Agri. Biol. J. Am. 1(1): 40-46. (2010).
2. Baharudin, B. and Shahrel, Mohd Lead and cadmium removal in synthetic wastewater using constructed wetland. Faculty of Chemical & Natural Resources Engineering Universiti, Malaysia Pahang. (2008).
3. ErakhrumenAgbontalor Andrew Phytoremediation: an environmentally sound technology for pollution prevention, control and remediation in developing countries. Educational Research and Review, 2 (7):151-156. (2007).
4. Babarinde, N.A.A., Babalola, J.O. and Sanni Adebowale, R.A Biosorption of lead ions from aqueous solution by maize leaf. International Journal of the Physical Sciences, 1(1):23-26.. (2006).
5. Miretzky Patricia, Saralegui Andrea and CirelliFerna´ndez Alicia Simultaneous heavy metal removal mechanism by dead macrophytes. Chemosphere, 62: 247–254. (2005).
6. Chong, Y., Hu, H. and Qian, Y Effects of inorganic nitrogen compounds and pH on the growth of duckweed. J. Environ. Sci., 24:35-40. . (2003).
7. David Tin Win, ThanMyintMyint and Tun Sein Lead removal from industrial waters by water hyacinth. AU J. T., 6(4):187-192. (2003).
8. Prassad, M.N.V., Malec, P., Waloszek, A., Bojka, M. and Strzallka, KPhysiological responses of Lemnatrisulca (duckweed) to cadmium and copper bioaccumulation. Plant Sci., 161: 881-889. (2001).
9. Cunningham, S. D. Huang, J. W. Chen, J. & Berti, W.R Abstracts of Papers of the American Chemical Society. 212, 87. (1996).
10. Nor, Y.M. The absorption of metal ions by Eichhorniacrassipes. Chemical Speciation and Bioavailability, 2: 85– 91. (1990).

ISBN: 978-93-8830-599-0

Hacking of Weather

Md Jalal Uddin and Mohammed Abdul Shahab

Asst Professor, Civil Dept

Abstract

In 200,000 years old human history of existence on planet earth humans have evolved and have been striving to get better every day. In this process humans have altered every natural method to match there purposes and satisfy their needs. Humans therefore have reworked every irrigation process, cultivation process, river systems ETC; and have moved towards much more productivity methods in every field of life but humans couldn't alter the weather according to their desire.

Weather has been an important factor for any process to occur on this magnificent and beautiful planet. The word weather roughly means "the state of the atmosphere at a particular place and time as regards heat, cloudiness, dryness, sunshine, wind, rain ETC". Of late humans have tried to alter the weather according to their desire to fulfil their needs since late 1990.

In our project we are trying to perceive the relation between "weather, irrigation and rainfall". This topic is a part of climate engineering, rainfall engineering and Irrigation engineering which are new streams of civil engineering domain.

This civil engineering project gives an overview of the modern methods which have been recently adopted by various developed countries to make weather or precipitation act as they desire for. Hence, it means they have hacked the natural process of precipitation and made it artificial.

Our project also tries to know the plight of farmers facing drought and floods due to irregularities in rainfall and steps taken by various central and state government organisations to rescue them from this crisis .But to our fate our government is least bothered about the farmers and hence leading to huge loss of population of this beautiful nation and its honest citizens by disasters which leave them helpless and hence in many cases farmers take extreme steps such as suicide and moving towards urbanisation.

Our project also consists of photographic references and various other data collected from meteorological department, forest department, irrigation department, land records department, Agriculture department and RTI.

Keywords: Climate engineering, rainfall engineering, environmental engineering, rainfall pattern droughts and floods.

In present day world due to pollution and various manmade faults such as storing more water than required from rivers which disturbs the natural course of rivers, Deforestation, Global Warming ETC and various other reasons there has been a huge difference in rainfall patterns .

As per new studies and research at MIT scientist have said that rainfall has patterns and this can be further classified in to major groups EL-NINO and LA-NINO effect which Is basically an oceanic cycles of warm water and cold water which maintain the temperature of earth.

Usually these cycles have a time period of 10 years but this may change according to place and geography of land. Further these cycles are main reasons behind occurrence of monsoon in any part of world.

In our project we try to understand the effects of irrigation on rainfall patterns and how we can tackle drought by modern methods of artificial precipitation.

Artificial precipitation is new technique followed by various developing and developed countries such as CHINA, SINGAPORE, AMERICA, RUSSIA ETC to satisfy there need of rains for farming so that the farmers don't run in to losses and farmers are satisfied.

INDIA is land of farmers and hence rainfall is an important factor for agriculture here .But due to drastic changes since last two decade there has been huge difference in monsoon patterns which has disturbed farmers life eventually. Due to huge industrial revolution going on in our country many farmers are leaving there farming lands and getting employed at industries for better employment and wages as they couldn't afford the cost of agriculture and mainly bore wells as there is no proper rainfall to irrigate the huge agricultural fields .

These problems are global issues now as every country has been affected by this crisis. Many countries such as America, china, japan, Russia, Chile, Singapore ETC have moved towards a new method ARTIFICIAL RAIN.

These ARTIFICIAL RAIN are new way of making precipitation possible further it enhances the chances of precipitation too, it also helps in satisfying the need of water for irrigation , and would further increase the ground water levels too.

ARTIFICIAL RAIN can be made possible by ICE BREAKING BOOM, RAIN ROCKER, THE ATMOSPHERE ZAPPER,SEEDING THE SKY, RIDING THE LIGHTING and various geographical

ISBN: 978-93-8830-599-0

changes to land mass to be irrigated or the place where precipitation is needed.

Our states Andhra Pradesh and Telangana have been 2nd in statics of 2012 suicide estimations .In recent two years 2015 and 2016 the number has gone very high and we are next to only Maharastra which also shares our fate of unseasonal rains and drought .This problems can be solved by these methods. And further restore the faith of farmers in irrigation and our government also.

We as civil engineers must feel responsible and try to implement these path braking and out of the box methods so that these we can be out of this crisis by some extent till an permanent solution of green future is made possible.

1.1   OBJECTIVE

The objective of our project is to better understand how irrigation can effect rainfall and also how we can plan an agriculture to feed our growing population .It will also allow us to "**MODEL , PREDICT AND ADAPT**" to changing climate. It also focuses on how artificial and natural methods of precipitation can make agriculture possible in drought hit zones.

2.0   PRECIPITATION

The objective of our project is mainly based on studying rainfall (or) in technical terminology it can be described as "PRECIPITATION". The history of precipitation goes back to 4.5 billion years ago when our planet broke from sun to form a land mass which was very hot back then but gradually cooled down which resulted in formation of gaseous clouds which ultimately lead to rains.

These rains (or) precipitation helped to bring down the temperature of earth and this made life possible on this planet. This precipitation must have lasted for many years as these rains are reasons for formation of oceans on earth. Since then our planet has under gone drastic changes and the land mass on our planet has become fertile but to irrigate the land mass we need fresh water where as 97% of water present on earth is saline which cannot cultivate the land mass. Only 3% of water is fresh water in which 68.7% is in form of glaciers and icecaps , Ground water 30.1% , remaining 1.2 % is in the form of surface water and other forms which includes rivers and lakes.

In many cases rivers don't exist near the land mass that is to be cultivated that means rain water and lakes becomes the only source for cultivation of land mass and now that show how rainfall is important and how precious is fresh water.

As mankind has grown from early man to modern man his methods have also changed from that day to present day so has been the scenario with patterns of clouds and rainfall all over the world. Studying these patterns helps us predict precipitation and better plan our irrigation schemes to enhance the productivity of agricultural lands.

2.1 CLOUDS

In meteorology, a cloud is an aerosol comprising a visible mass of minute liquid droplets or frozen crystals, both of which are made of water or various chemicals. The droplets or palrticles are suspended in the atmosphere above the surface of a planetary body. On Earth, clouds are formed by the saturation of air in the hemispheres (which includes the troposphere, stratosphere, and mesosphere). The air may be cooled to its dew point by a variety of atmospheric processes or it may gain moisture (usually in the form of water vapour) from an adjacent source. Nephology is the science of clouds which is undertaken in the cloud physics branch of meteorology.

ISBN: 978-93-8830-599-0

Cloud types in the troposphere, the atmospheric layer closest to Earth's surface, have Latin names due to the universal adaptation of Luke Howard's nomenclature. It was formally proposed in December 1802 and published for the first time the following year. It became the basis of a modern international system that classifies these tropospheric aerosols into five physical forms and three altitude levels or étages. These physical types, in approximate descending order of mean altitude range, include cirriform wisps and patches, stratocumuliform layers (mainly structured as rolls, ripples, and patches), cumuliform heaps and tufts, stratiform sheets, and very large cumulonimbiform heaps that often show complex structure. The physical forms are cross-classified by altitude levels to produce ten basic genus-types or genera. Some of these basic types are common to more than one form or more than one level, as illustrated in the stratocumuliform and cumuliform columns of the classification table below. Most genera can be divided into species, some of which are common to more than one genus. These can be subdivided into varieties, some of which are common to more than one genus or species.

Cirriform clouds that form higher up in the stratosphere and mesosphere have common names for their main types, but are sub-classified alpha-numerically rather than with the elaborate system of Latin names given to cloud types in the troposphere. They are relatively uncommon and are mostly seen in the polar regions of Earth. Clouds have been observed in the atmospheres of other planets and moons in the Solar System and beyond. However, due to their different temperature characteristics, they are often composed of other substances such as methane, ammonia, and sulfuric acid as well as water.

Classification of major types	Cirriform	Stratocumuliform	Cumuliform	Stratiform	Cumulonimbiform
Extreme level (meso-sphere)	Noctilucent				
Very high level (stratosphere)	Nacreous/non nacreous				
High-level	Cirrus	Cirrocumulus (layered)	Cirrocumulus(tufted)	Cirrostratus	
Mid-level		Altocumulus (layered)	Altocumulus (tufted)	Altostratus	
Low-level		stratocumulus	Cumulus (small)	Stratus	
Multi-level			cumulus		

Stratiform

Non-convective stratiform clouds appear in stable airmass conditions and, in general, have flat sheet-like structures that can form at any altitude in the troposphere. Very low stratiform cloud results when advection fog is lifted above surface level during breezy conditions. The stratiform group is cross-classified into the genera cirrostratus (high-étage), altostratus (middle-étage), stratus (low-étage), and nimbostratus (multi-étage).

Cirriform

Cirriform clouds are generally of the genus cirrus and have the appearance of detached or semi-merged filaments. They form at high tropospheric altitudes in air that is mostly stable with little or no convective activity, although denser patches may occasionally show buildups caused by limited high-level convection where the air is partly unstable.

Stratocumuliform

Clouds of this structure have both cumuliform and stratiform characteristics in the form of rolls or ripples. They generally form as a result of limited convection in an otherwise mostly stable airmass topped by an inversion layer.Thestratocumuliform group is cross-classified into layered cirrocumulus (high-étage),

ISBN: 978-93-8830-599-0

layered altocumulus (middle-étage), and stratocumulus (low-étage).

Cumuliform

Cumuliform clouds generally appear in isolated heaps or tufts .They are the product of localized but generally free-convective lift where there are no inversion layers in the atmosphere to limit vertical growth. In general, small cumuliform clouds tend to indicate comparatively weak instability. Larger cumuliform types are a sign of moderate to strong atmospheric instability and convective activity. Depending on their vertical size, clouds of the cumulus genus-type may be low or multi-étage. Tufted altocumulus and cirrocumulus genera in the middle and high étages are also considered cumuliform because they have a more detached heaped structure than their layered stratocumuliform variants.

Cumulonimbiform

The largest free-convective clouds comprise the genus cumulonimbus which are multi-étage because of their great vertical extent. They occur in highly unstable air and often have complex structures that include cirriform tops and multiple accessory clouds.

Étages and genera

The forms and resultant genus types can be grouped by étage. This is generally done for the purpose of cloud atlases, surface weather observations and weather maps. These maps are produced from information in the international synoptic code (or SYNOP) that is transmitted at regular intervals by professionally trained staff at major weather stations.

Non-vertical or single-étage genera are listed and summarised below in approximate descending order of the altitude at which each is normally based. Multi-étage clouds with significant vertical extent are separately listed and summarised in approximate ascending order of instability or convective activity.

AN INSIGHT INTO VARIOUS CLOUDS

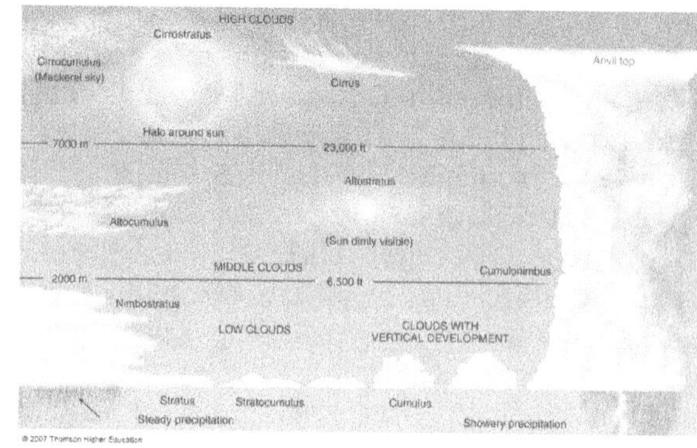

2.3   **HOW DO WE CALCULATE THE AMOUT OF WATER PRESENT IN CLOUD**

CLOUD TYPE	LWC(g/m^3)
Cirrus	0.03
Fog	0.05
Status	0.25-0.30
Cumulus	0.25-0.30
Stratocumulus	0.45
Cumulonimbus	1.0-3.0

The liquid water content (LWC) is the measure of the mass of the water in a cloud in a specified amount of dry air. It is typically measured per volume of air (g/m3) or mass of air (g/kg). (Bohren, 1998). This variable is important in figuring out which types of clouds are likely to form and is strongly linked to three other cloud microphysical variables: the cloud drop effective radius, the cloud drop number concentration, and the cloud drop size distribution. (Wallace, 2006). Being able to determine the cloud formations that are likely to occur is extremely useful for weather forecasting as cumulonimbus clouds are related to thunderstorms and heavy rain whereas cirrus clouds are not directly associated with precipitation.

Characteristics

The liquid water content of a cloud varies significantly depending on the type of clouds present in the atmosphere at a given location. The classification of the cloud is highly related to the liquid water content as well as the origin of the cloud. The combination of these two allows a forecaster to more readily predict the types of conditions that will be in an area based on the types of clouds that are forming or have already formed.

Relation to classification of clouds

Clouds that have low densities, such as cirrus clouds, contain very little water, thus resulting in relatively low liquid water content values of around .03g/m3. Clouds

ISBN: 978-93-8830-599-0

that have high densities, like cumulonimbus clouds, have much higher liquid water content values that are around 1-3g/m3, as more liquid is present in the same amount of space.

Maritime vs. continental

Maritime clouds tend to have fewer water droplets than continental clouds. The majority of maritime clouds have droplet concentrations between 100 drops/cm3 and about 200 drops/cm3. (Wallace, 2006). Continental clouds have much higher droplet concentrations ranging up to around 900 drops/cm3. (Wallace, 2006). However, the droplet radius in maritime clouds tend to be larger, so that the end result is that the LWC is relatively similar in both types of air masses for the same types of clouds. (Linacre, 1998).

Measuring techniques

There are several ways that can be used to measure the liquid water content of clouds .One way involves an electrically heated wire. The wire is attached to the power supply and is situated on the outside of the airplane. As it moves through a cloud, water droplets hit the wire and evaporate, reducing the temperature of the wire. The resistance caused by this is measured and is used to determine the power needed to maintain the temperature. The power can be converted to a value for the LWC. (Wallace, 2006).

Another way involves an instrument that uses scattered light from a large number of drops. This value is then converted to a value for the LWC. (Wallace, 2006).

A cloud chamber can also be used to simulate adiabatic ascent in the atmosphere through the decrease of pressure by removing air inside the chamber.

3.0   PROBLEMS WITH PRESENT DAY PRECIPITATION PATTERNS

As temperatures rise and the air becomes warmer, more moisture evaporates from land and water into the atmosphere. More moisture in the air generally means we can expect more rain and snow (called precipitation) and more heavy downpours. But this extra precipitation is not spread evenly around the globe, and some places might actually get less precipitation than they used to get. That's because climate change and global warming.

On average, the world is already getting more precipitation now than it did 100 years ago: 6 per cent more in the United States and nearly 2 per cent more worldwide.

This has a huge effect on farmers as many of agricultural schemes in our country and south Asia depend on monsoons and seasonal rains but due to global warming there has been huge change in patterns of rains since last 10 to 20 years.

As per recent studies scientist are trying to figure out patterns for example precipitation is expected to increase in higher latitudes and decrease in areas closer to the Equator. The northern United States will become wetter while the South, particularly the Southwest, will become drier.

Seashores have more probability of getting rainfall when compared to inlands

Places near to irrigation schemes have more probability of getting rainfall rather than lands far away from schemes in olden days this wasn't the case rain fall was evenly divided throughout and most as expected and mostly normal rainfall.

But as global warming has changed temperatures throughout earth surface it has become difficult to analyse rainfall patterns there are cases where 3 days of rainfall has actually given rain that exceed the limit of average rainfall of that place throughout year.

SOME BASIC CHANGES ON VARIOUS ASPECTS OF ENVIRONMENTS

To better understand the risks of climate change to development, the World Bank Group commissioned the Potsdam Institute for Climate Impact Research and Climate Analytics to look at the likely impacts of temperature increases from 2°C to 4°C in three regions. The scientists used the best available evidence and supplemented it with advanced computer simulations to arrive at likely impacts on agriculture, water resources, cities and coastal ecosystems in South Asia, South East Asia and Sub-Saharan Africa. Some of their findings for India include

Extreme Heat

4.0   HOW IRRIGATION HAS EFFECTED RAINFALL PATTERNS

Rainfall patterns have majorly changed in inlands and plain lands as these are places where most of irrigation schemes are constructed. These irrigations provide an huge surface of water for evaporation huge means huge moisture content in and heavier air this then flows in the direction of wind when this moisture in contact of cool air there is sudden downpour. These places are ending up to be forest and various other cool places where that rainfall can't be used properly.

These irrigation schemes have been providing moisture to air but can't capture that moisture to make rainfall possible. This is a major drawback as we are not able to cash on the moisture content at place required.

ISBN: 978-93-8830-599-0

And also at dams and reservoirs due to immense accumulation of water an strange type of bacteria has been noticed which increases the greenhouse gas levels while respiration .This is adds up to contributing 2% of global greenhouse gas levels.

5.0   HOW RAINFALL CAN BE MADE POSSIBLE BY VARIOUS ARTIFICIAL AND NATURAL METHODS

Precipitation has been something which has never been under human control traditionally we are taught that rain and weather are things that happen according to gods wish but as human race has been moving towards knowledge many studies have successfully helped in predicting the occurrence of rainfall at various places throughout this beautiful planet .

There have been references in history that rainfall at certain places had patterns of 10 years which is relatable to EL NINO and LA NINO effects so the more we put towards these new topics we can better understand and plan our irrigation , agriculture and various other things related to weather.

India has been an agricultural heaven which is known for its agricultural products and by products around the globe but due to global warming and industrial revolution going on in our country there has been and totally different scenario in agricultural sector when compared to business sector and industrial sector which are scaling new heights in global economy.

The most important part or the back bone of any type of agriculture in India has been rainfall as the water by rivers is not uniformly divided throughout the land mass and hence it becomes very difficult for farmers to irrigate the land.

Bore wells and Groundwater can be a better alternate but most of the farmers cannot afford these equipment's so it really gets difficult for them to get water to their farm lands.

This is not a problem that only we are facing but many countries globally are also facing them and many have even moved towards comfortable solution of artificial precipitation and natural methods of increasing the probability of rain.

ARTIFICIAL METHODS OF PRECIPITATION
1. CLOUD SEEDING
2. ICE BREAKING BOOM
3. RAIN ROCKETS
4. THE ATMOSPHERE ZAPPER
5. RIDING THE LIGHTING

These are some of the new artificial methods which are used by many developed countries and developed countries in order to get immediate rain fall from clouds.

1. CLOUD SEEDING

Cloud seeding, a form of weather modification, is the attempt to change the amount or type of precipitation that falls from clouds, by dispersing substances into the air that serve as cloud condensation or ice nuclei, which alter the microphysical processes within the cloud. The usual intent is to increase precipitation (rain or snow), but hail and fog suppression are also widely practiced in airports.

Cloud seeding also occurs due to ice nucleators in nature, most of which are bacterial in origin.

Methodology

The most common chemicals used for cloud seeding include silver iodide, potassium iodide and dry ice

ISBN: 978-93-8830-599-0

(solid carbon dioxide). Liquid propane, which expands into a gas, has also been used. This can produce ice crystals at higher temperatures than silver iodide. After promising research, the use of hygroscopic materials, such as table salt, is becoming more popular. When Cloud seeding increases snowfall takes place when temperatures within the clouds are between 19 and −4 °F (−7 and −20 °C).Introduction of a substance such as silver iodide, which has a crystalline structure similar to that of ice, will induce freezing nucleation.

In mid-latitude clouds, the usual seeding strategy has been based on the fact that the equilibrium vapour pressure is lower over ice than over water. The formation of ice particles in super cooled clouds allows those particles to grow at the expense of liquid droplets. If sufficient growth takes place, the particles become heavy enough to fall as precipitation from clouds that otherwise would produce no precipitation. This process is known as "static" seeding.

Seeding of warm-season or tropical cumulonimbus (convective) clouds seeks to exploit the latent heat released by freezing. This strategy of "dynamic" seeding assumes that the additional latent heat adds buoyancy, strengthens updrafts, ensures more low-level convergence, and ultimately causes rapid growth of properly selected clouds.

Cloud seeding chemicals may be dispersed by aircraft or by dispersion devices located on the ground (generators or canisters fired from anti-aircraft guns or rockets). For release by aircraft, silver iodide flares are ignited and dispersed as an aircraft flies through the inflow of a cloud. When released by devices on the ground, the fine particles are carried downwind and upward by air currents after release.

An electronic mechanism was tested in 2010, when infrared laser pulses were directed to the air above Berlin by researchers from the University of Geneva. The experimenters posited that the pulses would encourage atmospheric sulphur dioxide and nitrogen dioxide to form particles that would then act as seeds.

EFFECTIVENESS

New technology and research has produced reliable results that make cloud seeding a dependable and affordable water-supply practice for many regions. While practiced widely around the world, the effectiveness of cloud seeding is still a matter of academic debate. In 2003 the US National Research Council (NRC) released a report stating, "...science is unable to say with assurance which, if any, seeding techniques produce positive effects. In the 55 years following the first cloud-seeding demonstrations,

substantial progress has been made in understanding the natural processes that account for our daily weather. Yet scientifically acceptable proof for significant seeding effects has not been achieved.

Referring to the years 1903, 1915, 1919, 1944, and 1947 weather modification experiments, the Australian Federation of Meteorology discounted "rain making". By the 1950s, the CSIRO Division of Radio physics switched to investigating the physics of clouds and had hoped by 1957 to better understand these processes. By the 1960s, the dreams of weather making had faded only to be re-ignited post-corporatisation of the Snowy Mountains Scheme in order to achieve "above target" water. This would provide enhanced energy generation and profits to the public agencies that are the principal owners.

Cloud seeding has been shown to be effective in altering cloud structure and size and in converting super cooled liquid water to ice particles. The amount of precipitation due to seeding is difficult to quantify. There is statistical evidence for seasonal precipitation increases of about 10 per cent with winter seeding.

Clouds were seeded during the 2008 Summer Olympics in Beijing using rockets, to coax rain showers out of clouds before they reached the Olympic city so that there would be no rain during the opening and closing ceremonies, although others dispute their claims of success.

A 2010 Tel Aviv University study claimed that the common practice of cloud seeding to improve or induce rainfall, with materials such as silver iodide and frozen carbon dioxide, seems to have little if any impact on the amount of precipitation. A 2011 study suggested that airplanes may produce ice particles by freezing cloud droplets that cool as they flow around the tips of propellers, over wings or over jet aircraft, and thereby unintentionally seed clouds. This could have potentially serious consequences for particular hail stone formation.

Impact on environment and health

With an NFPA 704 health hazard rating of 2, silver iodide can cause temporary incapacitation or possible residual injury to humans and mammals with intense or continued but not chronic exposure. However, there have been several detailed ecological studies that showed negligible environmental and health impacts. The toxicity of silver and silver compounds (from silver iodide) was shown to be of low order in some studies. These findings likely result from the minute amounts of silver generated by cloud seeding, which are about one percent of industry emissions into the atmosphere in

ISBN: 978-93-8830-599-0

many parts of the world, or individual exposure from tooth fillings.

Accumulations in the soil, vegetation, and surface runoff have not been large enough to measure above natural background .A 1995 environmental assessment in the Sierra Nevada of California and a 2004 independent panel of experts in Australia confirmed these earlier findings.

"In 1978, an estimated 2,740 tonnes of silver were released into the US environment. This led the US Health Services and EPA to conduct studies regarding the potential for environmental and human health hazards related to silver. These agencies and other state agencies applied the Clean Water Act of 1977 and 1987 to establish regulations on this type of pollution."

Cloud seeding over Kosciuszko National Park a biosphere reserve is problematic in that several rapid changes of environmental legislation were made to enable the trial. Environmentalists are concerned about the uptake of elemental silver in a highly sensitive environment affecting the pygmy possum among other species as well as recent high level algal blooms in once pristine glacial lakes. Research 50 years ago and analysis by the former Snowy Mountains Authority led to the cessation of the cloud seeding program in the 1950s with non-definitive results. Formerly, cloud seeding was rejected in Australia on environmental grounds because of concerns about the protected species, the pygmy possum. Since silver iodide and not elemental silver is the cloud seeding material, the claims of negative environmental impact are disputed by peer-reviewed research as summarized by the international Weather Modification Association.

2   ICE BREAKING BOOM

In many cold countries there is scarcity of fresh water and in many cases due to hail storms and snow, crops get destroyed hence farmers incur huge losses so these kind of methods help in converting hail storms into rain which leads to fresh water for irrigation in places like these.

In this process and cannon is placed at a high altitude so that it is audible at every place another reason why it is kept at higher altitude is that it can be nearer to clouds as its main effect is on clouds and rains.

In many places clouds carry hail storms which eventually effects cultivation of fruits which are main crops to be sown in these places.

An cannon is placed at high altitude which is been filled with gun powder at lower part of cannon .This

gun powder is burnt which leads to an huge boom sound all over the place this shock wave disturbs the snow present in cloud and hence initiates rainfall .

EFFECTIVENESS

This is very effective in places where hails storms are common as these boom sounds can disintegrate huge masses of snow very effectively Hence they can cause rains immediately but these cannot be used in dry and warm places as this equipment is not effective there.

USES WORLDWIDE

At present France is the only main user of this particular technique but in some parts of United States of America this technique has been gaining popularity as this doesn't have any side effects on nature.

3.   RAIN ROCKETS

This method is similar to that of cloud seeding by aircraft but in this instead of aircrafts rockets are used as these doesn't require much of economical invest and is easily affordable.

Many countries have been using this method as these rockets have wide reach with minimal effect and these also can be transported easily without any hassle.

EFFECTIVENESS

This is very effective and a very practical perspective of squeezing out clouds so that drought hit zones can be cultivated by rain water. Everything about this method is similar to that of cloud seeding by aircraft just the medium through which silver nitrate is placed in to clouds is different.

USES WORLDWIDE

Many developed countries worldwide have been using this methods mainly CHINA, JAPAN, RUSSIA, U.S.A and some parts of Middle East Asia.

CHINA has a dedicated Govt. organisation which works on identifying these clouds and areas which should be seeded with this rain rockets to match the needs of farmers and industries.

Of late Dubai and some of Middle East Asian countries are also using these to attract tourist which increases there income through tourism.

4   THE ATMOSPHERE ZAPPER

This is a new technic which has been introduced Dubai and is still under testing and has been giving fruitful results since its application .Basically in this method rain is pulled by electricity. The theory is that electrified, umbrella-shaped towers can send negatively charged particles into the air, increasing the chance that super cooled droplets will collide with freezing nuclei, thus becoming rain. Experts, however, are highly sceptical.

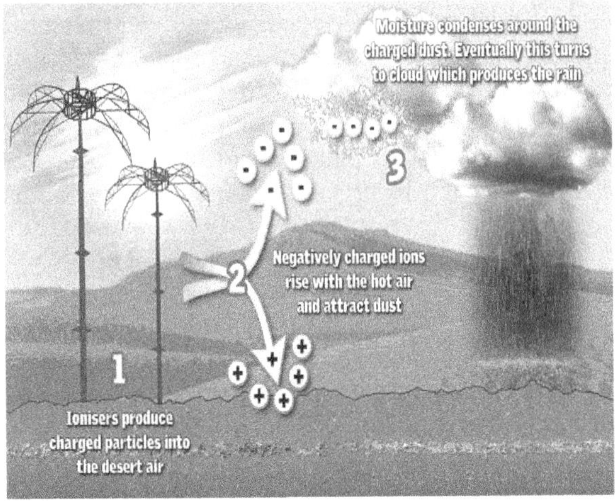

EFFECTIVENESS

This method is only used in Dubai by this time but as fruitful results are been yielded by method could be further used in various other countries. But this requires a lot of skill and electricity to be implemented.

5   RIDING THE LIGHTING

As unlikely as it sounds, researchers around the world have tried to redirect lightning by taking an long wire, tying one end to the ground and the other to the rocket , then firing the rocket in to thunder storm more recent promising developments include research at the university of Arizona in Tucson that shows high intensity lasers could redirect lighting

NATURAL METHODS

Naturally there are is no certain method that leads to immediate rain as of artificial methods but these natural methods guarantee a non-contaminated and pure rain fall.

Basically afforestation tops the list of natural methods of rainfall then comes planning of agricultural and irrigation simultaneously so that rainfall can be effectively captured. Then there is increasing number of water bodies so a well-balanced eco system can be created. This all methods contribute immensely towards creating an ecosystem which creates an chances of rainfall.

6.0   PROBLEMS DUE TO FLOODS AND DROUGHTS

Compartmentalised responses are unlikely to be adequate to address the current water crises.

Children watch the view of almost dry river Tawi, in Jammu, 17 May 2016. Credit: EPA/Jaipal Singh.

India is facing one of its most serious droughts in recent memory – official estimates suggest that at least 330 million people are likely to be affected by acute shortages of water. As the subcontinent awaits the imminent arrival of the monsoon rains, bringing relief to those who have suffered the long, dry and exceptionally warm summer, the crisis affecting India's water resources is high on the public agenda.

ISBN: 978-93-8830-599-0

Unprecedented drought demands unconventional responses and there have been some fairly unusual attempts to address this year's shortage. Perhaps the most dramatic was the deployment of railway wagons to transport 5,00,000 litres of water per day across the Deccan plateau, with the train traversing more than 300 kilometres to provide relief to the district of Latur in Maharashtra state.

The need to shift water on this scale sheds light on the key issue that makes water planning in the Indian subcontinent so challenging. While the region gets considerable precipitation most years from the annual monsoon, the rain tends to fall in particular places – and for only a short period of time (about three months). This water needs to be stored and made to last for the entire year.

In most years, it also means that there is often too much water in some places, resulting in as much distress due to flooding as there currently is due to drought. So there is a spatial challenge as well – water from the surplus regions needs to reach those with a shortfall, and the water train deployed in Maharashtra is one attempt to achieve this.

Grand ambitions

The current crisis has led the Indian government to announce that it hopes to resurrect an ambitious plan to try and link the major river basins of the country, under the Interlinking of Rivers Project. The scale and magnitude of this exercise, both financial (it is estimated to cost more than £100 billion) and in engineering terms (involving the transfer of 174 billion cubic metres of water annually) is unprecedented.

Fishing boats on the banks of the Brahmaputra river in Guwahati, Assam, India. EPA/Stringer.

Critics suggest that it is unlikely to work and is likely to create further ecological and social disruption, especially due to the uncertainties in weather and precipitation patterns due to climate change. There is a risk that other alternatives, perhaps less dramatic in their scope, might be neglected in the rush for the big headline-grabbing schemes.

A specific way forward might be to work more directly with natural processes to secure the regeneration of water sources at the local level. In the dry plains, this involves the revitalisation of aquifers and the replenishment of groundwater through recharge during the monsoon, as has been attempted already in some regions. In the hilly areas, there is considerable scope for investment in spring recharge and source sustainability, as has been undertaken on a significant scale in the Himalayan state of Sikkim.

Our current research is examining the need to invest in source protection and sustainability in detail, especially in the Himalayas, which have been described as the "Water Towers of Asia". Urbanisation trends in the region suggest that there will be a growing number of small towns and settlements that will need water infrastructure to meet their needs – and there is a critical need to secure these water sources.

Deforestation, land conversion and degradation, as well as urban encroachment due to illegal construction, pose major threats to the water bearing capacity of the Himalayan landscape. There is an urgent need to invest in the identification, protection and restoration of these "critical water zones".

Potential for conflict

The Himalayan context also demonstrates the transboundary nature of the water issue. The Hindu Kush Himalayan region extends across eight countries, from Afghanistan to Myanmar, and supports ten major river systems, potentially affecting the lives of more than 1.5 billion people. Cooperation across political boundaries is vital to manage these fragile resources, further threatened by the uncertain impacts of climate change.

Introduction

We know that water is necessary, both for sustainable human development and for the healthy functioning of the planet's ecosystem. Availability of freshwater globally however, is limited. Out of the 2.7 per cent of a total amount of 1 400 million km3 of freshwater, the major portion occurs in the form of permanent snow cover or deep aquifers and only a small fraction is available for use.

Although India has to support 16 per cent of the world's population and 15 per cent of livestock, we have only 2.4 per cent of the land and 4 per cent of the water resources of the world. Out of about 4 000 km3

ISBN: 978-93-8830-599-0

of precipitation in a year, as much as 3 000 km3 comes as rainfall in a short monsoon period of three to four months from June to September. The distribution of the water thus available is not uniform and is highly uneven in both space and time. The average annual water resource potential of the country is estimated to be 1 869 km3. Due to hydrological, topographical and geological limitations, however, only 690 km3 of surface water can be utilized by conventional storage and diversion structures. The annual recharge of groundwater is 433 km3.

Two major problems faced by the country are drought and floods, which are discussed in the succeeding paragraphs:

Floods and drought

Today, droughts and floods are a common feature and their co-existence poses a potent threat, which cannot be eradicated but has to be managed. Transfer of the surplus monsoon water to areas of water deficit is a potential possibility. This would also help create additional irrigational potential, the generation of hydropower, as well as overcoming regional imbalances.

The recurrence of drought and famines during the second half of the 19th century necessitated the development of irrigation to give protection against the failure of crops and to reduce large-scale expenditure on famine relief.

Floods in India

Floods are recurrent phenomena in India. Due to different climatic and rainfall patterns in different regions, it has been the experience that, while some parts are suffering devastating floods, another part is suffering drought at the same time. With the increase in population and development activity, there has been a tendency to occupy the floodplains, which has resulted in damage of a more serious nature over the years. Often, because of the varying rainfall distribution, areas which are not traditionally prone to floods also experience severe inundation. Thus, floods are the single most frequent disaster faced by the country.

Flooding is caused by the inadequate capacity within the banks of the rivers to contain the high flows brought down from the upper catchments due to heavy rainfall. Flooding is accentuated by erosion and silting of the river beds, resulting in a reduction of the carrying capacity of river channels; earthquakes and landslides leading to changes in river courses and obstructions to flow; synchronization of floods in the main and tributary rivers; retardation due to tidal effects; encroachment of floodplains; and haphazard and unplanned growth of urban areas. Some parts of the country, mainly coastal areas of Andhra Pradesh, Orissa, Tamil Nadu and West Bengal, experience cyclones, which are often accompanied by heavy rainfall leading to flooding.

Area prone to flood

In 1980, RashtriyaBarhAyog (National Commission on Floods) assessed the total area liable to flooding in the country as 40 million hectares (ha), which constitutes one-eighth of the country's total geographical area. The Working Group on Flood Control Programme set up by the Planning Commission for the Tenth Five Year Plan put this figure at 45.64 million ha. About 80 per cent of this area, i.e. 32 million ha, could be provided with a reasonable degree of protection.

Damage from floods

More significant than the loss of life and damage to property is the sense of insecurity and fear in the minds of people living in the floodplains. The after-effects of flood, such as the suffering of survivors, spread of disease, non-availability of essential commodities and medicines and loss of dwellings, make floods the most feared of the natural disasters faced by humankind.

Flood damage

	Maximum	Average
Area affected	17.5 million ha (1978)	7.63 million ha
Crop area affected	10.15 million ha (1988)	3.56 million ha
Population affected	70.45 million (1978)	32.92 Million
Houses damaged	3 507 542 (1978)	1 234 616
Heads of cattle lost	618 248 (1979)	91 242
Human lives lost	1 1316 (1977)	1 560
Damage to public utilities	US$ 705 million (1998)	US$ 126 million
Total damage	US$ 1 255 million (1998)	US$ 307 million

Heavy flood damage was inflicted during the monsoon of 1955, 1971, 1973, 1977, 1978, 1980, 1984, 1988, 1989, 1998, 2001 and 2004. Highlights of the damage are given below:

Flood-prone areas are shown in the map.

ISBN: 978-93-8830-599-0

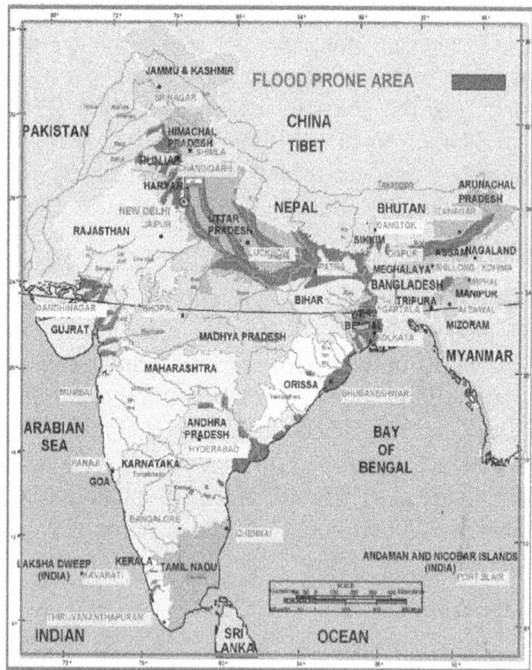

Flood-prone areas

Drought

Drought is a recurrent natural feature which results from the lack of precipitation over an extended period of time (e.g. a season or several years). It is a temporary deviation of rainfall and moisture conditions from the mean, thus differing from aridity and seasonal aridity. It is a creeping phenomenon and, unlike other hazards, can last for months and, in severe cases, years. Drought affects virtually all climatic regions and more than one-half of the Earth is susceptible to droughts every year. Regions with higher variability of rainfall and runoff are more vulnerable. Depending on the likely impact, the phenomenon of drought can be categorized in several ways, such as meteorological, hydrological and agricultural. The spatial extent of drought is much greater than for any other hazard and is not limited to basin or political boundaries. Long- lasting droughts lead to degradation of soil, plant and animal habitats and social disruption.

During a severe drought in 1917/1918, the Jhelum River in Kashmir dried up completely. Out of the 328 million ha geographical area of India, 107 million ha (nearly one-third), spread over administrative districts in several states, is affected by drought. It includes about 39 per cent of cultivable land and about 29 per cent of our population. India has experienced 22 major droughts during the last 131 years. The 2002 drought, one of the severest in India, affected 56 per cent of its geographical area, the livelihoods of 300 million people and 150 million cattle in 18 states. The Government of

India had to provide relief amounting to about US\$ 4500 million.

Water-resources development in India

India's population of about 1 billion (2001 census) is expected to stabilize at about 1.6 billion by 2050. This would require some 450 million tones of food grain annually. The basic needs for water of rural and urban populations and cattle, as well as industry and environment and ecological management, also have to be met, taking into account land-use policies, degradation of water resources, depletion of aquifers, etc. To this end, long-term planning of the utilization of the country's water resources is required to meet the various competing demands on a sustainable basis. The strategy for mitigating the effects of drought and floods is the optimal development of scarce water resources.

After independence, planned development of water resources was taken up mainly through the creation of storage projects, as well as extension, renovation and modernization of existing projects. So far, storage capacity of some 213 billion m3 has been created in the country and projects under construction will increase this to 289 billion m3. A further 108 billion m3 is expected to be created through projects under consideration.

Institutional arrangements

At the central level, the Union Ministry of Water Resources is responsible for development, conservation and management of water as a national resource, i.e. for general policy on water resources development and for technical assistance to the states on irrigation, multipurpose projects, groundwater exploration and exploitation, command area development, drainage, flood control, water-logging, sea-erosion problems, dam safety and hydraulic structures for navigation and hydropower. It also oversees the regulation and development of inter-state rivers. These functions are carried out through various central organizations. Urban water supply and sewage disposal is handled by the Ministry of Urban Development, while rural water supply comes under the purview of the Department of Drinking Water under the Ministry of Rural Development. The subject of hydro-electric and thermal power is the responsibility of the Ministry of Power. Pollution and environment control comes under the purview of the Ministry of Environment and Forests.

Water being a state subject, the state government has primary responsibility for use and control of this resource. The administrative control and responsibility for development of water rest with the various state departments and corporations.

ISBN: 978-93-8830-599-0

National water policy

The National Water Policy adopted by the National Water Resources Council in April 2002 highlights the provisions for project planning, surface- and groundwater development, irrigation and flood control.

Irrigation plays a major role in increasing the production of food grains. The policy provides following directives for irrigation management:

- Irrigation planning either in an individual project or in a basin as a whole should take into account the irrigability of land, cost-effective irrigation options possible from all available sources of water and appropriate irrigation techniques for optimizing water- use efficiency. Irrigation intensity should be such as to extend the benefits of irrigation to as large a number of farming families as possible, keeping in view the need to maximize production;
- There should be close integration of water- and land-use policies.
- Water allocation in an irrigation system should be done with due regard to social equity and justice. Disparities in the availability of water between head-reach and tail-end farms and between large and small farms should be obviated by adoption of a rotational water distribution system and supply of water on a volumetric basis subject to certain ceilings and rational pricing;
- Concerted efforts should be made to ensure that the irrigation potential created is fully utilized. For this purpose, the command area development approach should be adopted in all irrigation projects.

The following provisions exist in National Water Policy 2002 as regards flood control and moderation:

- There should be a master plan for flood control and management for each flood prone basin;
- An adequate flood cushion should be provided in water-storage projects, wherever feasible, to facilitate better flood management. In highly flood- prone areas, flood control should be given overriding consideration in reservoir-regulation policy, even at the cost of sacrificing some irrigation or power benefits;
- While physical flood-protection works like embankments and dykes will continue to be necessary, increased emphasis should be laid on non-structural measures such as flood forecasting and warning, floodplain zoning and flood-proofing in order to minimize losses and reduce recurring expenditure on flood relief.

Irrigation developments and protection against droughts

River-valley projects serve a basic necessity of a country whose economy is based largely on agriculture. Irrigation and power are crucial inputs for increased productivity. A high priority has, therefore, been given in the national planning process to the creation of river-valley projects ever since India gained independence. The prominence given to river-valley projects in India's successive five-year plans is consequent to the contribution to the prosperity of the country by water-resources projects.

As stated earlier, of all natural disasters, drought affects the greatest number of people in the world, especially in India. The occurrence of drought cannot be prevented but being well prepared for its likely occurrence can lessen its impacts. Droughts have two basic components: climatic (decrease in precipitation) and demand (use of water). In responding to droughts, governments tend to concentrate most of their efforts on reducing the demand for water, although there are limited options for controlling the climatic component. Thus, drought-planning strategies should have a clear objective and purpose; involve stakeholder participation; have a good inventory of resources; identify groups at risk; be able to integrate science and technology with policy; publicize the proposed plan and invite public responses; and have an appropriate education programme. The long-term measures for drought mitigation would normally include creation of ground- and surface-water storages, integration of small reservoirs with major reservoirs, integrated basin planning; inter-basin transfer of water, etc. Long-term adaptation involves the development of community-based natural resources management plans, developed and implemented through a participatory approach, and making full use of traditional knowledge.

Water used for irrigating fields near Maniyampadi. The water comes from a diesel engine driven pump and is also used for drinking and cooking.

ISBN: 978-93-8830-599-0

Thus, most measures focus on management, re-allocation and distribution of existing water resources and on establishing priorities accordingly for different uses. Commonly known means adopted for combating drought and promoting development are:

- Improving national capabilities, including training and human resource development, for assessing water resources and determining water use on a continuing basis and for the planning and management of these resources;
- Conserving water resources and optimizing their use;
- Augmenting the supply of water locally by exploiting surface- and groundwater, taking into account long-terms trends, the future demands of local communities and other needs;
- Augmenting the supply of water by transfers from more permanent surface-water sources (lakes and rivers) and from groundwater resources in arid and semi-arid areas and/or long-distance transfers from humid areas if practically and economically possible—and environmentally acceptable.

Many major, medium and minor water-resources projects have been constructed over the past six decades. As a result of this development, the irrigation potential of India increased from about 23 million ha in 1951 to about 102 million ha in 2006, resulting in increased production of food grains from 51 million tonnes to about 212 million tonnes, making the country self-sufficient.

The cultivable area of the country is estimated to be about 186 million ha, of which about 142 million ha is under cultivation. With growing population and industrialization putting pressure on land, it is expected that cultivated area will stabilize at about 140-145 million ha. As irrigated agriculture is more productive than non-irrigated agriculture, it is imperative to irrigate more land in order to meet the country's future needs for food and fibre.

Sustained efforts so far have ensured considerable progress, but there is still a long way to go to ensure water for all in adequate quantity and quality. The problems are further compounded by fresh issues that have surfaced over the last few years.

Approach to drought management

The behaviour of the monsoon is usually erratic and uncertain in India. Kharif (summer crop) production depends on the quantum and distribution of rainfall. The behaviour of monsoon is broadly classified as:

- Normal season with normal onset, cessation and distribution of the monsoon;
- Delayed onset of the monsoon;
- Normal onset but early withdrawal of the monsoon;
- Normal onset and cessation but prolonged drought period in between (inter-spell dry period;
- Flood/excess rains;
- Uneven distribution of rain.

The preparations for dealing with such situation, which is necessary to maintain from year to year, are:

- Early warning;
- Early response;
- An efficient intelligence system;
- Timely maintenance of the irrigation system and adoption of a crop stabilization strategy;
- An effective programme of relief works by advance shelf of projects of the works by different departments
- Pre-positioning of adequate foodstuff and their delivery;
- Alternate arrangements for drinking-water supply;
- Construction of deep wells and bore wells and repair of those which are defunct and continuous repair of hand pumps.

Initiative taken for drought management

From 1900 to 2002, droughts in India resulted in 2 750 430 deaths and affected some 900 million people, apart from huge financial losses. It is the creeping effect of drought over long periods and its severity that sensitized the Government of India to treat the problem from several angles—scientific, technological, economic, social and environmental. Some of the initiatives taken for drought management by the Government are:

- Enhancement of the capabilities of long-range forecasts to climate modelling and weather forecasting;
- In 1989, the National Centre for Medium Range Weather Forecasting started to forecast weather on a medium-term basis (3-10 days in advance);
- Monitoring of storage position of reservoirs: 76 important reservoirs of the country having a total live storage capacity of 131.22 billion m3 are being monitored. A further 49 have also been identified for inclusion in the monitoring system, which will increase storage capacity of the monitored reservoirs to 156.69 billion m3, i.e. about 74 per cent of the total capacity of 213 billion m3 created so far;

ISBN: 978-93-8830-599-0

- Efforts are under way to improve the efficiency of the irrigation system;
- The National Agricultural Drought Assessment and Monitoring System became operational in 1989;
- The National Centre for Disaster Management was set up in 1995 to undertake human-resource development, research, building a database and providing information services and documentation on disaster management;
- Many programmes to prevent/ mitigate drought in the long term;
- Supporting research to provide solutions to drought-related problems;
- Setting-up of a National Data Bank under the All India Co-ordinated Project on Agrometeorology at the Crop Research Institute for Dry Land Agriculture, Hyderabad;
- Setting-up of a National Disaster Management Authority.

The new Drought Risk Management Programme under formulation aims to build on the previous Programme's experience to reduce the vulnerabilities of communities to drought through community-based approaches and appropriate risk management and better decision-support systems at state and district levels.

Approach to flood management

Approaches to dealing with floods may be any one or a combination of the following available options:

- Attempts to modify the flood
- Attempts to modify the sus-ceptibility to flood damage
- Attempts to modify the loss burden
- Bearing the loss.

The main thrust of the flood protection programme undertaken in India so far has been an attempt to modify the flood in the form of physical (structural) measures to prevent the floodwaters from reaching potential damage centres and modify susceptibility to flood damage through early warning systems.

Structural measures

The following structural measures are generally adopted for flood protection:

- Embankments, flood walls, sea walls
- Dams and reservoirs
- Natural detention basins
- Channel improvement
- Drainage improvement
- Diversion of flood waters.

Of these measures, embankments are the most commonly undertaken in order to provide quick protection with locally available material and labour. The major embankment projects taken up after independence are on the rivers Kosi and Gandak (Bihar), Brahmaputra (Assam), Godavari and Krishna (Andhra Pradesh), Mahanadi, Brahmani, Baitarni and Subarnarekha (Orissa) and Tapi (Gujarat). These embankments play an important role in providing reasonable protection to vulnerable areas. Realizing the great potential of the reservoirs in impounding floods and regulating the flows downstream for flood moderation, flood control has been sought to be achieved as one of the objectives in multipurpose dams. Reservoirs with a specifically allocated flood cushion have been constructed on the Damodar system in Jharkhand and the Hirakud and Rengali dam in Orissa. However, many other large storage dams, e.g. Bhakra dam, without any earmarked flood storage, have also helped in flood moderation.

During the post-independence period, multi-purpose projects such as the Damodar Valley Corporation (DVC) reservoirs, the Bhakra-Nangal project, Hirakud dam, NagarjunaSagar project etc., have been constructed to increase food production, energy generation, drinking-water supply, fisheries development, employment generation, flood moderation, etc. These large dams have played a significant role in reducing damage by way of flood moderation. One of the important flood moderation examples achieved by dams is that of Damodar Valley, where four reservoirs were constructed with flood management as one of the objectives. During the 2000 monsoon, DVC reservoirs saved the life and property of people from a possible disaster through flood moderation.

Up to 2005, 34 398 km of new embankments and 51 318 km of drainage channels were constructed. In addition, 2400 town protection works were completed and 4 721 villages were raised above flood levels. Barring occasional breaches in embankments, these works gave reasonable protection to an area of some 16.5 million ha.

Non-structural measures

Non-structural measures include:

- Flood forecasting and warning
- Floodplain zoning
- Flood fighting
- Flood proofing
- Flood insurance.

ISBN: 978-93-8830-599-0

A brief description of the most important measure, i.e. flood forecasting, and the progress made so far is given below.

Flood forecasting and warning network in India

Of all the non-structural measures for flood management which rely on the modification of susceptibility to flood damage, the one which is gaining increased/ sustained attention of planners and acceptance by the public is flood forecasting and warning, which enable forewarning as to when the river is going to use its floodplain, to what extent and for how long. As for the strategy of laying more emphasis on non-structural measures, a nationwide flood forecasting and warning system has been established by the Central Water Commission.

Flood forecasting and flood warning in India commenced in a small way in the year 1958 with the establishment of a unit in the Central Water Commission, New Delhi, for flood forecasting for the river Yamuna at Delhi. This has now grown to cover most of the flood-prone interstate river basins. The Central Water Commission is currently responsible for issuing flood forecasts at 173 stations, of which 145 are for river stage forecast and 28 for inflow forecast. On average, about 6 000 flood forecasts are issued every year with a maximum of 7 943 forecasts in 1998. The forecasts issued by the Central Water Commission have been consistent with about 96 per cent accuracy as per the present norms of the Central Water Commission. A forecast is considered to be reasonably accurate if the difference between forecast and corresponding observed level of the river lies within ±15 cm. In the case of inflow forecasts, variations within ±20 per cent are considered acceptable, as a result of which the flood-forecasting and warning services have rendered immense benefit to those in flood-prone areas.

Modernization of flood forecasting services

The Central Water Commission is making a constant endeavour to update and modernize forecasting services on a continuous basis to make flood forecasts more accurate, effective and timely. Initiatives being taken for modernizing flood forecasting services are:

- The establishment and modernization of the flood forecasting network, including inflow forecast through automated data collection and transmission; use of satellite-based communication systems through very small aperture terminals; and improvement of forecast formulation techniques using computer-based catchment models;

- Development of a decision-support system for flood forecasting and inundation forecast model for the Mahanadi basin and flash flood forecasting for Sutlej basin;

- Development of a real-time flood-forecasting system for the Brahmaputra and Barak basin, envisaging data collection through automatic sensors and transmission through satellite and forecast formulation using a computer-based mathematical model.

Disaster management in India

India has traditionally been vulnerable to natural disasters on account of its unique geoclimatic conditions. Floods, droughts, cyclones, earthquakes and landslides have been recurrent phenomena. About 60 per cent of the landmass is prone to earthquakes of various intensities; over 45 million ha are prone to floods; about 8 per cent of the total area is prone to cyclones and 68 per cent of the area is susceptible to drought. In the decade 1990-2000, an average of about 4 344 people lost their lives and 30 million were affected by disasters every year. The loss in terms of private, community and public assets was astronomical.

Over the past couple of years, the Government of India has effected a paradigm shift in its approach to disaster management. The new approach derives from the conviction that development cannot be sustainable unless disaster mitigation is built into the development process. Another cornerstone of the approach is that mitigation has to be multi-disciplinary, spanning all sectors of development. The new policy also emanates from the belief that investments in mitigation are much more cost-effective than expenditure on relief and rehabilitation.

Disaster management occupies an important place in this country's policy framework, as it is the poor and the underprivileged who are worst affected by calamities/disasters.

The steps being taken by the Government emanate from the approach outlined above. This has been translated into a National Disaster Framework (roadmap) covering institutional mechanisms, a disaster prevention strategy, early warning systems, disaster mitigation, preparedness and response and human resource development. The expected inputs, areas of intervention and agencies to be involved at the national, state and district levels have been identified and listed. There is now, therefore, a common strategy underpinning the action being taken by all the participating organizations/stakeholders.

ISBN: 978-93-8830-599-0

assured irrigation over a period of four years (2005-2009) through completion of ongoing major, medium and extension renovation and modernization projects, the repair, renovation and restoration of water bodies and groundwater development for irrigation has been taken up by the Government of India.

7.0   IMPLEMENTATIONS USES WORLDWIDE ASIA CHINA

The largest cloud seeding system is in the People's Republic of China. They believe that it increases the amount of rain over several increasingly arid regions, including its capital city, Beijing, by firing silver iodide rockets into the sky where rain is desired. There is even political strife caused by neighbouring regions that accuse each other of "stealing rain" using cloud seeding. About 24 countries currently practice weather modification operationally.China used cloud seeding in Beijing just before the 2008 Olympic Games in order to clear the air of pollution. In February 2009, China also blasted iodide sticks over Beijing to artificially induce snowfall after four months of drought, and blasted iodide sticks over other areas of northern China to increase snowfall. The snowfall in Beijing lasted for approximately three days and led to the closure of 12 main roads around Beijing. At the end of October 2009 Beijing claimed it had its earliest snowfall since 1987 due to cloud seeding.

INDIA

In India, cloud seeding operations were conducted during the years 1983, 1984–87,1993-94 by Tamil Nadu Govt. due to severe drought In the years 2003 and 2004 Karnataka government initiated cloud seeding. Cloud seeding operations were also conducted in the same year through US-based Weather Modification Inc. in the state of Maharashtra. In 2008, there were plans for 12 districts of state of Andhra Pradesh.

INDONESIA

In Jakarta, cloud seeding was used to minimize flood risk in anticipation of heavy floods in 2013, according to the Agency for the Assessment and Application of Technology.

SOUTHEAST ASIA

In Southeast Asia, open burning haze that pollutes the regional environment. Cloud-seeding has been used to improve the air quality by encouraging rainfall.

On 20 June 2013, Indonesia said it will begin cloud-seeding operations following reports from Singapore and Malaysia that smog caused by forest and bush fires in Sumatra have disrupted daily activities in the neighbouring countries. On 25 June 2013, hailstones were reported to have fallen over some parts of Singapore. Despite denials, some believe that the hailstones are the result of cloud seeding in Indonesia.

In 2015 cloud seeding was done daily in Malaysia since the haze began in early-August. Thailand started a rain-making project in the late-1950s. Its first efforts scattered sea salt in the air to catch the humidity and dry ice to condense the humidity to form clouds. The project took about ten years of experiments and refinement. The first field operations began in 1969 above KhaoYai National Park. Since then rainmaking has been successfully applied throughout Thailand and neighbouring countries. On 12 October 2005 the European Patent Office granted to King BhumibolAdulyadej the patent Weather modification by royal rainmaking technology.

KUWAIT

To counter drought and a growing population in a desert region, Kuwait is embarking on its own cloud seeding program, with the local Environment Public Authority conducting a study to gauge its viability locally.

UNITED ARAB EMIRATES

The UAE is one of the first countries in the Persian Gulf region to use cloud seeding technology. It adopted the latest technologies available on a global level, using sophisticated weather radar to monitor the atmosphere of the country around the clock.

In the United Arab Emirates, cloud seeding is being conducted by the weather authorities to create artificial rain. The project, which began in July 2010 and cost US$11 million, has been successful in creating rain storms in the Dubai and Abu Dhabi deserts.

The United Arab Emirates has an arid climate with less than 100mm per year of rainfall, a high evaporation rate of surface water and a low groundwater recharge rate. Although rainfall in the UAE has been fluctuating over the last few decades in winter season, most of that occurs in the December to March period. During the summer months, the prevailing Indian Monsoon drought effect leads to a build-up of cumulus clouds especially along the mountainous terrain in the eastern UAE.

The UAE cloud-seeding Program was initiated in the late 1990s. By early 2001 the Program was being conducted in cooperation with the National Centre for Atmospheric Research (NCAR) in Colorado, USA, the Witwatersrand University in South Africa and the US Space Agency, NASA.

In 2005, the UAE launched the UAE Prize for Excellence in Advancing the Science and Practice of Weather Modification in collaboration with the World Meteorological Organization (WMO). This prize was thereafter reshaped into the International Research Program for Rain Enhancement Science.

It subsequently became the UAE Research Program for Rain Enhancement Science in January 2015. The Program for Rain Enhancement Science is an initiative of the United Arab Emirates Ministry of Presidential Affairs. It is overseen by the UAE National Centerof Meteorology & Seismology (NCMS) based in Abu Dhabi. Among its key goals are advancing the science, technology and implementation of rain enhancement and encouraging additional investments in research funding and research partnerships to advance the field, increasing rainfall and ensuring water security globally.

The UAE now has more 75 networked automatic weather stations distributed across the UAE, 7 air quality stations, a sophisticated Doppler weather radar network of five stationary and one mobile radars, and six Beechcraft King Air C90 aircrafts for cloud seeding operations. Natural salts such as potassium chloride and sodium chloride are used in these operations. At present, the UAE mostly seed with salt particles in the eastern mountains on the border to Oman to raise levels in aquifers and reservoirs.

Forecasters and scientists have estimated that cloud seeding operations can enhance rainfall by as much as 30 to 35 per cent in a clear atmosphere, and by up to 10 to 15 per cent in a turbid atmosphere.

A total of 187 missions were sent to seed clouds in the UAE in 2015, with each aircraft taking about three hours to target five to six clouds at a cost of $3,000 per operation

EUROPE

Cloud seeding began in France during the 1950s with the intent of reducing hail damage to crops. The ANELFA project consists of local agencies acting within a non-profit organization. A similar project in Spain is managed by the Consorciopor la LuchaAntigranizo de Aragon. The success of the French program was supported by insurance data; that of the Spanish program in studies conducted by the Spanish Agricultural Ministry.

The Soviet Union created a specifically designed version of the Antonov An-30 aerial survey aircraft, the An-30M Sky Cleaner, with eight containers of solid carbon dioxide in the cargo area plus external pods containing meteorological cartridges that could be fired into clouds. Soviet military pilots seeded clouds over the Belorussian SSR after the Chernobyl disaster to remove radioactive particles from clouds heading toward Moscow. At the July 2006 G8 Summit, President Putin commented that air force jets had been deployed to seed incoming clouds so they rained over Finland. Rain drenched the summit anyway. In Moscow, the Russian Airforce tried seeding clouds with bags of cement on June 17, 2008. One of the bags did not pulverize and went through the roof of a house. In October 2009, the Mayor of Moscow promised a "winter without snow" for the city after revealing efforts by the Russian Air Force to seed the clouds upwind from Moscow throughout the winter.

In Germany civic engagement societies organise cloud seeding on a region level. A registered society maintains aircraft for cloud seeding to protect agricultural areas, for example in wine growing areas, in the district Rosenheim, the district Miesbach, the district Traunstein (all located in southern Bavaria, Germany) and the district Kufstein (located in Tyrol, Austria). Another society for cloud seeding operates in the district of Villingen-Schwenningen.

AUSTRALIA

In Australia, the activities of CSIRO and Hydro Tasmania over central and western Tasmania between the 1960s and the present day appear to have been successful. Seeding over the Hydro-Electricity Commission catchment area on the Central Plateau achieved rainfall increases as high as 30 percent in autumn. The Tasmanian experiments were so successful that the Commission has regularly undertaken seeding ever since in mountainous parts of the State.

In 2004, Snowy Hydro Limited began a trial of cloud seeding to assess the feasibility of increasing snow precipitation in the Snowy Mountains in Australia. The test period, originally scheduled to end in 2009, was later extended to 2014. The New South Wales (NSW) Natural Resources Commission, responsible for supervising the cloud seeding operations, believes that the trial may have difficulty establishing statistically whether cloud seeding operations are increasing snowfall. This project was discussed at a summit in Narrabri, NSW on 1 December 2006. The summit met with the intention of outlining a proposal for a 5-year trial, focusing on Northern NSW.

ISBN: 978-93-8830-599-0

The various implications of such a widespread trial were discussed, drawing on the combined knowledge of several worldwide experts, including representatives from the Tasmanian Hydro Cloud Seeding Project however does not make reference to former cloud seeding experiments by the then-Snowy Mountains Authority, which rejected weather modification. The trial required changes to NSW environmental legislation in order to facilitate placement of the cloud seeding apparatus. The modern experiment is not supported for the Australian Alps.

In December 2006, the Queensland government of Australia announced a $7.6 million in funding for "warm cloud" seeding research to be conducted jointly by the Australian Bureau of Meteorology and the United States National Center for Atmospheric Research. Outcomes of the study are hoped to ease continuing drought conditions in the states South East region.

AFRICA

In 1985 the Moroccan Government started with a Cloud seeding program called 'Al-Ghait'. The system was first used in Morocco in 1999, It has also been used between 1999 and 2002 in Burkina Faso and from 2005 in Senegal. For this program two aircraft were equipped with special instrument.

8.0   CONCLUSION

As the world is speeding ahead in the field of artificial rains we INDIA an agricultural country which is mainly dependent on rains for water need to implement these modern techniques so that drought and floods could be avoided.

As these methods are economical and non-hazardous they can be easily implemented and would help in serving the purpose of rains all over country and also drought effected areas.

ISBN: 978-93-8830-599-0

Municipal Solid Waste Management of Indian Cities

Vaibhav Sapkal[1]*, Bharat B Jindel[2] and Rahul Sharma[3]
[1,2,3]School of Civil Engineering, Shri Mata Vaishno Devi University, Katra, India
vrsapkal@gmail.com

Abstract

India (officially the Republic of India) located in South Asia. It is the seventh-largest country by geographical area. Its Area is 3287263 km^2. Urbanization is now becoming a global phenomenon, but its ramifications are more pronounced in developing countries. Due to rapid urbanization and uncontrolled growth rate of population, Solid Waste Management (SWM) has become acute in India. The waste characteristics are expected to change due to urbanization, increased commercialization and standard of living. Waste disposal is one of the major problems being faced by all nations across the world. It is more than a menace in our country. In Indian cities the waste is generally not weighed. It is measured by volume to determine the quantity of waste disposed. Several studies conducted by NEERI and other consultants have shown that the waste generation rates are low in smaller towns whereas they are high in cities over 20 lac population. The range is between 200 gms. per capita / day and 500 gms / capita / day. Thus a small town with a population of 100000 would generate about 10 to 50 tones of waste daily. Metropolitan cities like Mumbai, Chennai, Delhi and the likes generate staggering amounts of waste. If we carefully analyze this waste we will realize that more than 60% of it is biodegradable. There are different types of solid waste like Municipal solid waste, Agricultural solid waste, Hazardous solid waste, Industrial solid waste, Bio medical solid waste, E-waste solid waste, Radioactive solid waste etc. Honorable Supreme Court of India has given directions to MOEF regarding management of MSW. According to directions of supreme court of India MOEF published the MSW 2000 management and handling rules in the gazette of India and it is mandatory for all the municipalities to follow the rules.

1. INTRODUCTION

India (officially the Republic of India) located in South Asia. It is the seventh-largest country by geographical area. Its Area is 3287263 km^2. Urbanization is now becoming a global phenomenon, but its ramifications are more pronounced in developing countries. Natural growth of population, reclassifications of habitation and migration trends are important in urban population in India. Due to rapid urbanization and uncontrolled growth rate of population, Solid Waste Management (SWM) has become acute in India. The waste characteristics are expected to change due to urbanization, increased commercialization and standard of living. Waste disposal is one of the major problems being faced by all nations across the world. It is more than a menace in our country. In Indian cities the waste is generally not weighed. It is measured by volume to determine the quantity of waste disposed. Several studies conducted by NEERI and other consultants have shown that the waste generation rates are low in smaller towns whereas they are high in cities over 20 lac population. The range is between 200 gms. Per capita / day and 500 gms / capita / day. Municipal solid waste generation rate in different states in India is shown in Table 1.

Table 1 Municipal solid waste generation rate in different states in India

Sr No	Name of the state	Municipal population (according to census of India 2011)	Municipal solid (tones/day)	Per capita generated (kg/day)
1	Andhra Pradesh	10,845,907	3943	0.364
2	Assam	878,310	196	0.223
3	Bihar	5,278,361	1479	0.280
4	Gujarat	8,443,962	3805	0.451
5	Haryana	2,254,353	623	0.276
6	Himachal Pradesh	82,054	35	0.427
7	Karnataka	8,283,498	3118	0.376
8	Kerala	3,107,358	1220	0.393
9	Madhya Pradesh	7,225,833	2286	0.316
10	Maharashtra	22,727,186	8589	0.378
11	Manipur	198,535	40	0.201
12	Meghalaya	223,366	35	0.157
13	Mizoram	155,240	46	0.296
14	Orissa	1,766,021	646	0.366
15	Punjab	3,209,903	1001	0.312

ISBN: 978-93-8830-599-0

Contd….

Sr No	Name of the state	Municipal population (according to census of India 2011)	Municipal solid (tones/day)	Per capita generated (kg/day)
16	Rajasthan	4,979,301	1768	0.355
17	Tamil Nadu	10,745,773	5021	0.467
18	Tripura	157,358	33	0.210
19	Uttar Pradesh	14,480,479	5515	0.381
20	West Bengal	13,943,445	4475	0.321
21	Chandigarh	504,094	200	0.397
22	Delhi	8,419,084	4000	0.475
23	Pondicherry	203,065	60	0.295
	INDIA	**128,113,865**	**48,134**	**0.376**

2. MUNICIPAL SOLIS WASTE MANAGEMENT IN INDIA

We all are aware about Air, Water and Noise pollution created due to industrial activities and various steps taken by industries to control this pollution. However, the day to day waste generated from our residential and business activities called as Municipal Solis Waste (MSW), which is responsible for creation of Air, Water and Soil pollution, is not being handled scientifically in India. The quantity and quality of Municipal Solis Waste generation are among the major criteria to decide richness of city. More the quantity of MSW the richer is the city. As our richness increases the lifestyle changes, resulting in generation of more quantity of MSW. The Hon'ble Supreme Court of India constituted a Committee for suggesting improvements in SWM practices in Class I cities in India. The Hon'ble Supreme Court of India entertained the Writ Petition No. 888 of 1996 and after several hearings felt it appropriate to constitute a Committee to look into all aspects of Solid Waste Management in class I cities of India.

2.2   Guidelines given to Municipalities according to MSW 2000 Rules

Handling of Municipal Solid Wastes has been considered as one of the critical issues in environmental management. Implementation of Municipal Solid Wastes (Management & Handling) Rules, 2000, has been very slow owing to inadequate financial resources and to some extent lack of technical/management experience with the Urban Local bodies (ULBs).' Ministry of Environment & Forests (MoEF) has stressed on formulation of time-bound action plan for Municipal Solid Wastes Management (MSWM) for metro cities and State capitals to begin with. However, formulation of action plan would need reliable data for each local body and considering the availability of resources, realistic time-targets should be set.

3.   MSWM IN INDIA: AT A GLANCE

Objectives of MSW Management

   (a) Improve the quality of waste management process

   (b) Improve recycling and waste collection infrastructure on campus
   (c) Reduce the total amount of waste generated each year
   (d) Divert waste from landfill through re-use and recycling initiatives
   (e) Consider waste as a resource and realize its value wherever possible
   (f) Reduce the cost of waste disposal
   (g) Encourage and influence all campus users (staff, students and visitors) to follow the
   (h) principles of sustainable waste and resource management
   (i) Provide opportunities for curricular and extra-curricular activities on campus with due attention to environment friendliness.

In India, the community bin collection system is the main practice used for waste collection. In this system, residents deposit their waste into the nearest community bins located at street corners at specific intervals. Waste segregation at the source is minimal. Segregation of MSW into dry and wet wastes is carried out only in limited areas of a few cities, and in these areas, separate containers are used for collection of dry and wet wastes. Waste generated in households is generally accumulated in small containers (often plastic buckets) and then disposed of into community bins. Containers for household storage of solid wastes are of many shapes and sizes, and are fabricated from a variety of materials. Residents usually store waste in 16–20 l plastic buckets. The type of container generally reflects the economic status of the waste generator. The containers generally are constructed of metal, concrete, or a combination of the two. Wherever the Common Municipal Solid waste Management Facilities (CMSWMF) proposed to be implemented through an agreement in between two or more local bodies, prior environmental clearance from MoEF as per EIA notification dated 14th September, 2007 is required. The various treatment and disposal options os MSW are shown in Fig. 1.

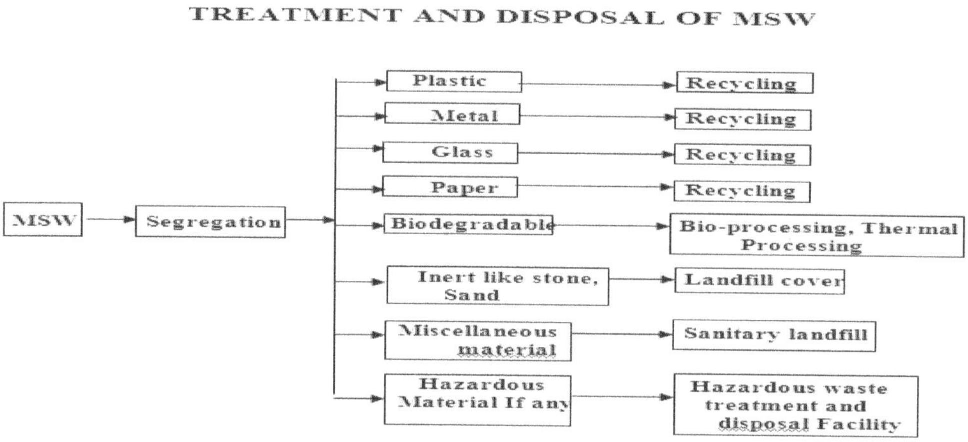

TREATMENT AND DISPOSAL OF MSW

MSW → Segregation →

- Plastic → Recycling
- Metal → Recycling
- Glass → Recycling
- Paper → Recycling
- Biodegradable → Bio-processing, Thermal Processing
- Inert like stone, Sand → Landfill cover
- Miscellaneous material → Sanitary landfill
- Hazardous Material If any → Hazardous waste treatment and disposal Facility

Fig. 1 Various treatment and disposal options of MSW

4. CONCLUSION

Management of solid waste continuous to remain one of the most neglected area of urban development in India. In Indian cities the waste is generally not weighed. It is measured by volume to determine the quantity of waste disposed. Several studies conducted by NEERI and other consultants. MSW management pertain mainly to inadequate MSW management solid waste management activity like collection, transportation, processing, disposal, initiation of house-to-house collection system, adoption of large variety of community bins in cities having community bin system, adoption of front end loaders for mechanical lifting of waste at open collection spots. adoption of source specific collection system emphasis on segregation of dry and wet waste at source introduction of spot fining system ensuring minimum manual handling of waste.

Emphasis on improvement of existing disposal site keeping in view of MSW Handling Rules, 2000 Avoiding open dumping of MSW Creation of sanitary landfills Selection of landfill site as per the guidelines of CPCB/NEERI Monitoring of management of hazardous wastes in keeping with the legislation Adoption of separate treatment and disposal facilities for hazardous waste.

Discontinuing the mixing of biomedical waste with the MSW Ensuring proper segregation of various categories of biomedical waste as also treatment and disposal of these wastes as per legislation. To overcome the deficiencies in the existing MSWM systems, an indicative action plan incorporating strategies and guidelines has been delineated. Based on this plan, municipal agencies can prepare specific action plans for their respective cities. A need also exists to strengthen existing monitoring mechanisms, particularly from the point of view of implementation of provisions made in MSW (Management and Handling) Rules, 2000.

ISBN: 978-93-8830-599-0

A Study on Global Ecological Damage Due to Fast Fashion Apparel in Textile Industry

J. Lakshmi Samhitha

B. Tech (Civil Engineering) II Year, CMR College of Engineering and Technology, Kandlakoya, Medchal, Telangana.

Abstract

The textile industry, being the third largest source of pollution in the world, is facing criticism due to extensive water consumption and usage of hazardous chemicals from material growth processes to disposal phase. This industry is responsible for causing land, air and water pollution, affecting human health and damaging the ecosystem. Spike in production due to fast fashion is creating exponential damage in ecological elements, at a rapid rate. Chemicals of toxic nature are absorbed through skin, pollute the air we breathe, seep into natural water bodies and enter our food chain. This paper discusses key environmental issues involved in the textile industry, focusing on the ongoing production pattern called fast fashion. In order to create sustainable textiles, the main change aspect is categorization of textile fabrics, based on various factors and grading these fabrics. This paper also emphasizes on the recycling of textiles for further varied uses and contributes to cleaner production and sustainability in the textile industry by initiating a discussion on the opportunities for change in the manufacturing of textiles in accordance with the laws.

Keywords: Textile industry, environmental issues, sustainability, pollution, fast fashion, recycling, hazardous chemicals, water consumption.

I. INTRODUCTION

Cotton textile industry with an annual production of about 400 million meters of cloth and approximately 1000 million Kg of yarn is one of the biggest industries in the country. A large volume of waste water is originated from the different process in the mill.Water is used as the principal medium to apply dyes and various finishes. Water mass balances have to be considered per typical category of industries. Dye bath wastewater generated by textile mills is often rated as the most polluting among all industrial sectors. The pollution load is characterized by high color content, suspended solids, salts, nutrients and toxic substances such as heavy metals and chlorinated organic compounds. Many textile mills in the state currently discharge their wastewater to local wastewater treatment plants with minimum treatment such as pH neutralization. This process removes much of the residual dye colour. Larger mills can discharge more than 2 million gallons of wastewater of this kind per day.

II LITERATURE REVIEW

Enormous quantities of wastewater characterized by high concentrations of chemical oxygen demand (COD), suspended solids, heavy metals and salts are produced in the textile industries (Yurtsever et al., 2015). The highly variable nature of textile wastewater is due to the use of multiple processes such as dyeing, finishing, sizing, several washing, and rinsing cycles (Khandegar and Saroha, 2013; Prigione et al., 2008; Sahinkaya et al., 2008).Reactive dyes are the most widely used with a market share of 60–70% (Çinar et al., 2008; Sen and Demirer, 2003a) and about 20–50% of the applied dyes remain in the aqueous phase during dyeing which eventually leads to the colorization of the flow. These reactive dye molecules, which are often highly water soluble, generally pass through the conventional system of activated sludge or municipal wastewater treatment systems, virtually unchanged (Willmot et al., 1998). The qualities of the products required by the textile industry and consumers, namely high wash stability and wide range of colors, require the use of these reactive dyes even if they are not biodegradable (Cooper, 1998). In the textile industry, the dyeing processes are an essential link in the production chain. The effluents discharged by these processes contain products which give them a toxic character, notably dyes, detergents, insecticides, fats, oils, sulphates, solvents, and fibers (Gebrati et al., 2011; Balanosky et al.,2000).According to Mountassir et al. (2013) and Kacha et al. (1997), this wastewater is also characterized by high pH, high temperature, and organic load as well as extreme colors. These carcinogenic and mutagenic pollutants are resistant to biodegradation and possess a potential for inhibiting microbial activity.

III METHODOLOGY

The primary data was collected from the households from the Residential areas in Hyderabad using Questionnaire as the research instrument. A sample of 40 respondents was taken from Twin cities of Hyderabad and Secunderabad randomly.

ISBN: 978-93-8830-599-0

IV   DATA ANALYSIS

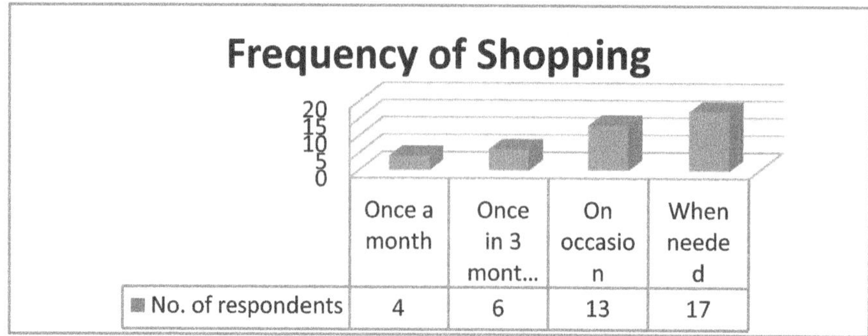

Fig. 1

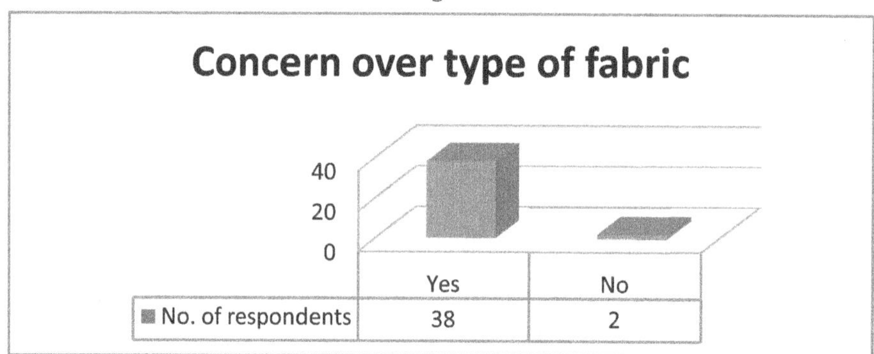

Fig. 2

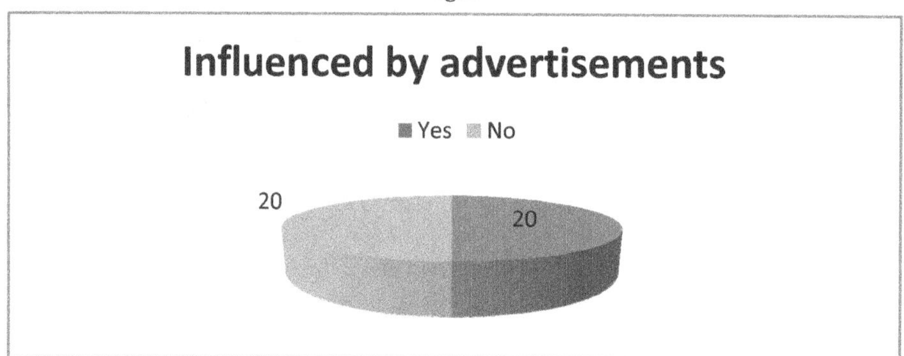

Fig. 3

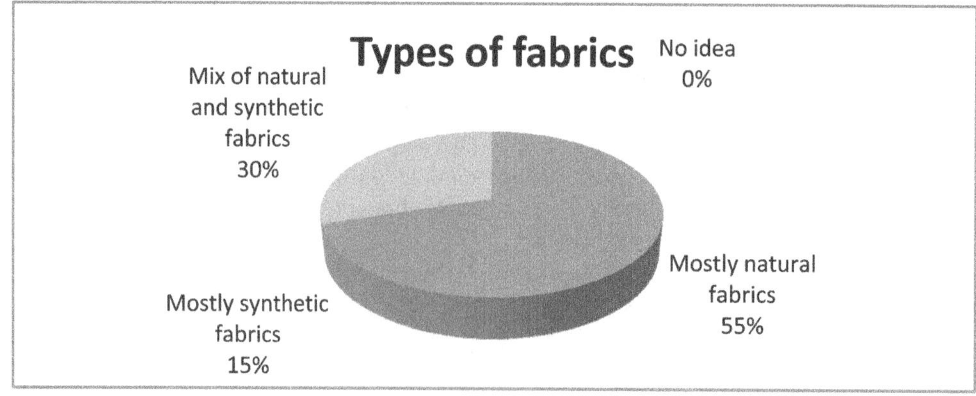

Fig. 4

ISBN: 978-93-8830-599-0

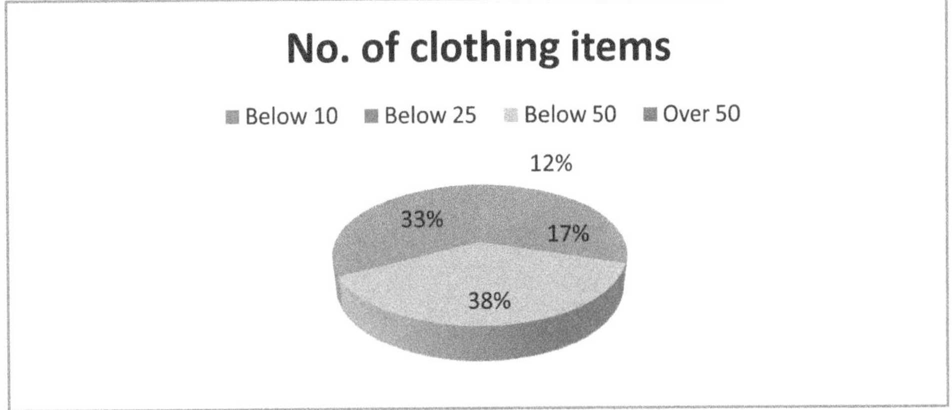

Fig. 5

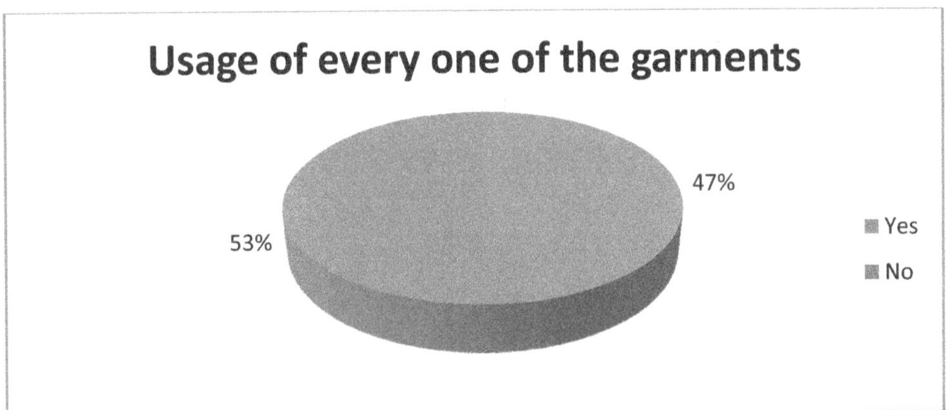

Fig. 6

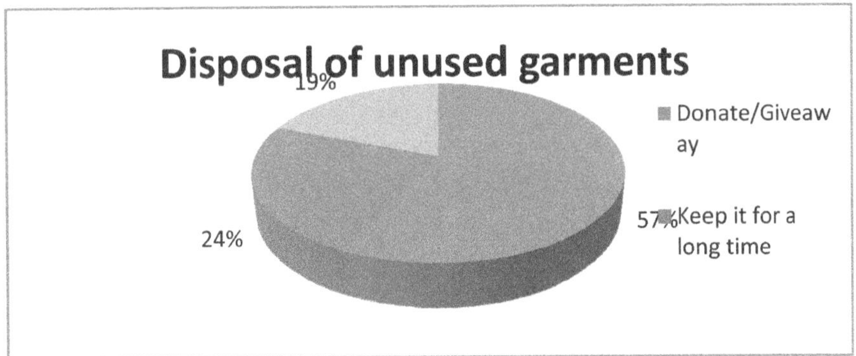

Fig. 7

ISBN: 978-93-8830-599-0

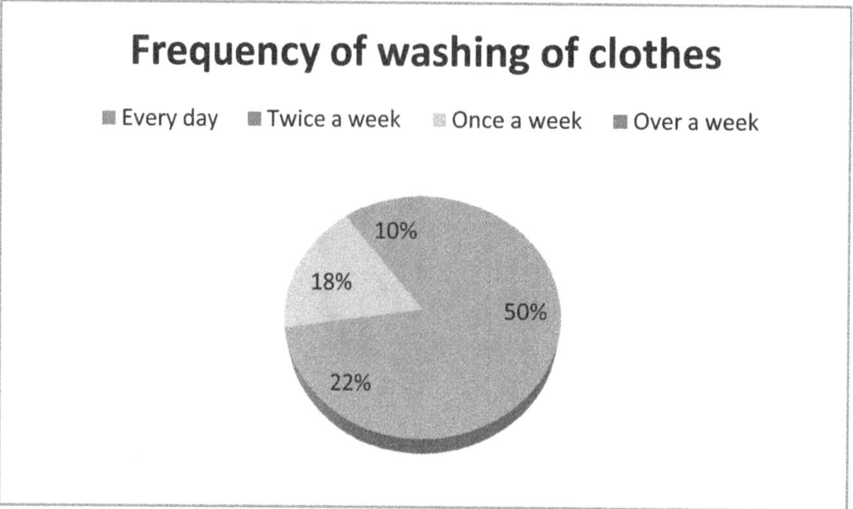

Fig. 8

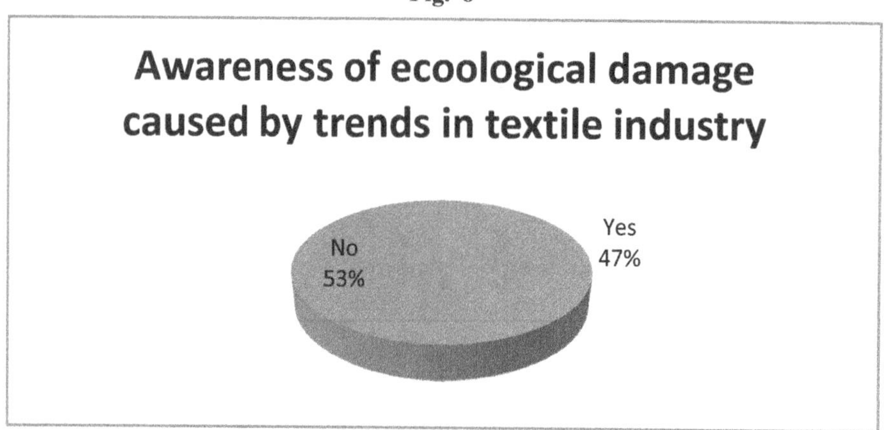

Fig. 9

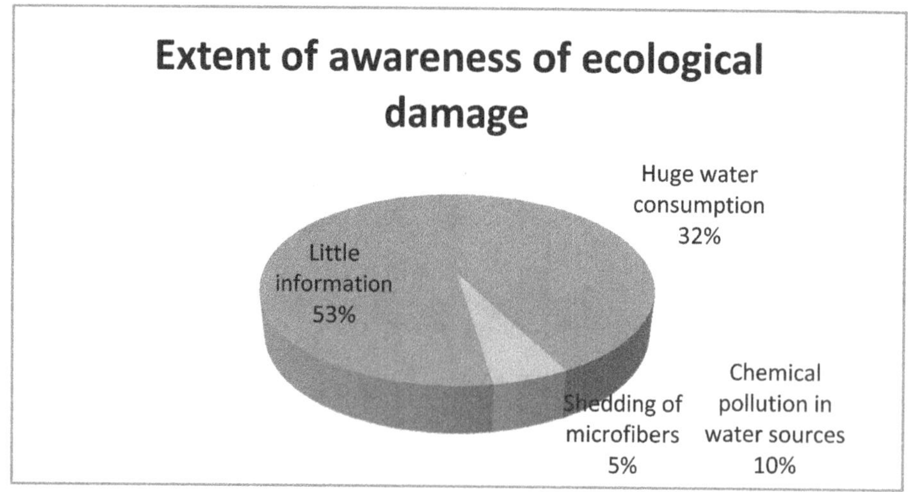

Fig. 10

ISBN: 978-93-8830-599-0

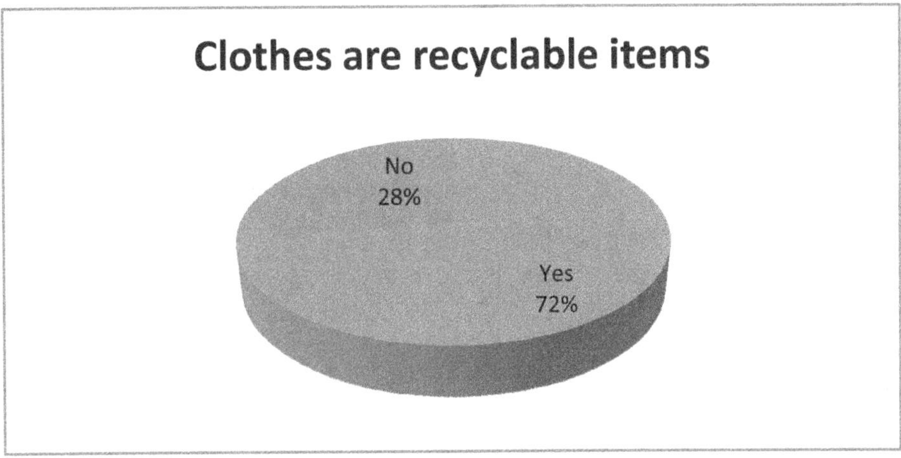

Fig. 11

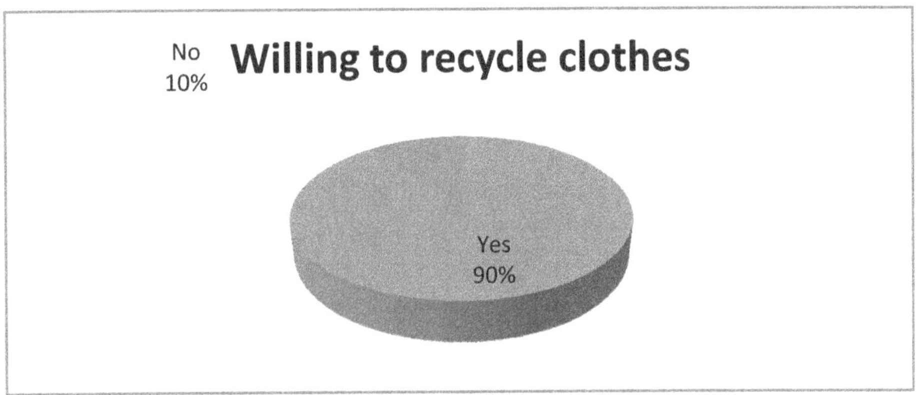

Fig. 12

V. DATA FINDINGS

1. Among total 40 respondents, 4 persons buy clothing items once a month, 6 persons buy once in three months. 13 persons buy on occasion or seasonally and 17 persons buy clothes only when needed.

2. Among 40 respondents, 38 show concern over the type of fabric they buy, while shopping. 2 are not concerned about type of fabric.

3. Among the 40 respondents, 20 respondents agree to being influenced by advertisements and offers issued by clothing retailers, while 20 are not influenced.

4. Among 40 respondents, 22 respondents possess mostly natural fabrics, 6 respondents possess mostly synthetic fabrics. 12 own clothing of a mix of natural and synthetic fabrics and no respondent is without an idea of the type of fabrics they own.

5. Among 40 respondents, 5 respondents possess below 10 items of clothing, 7 possess below 25, 15 possess below 50 clothing items and 13 possess over 50 items of clothing.

6. Among 40 respondents, 19 respondents wear/use every one of the garments they own while 21 respondents do not use every garment they own.

7. Among 40 respondents, 21 respondents had answered No when they were asked if they wear/use every one of the garments that they own.

8. Among 40 respondents, 21 respondents, 12 respondents donate/give away the garments, 5 respondents keep the garments for a long time and 4 reuse for household purposes.

9. Among 40 respondents, 20 respondents wash their clothes every day, 9 respondents wash their clothes twice a week and 7 wash once a week. Only 4 of the respondents wash their clothes over a week.

10. Among 40 respondents, 19 respondents have basic awareness of ecological damage caused by various trends in the textile industry while 21 respondents are not aware.

11. Among 40 respondents, 19 respondents are aware of the ecological damage caused by various trends in the textile industry. Of these 19, 6

ISBN: 978-93-8830-599-0

respondents are aware of the huge water consumption, 2 are aware of the chemical pollution in water sources, 1 is aware of the shedding of microfibers in water and 10 respondents have little information/knowledge on this issue.

12. Among 40 respondents, 29 respondents agree to the statement –"Clothes are recyclable items." 11 respondents do not agree.

13. Among 40 respondents, 29 agree that clothes are recyclable items. Of the 29, 26 respondents are willing to recycle their clothes in order to reduce ecological damage while 3 are not willing.

VI. CONCLUSION

Textile Industry is being forced to consider water conservation for many reasons. Since a significant portion of the equipment and machinery is highly customized, it is difficult to recommend new ways of saving water. The primary reasons being the increased competition for clean water due to declining water tables, reduced sources of clean waters, and increased demands from both industry and residential growth, all resulting in higher costs for this natural resource. Water and effluent costs may in the more common cases; account for as much as 5% of the production costs. The equipment used in a water conservation program is relatively inexpensive, consisting in most cases of valves, piping, small pumps, and tanks only. The operating costs for these systems are generally very low. Routine maintenance and, in some cases, electricity for the pumps, would be the major cost components.

REFERENCES

1. Balanosky, E., Herrera, F., Lopez, A., Kiwi, J., 2000. Oxidative degradation of textile waste water. Modeling reactor performance. Water Res. 34, 582–596.

2. Çinar, Ö., Yasar, S., Kertmen, M., Demiröz, K., Yigit, N.Ö., Kitis, M., 2008. Effect of cycle time on biodegradation of azo dye in sequencing batch reactor. Process Saf. Environ. Prot. 86, 455–460. https://doi.org/10.1016/j. psep.2008.03. 001.

3. Cooper, P., 1998. Ecotextile. Conference, Proc. of the 2nd Int. Textile Env. Conf., Bolton, UK. 7–8th April, p. 57.

4. Gebrati, L., Loukili, Idrissi L., Mouabad, A., Nejmeddine, A., 2011. Use of Daphnia test for assessing the acute toxicity of effluents from a textile industry in Marrakech (Morocco). Phys. Chem. News 60, 133–140.

5. Kacha, S., Ouali, M.S., Elmalih, S., 1997. Élimination des colorants des eaux résiduaires de l'industrie textile par la bentonite et des sels d'aluminium. Rev. Eau 10, 233–248.

6. Khandegar, V., Saroha, A.K., 2013. Electrocoagulation for the treatment of textile industry effluent – a review. J. Environ. Manage 128, 949e963..

7. Prigione, V., Tigini, V., Pezzella, C., Anastasi, A., Sannia, G., Varese, G.C., 2008.

a. Decolourisation and detoxification of textile effluents by fungal biosorption. Water Res. 42, 2911–2920.

8. Mountassir, Y., Benyaich, A., Rezrazi, E., Gebrati, L., 2013. Wastewater effluent characteristics from Moroccan textile industry. Water Sci. Technol. 67 (12),2791–2799.

9. Sandra, G.M., Renato, S.F., Nelson, D., 2000. Degradation and toxicity reduction of textile effluent by combined photocatalytic and ozonation processes. Chemosphere 40, 369–373.

10. Sen, S., Demirer, G.N., 2003a. Anaerobic treatment of real textile wastewater with a fluidized bed reactor. Water Res. 37, 1868–1878. https://doi.org/10.1016/ S0043-1354(02)00577-8.

11. Yurtsever, A., Sahinkaya, E., Aktas, Ö., Uçar, D., Çınar, Ö., Wang, Z., 2015,

a. Performances of anaerobic and aerobic membrane bioreactors for the treatment of synthetic textile wastewater. Bioresour. Technol. 192, 564–573. https://doi.org/10.1016/j.biortech.06.024.

12. Willmot, N., Guthrie, J., Nelson, G., 1998. The biotechnology approach to color removal from textile effluent. JSDC 114, 38–41.

ISBN: 978-93-8830-599-0

Review on Groundwater Recharging along the Plastic Roads

Vasala Sai Charan[1], B. B Kori[2], Rajarshi Saha[3] and L. Ravi[4]
[1]M.Tech Student, JNTUH-IST [2]Professor, Guru Nanak Dev Engineering College, Bidar
[3]Scientist, NRSC Hyderabad, [4]Research scholar, CSIT, JNTUH-IST

Abstract

Plastic widely used in our daily life for packing, protecting, carrying and for disposing other waste also. Plastic is a non-biodegradable. It is the main constituent of municipal waste. Plastic has long life time. If we burnt the plastic it will release toxic pollutants to atmosphere. Sometimes some wild animals will die by consuming plastic waste. To reduce the plastic waste we may use it for laying of roads by adding plastic waste with bituminous that is plastic roads.

In Present scenario people facing problems like lack of water, to overcome the water scarcity we need to increase the groundwater levels. It is possible if effective using of rainwater come into account that is groundwater recharging.

Keywords: Muncipal waste, plastic waste, road laying, plastic road, bituminous, water scarcity, ground water, rain water, ground water recharging.

I. INTRODUCTION

Nearly 15,342 tonnes of plastic wastes are coming per day in India. In India 30-40% of villages don't have proper road transportation; most of the bitumen roads undergo maintenance twice a year because of improper laying and using cheap quality materials for low cost roads. To reduce the cost of road laying and minimize the maintenance cost such as repairing patches etc., it is better to use plastic waste which is already used, low cost, and foremost to save the environment. Waste plastic may be used as binder and modifier. If we use plastic waste we may control air pollution (if we burnt plastic), land- pollution (land filling), and water pollution (as per new researches seas and oceans contains more than 33million tonnes of plastic wastes). We may also decrease municipal solid waste contents; it is indirectly helpful for our prime minister's prestigious mission "Swachh Bharath" (because beside our roads and residencies we saw huge quantities of plastic wastes).

Even though we are drilling more than 200ft for bore well, we are not in a position to find water levels in the present situation,   due to deforestation, urbanization, improper consuming of water, enamorous growth of global population and industrialization. If we effectively use the rainwater there may be chances of increasing the ground water levels. To implement this we have to construct a path for rainwater flow besides footpaths with a collecting ground water recharger.

II.   EASE OF USE

A.   Materials used and methodology

2.1 For plastic-bituminous road we have to use following materials.

- Fine aggregate   : sand, rock dust
- Filler   : cement lime, rock dust
- Binder   : bitumen with plastic waste
- Coarse aggregate : granite

2.1.1   Plastic and different types of plastic wastes and their origin:

The pure material which is obtained from the process of polymerisation is known as polymer. After adding fillers, resins, plasticisers, stabilizers that polymer is termed as plastic.

Different types of plastics and their origin are given below.

(a) Low density polyethylene (LDPE): bags, bin lining etc.

(b) High density polyethylene (HDPE): pharmaceutical bottles, milk, bottle caps etc

(c) Polystyrene : clear egg packs, bottle caps

(d) Polypropylene: film wrapping for biscuits, microwave trays for ready-made meals etc

(e) Foamed polystyrene: food trays, disposable cups, protective packing

(f) Polyvinyl chloride: toys, pipes, pens etc

Etc are different types of plastic wastes

2.1.2 Plastic waste can be used in plastic roads after doing   following processes:

1. Segregation: we have to segregate different types of plastics as prescribed above.
2. Cleaning process: In this we clean the plastic waste because it is a municipal waste.
3. Shredding process: In this we have to reduce the size of plastic waste. Generally waste particle not greater than the 2-3mm in width and 8 mm in length.
4. Collection process: collect plastic waste after shredding process. This plastic waste is used to mix with bitumen.

ISBN: 978-93-8830-599-0

2.1.3 Bitumen

It is a black, sticky and high viscous liquid. It is a   by-product of fractional distillation of crude petroleum. It has good cohesive and adhesive property. Commonly used bitumen grades are 60/70 and 80/100.

2.2 Ground water recharger:

For ground water recharger we need follow below given developments:

(a) Recharger: Dig a hole as much as required depth and diameter for every 1K.M. based on soil permeability, porosity, average rainfall intensity and ground water levels

(b) Sand and pebbles layers

(c) Flow path: which is used to create a way to rain water

(d) Screening: They are the separators of solid waste, which is used to allow only water to the hole.

(e) Pipe line from recharger to artificial lake:

In this if water flow is very high and porosity and impermeability of soil is high water may be outflows so if we arrange a pipeline from dig to an artificial lake the water utilization is more. By artificial lake we can use water for our daily-life usages.

III.   PREPARATION OF PLASTIC ROADS AND GROUND WATER RECHARGING

3.1 There are two processes for laying of plastic roads:
1. Dry process
2. Wet process
    (a) Dry process: In this aggregate heated to 170-180^{0}c in the mini hot mix plant, the shredded waste plastic is added in equal proportions, immediately the hot bitumen of required grade- (i.e. 60/70 or 80/100, 160^{0}c) is added. The mixture is added to road and the road is laid.
    (b) Wet process: In this method the plastic waste is directly mixed with hot bitumen at a temperature   of 560^{0}c and stirred with mechanical   stirrer,   after   that   some stabilizers are added and treated with proper cooling.

3.1.1   REQUIRED PROPERTIES OF GOOD PLASTIC ROADS:

Pavement surface layer of road is ready to withstand weather and tyre. Pavement must be exhibit characteristics like friction, strength, noise and impermeable to surface water.

1. Moisture absorption and void measurement:

The aggregate is chosen on the basis of is moisture absorption capacity, porosity, and strength as per IS codes. Aggregate coated with plastic waste improved the quality of roads with respect to the voids, moisture absorption and soundness.

2. Soundness test:
    It is used to study the resistance of aggregate to weathering action.

3. Los Angel's Abrasion Test:
    This test is used to determine the wear and tear on aggregate caused by iron wheel or rubber tire.

4. Marshall stability test:
    This test is used to determine resistance of bituminous material to distortion, displacement, rutting and shearing stress.

5. Softening point test by RING AND BALL APPARATUS:
    This test is used to find the softening point of temperature at which the material attains particular degree of softening.

6. Ductility index test:
    This test is used to find how much distance plastic waste-bitumen will elongate before breaking.

7. Flash and fire point:
    This test is used to find the minimum flash point limits to prevent the inclusion of highly inflammable volatile fractions in kerosene distillates.

Other tests are skid resistance test, sand-patch   test, benkel man deflection test etc.

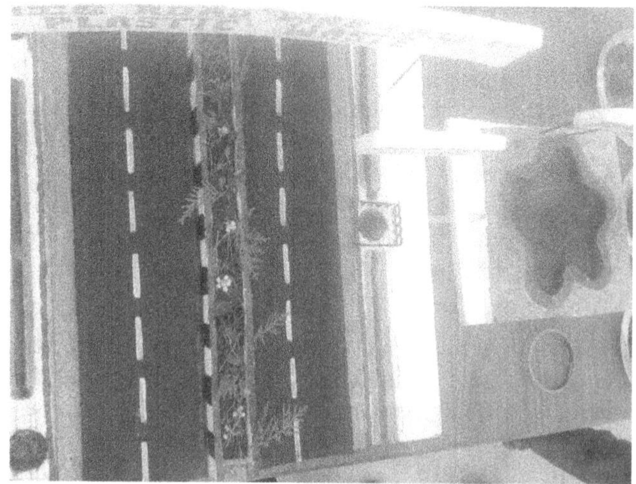

Fig. 1 Shows model representation of our theme:

ISBN: 978-93-8830-599-0

COMPARISION OF ADVANTAGES BETWEEN DRY AND WET PROCESSES:

By using dry process we may increase the percentage of waste plastics, this process doubles the binding property of aggregates, no new or special equipment are used that is no usage of stirrer, we may also decreases the cost of the project and there is no chances of maintenance of roads up to 6-7 years.

By using wet process main advantage is we can use any type, size, and shape of waste plastic.

COMPARISON BETWEEN ORDINARY ROAD AND PLASTIC ROAD:

S. no	Properties	Plastic road	Ordinary road
1	Marshall stability value	high	Low
2	Binding property	Better	Good
3	Softening point	Less	More
4	Penetration value	More	Less
5	Tensile strength	High	Less
6	Rutting	Less	More
7	Stripping	No	More
8	Seepage of water	No	Yes
9	Durability of the roads	Better	Good
10	Cost of pavement	Less	Normal
11	Maintenance cost	Nil	More
12	Eco-friendly	Yes	No

3.2 Ground water recharger

Process of making ground water recharger:

To prepare a ground water recharger we have to make a hole of required depth and diameter based upon the average rainfall intensity, runoff, ground- water levels of particular place, water table, porosity and permeability of that soil. After that make layer of sand particles of size 1-3mm, then after make a layer of pebbles of size greater than 4.75mm.

> Make convey (flow path) besides the roads for flowing of rain water. That flow path must be indicated with the sign boards to alert vehicles and public.

> Use screenings to avoid solid particles like plastics, sand, etc.

> Make a pipeline to send the water to artificial lake if levels of water high compare to recharger.

> The water in artificial lake may be utilised for domestic use. Artificial lake is created for every 2-3 rain water rechargers. If this arrangement system was constructed beside of roads there is much more benefit, because water scarcity decreases.

This whole set-up is collectively known as ground water recharger.

IV. ADVANTAGES OF PLASTIC ROADS AND GROUND WATER RECHARGER

Advantages of plastic roads:

- Increase the binding property.
- No stripping and no potholes.
- Have a high resistance towards the rainwater and water stagnation.
- No leaching of plastic occurs.
- The strength of roads is increased by 100%.
- The load withstanding property increases.
- The cost of road construction is also decreased by 20-25%.
- Maintenance cost of road is almost zero.
- Disposal of waste plastic will no longer be a problem.
- Air, water and land Pollutions minimized.
- Skid resistance also improved.
- Municipal waste levels are decreases.

ADVANTAGE OF GROUND WATER RECHARGER:

The major benefit of ground water recharger is the increasing of groundwater levels. Decrease the problem of scarcity of water.

V. CASE STUDIES

- The performance of the road stretches constructed using waster plastic in Karnataka is found to be satisfactory
- In Tamil Nadu, 1000KM roads in various places were constructed using waste plastic as an additive in bituminous mix under the scheme "1000KM Plastic- Tar Road", and found that, the performance of all the road stretches are satisfactory.
- Laboratory studies were carried out at the Centre for Transportation Engineering of Bangalore University, in which the plastic was used as an

additive with heated bitumen and different proportions (ranging from zero to 12% by weight of bitumen) the results of the laboratory investigation indicated that, the addition of processed plastic of about 8.8% by weight of bitumen, helps insubstantially improving the stability, strength, fatigue life and other desirable properties of bituminous concrete mix, even under adverse water-logging conditions. The addition of 8.0% by weight of processed plastic for the preparation of modified bitumen results in a saving of 0.4% bitumen by weight of the mix or about 9.6% bitumen per cubic meter of BC.

VI. CONCLUSION

If plastic roads are laid, disposal of plastic wastes is no more problems, no problem of dirtiness nearer residences, public places, and roads because of plastic waste. Lifetime of roads also increases and cost of the laying also decreases by 15-25%. New employment sector of plastic waste collection, segregation, etc are going to be established. And finally it is an eco-friendly construction. Ground water recharger along the roads is a one of the best way to save Rain water and increase the levels of ground water.

REFERENCES

1. R.Vasudevan , "Utilization of Waste Polymers for Flexible Pavement and Easy Disposal of Waste Polymers".
2. R.P Mani, K.N Mishra, B. Rama Devi, V.R Reddy "Engineering Chemistry 3e", " polymers", 1st impression,   2014, pp. 109-110.
3. Amit Gawande, G. Zamare, V.C. Renge, Saurabh Tayde, G.Bharsakale, "An Overview on Waste Plastic Utilization in Asphalting of Roads", vol.3/issue 2/april-june.2012/01-05.
4. J.M.Mauskar "Performance Evaluation Of Polymer Coated Bitumen Built Roads", PROBES/122/2008-2009.
5. S.S.Varma, "Roads from Plastic Waste", Indian Concrete Journal, November 2008.
6. Google search on plastic waste generated in India per day and municipal plastic waste.
7. "Utilization of waste plastic in bituminous mixes for road constructions", unknown author.
8. Dr. P.K. Jain, "Plastic Waste Modified Bituminous Surfacing For Rural Roads".
9. Araceli Magsino Monsada, "waste plastic recycling technologies in the Philippines", march 01-2004.

Utilization of Plastic Waste in Bituminous Pavements

B. Mahesh

Assistant Professor Department of Civil Engineering, Vignan Institute of Technology and Science, Deshmukhi

Abstract

Plastics are characterized as the substances which are non-biodegradable in nature that show negative impacts, for example, debasing our condition. A noteworthy issue these days confronted is the transfer of such risky plastic wastes. These wastes are non-bio-degradable (may not decompose for a few or many years) in nature causing ecological contamination and cleanliness issues. So the major problem due to these plastic wastes can be eliminated by the process of recycling or reusing them for various purposes such as usage of plastic waste in the construction of bituminous (flexible) pavements. The usage of plastic helps in producing plastic coated aggregates which shows a major impact on the durability of the pavements. The main advantage of using plastic in flexible pavements is that it improves the life span of the pavement by reducing the abrasion, wear and tear. These polycoated aggregates will not only increase the life span but also increases the stability, binding nature of the aggregates.

1.0   INTRODUCTION

Defilement on account of plastic is characterized as the strategy of obtaining of plastic waste items, (for instance, plastic containers and plastic sacks) in the Environment that goes about as a peril to normal life, and individuals. They are characterized basically into scaled down scale, meso, or full scale rubbish, in perspective on size. Plastic being a shoddy and extreme material is the principle reason of high plastic age by individuals. Plastic waste collection can filthy land, courses, and oceans. It is assessed that 1.1 to 8.8 million metric tons (MT) of plastic waste enters the ocean from waterfront territories consistently. Living creatures, expressly marine animals, can be influenced either by mechanical effects, for instance, trap in plastic articles or issues related to contamination of water as a result of plastic waste or through prologue to compound substances inside plastics that intrude with their physiology. Individuals are also impacted by plastic defilement. Beginning at 2018, around 380 million tons of plastic waste is made worldwide consistently. From the 1950s up to 2018, a normal 6.3 billion tons of plastic have been conveyed far and wide, of which a normal 9% has been reused and another 12% has been singed. In the UK alone, more than 5 million tons of plastic are used each year, of which only a normal one-quarter is reused, with the remainder of to landfills. From now on it is essential to reuse plastic waste by using it amid the assembling of bituminous asphalts pavements.

1.0   PLASTIC ROADS

Plastic waste accept a critical activity in growing the nature of an asphalt, decreasing asphalt weariness. These asphalts have better resistance towards various biological effects, for instance, water and cool atmosphere. Since a ton of plastic waste is required for laying somewhat stretch of street, the proportion of waste plastic to be singed will decrease. The liquid type of plastic waste materials indicates extraordinary attachment property. In the wake of cooling, the blend was tried for pressure and twisting minutes, and the outcomes have announced that the plastic can be utilized as a decent restricting specialist in the bituminous asphalts development.

2.0   OBJECTIVES

(a) To comprehend the procedure associated with the utilization of waste plastic in bituminous asphalts make.

(b) To figure the conduct of the stability of blend under different rates of included plastic.

(c) To know the stability of the plastic asphalt and typical asphalts.

(d) To examine the ideal amount of plastic that can be used in the adaptable asphalts.

3.0   Advantages

(a) Plastics soften at around 130°C.

(b) No toxic gasses are evolved from plastic in the temperature range of 130-180°C.

(c) Plastic when in molten state shows adhesion property hence used as a binder.

(d) To reduce usage of bitumen content in pavement construction.

(e) Mixed with a binder like bitumen to enhance their binding property.

(f) This methodology helps to increase the strength and life span of the road.

4.0   Scope of the project

(a) The behavior of bituminous pavements under various other types of polymers.

ISBN: 978-93-8830-599-0

(b) The behavior of bituminous pavements under addition of more amount of plastic content to the aggregates.

(c) Usage of various other additives to improve pavement stability.

5.0   Literature review

Dr. R. Vasudevan (2007) – stated that the usage of plastic in the construction of pavement have shown several advantages such as good binding effect with the aggregates and the stability of the pavement can be improved by it.

Anzar Hamid Mir: Use of Plastic Waste in Pavement Construction: An Example of Creative Waste management - stated that the stability of the bitumen aggregate mix has increased on the addition of plastic to the aggregate.

R.Manju; Sathya; Sheema K: Use of plastic waste in the bituminous pavement – states that polymer coated aggregates reduce voids and moisture absorption and the use of plastic mix will reduce bitumen content by 10%.

6.0   Tests on bitumen

Following tests are performed on the bitumen

1. Penetration test on bitumen

Penetration value test on bitumen is done to measure the hardness or consistency of the bituminous material. Penetration value of bitumen is defined as the vertical distance penetrated by a standard needle into the bituminous material under standard conditions of load, time and temperature. This distance is measured in millimeter. The penetration test is used for evaluating the consistency of bitumen and helps in determining the grade of Bitumen.

Tabulations and Calculations

Table 1 Penetration test values

Sample no	Penetration Reading(mm)
Sample 1	50
Sample 2	55
Sample 3	55
Average Penetration reading	53

Results: The penetration reading of the Bitumen used is 53 And the grade of the Bitumen is S55 S55 indicates that the penetration values of the specific bitumen sample lie between penetration values 50 to 60.

2. Softening point test on bitumen

Softening point demonstrates the inclination of a material to stream at expanded temperature experienced in administration. Ring and ball device is utilized to decide the softening purpose of bitumen, black-top and coal tar. Softening point demonstrates the temperature at which cover have a similar thickness. Higher softening point demonstrates that they won't stream amid administration.

TABULATION AND CALCULATIONS

Table 2 Softening point test values

Sample No	Softening point readings
Sample 1	$80^\circ C$
Sample 2	$80^\circ C$
Sample 3	$85^\circ C$
Average Softening point	$81.6^\circ C$

RESULTS

The softening point of the Bitumen sample is $81.6^\circ C$

3. Flash and fire point test

Flashpoint is defined as the lowest temperature at which the bitumen material starts to flame up or catch fire. The fire point is defined as the lowest temperature of the liquid at which vapor combustion and burning commences. Pensky- Martens apparatus is used to determine the flash and fire point of the bitumen sample.

Tabulations and calculations

Sample no	Flash point	Fire point
Sample 1	180°C	210°C
Sample 2	178°C	200°C
Sample 3	175°C	200°C

RESULTS

Flash point = 180°C Fire point = 203°C

4. Ductility test

The ductility test is used to measure the adhesive property of bitumen and its ability to elongate. In the flexible pavement design, it is necessary that binder should form a thin ductile film around aggregates so that physical interlocking of the aggregates is enhanced. The results are measured in cm.

ISBN: 978-93-8830-599-0

Table 3 Ductility test values

Sample No.	Ductility value (cm)
Sample 1	40
Sample 2	45
Sample 3	43
Average ductility value	42.6

RESULTS

The distance at which the thread of each bitumen sample breaks is recorded (in cm) to report as ductility value. The average penetration value of 3 tests of a given bitumen sample is recorded as 42.6cm.

7.0   Marshall Stability test

Marshall test is for the most part utilized in routine test programs for clearing occupations. The stability of the blend is characterized as a greatest burden conveyed by a compacted example at a particular test temperature of 600 °C. The stream is estimated as the disfigurement in units of 0.25 mm between states of no heap and most extreme burden conveyed by the example amid stability test.

Tabulations and Calculations

Table 4 Marshall Stability Test Values

Sample No.	Bitumen content (%)	Plastic content (%)	Marshall stability(kg)	Marshall stability (KN)	Flow value (mm)
1.	10	0	1240	12.16	3.6
2.	9	10	1720	16.86	4.5
3.	8.8	12	1830	17.94	4.7
4.	8.6	14	1940	19.02	4.9
5.	8.4	16	1890	18.53	4.8

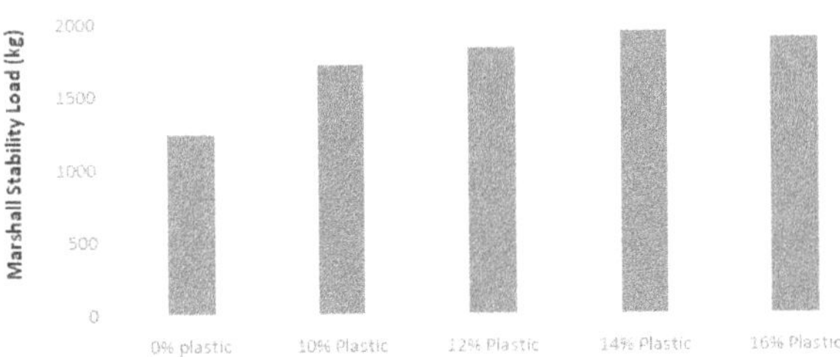

Fig. 1(a) Effect on load carrying capacity

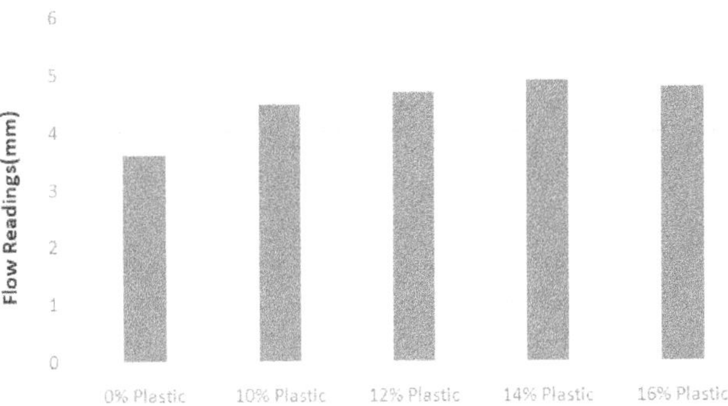

Fig. 1(b) Effect on Flow readings

8.0   CONCLUSIONS

The plastic blended with bitumen and aggregates is utilized for the better execution of the roads. The polymer covered on aggregates diminishes the voids and dampness ingestion. This outcomes in a decrease of trenches and there is no pothole arrangement. The plastic asphalt can withstand overwhelming traffic and are strong than adaptable asphalt. This new innovation is eco-accommodating.

(a) The stability of the blend is expanded with an expansion in plastic substance.

(b) The load conveying limit of the example with plastic substance is more when contrasted and the aggregate and bitumen test.

(c) 8.6% Bitumen and 14% of plastic are the rates that give the most astounding qualities to the asphalt.

(d) Further increment in plastic rate has demonstrated a decline in quality because of diminished Bitumen content prompting a decrease of restricting property between materials.

(e) Hence most extreme quality with a decrease in bitumen level of about 2.4% when contrasted with common example can be accomplished at the expansion of 14% plastic as it were.

9.0   REFERENCES

1. Use of Plastic Waste in Pavement Construction: An Example of Creative Waste management.

2. Author: Anzar Hamid Mir Use of waste plastic materials for road construction in Ghana

3. Author: Johnson Kwabena Appiah, Victor NanaBerko-Boateng, Trinity AmaTagbor S. Kumar and S. A. Gaikwad, Municipal Solid Waste Management in Indian Urban Centres: An

4. Approach for Betterment", in K. R. Gupta (Ed): Urban Development Debates in the New Millennium,

5. Atlantic Publishers and Distributors, New Delhi (2004) pp. 100-111.

ISBN: 978-93-8830-599-0

6. Manual on Municipal Solid Waste Management, Table 3, 6 (2000).

7. Priya Narayan, Analyzing Plastic Waste Management in India: a Case study of Polybags and PET bottles, IIIEE, Lund University, Sweden (2001) pp. 24-25.

8. CPCB report on Assesment of Plastic Waste and its Management at Airport and Railway Station in Delhi, December (2009) p.

9. National Plastic Waste Management Task Force (1997).

10. Plastics for Environment and Sustainable Development, ICPE, 8(1) (2007).

11. S.S. Verma, Roads from plastic waste, The Indian Concrete Journal, Nov.2008,43-44

12. S.Rajasekaran, Dr. R. Vasudevan2, Dr. Samuvel Paulraj, Reuse of Waste Plastics Coated AggregatesBitumen Mix Composite For Road Application – Green Method, American Journal of Engineering Research,2013, Volume-02, Issue-11

ISBN: 978-93-8830-599-0

Highway Safety Assessment Using Intellectual GIS-Based Road Accident Investigation and Real-Time Monitoring Automated System Using GIS/GPRS

Md Jalal[1], K. Jagdeesh[2] and V. Sireesha[3]

Asst Professor, Civil Engineering Dept

Abstract

All road safety audit process done for Indian condition are effective, but tedious and also time consuming. This paper is intended to illustrate applications of Geographic Information System (GIS) in transportation planning for which authors have made an attempt to simplify road safety audit process by integrating it with a GIS based software "Gram ++". This technique will help the road auditors to differentiate different sites of a road segment having higher accident frequency with sites having low accident frequency which should be potentially improved in a cost effective manner. It will also help road safety audit experts to identify and focus on the real problems at specific site of road segment right from planning stage for which authors have presented the procedure & discussed how it can be applied so that all road features can be considered simultaneously.

In India with more than one lakh fatalities per annum accounts for about 10 % of total world's road fatalities. The share of National Highways and State Highways in the total road network is just 6 % but these cater to 70 to 75 % of total road traffic in India. However, the National Highways, which constitute less than 2 % of the total road network, account for 20 % of total road accidents and 25 % of total road traffic fatalities occurring on Indian roads. Further, the severity of road traffic accidents on National Highways is more because of higher speeds as compared to other roads. The road safety situation in India is worsening. Accidents, fatalities and casualties have been increasing dramatically over last 20 years – about 5 % growth rate over last two decades - partly due to exponential growth of vehicles. The death rate per vehicle is 10 to 20 times higher in India as compared to high-income countries like Sweden, Norway, Japan, Australia, UK and USA. It is much higher even when compared to many low-income countries like Brazil, Mexico and Malaysia. Pedestrians, bicyclists and motorized two-wheeler riders are the Vulnerable Road Users (VRU), which constitute 60-80 % of all traffic fatalities in India. This seems logical as this class of road users forms the majority of those on roads. On highways, the proportion of VRU and other motor vehicle occupants are 32 % and 68 % respectively. In addition, they sustain relatively more serious injuries even at low velocity crashes, unlike car occupants who are protected by impact absorbing metallic body of the vehicles.

Accidents have been a major social problem in the developed countries of world for over fifty years. It is only in the past decade that developing countries like India have began to experience large increase in the number of road accidents taking place and have found it necessary to institute road safety programs. It is strongly felt that most of the accidents, being a multi factor event, are not merely due to drivers fault on account of driver's negligence or ignorance of traffic rules and regulations, but also due to many other related factors such as abrupt changes in road conditions, flow characteristics, road user's behavior, climatic conditions, visibility and absence of traffic guidance, control and management devices. In the above context an attempt is made to study various types of accidents including causative factors and blackspot identification in Cyberabad area. The study involves collection of accident data from various police stations for the period of four years

On the other hand a GIS (Geographic Information System) technique used for development of accident black spot map with same data. Based on the data collected, the data analysis, Black spot identification has done. Based on the spatial relationship between traffic accidents and road network elements, two-way association relationship is defined by spatial relationship computation. An attribute table is created to give information about various fields involved in black spots. Maps are made also with GPS (Global Positioning System) data and compared accuracy with traditional way of accident records. This attempt will show how GPS & GIScombines to give accurate black spot identification, rather than relying on assumed data for the location.

Keywords: Road safety Accidents, GIS, Fatal and non Fatal, GPS., GPRS

1. Introduction

Road traffic accident is complicated to analyze as it crosses the boundaries of engineering, geography, and human behavior. Therefore, there is a need for a more systematic approach, which can automatically detect statistically significant spatial accident clusters and offering repeatable results. To implement traffic accident countermeasures effectively and efficiently, it is important to identify accident-prone locations and to analyze accident patterns so that the most appropriate measures can be taken for each specific location.

Research Motivation

The research undertaken intends to find a completed solution to covering all aspects in managing and monitoring accident data. The system is based on GIS and telecommunications infrastructures technologies. In this case, IRAS (Intelligent Road Accident System) is used to get the better results from accident data, which includes the most effective and useful queries, reports, charts and advanced graphical user interface. The work evaluates for performance of a GIS-based solution in order to integrate infrastructure communication systems such as WiMAX and GPRS to develop a multiplatform middle-ware for real-time monitoring, automated services. The development is based on aspect and object oriented software design. In addition, the other objective is to be established a dataware house for the real-time smart decision support system for automated services and analysis. Also, this system offers Location Based Services (LBS) for Clients' cell phone, PDAs, smart phones and laptops.

System Architecture

3.1. Enabling Telegeoinformatics

Telegeoinformatics is enabled through advanced in such fields as geopositioning, mobile computing, and wireless networking. There are different architectures possible for Telegeoinformatics, but one that is expected to be widely used is based on a distributed mobile computing environment where clients are location aware, that is capable of determining their location in real-time, and interconnected to intermediary servers via WiMAX/GPRS or even wired networks. Telegeoinformatics can be based on different architectures to meet different requirements of applications, where clients and servers are connected via wireless networks. The middleware is used for performing many computations and activities and linking the different components of Telegeoinformatics. One of the key responsibilities of the middleware is ensuring interoperability among heterogeneous data, software, and functions. Figure 1 shows an ideal middleware for Telegeoinformatics.

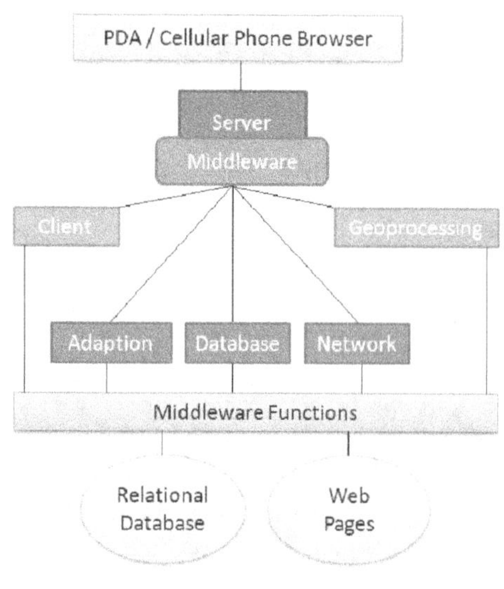

Fig. 1

Interoperability of Telegeoinformatics

In order to make Telegeoinformatics interoperable, the middleware must be based on special mechanism and protocols. Insomuch as geospatial data and geoprocessing are central to Telegeoinformatics, providing geospatial interoperability, eg., in Location-Based Services (LBSs) led by the Open GIS Consortium (OGC), should be one of the objectives of the middleware (OpenLS, 2001).

Strategies and Adaptation

One adaptation strategy in Telegeoinformatics is to adapt to the client machine the user would use to access the system. There are several client variations: PDA, Desktop PC, laptop, cell phone. Differences include storage capacity, processing power and user interface. Telegeoinformatics should have knowledge about these client's limitations with respect to the output interface and adjust its output information presentation to the capabilities available. Heineman (1999) has analyzed and evaluated various adaptation techniques for software components. On such technique is active interface. This technique acts on port requests between software components, which is where method request are received. Another technique is Automatic Path Creation (APC), which is a data format, and routing technique that allows multi data format adaptation and adapts to current network conditions (Zao and Katz, 2002). Adaptation is not the modification of components by system designers; this is considered component evolution. Adaptation should be accomplished automatically by the system with little user intervention. In addition, adaptation should be considered a design capable of adapting to users with respect to the user's needs (Stephanidis, 2001).

Terminal-Centric Positioning

The Terminal-Centric methods rely on the positioning software installed in the mobile terminal. The method, which is used in this research, is Network Assisted GPS (A-GPS); this method can also be used in the network-centric mode, according to Andersson (2001). A-GPS uses an assisting network of GPS receivers that can provide information enabling a significant reduction of the time-to-first-fix (TTFF) from 20-45 s, to 1-8 s, so the receiver does not need to wait until the broadcast navigation message is read. It only needs to acquire the signal to compute its position almost instantly. For the timing information to be available through the network, the network and GPS would have to be synchronized to the same time reference. According to Andersson (2001) the assistance data is normally broadcast every hour, and thus it has a very little impact on the network's operability.

ISBN: 978-93-8830-599-0

Network-Centric and Hybrid Positioning

Cell Global Identity with Timing Advance (CGI-TA) is one of the network-centric and hybrid positioning methods, which is used in this research. CGI uses the cell ID to locate the user within the cell, where the cell is defined as a coverage area of a base station (the tower nearest to the user). It is an inexpensive method, compatible with the existing devices, with the accuracy limited to the size of the cell, which may range from 10-500 m indoor micro cell to an outdoor macro cell reaching several kilometers (Andersson, 2002). CGI is often supplemented by the Timing Advance (TA) information that provides the time between the start of the radio frame and the data burst. This enables the adjustment of a mobile set's transmit time to correctly align the time, at which its signal arrives at the base (Snap Track, 2002).

4. Network-Based Service

The deployment of wireless-based LBSs relies on a common standards-based network infrastructure. The telecom market is currently experiencing a transition in the delivery of services from proprietary and network closed implementations to an open, IP-based service environment. The positioning service will be an important component in this open, IP-based service environment. The building of positioning information with LBSs, personalization, security, and messaging will be key for any operator when offering service packaging to the clients. The LBSs request/response flow is generally carried out within a wireless carrier network that includes mobile phone, the wireless network, the positioning server, gateway servers, geospatial server and the LBS application. The mobile phone provides a keypad for query and either a numeric or graphical interface for display. The positioning server, usually embedded in the wireless carrier's infrastructure, calculates the position of the device using one or more positioning approaches. The various wireless location measurement technologies fall into two broad categories: network-based and handset-based solutions (Campbell, 2001). Network-based solutions rely on base station radios to triangulate the position of a roaming mobile device, either with received radio signals or with transmitted synchronization pulses. The advantage of this approach is that it enables every user to access LBS without the need to upgrade the handset. Handset-based solution are systems that incorporate the measuring and processing of the location information within the handset. GPS is the principal technology and has been further enhanced by the development of A-GPS, which allows faster and more accurate service (Moeglein, 2001).

5. Multi Criteria Decision Making (MCDM)

The MCDM tool, which is embedded in server-side application, indicates AHP (Analytical Hierarchy Process) method, which can be used for decision making in GIS-based solution using input numeric values by the user. This tool is based on pairwise comparison method that developed by Tomas Saaty in 1970 in context of MADM (which is refer to attributes) method. It represents a theoretically founded approach to computing weights that are representing the relative importance of criteria. In this technique, weights are not assigning directly, but represent a "best fit" set of weights derived from the eigenvector of square reciprocal matrix. The objective of the AHP is to ensure that evaluation of weighting is consisted or not. The AHP relies on three fundamental assumptions:

(i) Preferences for different alternatives depend on separate criteria, which can be reasoned about independently and given numerical scores.

(ii) The score for a given criteria can be calculated from sub-criteria. That is, the criteria can be arranged in a hierarchy and the score at each level of hierarchy can be calculated as a weighted sum of the lower level scores.

(iii) Suitable scores can be calculated from only pairwise comparisons.

AHP is a mathematical decision making technique that allows consideration of both qualitative and quantitative aspects of decisions. It reduces complex decisions to a series of one-on-one comparisons, and then synthesizes the results. Compared to other techniques like ranking and rating, the AHP uses the human ability to compare single properties of alternatives, it not only helps decision makers choose the best alternative, but also provides a clear rational for the choice.

Methodology

The develop system consist of the main module that operate as a real-time monitoring system. The system started with a field data collection which perform as client. All data which are related to accident, can be categorizes into spatial data and non-spatial data which are merged into one database. The merged data is transferred to the dataware house via internet by using WiMAX/GPRS. Dataware house will then be established with two different reference of data includes Road Networks (map data) and Datacenter (description data). The main system uses integrated data from the dataware house and analyzes on them to achieve various types of statistical reports, these reports can be customized and improved with different elements by interaction through the designed GUI. The main system

ISBN: 978-93-8830-599-0

will be able to recycle the outcome of the primary analyses into the system and pass the statistical reports to MCDM unit (Multi Criteria Decision Making). Under the module, AHP (Analytical Hierarchy Process) technique is used for decisions making process. The results will then be used in the next module, which is SDM (Smart Decision Maker). In this module, reports can be compared together, make the best decision for diagnosis, and define the proposed solution related to current situation. SDM unit has potential to modeling suggested solution based on GIS interface. Figure 2 shows the process and relations between main system and the other parts to provide the proposed solution.

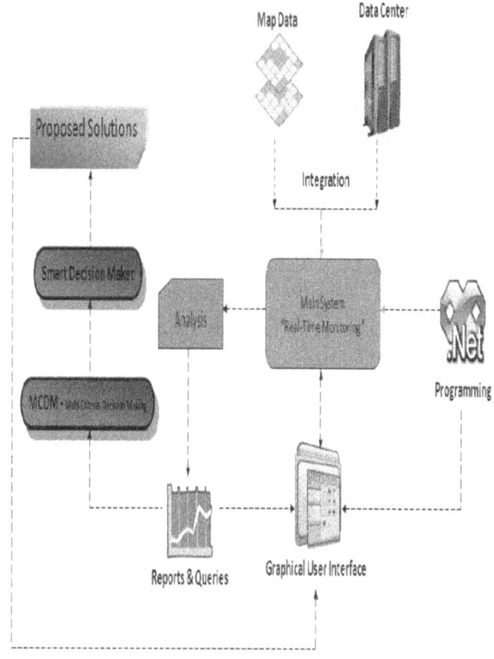

Fig. 2 The relationship between main system and MCDM and SDM modules

There are two types of clients that should be defined to the system, Police and end users (citizens). After each accident, the clients (citizens) can provide necessary information to the main system by using their Mobile Phone (SMS) or using their PDA (WiMAX/GPRS). The main system automatically will search and announce the nearest police vehicle via internet around the accident location and suggest the best path to get to the accident location based on the traffics information, time and distance. In this case, police officers using PDA via internet directly to the main system will key all accident information in and then the system will automatically send data to the dataware house and update the database. Also at the same time, system will be able to inform other involved organizations and companies such as medical emergency services, insurance and car service companies. Figure 3 shows the relationship between main system and clients.

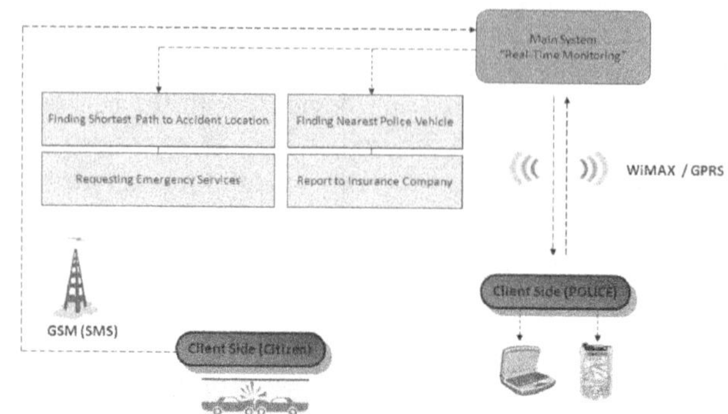

Fig. 3 The relationship between main system and clients

Results and Discussion

The system being developed manage to support online communication through the GSM/GPRS/WiMAX services. User who wants to use this system must have a few equipments on his/her car which call On- Board Unit (OBU). These equipments consist of GPS and wireless communication devices. The main system also consists of Auto Alarming System (AAS) to inform the system for any accident events throughout the city. In this case, when an accident occurs, the location of the accident sends to the main system by client (citizen) using GPS devices installed on the vehicle, this task is done by calling OBU which, automatically or manually send the accident position to the server via internet using WiMAX or GPRS. From the signal received, the system will find the nearest police vehicle according to the accident coordinate system and sends the primary information to client (police officer) in their PDAs via internet. Then, accident data will be send to data center and store on the system and dataware house. This system can send online data to road accident database using PDA or Smartphone by police officers in real time access. In addition, Smart Decision Making (SDM), Multi Criteria Decision Making (MCDM) and Automatically Proposed solution system (APSS) units can be applied professionally and analytically for making the particular reports and queries in a proper manner.

Conclusion

The research outcome is a comprehensive system to cover accident management and analysis, smart automated service for accident locations, accident and service diagnosis, reducing the number of accidents,

ISBN: 978-93-8830-599-0

increasing the level of road safety and fast delivery services such as insurance companies and emergency services. For further works, the system should also include the ability to inform clients about the risk zones in all over the city using LBS services, based on the main system reports, queries and real time traffic monitoring. The system classifies and categorizes the road networks into four zones, so when a vehicle enter to each zone, the system will automatically send the risk message to alert the accident risk on that particular zone or location of the city. In addition, the system shall offer some services for clients such as air download client version (for PDAs or Smartphone) of application for using these LBS services. All LBS data, which will be shown on client phone, are generated by the main system with real time updates via internet, so the clients have real time information about the traffic conditions and accident risk zones around them

ISBN: 978-93-8830-599-0

India's Initiative Approach for Disaster Management Programmes

Zainab Fatima Ali[1] and Khaja Mohiuddin[2]
[1]B.tech Nawab Shah Alam College of Engg & Technology, JNTUH, Hyderabad, India
M.Tech, Construction Technology, VTU,Belgaum, Mumbai, India
fathima98ali@gmail.com; kmohiuddin88@gmail.com

Abstract
Disaster has been occurring in this earth since time immemorial loss of millions of lives and huge properties due to disaster like earthquakes, volcanoes, tsunami has been occurring time to time to show the furry of nature. However due to technology it is possible to initiate disaster management program at international and national level to reduce the effect of such disaster to an extent. This paper gives an idea about the India's initiative for disaster management programs.

As India is prone to disaster due to its geological and geographical area the objective of this study is to give India's approach to the disaster management program with respect to its different support programs, tool & techniques adopted to carry out early forecasting of disaster, various agencies involved in mitigation, preparedness, response, and recovery operation and the various rules policies and regulation governing the disaster management.

Keywords: India, Disaster Management, Mitigation, Preparedness, Response, Recovery, NDMA,NDRF , Capacity Development.

I. INTRODUCTION

Disaster is a catastrophic situation where the normal life is affected with great loss to the eco system with emergency like situation to save lives and environment[1]

Disaster can be classified as

1. Natural Disaster – which is beyond the control of human and have an immediate impact ex floods, earth quakes etc.
2. Human Instigated/ Manmade Disaster: These are emergency which may occur due to major happenings such as fire, war.[5]

Disaster are further classified as

Water and Climate related Disasters	Geologically related Disasters	Chemical Industrial and Nuclear	Accident related Disasters	Biologically related Disasters
floods, cyclones, tornadoes, hailstorm, cloud burst heat wave and cold wave, snow avalanches, Droughts ,sea erosion, thunder and lightning, tsunami	Landslides and mudflows, earthquakes, dam failures/ dam bursts, mine fires	chemical and industrial disasters ,nuclear disasters	forest fires, urban fires, mine flooding, oil spill, major building collapse, serial bomb blasts, festival related disasters, electrical disasters and fires air, road and rail accidents, boat capsizing	Biological disasters and epidemics, Pest attacks, Cattle epidemics, Food poisoning

Disaster in India

There is no country in the world that is total free from disaster, so do India as disaster is not only natural but can also be human instigated.

Due to its geographical location and geological formation India is highly prone to disaster. According to NDMA report about 40 million hectare of total area in India is prone to floods. 8% of total coastline area (5700 km) is prone to cyclone.

68 % of total area (116 districts area) is prone to drought. 8% of total area is prone to landslides.

55% of area is prone to seismic zone III & IV. 65% of forest is under potential fire threat.

East coast, part of west coast and Andaman and Nicobar island are prone to tsunami.

Table 1 History of some disaster in India over a decade.[1]

Event	Year	Area
Maharashtra Floods	2005	Maharashtra
Tsunami	2004	TN,AP, Andaman& Nicobar islands, Pondicherry

Contd…

Floods	2014	Jammu and kashmir
Cyclone Vardah	2016	Chennai
Cyclone Nilam	2012	Tamil Naidu
Drought	2009	More than 10 States

*Source: National Policy on DM 2009 (Govt of India)

Disaster Management is a well planned strategy or an efforts to reduce the overall hazards caused by disaster, with a focus on formulation of a plan to reduce the effect of disaster, no disaster management plan can avert or eliminate the disaster but this can minimize the effect of disaster to an extent.DM has four phases.[2]s

Mitigation	:	Is Minimizing or reducing the effects of disaster
	:	Includes any activities that prevent an emergency or reduces the chance of an disaster happening
	:	Mitigation activities take place **before** and **after** disaster.
Preparedness	:	Prior planning on how to respond to a disaster
	:	Includes emergency exercises, training and preparedness plan
	:	Preparedness activities take place **before** an disaster occurs.
Response	:	Response is putting your preparedness plans into action, So as to minimize the hazards created by the disaster
	:	Activities such as rescue, emergency, relief and so on
	:	Response activities take place **during** an emergency.
Recover	:	Includes actions taken to return to a normal or an even safer situation following an disaster
	:	Includes a temporary housing, claims, medical care and so on
	:	Recovery activities take place **after** an emergency.

History and Frame work of Disaster Management in India.[3]

- Until the year of 2001 Agricultural Ministry was responsible for disaster management in India.
- After Gujarat earthquake under the chairmanship of PM. All Party National Committee was constituted in Feb 2001.
- Then the Responsibility of DM was transferred to Ministry of Home Affairs in June 2002.

- Sept 2005 establishment of NDMA and the Disaster Management Act was passed in 2005 Dec.

India's Approach

India's approach to disaster management has shifted from relief centric from integrated and holistic approach.[1] The shift has been taken into account the prevention, mitigation and preparedness after introduction of Disaster Management act of 2005.

In India, National Disaster Management Authority (NDMA) is the regulatory authority which lays all the rules and policies regarding Disaster Management.

Objective of NDMA
- Regulates rules policy and guidelines for DM plan.
- Implementation coordination and enforcement of policy   and plans.
- Take Measures for Prevention , Mitigation , Preparedness , Awareness Rehabilitation and Recovery
- Response
  National Executive Committee – low level disaster
  Execution by NEC and
  National Crisis Management   - Grave Disaster
  Committee (NCMC)
  Coordination by NDMA

In Additional
- National Disaster Response Force (NDRF) : Superintendence, Direction and Administrative Control with NDMA.
- National Institute of Disaster Management (NIDM):to work under the guidelines of NDMA.

At state level the State Disaster Management Authority (SDMA's) which is headed by chief minister of state, along with chairperson of state executive committee it consists of eight member to be nominated by chief minister and chairperson & will lay down policies and state plan in according to the guidelines of NDMA.

State Government will establish District Disaster Management Authority (DDMA) in each district which will regulate the planning co-coordinating and implementing body of DM in according to the policy and guidelines of NDMA, with total members not exceeding seven the DDMA is headed by District Magistrate.

Other important institutional arrangement includes Armed Forces, Paramilitary Forces, States Police Forces, State Fire Services, Civil Defense Forces, Home guards , NCC Etc.

Disaster Management Framework in India. [3]

- PM is the chairperson of NDMA.
- Structure of Framework is for all the three levels
    - (i) National
    - (ii) State
    - (iii) District.
- National Executive Committee (NEC) Secretaries of 14 ministries and chief of integrated defense staff function as executive committee of NDMA.

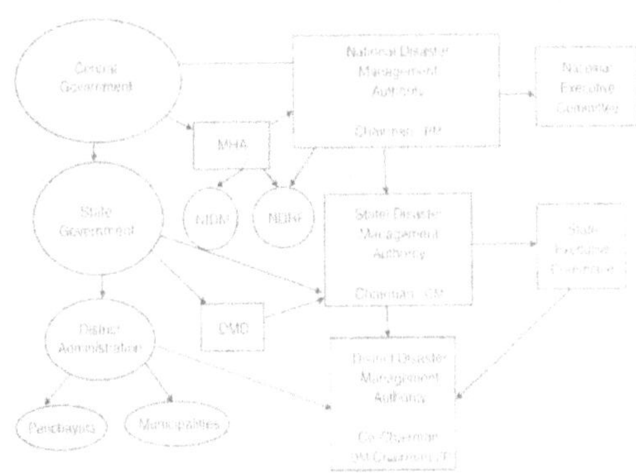

Fig. 1 Disaster Management Frame work

Nodal Ministries responsible for Disaster Management

- Natural Disasters (Flood, Tsunami, Cyclone, Earthquake etc.)- Ministry of Home Affairs (MHA)
- Drought-Ministry of Agriculture
- Biological Disasters-Ministry of Health and Family Welfare
- Chemical Disasters-Ministry of Environment & Forests (MoEF)
- Forest related Disasters- MoEF
- Nuclear Disasters-Ministry of Atomic Energy
- Air Accidents-Ministry of Civil Aviation
- Railway Accidents-Ministry of Railways.

National Executive Committee (NEC):

- The NEC is the executive committee of the NDMA.

- The NEC comprises the Chairman, Union Home Secretary & various other ministries Secretary.

Financial Arrangements

- Present funding mechanism for disaster relief (CRF/NCCF) to continue.
- Act provides for constitution of Disaster Response Fund and Disaster Mitigation Fund at National, State and District level.
- Each Department of Central and State Governments to make provision in annual budgets for implementation of District Plan.
- Provides penalties for obstruction, false claims, misappropriation, false warnings etc.

Institutional Steps to be taken at various level in case of severe disaster

National level:

- ➢ Under cabinet secretary the National Crisis Management Committee gives direction and plan to the crisis management group.
- ➢ Crisis Management group in MHA reviews the situation coordinates with various ministry to support the effected state.
- ➢ Task force may be set to coordination of relief measures in case of severe disaster by union cabinet.

State level:

- ➢ Under Chief Secretary a state National Crisis Management Committee comprising of all functionaries from various department and central official representatives.
- ➢ State disaster management commissioner is the head for ball coordinating and relief operation.

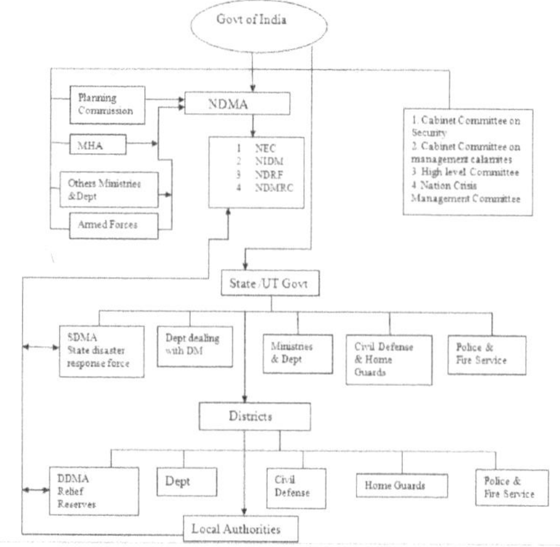

Fig. 2 Structure of DM in India

ISBN: 978-93-8830-599-0

National Institute of Disaster Management (NIDM)

- Central Govt shall constitute the NIDM as per chapter 7 of DM Act of 2005.

Objectives:
- ✓ Promote training and research.
- ✓ Networking and documentation of disaster management policies and mitigation measures.

National Disaster Response Force (NDRF)

- A highly technical, multi skilled, multidisciplinary force capable of handling all type of disaster
- Dedicated exclusively for disaster response as per Section 44 (1) , section 44 (2) , section 45.

Composition of NDRF Battalion
- ✓ In total there are 10 battalion with 8 existing and 2 raising
- ✓ Each Battalion have 1149 personal with 18 specialist team of 45 member in each battalion
- ✓ Each team to have 2 engineers, 2 paramedics, 1 technician, 1 electrician , 1 dog squad, 4 SAR teams, 1 diving & 1 water rescue team.
- ✓ Total number 18teams x 8battalions = 144 teams

Various Agencies Providing Warning & Forecasting System in India

Indian Meteorological Department (IMD)
- IMD provides cyclone warnings from the Area & Cyclone Warning Centre (ACWCs). It has developed necessary infrastructure to originate and disseminate the cyclone warnings at appropriate levels.
- Indian Meteorological Department (IMD) It has made operational a satellite based communication system called Cyclone Warning Dissemination System for direct Forecasting dissemination of cyclone warnings to the cyclone prone coastal areas.
- IMD runs operationally a Limited-area Analysis and Forecast System (LAFS), based on an Optimal Interpretation (OI) analysis and a limited area Primitive Equation (PE) model, to provide numerical guidance.

National Remote Sensing Agency (NRSA)
- Long term drought proofing programmes on the natural resources of the district have been greatly helped by the use of satellite data obtained by NRSA.
- Satellite data can be used very effectively for mapping and monitoring the flood inundated areas, flood damage assessment, flood hazard zoning and past flood survey of river configuration and protection works.
- Seismological Observations: Seismological observations in the country are made through national of 36 seismic stations operated by the IMD, which is the nodal agency. These stations have collected data over long periods of time.

Warning System for Drought :
- The National Agricultural Drought Assessment and Management System (NADAMS) has been developed by the Department of Space for Department of Agriculture and Cooperation, and is primarily based on monitoring of vegetation status through National Oceanic and Atmospheric Administration (NOAA) Advanced Very High Resolution (AVHR) data.
- The drought assessment is based on a comparative evaluation of satellite observed green vegetation cover (both area and greenness) of a district in any specific time period, with that of any similar period in previous years.

Flood Forecasting :
- Flood forecasts and warning are issued by the Central Water Commission (CWC) & Ministry of Water Resources.
- These are used for alerting the public and for taking appropriate measures by concerned administrative and state engineering agencies in the flood hazard mitigation. Information is gathered from the CWCs vast network of Forecasting Stations on various rivers in the country.

Cyclone Tracking:
- Information on cyclone warnings is furnished on a real-time basis to the control room set up in the Ministry of Agriculture & Government of India.
- High-power Cyclone Detection Radars (CDRs) that are installed along the coastal belt of India have proved to be a very useful tool to the cyclone warning work. These radars can locate

ISBN: 978-93-8830-599-0

and track approaching Tropical Cyclones within a range of 400 km.

- Satellite imagery received from weather satellite is extensively used in detecting the development and Warning & movement of Tropical Cyclones over oceanic Forecasting regions, particularly when they are beyond the range system of the coastal radars.
- The existing mode of dissemination of cyclone warnings to various government officials is through high priority telegrams, telephones, telex and fax.

Space Based Support and Services For Disaster

Department of Space (DOS) and Indian Space Research Organization (ISRO) are responsible for providing space based support and service for disaster management in India.

Disaster Management Support Program (DMSP) of DSO and ISRO.

Disaster Management support program is a aerospace system program which include both imaging and communication of ISRO & DSO provide timely support and services.

Disaster Management Support Program(DMSP)

- Addresses zoning of hazards with monitoring and assessment of major natural disaster by use of GIS based database system.
- Forecasting information regarding disaster such as Floods, Cyclones ,Landslides, Earthquakes, Etc
- Developing tools and techniques and also a satellite based communication system.
- Early warning of disaster by conducting development and research programmes.

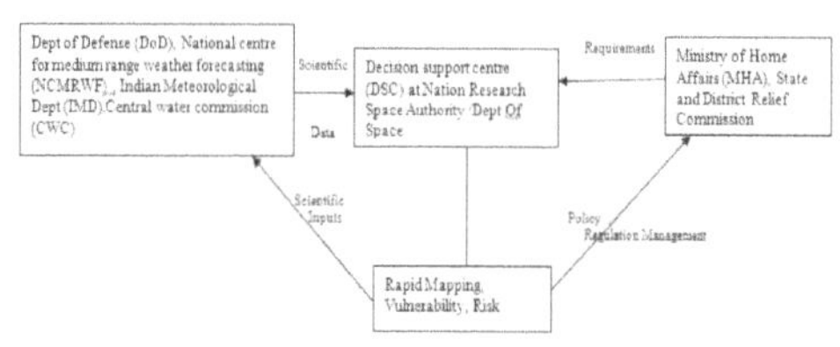

Mechanism of Space Enable Services for Disaster Management

Disaster Prevention & Mitigation

As Disaster cannot be fully eliminated but its effect can be reduced by proper planning and assessment in risk prone areas. A multi pronged approach is been adopted with following measures .Taking mitigation measure with help of Central Ministries in high sensitive areas. Encouraging and assistance of state level mitigation programs.

- Risk assessment & Vulnerability mapping
  - ✓ Use of Geographical Information system (GIS), remote sensing application and application of Global positioning system in disaster management.
- Disaster in urban area:
  - ✓ Of late due to urbanization there is an alarming increase in such type of disasters. This can be minimize by preparing an action plan on the unplanned urbanization and by ensuring a safe human habitat against all forms of disaster.
- Critical Infrastructure:
  - ✓ Infrastructure such as Dams, Bridges, Power station, Flyovers, Town should be properly monitored with respect to its safety standard to avoid and untoward mishap as there structure of vulnerable to any disaster.
- Environmental Approach & Climatic Change Adoption:
  - ✓ A disaster management program must be ecologically sustainable. The preservation of river, costal area, agricultural and soon should be well balanced period zonal regulation should be up gradation for natural habitats.
  - ✓ Due to climate change there is increase in natural disaster and hence to meet there challenges shall be carried out in sustained and effective manner.

The National Policy on Disaster Management (2009) has been finalized and approved by Home Inter-Ministerial consultation process has been completed. With the disaster management plan focuses on enhancing the mitigation capabilities for any type of multiple hazards. The five areas of action for reducing the risk are

- Understanding Risk (Priority-1).
- Interagency Coordination (Priority-2).
- Inverting in DRR – Structural Members (Priority-2).
- Inverting in DRR – Non Structural Member (Priority-2).
- Capacity Development (Priority-3).

Preparedness:

1. Forecasting and early warning system:

ISBN: 978-93-8830-599-0

- ✓ Upgrading or modernization of forecasting and early warning system
- ✓ Partnership with international agency like World Metrological Organization (WMO), Pacific Tsunami Warning System.

2. Information technology and communication support:
   - ✓ Up gradation and Monitoring of National Emergency Communication network , with reliable data and faster sharing with up to date data and faster sharing information with setting up of Infrastructure Emergency Centre

3. Strengthening of emergency operation centre.
   - ✓ At Central, State and District level with all the infrastructure equipped will be advantageous.

4. Preparedness with respect to medical:
   - ✓ For any DM Plan medical preparedness is an crucial component Ministry of Health and Family Welfare, State & Premier Medical Research will formulate medical research in coordination with the NDMA and in case of mass casualty the state and district will formulate the same as per the guidelines and policy in any disaster phase.
   - ✓ This plan will also include post disaster diseases surveillances system accessing services such as blood banks ambulance.
   - ✓ Creation of Heliambulance , mobile surgical team, railways dept accident relief medical van (ARMV) at every 100km.[1]
   - ✓ Proper and speedy disposal of dead bodies and animal carcasses should be given the due weight age.[1]

5. Training, Mock drill and Stimulation :
   - ✓ With assistance from NDMA the State and UT's will conduct mock drill in different parts of countries ,for efficacy of plans and standard operating procedure (SOP's) & refined through, trainings, seminars and mock drill and so to check the response of stakeholders.

Partnership for Mitigation and Preparedness:
1. Community based preparedness.
2. Stake holder participation like SDMA's, DDMA's, Civil defense.
3. Corporate social responsibility media partnership.
4. Media play a key role during the disaster Management will passing on the information of all phases of Disaster.

5. Corporate section has been the supporting the relief and rehabilitation activities.

Response
- ✓ Role of NEC :It will co-ordinate in response to an event and of disaster as per guide lines of NDMA.
- ✓ Role of nodal and central ,states, locals and district authority: Nodal ministry will chart out a response plane as per the National response plan all the authority will Co-ordinate as per the disaster specific guidelines.
- ✓ Standard Operating Procedure (SOP's): All the related authority will prepare a sop, which includes resource, medicals, assistant.
- ✓ Level Of Disaster: The level of disaster is determined by SOP's and a message issuing alert is circulated to various agencies about disaster. These SOP's will be reviewed periodically for disaster response.
- ✓ Incident Command System (ICS):A traditional structure exists which manages disaster in India in an administrative hierarchy system.
- ✓ Medical Response: It should be quick and effective with adequate medical resource at an site of disaster. Post disaster medical response is crucial to prevent out break of epidemics.
- ✓ Animal Care:Both wild and domestic animals are exposed to effect of disaster.A necessary and appropriate measures to be taken to provide them shelter after math of disaster should be available.

Relief

 Relief is no longer an assistance or provision of emergency relief supplies on time but viewed as an comprehensive system of facilitation of assistance of the victims of disaster for their rehabilitation.

With the following Objectives:
- ✓ Management of Relief Supplies
- ✓ Setting of temporary relief camp
- ✓ Review standard of Relief.
- ✓ Temporary livelihood
- ✓ Option and social economic rehabilitation.

Rehabilitation and Recovery [1]

The approach to the reconstruction process has to be overarching so as to convert this difficult event into opportunity. The administration, the stakeholders and the communities need to stay focused on the needs of this phase as with the passage of time, the sense of urgency gets diluted. This phase require the most patient effort. The appropriate choice of technology and

ISBN: 978-93-8830-599-0

project impact assessment needs to b e carried out to establish that the projects contemplated do not create any side effects on affected area.

With the following Objectives
- ✓ Speedy and owner driven
- ✓ Livelihood restoration
- ✓ Liking recovery with safe development.

Capacity Development

It's a process which comprises of awareness generation, training ,education, research and development , is effective with only active participation of stakeholders.

Recent Capacity Development Programs initiated in India[4]

1. Capacity building project (in partnership with (IGONU) and ministry of rural development.
   a. States. : 11
   b. Districts. : 55
   c. Training of PRI / ULB representatives. : 12,375 (225 per district).
   d. Training of Government Functionaries. : 4,125 (75 per district).

2. Under the Ministry of Panchayati Raj - training in state institutes of rural development and national institute of rural development for is under active planning
   a. Awareness generation and preparedness through PRI training institutes.
   b. Disaster resilience ensured in their projects.
   c. Similar consultation is in progress with ministry of urban development for training of urban local bodies.
3. Strengthening of Fire Services.
   a. Present shortage : 91%.
   b. Weakest link presently in dm apparatus of the country – matter of grave concern.
   c. equipment needs immediate upgradation to meet operational challenges
   d. Fire services need to be upgraded into fire and emergency services.
   e. Discussion carried out with finance commission for necessary allocation of funds. Presently the fire services though most valuable are the weakest link.
   f. Recently GoI approved rs 200 cr.

13th finance commission award fire services and capacity building[4]

a. Finance commission has recognized the need for modernization of the fire services.
b. Finances allotted to ulbs for revamping and upgradation of fire services in cities over one million population.• compliance to be confirmed through gazette notification.
c. In addition to the above, finance commission has allotted specific funds for revamping fire services to Andhra Pradesh , Haryana , Orissa ,West Bengal and Uttar Pradesh.
d. For capacity building and up gradation of SDRF of Rs. 525 corers allotted.
e. For national disaster response reserves rs. 250 cr allotted.

CONCULSION

- Use of IT infrastructure network with remote sensing technology, GIS, duly empowered with a well efficient DM structure will help to realize the fully save vision .
- A separate financial arrangement under a capable national leadership should be allocated annually not only if disaster take place in an area but for pre disaster and post disaster measures and programs.
- Various rules, regulation, policies, formation of related institution are only steps to minimize the disaster. As disaster cannot be fully eliminated its effect can be minimize to some extent & hence its duty of each and every citizen of India to actively act as stake holder in disaster related activates. A destination has been described and hopefully direction is shown. The stage is has been set and road maps now need to be rolled out.

Acknowledgment

We gratefully, acknowledge the overall assistance & guidance received to carry out this case study, from National Policy on Disaster Management 2009, published by NDMA.

References

1. National Policy on Disaster Management 2009 & (NDMP)2018 , published by NDMA. .

2. Disaster Management Cycle-A theoretical approach by Himayatullah khan, Laura birea & Asmatullah Khan.

3. Disaster Management Concept and Framework- Dr A D Kashik (NIDM) Govt of India , New Delhi.

4. Disaster Management intiative policy, perpective and effective response mechanism in India, march 2003 at 28 ALNAP meeting –By Dr Muzzafar Ahmed.

5. Central Board of Secondary Education (CBSE) , Natural Hazards and Disaster Management Delhi 2006. S

"Comparative Study of India's 2018-Flood Risk Evaluation Using Digital Image Processing & Geo Spatial Analysis"

Y. S. Prasanna[1] and B. Pratyusha[2]
[1]Department of Mechanical Engineering and [2]Department of Civil Engineering, Hyderabad.
ysp1106@gmail.com and Prathyushacivil09@gmail.com

ABSTRACT

2018, the year in which Indian monsoon floods hit the southern Indian states of Kerala beginning july 2018 left nearly 500 dead and around 1.5 million displaced from their native to other places continuing the count with the cyclonic Titli floods in October 2018. It is noticeable that since many years monsoon weather analysis in India tells us clearly that there is a huge effect of cyclonic floods in between july & December of every year, but still exact evaluation & qualitative measures are not under practice. Hence there exists a large need to evaluate clearly and comparatively the floods structure & analysis in the country. Also risk assessment & preparative precautionary techniques & methods have to be framed so as to decrease the damage rate caused in the country because of these natural disasters.

In this paper we try to present a clear analysis of floods risk evaluation in the society with the support of emerging scientific and technological assessment techniques like DIP, Geomorphological Studies, and hydrological methods using satellite data. And government responsibility in encouraging military & Para military forces indigenous efforts have indeed yielded results in increasing the safe life rate at the end of the floods occurrence. Also there exists a need to encore public responsibilities every year in the country.

Keywords: Kerala floods, titli, DIP, geomorphology, Hydrology, risk assessment, evaluation

I. INTRODUCTION

India's economic growth has always been under the caprices of the weather, especially on extreme weather events. Besides heavy agricultural losses, such extreme events also result in huge losses of life, property, and disruption to economic activities, especially when highly urbanized and populated regions of the country are in the eye of the extreme weather events. Mumbai floods of 2005 (Gupta 2007) and 2015 (Sarkar and Singh 2017), and Kedarnath floods of 2013 (Rao 2014; Sati and Gahalaut 2013) cyclonic titli floods and kerala floods in 2018 are recent examples of flood-related life and property losses in India.

A. Floods

Rain-water falls on big mountains as well as upon plains. As the river bed is lower in level, all the rain-water flows to low lands and rivers. When the rainfall is heavy, rivers, streams and other channels cannot contain all the rain-water in their beds. Then the water overflows to their banks. This is called a flood. At times flood comes all on a sudden. People sleep at night in their houses peacefully. They wake in the morning to see their houses surrounded with water. This cause is an untold suffering and creates immense miseries to the people. Their belongings are washed away. Houses collapse and people are rendered homeless. Many men, women, children and cattle are swept away by the force of the river. Floods destroy crops standing in the fields and cause famine. They also cause epidemics. Many people lose their resources and tools and thus lose their occupation. It is a very pitiable sight to see cattle and men being washed away by the current of the river during the floods. Often entire families are washed away and they drift on the thatches and frail barges, aimlessly. The flood also damages railway lines and makes the running of trains impossible.

At the time of flood the Government of India adopts various measures to alleviate the sufferings of the people. It arranges for both long-term and short-term relief. The short-term relief means immediate relief. This includes distribution of food, cloth, medicines etc. The long-term relief means free distribution of seedlings, remission of rent, grant of loan for reconstructing houses and for reclaiming land, etc. Even private relief societies are organizes by the people. They also render much help to the flood-stricken people.

B. Flood Hydrology

Every year about six to eight major floods occur in one or the other part of the country (Dhar, 1986). Although the rivers have two to three months of high flows and floods can occur during any part of the monsoon season, the annual flood peaks are generally recorded in the months of August and September. The annual peak floods on large rivers range between 6,000 and 80,000 m^3/s. These are approximately 1.3 to 4 times greater than the mean annual flood discharges (Table 1) and rank among the global highest for their respective drainage areas.

ISBN: 978-93-8830-599-0

Whereas single-peak flood events during the monsoon season are a common feature in case of most of the rivers, sometimes successive flood-generating meteorological conditions may give rise to multiple peak events. The 1970 floods on the Narmada River, for example, were associated with such a situation. Multiple floods are also relatively frequent in the Ganga plains. Single-peak events are commonly observed on smaller rivers or on rivers flowing mostly through arid and semi-arid areas.

Short-lived, but highly damaging, flash-floods are generally produced by intense convectional storms during the post- and pre-monsoon seasons, and sometimes during the monsoon season. Drainage basin characteristics (steep slopes, high drainage density) also play an important role in the generation of flash-floods. There is considerable interregional variability in both the annual monsoon flows and the peak flood discharges (Table 1). This is primarily due to the marked variation in the regional pattern of monsoon rainfall, with some basins receiving higher rainfall in the catchment areas than the others. As a result of this there is great diversity in hydrological characteristics within the monsoonal region, with some rivers revealing ephemeral characteristics.

Table 1 Annual peak discharge characteristics of some large rivers

River	Site	Mean monsoon flood m^3s^{-1}	Maximum monsoon flood m^3s^{-1}	Standard deviation
Ganga	Farakka	55 776	72 915	7 979
Godavari	Dowleswaram	29 207	79 990	12 378
Krishna	Vijayawada	14 775	30 040	4 516
Narmada	Garudeshwar	25 399	69 000	12 781
Tapi	Kathore	9 698	41 700	5 883
Satluj	Bhakra	3 941	9 203	1 664
Mahanadi	Naraj	29 269	42 334	8 731
Damodar	Rhondia	6 328	14 761	3 103
Kosi	Sunakhambi-Kola	8 178	17 445	2 834
Brahmaputra	Pandu	48 160	72 748	8 984

C. Causes of floods

There is substantial inter-annual variability in the monsoon rainfall over the Indian region, with high variability in monsoon rainfall there are inevitable droughts or floods. Therefore, it is not surprising that India is often referred to as the land of droughts and floods. it has now been well established that this inter annual variability in the monsoon rainfall is connected to large scale phenomena. even during a normal monsoon year some parts of the country receive more rainfall than the others. therefore, forgive near the current and distribution of large plates are determined by the amount and distribution of heavy to excessive rainfall.

Large floods on most Indian rivers are a direct result of intense cyclonic Storms and depressions. in general exceptionally heavy rains are associated with slow moving and stationary low pressure systems. The flood generating rainstorms are mainly confined to two measure zone. One zone is located over the Ganga basin in the Punjab plains second zone includes the central India as well as Northern half of the Peninsular India.

D. Flood Forecasting Techniques

According to the various concepts used in developing models the models can be classified into five categories

i.   Based on correlation/coaxial diagrams between two variables or even more,

ii.   Mathematical equations developed using regression/ multiple linear regressions mix which combine independent variable more than one variable

iii. Hydrological models such as rainfall runoff model and routing technique's. The rainfall runoff model is constituted with Quasi distributed and distributed models whereas routing techniques and distributed ones which are used for the floods acquiring in the country.

iv.   A hydraulic model such as the dynamic wave routing is one of the precise forecasting techniques used in order to find flood data.

v.   Data-driven hydrological models are classified as artificial neural networks farmer using expert system design on FF, adaptive neuro-fuzzy interference system models Police Stop

E. Flood management problems

Floods constituted as natural hazards in India therefore, several programs are launched in order to control floods in our country. The programs primarily included engineering structural measures and in the last few decades as a result of flood control programs for about 14.4 million hectares of land has benefited.

It is also very important to note that flood damage is not just excess rainfall but also a complex process which require a deeper understanding. And inspite of many efforts to control and manage floods, flooding continues to be the single most destructive natural

ISBN: 978-93-8830-599-0

Hazard in the country. In this context the primary problems are;

(i)   lack of long continuous and reliable hydrological reports to analyse and evaluate the flood characteristics of the monsoon rivers.

(ii)   Difficulty in understanding the complex flood process due to enormous variability of river types.

(iii)   lack of quantitative data on the flood phenomena and sediment transport during extreme floods with annual exceedence probabilities of less than 10^{-2} .

(iv)   lack of multivariate data on geomorphic and environmental performance of the flood protection works, to evaluate the consequences of flood control structures.

II. 2018-KERALA FLOODS

In 2018, Kerala is left with more than three hundred dead thousands missing millions left homeless as Kerala floods have become one of the worst disasters in a century. The state experienced unprecedented rainfall in the couple of weeks of August, more than twice that of what is the average around this time of the year. It led flooding of rivers, overflowing of dams all the disaster one could anticipate. Questions begin to rise as images of nature's bury and safety operations begin to flow as the scale of destruction being clear.

Was this a natural calamity or manmade disaster happened? Could we have prevented such a massive tragedy? The shear Amount of rainfall in first 3 days of August broke previous records. The first affected region was the Idukki district owing to several dams including the famous Idukki dam. It received more than 1400 millimetre of rainfall in August 2018.

In the first week of August the rainfall was 770.9 mm till 20th August against the normal 300.4 mm getting 157 percent of higher rainfall. if we look at how much it rained in Kerala in August 2018 overall the state it recorder 2018 7.67 mm of rainfall where as in November 2015 in Chennai it is 12 18.6 mm within 4 days cost Chennai floods. In June 2005 Mumbai was affected with 944 mm rain in just 24 hours.

when we compare amount of rainfall over a wide region the result could be only one. Rivers began to swept, dams were overflown to their strength and authority did something they had not in a hundred years; they open the Gates to release massive volume of water to ease pressure on the dams. not one or two but 35 dams out of 39 in the state begin to release in the same time. All the 14 districts were traced under red alert. The maximum capacity of Idukki Dam to put water was kept at 2403

feet when authorities decided release water. At one point 800 thousand litres of water per second poured out of the Dam.

Then questions were raised why water from the dams not released as a precaution to avoid what was also perceived as a manmade disaster. State government claims they never received The Warning from the Meteorological department.

Some of the environmental list blame it's the government failure on the environment policy drafting. an ecologist renowned scientist from IISC professor Madhav Gadgil said flood caused due to unjustified human interventions in natural process. Major storage dams released water suddenly affecting landslides due to illegal encroachment of forest land.

This could have been avoided releasing water in advance rather than waiting for dams to fill up.

Secondly the extensive quarrying, mining, mushrooming of high Rises on the Hillside part of tourism are adding to the flooding misery in Kerala.

In 2011, Gadgil went getting a panel on Western Ghats ecology and recommended stringent steps to ensure such disasters are prevented also The panel have recommended that the entire Western Ghats stretching over 1500 kilometres passing through 6 States including Kerala should draft several measures including strict regulation on construction of lands, housing on hillsides etc.

A. Rainfall forecasting

Indian meteorological department released a data sheet of rainfall forecast like this on august 19th, 2018. Rainfall over Kerala during southwest monsoon season 2018 (1 June to 19 August, 2018) has been exceptionally high. Kerala so far received 2346.6 mm against normal of 1649.5 mm (above normal by 42%). The spatial distribution of district-wise seasonal rainfall is shown in table 2. It indicates that highest excess rainfall is recorded over Idukki District (92% above normal) followed by Palakkad (72% above normal).

Table 2 The over rainfall in Kerala during June , July and August-2018

Month	Actual Rainfall (mm)	Normal Rainfall (mm)	% Departure from normal
June, 2018	749.6	649.8	15
July, 2018	857.4	726.1	18
1-19th august,2018	758.6	287.6	164
1 June to 19 August, 2018	**2346.6**	**1649.5**	**42%**

ISBN: 978-93-8830-599-0

Fig. 1 Kerala aerial view on 20.8.2018

If we observe the above figure, one can easily feel the extent of how the rainfall prone disaster effected the state kerala. the dams overflow situation had led sudden water thrust on banks of rivers, which made a way through for water to flow over the villages located around the dam.

B. Kerala Flood Forecasting

After experiencing continuous rainfall in some of the areas near the dams in Kerala, IMD(Indian Meteorological Department) released flood forecast report in due time.

Specially, there were two consecutive active spells with above normal rainfall peaking around 14 & 20 June respectively. Another peak rainfall activity was experienced around 20 July. The week-wise rainfall departure over Kerala is shown below. Heavy rainfall observed (7 cm or more) during 8 August to 21st August. Due to the above rainfall scenario prevailed till end of July, 2018 over Kerala, all of major 35 odd reservoirs storage was close to the full reservoir level (FRL) and had no buffer storage to accommodate the heavy inflows from 10th August. The continued exceptional heavy rainfall in August (with 170% above normal so far) in the catchment areas had compelled the authorities to resort to heavy releases downstream into the rivers. Such a scenario that continued for almost a week now has caused overflowing of all river banks leading to widespread flooding almost all over the state. A fresh low pressure area has formed over northwest Bay of Bengal on 19th August 2018 morning, however it will not have any significant impact over Kerala.

Since there was continuous rainfall with 758mm approx.; by the time forecasting analysis report reached the authority the dams were completely filled with water. Therefore for the aid of dam safety precaution, the authorities of the concern dam need to open the gates so as to let go the stagnated water to other side of the dam. This led to the dangerous outcome driving towards flooding of delta regions of Kerala completely.

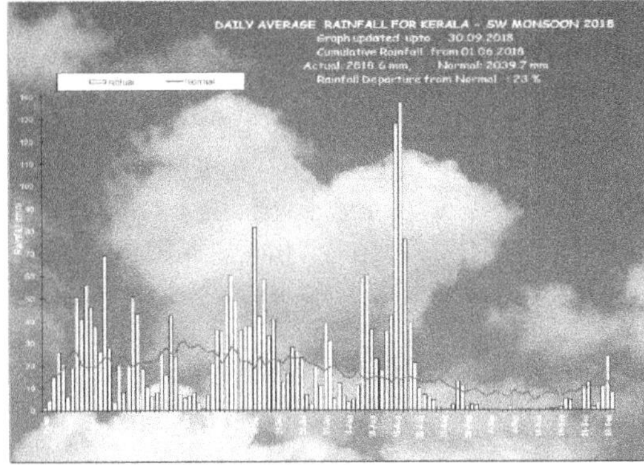

Fig. 2 Average Rainfall in Kerala

The figure 2 shows the clear image of average rainfall for Kerala from June 2018 to August 2018. The average rainfall reaches high or to the critical point in the month of August 2018 from 11th to 20th. This period lead to the devastating natural disaster like floods.

C. Flood Data and Method

From one of the experimentation we obtained the following information .The daily observed average rainfall data is obtained from India Meteorological Department (IMD) at 0.25° (till 21st August. IMD rainfall dataset is developed using more than 6000 observing gauge stations across India, and a20 substantial number of stations are located in Kerala. Rainfall observations of each station were interpolated using Inverse Distance Weighting interpolation. Using the gridded data of rainfall the mass curve (the relationship between cumulative rainfall and time) and depth-duration- frequency (DDF) curves are used to evaluate the severity and intensity of extreme rainfall during August 2018 in Kerala.

DDF curves are similar to the intensity-duration-frequency (IDF) curves but with rainfall depth instead of intensity. The mass curve and DDF curves are widely used to extract the information of a storm event. Generalized Extreme Value (GEV) distribution is used to analyze extreme hydro-climatic events. The next step is fitting of the GEV distribution to annual maximum rainfall for the selected durations. The GEV distribution has three parameters (location, scale, and shape) that were estimated using the Maximum Likelihood method. Using the GEV distribution, DDF curves are developed for 1-15 days extreme rainfall duration for 20, 50, 100, 200, 5 and 500 years return period. the goodness of fit of GEV distribution using QQ plots and Chi-Square test are tested, and found that the GEV distribution passed

ISBN: 978-93-8830-599-0

the goodness-of-fit test for the most of the rainfall durations .

To analyze the reservoir condition before the flood event, reservoirs storage information for the seven major reservoirs (Idukki, Idamalyar, Kakki, Kallada, Malampuzha, Parambikulam, and Periyar) is obtained from India Water Resources Information System (WRIS) (IWRIS, www.india-wris.nrsc.gov.in) for the period 2007-2018. Using the IWRIS data, the abnormality in the reservoir storage before the flood in Kerala is analysed. There is a number of small and medium size reservoirs in Kerala; however, as their storage information is not available, we mainly focus on the major reservoirs.

D. *Geo spatial Analysis*

Kerala received actual rainfall of 2394.1 mm from June 1 to Aug 22, 2018, which exceeds normal rainfall of 1701.4 mm by 41 % (IMD). Rainfall data of IMD-GPM at 0.25 degree area shown. Idukki district received highest rainfall of 3555.5 mm during same time, with two major spells of rainfall during Aug 07-10 and Aug 14-18, 2018. Heavy rainfall resulted in high surface runoff in major river basins of Kerala, filling of all dams and subsequent opening of these dams; caused widespread flooding in downstream areas, low lying areas near coast and backwaters of Kerala.

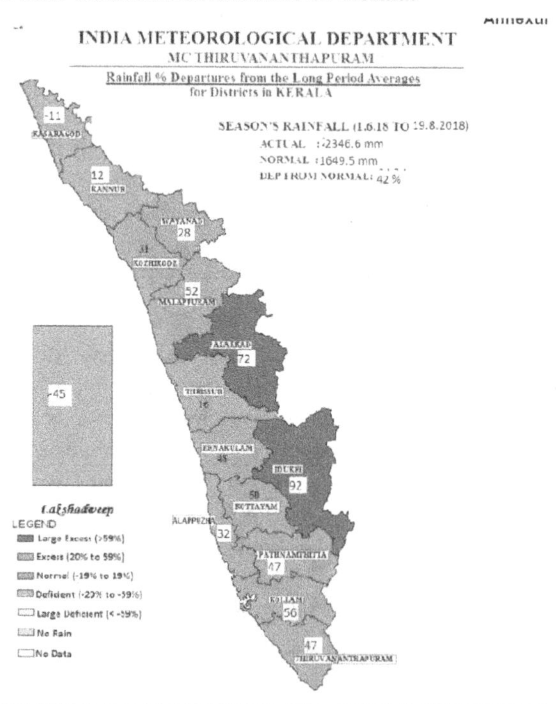

Fig. 3 Average rainfall in kerala disticts

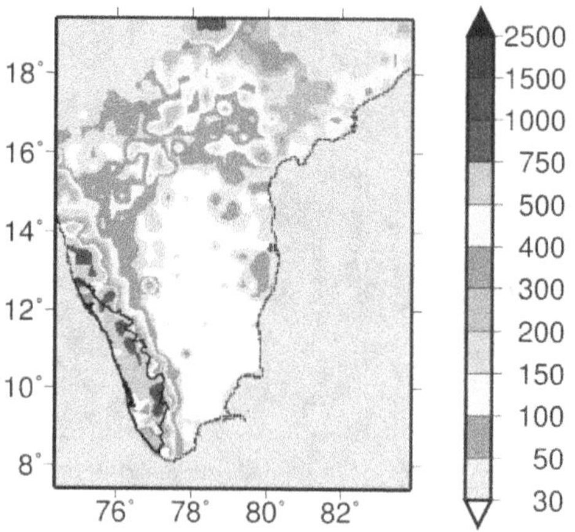

Fig.4 Cummulative rainfall in kerala(red-highest)

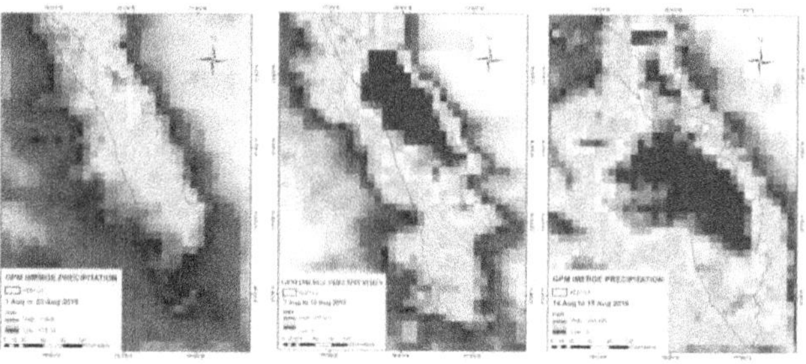

Fig. 5 Geospatial distribution of rainfall in Kerala

E. *Hydrological Modelling*

There are 12 river basins which are majorly flood effected and are studied for detailed basin wise DEM based hydro data processing and virgin hydrological simulations using Hydrological Modelling System (HMS) . the IMD-GPM gridded data used as input met data, SCS method for loss, and SCS unit hydrograph as runoff transformation, Muskingum-Cunge as routing & constant monthly baseflow, in all basins simulations. The detailed hydrological modelling was done for Periyaar basin with outlet at Alluva Manappuram Rail Bridge, Periyaar Nagar using HEC-HMS model. Basin has 47 sub-watersheds, with total area of 4276.9 km2 and 527.16 km of drainages. Rainfall data from IMD product at 0.25 degree was used for simulating the river flows. Rainfall at various sub watersheds and simulated river flood flow hydrographs are shown in figures. Two flood peaks are seen on 10 and 16 August 2018.

ISBN: 978-93-8830-599-0

Table 3 Major flood affected river basins

S. No	Basin name	Area (km^2)	Total Drainage* (km)	No. of sub-watersheds*	Mean Tc (hours)**	No of major dams/reservoirs
1	Periyar	4276.91	527.16	47	6.92	12
2	Pamba	2486.7	320.37	17	12.14	4
3	North Region	1008.4	114.73	11	11.21	2
4	Muvattupuzha	1560.7	196.15	15	13.35	2
5	Mannar	1234	155.83	7	20.48	0
6	Kambini	6853	834.87	89	9.75	5
7	Kadalundi	886.79	109.5	8	14.54	0
8	Chaliyar	2915.73	329.93	37	7.16	1
9	Chalakudy	1309.84	187.04	15	5.94	4
10	Bharathapuzha	5884.24	782.92	73	9.88	6
11	Kodoor	1045.28	107.15	14	10.69	0
12	Puzhakal	1523.473	176.7	19	12.4	3

The watersheds are delineated using SRTM-30 m DEM, and minimum watershed area threshold for defining a stream is given as 50 km^2. The derived drainages and number of sub watersheds are based on this threshold. The total drainage length and number of sub-watershed will increase if we decrease this threshold. The Time of concentration is calculated for each sub watershed of each river basin with 50 km^2 area threshold and TR-55 method. The actual time of travel for entire basin will be higher than the mean Tc.

F. Flood Management Problem

Around 8.70 lakh hectares was estimated as flood prone area out of 38.90 lakh hectares of geographical area in Kerala. Its estimation of flood prone areas was taken up by regional committees for scientific assessment.

Kerala High Court appointed amicus curiae for the investigation on kerala floods, whose report shows that Dam water management seemed to have aggravated the damages caused by Kerala floods in 2018. due to this around 450 people were killed and properties worth crores of Rupees in the massive floods in August 2018. The flood has affected 5.5 million people overall in which 4.5 million displaced to other. After investigation amicus curiae stated that the simultaneous opening of many dams is the cause for the disasters. The impact of disaster could have been reduced if the Dam gates opened and water were released in a regulated manner. also the report presented by Amicus curiae states that none of the 79 States in the state Kerala were maintained or used for flood Control Programme. According to the report dams in the state had not maintained effective flood control zone and Flat cushion.

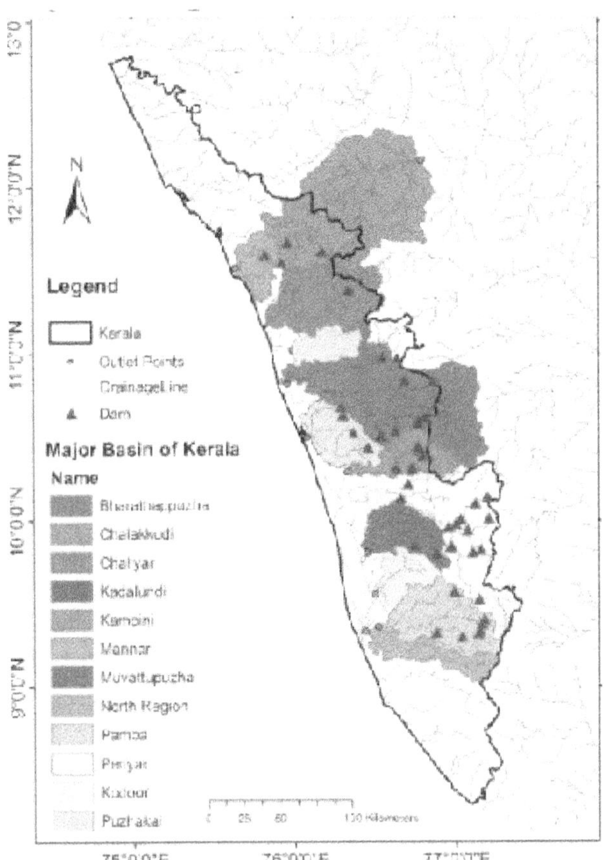

Fig. 6 Hydrological data of major river basins

Fig. 7 Idukki Dam

II. TITLI CYCLONIC FLOOD

Immediately after experiencing Kerala floods, in October 2018 India was affected with very severe cyclonic storm called TITLI over Bay of Bengal. a Low pressure area is formed in the Andaman sea on October 6th and within the next two days the disaster entered the Bay of Bengal and became a depression. on October 8 a very severe cyclonic storm between 4:30 a.m. and 5:30 a.m. rised namely "Titli" near Palasa Andhra Pradesh a peak intensity.

ISBN: 978-93-8830-599-0

Titli killed at least 77 people in Odisha and left a couple of others missing and displaced. due to flooding and landslides another 8 went to death in Andhra Pradesh. its Weakened into depression before entering West Bengal and also in some parts of South Bengal and bringing Torrential rain and Gale force wind. Floods caused by Titli totally damaged around rupees 3673 crores in Andhra Pradesh and 3000 crores in Orissa.

A. Geo-Spatial Analysis

Using satellite data to visualize the extent, magnitude, and height of flooding the occurrence of floods is mainly due to of heavy rainfall (~2,377 mm) which meant that the area was under a dense cloud cover. Satellites <u>Sentinel</u> 1A and 1B contains a synthetic-aperture radar (<u>SAR</u>) instrument that can collect data in any weather, day and night. Sentinel data is processed by the Sat Sure to generate Digital Elevation Models (DEMs) of the area, using which creation of interactive web map is done which is further equipped to intelligently display such a heavy dataset.

As per Government of Kerala's order performance of spatial analysis is made to divide Kerala into zones of high, medium, and low flood risk. Combining our spatial data with weather and administrative data, we delineated Kerala for a greater understanding of the floods.

The application of geo spatial analysis is performed on the wake of Cyclone Titli on the eastern coast this October. The tropical cyclone from the Bay of Bengal is categorised as a Category 2 hurricane on the Saffir-Simpson scale (SSHWS) which is called as a Very Severe Cyclonic Storm. Titli has primarily affected the states of Odisha and Andhra Pradesh, causing landslides and heavy flooding. With this effect mapping of Ganjam district of Orissa and Srikakulam district of Andhra Pradesh is made. Using Sentinel SAR imagery of October, mapping of the flood extent for 10th and 17th October is made and visualised it on the web map. Along with this, the potential crop damage to rice against the actual acreage is analysed through mapping from GIS. The difference from mapping Kerala was that the cyclone is made from Satsure satellite system.

B. Rainfall Forecasting

The rainfall forecasting was made for the states Orissa and Andhra Pradesh due to the high intensity wind formations. The table 3. Shows the water level in the states of Orissa and Andhra Pradesh. Average rainfall of the state recorded on 13th October 2018 for the last 24 hours is 17.5mm.   4 districts have received an average rainfall of 50mm and 100mm. The districts are balasore-86.9mm, mayurbhanj-78.8mm, bhadrak-68.8mm, jajpur-61.9mm.due to heavy rainfall the flood situation is raised in the rivers Rushikulya and vanshadara.

Fig. 8 GIS image Cyclone TITLI

Fig. 9 Floods due to TITLI in Orissa

REFERENCES

1. Kumar, S.: Rapidly Assessing Flood Damage in Uttarakhand, India. [online] Available from: http://documents.worldbank.org/curated/en/724891468188654600/Rapidly-assessing-flood-damage-in-Uttarakhand-India, 2013.

2. The Kerala flood of 2018: combined impact of extreme rainfall and reservoir storage. Vimal Mishra1*, Saran Aadhar1, Harsh Shah1, Rahul Kumar1, Dushmanta Ranjan Pattanaik2, Amar Deep Tiwari, Hydrol. Earth

Syst. Sci. Discuss., https://doi.org/10.5194/hess-2018-480

3. Kumar, A., Dudhia, J., Rotunno, R., Niyogi, D. and Mohanty, U. C.: Analysis of the 26 July 2005 heavy rain event over Mumbai, India using the Weather Research and Forecasting (WRF) model, Q. J. R. Meteorol. Soc., 134(636)(October), 1897–910 [online] Available from: https://agupubs.onlinelibrary.wiley.com/doi/abs/10.1002/qj.325, 2008.

4. Indian Meteorological Department, dated August and October 2018, (online) http://www.imd.gov.in/press_release/20181016_pr_347.pdf.

5. IMD Regional specialized Meteorological Centre for cyclones http://www.rsmcnewdelhi.imd.gov.in/index.php?lang=en

6. The Kerala flood of 2018: combined impact of extreme rainfall and reservoir storage Vimal Mishra1*, Saran Aadhar1, Harsh Shah1, Rahul Kumar1, Dushmanta Ranjan Pattanaik2 , Amar Deep Tiwari.

7. https://www.freepressjournal.in/editorspick/kerala-floods-kerala-must-look-within-t-r-ramachandran/1340226

8. Situation report on very severe cyclone TITLI, SEOC, Orissa.

9. The spatio temporal aspects of monsoon floods in India; Disaster Management , Harsh K Gupta.

10. Disaster Mitigation, sahni-dhameja-medur.

ISBN: 978-93-8830-599-0

Disaster Management in India

Mohammed Jalal uddin Mohammed Ghouse and Quayyum Md
Assistant professor, Department of Civil Engineering, Mahbubnagar, India

Abstract
India has been traditionally vulnerable to natural disaster on account of its unique geo-climate conditions. Floods, droughts, cyclones, earthquakes, and landslides have been recurrent phenomena. About 60% of the landmass is prone to earthquake of various intensities; over 40 million hectares is prone to floods; about 8 % of total area is prone to cyclones and 68% of the areas is susceptible to drought. In the decade 1990-2000, an average of about 4344 people lost their lives about about 30 million people were affected by disaster every year. The loss in terms of private, community and public assets has been astronomical. At the global level, there has been considerable concern over natural disaster. Even as substantial scientific and material progress is made, the loss of life and property due to disaster has not decreased. In fact human toll and economic losses have mounted. It was in this background that the UN general assembly in 1989 declared 1990-2000 as the International decade of natural disaster reduction with the objective to reduce loss of lives and property and restrict socioeconomic damage through concerted international action. The Government of India have adopted mitigation and prevention as essential components of their development strategies. The Tenth Five Year Plan documents has a detailed chapter on Disaster Management.

The plan emphasizes the fact that development can not be sustainable without mitigation being built into development process. Each State is supposed to prepare a plan scheme for disaster mitigation in accordance with the approach outlined in the plan. In brief, mitigation is being institutionalized into development planning. The Finance Commission makes recommendation with regard to devolution of funds between Central Government and State Government as also outlays for relief and rehabilitation. The Government of India have issued guidelines that where there is a self of projects, projects addressing mitigation with be given priority. It has also been mandated that each projects in a hazard prone area will have disaster prevention /mitigation as a term of reference and the project documents has to reflect as to how project addresses that term of reference. In the sections are discussed the measures shortcoming, measures taken for the mitigation of the disaster.

Keywords: Disaster, Management, Mitigation, Rehabilitation, Prevention, Relief

1. Introduction

It is really an unfortunate and undesirable situation that in our country where more than 6 crore people are affected by disasters every year. Statistics is shown in figure,

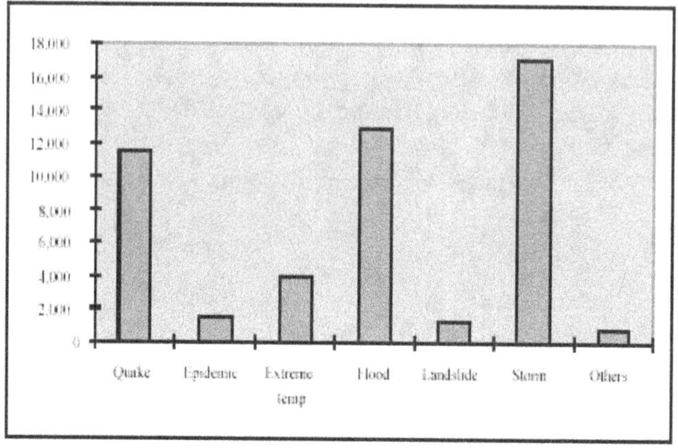

Fig. 1 Mortality due to natural hazards (1990-2000)

we have no policy on systematic disaster Management. It is only after a disaster strikes that the wheels of the government, both at the centre and at the states, move and that too slowly. Despite the need to build up capabilities to meet the challenges of disasters, the thrust has unfortunately been on alleviation and relief. Even the relief has not been quick and adequate, as few disasters such as Orissa super cyclone, Tsunami of 2004, Gujarat earthquake etc experiences has shown. India's response to and tackling of this two major disasters has thrown up the following weakness in our disaster management efforts.

(a) Inadequate Early Warning System

Though, the forecasting, monitoring and warning mechanisms are beautifully articulated on paper in practice, the warnings are not early enough and they do not reach all those likely to be affected. In case of Tsunami, 2004; Bhuj earthquake etc for example, communication facilities which could have resulted in better co-ordination of warning and reduction of damage to life and property were inadequate.

ISBN: 978-93-8830-599-0

Tsunami of 2004

Bhuj(2001) earthquake

DMT members rescue affected people on hand made banana rafts

Orissa super cyclone

Bihar Flood 2009

(b) Lack of Pre-disaster Preparedness

With disasters striking India with increased regularity, there should be a plan in place to tackle the disaster and reduce its impact. On the contrary, people are caught unaware time and again. There is not planned information system as to what needs to be done when faced with a calamity. For example, during Tsunami, 2004, dead body laid floating in the water for many days due to the unavailability or lack of required equipment to meet the need of the time/emergency.

(c) Inadequate and Slow Relief

Relief is an important aspect of the disaster management to provide help to the affected people. The relief operations are often handled in ad hoc and haphazard manner. How efficiently to provide food, medicine, to reduce the suffering of the affected people etc are addressed and met improperly. Even days after the Bhuj earthquake, and Tsunami, 2004,, many people could not be provided with safe drinking water, temporary shelter, and medicines. Such a scenario gives rise to law and order problem- looting of the relief materials and outbreak of the epidemic due to rotting dead bodies on the other hand.

(d) Lack of Co-ordination

Disaster management requires concerted efforts from Central Government, State Government, NGOs, International agencies and private sectors etc. Because of the lack of the co-ordination, relief material is not property distributed among the people. Even worst happens when they are mis-utilized and are not distributed uniformly.

(e) Slow Rehabilitation and Reconstruction

While immediately after a disaster strikes, there is hectic relief and rescue mission, mainly aimed at feeding the people and stalling the outbreak of an epidemic, relief and rescue can not go on endlessly and rehabilitation and reconstruction should be given proper attention. However, this is an area which is often ignored and progressed is slow once the initial attention fades away. Restoration of infrastructure, hospitals, schools, houses, and sources of living of the people needs to be given proper attention.

(f) Proper Administration

A quick assessment of the extent of the damage is necessary so that relief and rehabilitation work can be properly planned. However, it was seen that even many months after the Bhuj earthquake and Tsunami of 2004, the government was yet to finish the preliminary survey of assessing the total impact of the damage. Apart from this, poor administration frustrated the best intentions and efforts of private initiatives. After the quake, Gujarat government was too slow and indecisive on some of the best rehabilitation plans proposed by the NGOs and Corporate.

(g) Poor Management of Finances for Post-disaster Relief

Mostly relief and rehabilitation work suffers from the lack of co-ordination, proper management, and supervision at all levels and indicated the absence of adequate planning and preparedness to meet any

ISBN: 978-93-8830-599-0

emergency. Consequently, the funds are mis-utilized and relief measures were tardy and inadequate, providing scope for pilferage of relief and rehabilitation remained unutilized and there is huge shortfall in distribution of emergency relief, shelter material cloths, house building assistance etc. There have also been reports of relief and rehabilitation funds being utilized for paying salary arrears of the state government employees.

(h) Symbolism Rather than Relief

It has been a recurrent experience that rather than making a serious effort at planning and management for tackling frequent disasters, our government adopts symbolic gestures like helicopter survey of disaster affected areas. The politics of relief works in a manner that tall claims are made by the Government other than the affected state to help the affected districts and by sending huge financial help but these claims prove hollow once the calamity recedes.

(i) No Instruction for Pre-seismic Period

There is no instruction for the pre-seismic period. Unfortunately, in the present administrative set up, no official will visit the pople during pre-seismic period to tell them about an eminent earthquake. But, during the post-seismic period, a large number of officials will visit the affected people with food, tents, medicine, cloths and compensation funding to the relatives of the dead. This scenario has been repeated after Latur (1993), Jablpur (1997), Bhuj (2001), Andman (2004), and Kashmir(205) earthquakes. This pathetic situations has to be changed at the earliest. The issue need to be seriously pondered at the national level. The sole reason for this is the lack of knowledge about earthquake precursors and earthquake prediction. Most of the earthquake disaster management experts, agencies, and offices have a strong conviction that an earthquake can not be predicted. They are correct to some extent. Till now, there was onle one case of successful prediction in China. Earthquake prediction has almost become a taboo in most of the disaster management offices. The relevant rules also are empowered to take penal action against anyone who talks about earthquake prediction. As a result, an impression is inadvertently created in the society that moist of the disaster management agencies come in the picture during post-seismic period to clear the debris and the corpses. The present situation is skewed. On one hand, it is accepted that a large-magnitude earthquake is due and it may occur anytime; On the other hand, most of the disaster management agencies feet that an earthquake can not be predicted. If we want to protect people form an earthquake, it is essential that a suitable precursory warning is issued even in case of moderate scale of earthquake as it makes the people of the region aware the region is prone to disaster, and they should be careful.

Measures/Facts Taken to Improve Disaster Management in India

Central Level

At the central or national level, Ministry of Home affairs is entrusted with the nodal responsibility of managing disaster. At the apex level, there are tow cabinet committees viz. cabinet committee on national calamity and cabinet committee on security. All the major issues concerning natural disasters are placed before cabinet committee on natural calamity whereas calamities which can affect internal security or which may be caused due to use of nuclear, biological or chemical weapons etch are placed before cabinet committee on security. The NCMC (National Crisis Management Committee) is the next important functionary. The cabinet secretary heads it. It includes secretaries of concerned department/ministers. Its main function is to give direction to Crisis Management Group (CMG) and any minister/department for specific action needed for meeting the crisis situation. CMG lies below the NCMC. The Central Relief Commissioner is its chairman. His primary function is to coordinate all the relief operations for natural disaster. Apart from coordinating the relief operations, it reviews the contingency plans formulated by Central Ministers/Department and measures required for dealing with natural disaster. CMG meets every six months however in event of any disaster it frequently meets to review the relief operation and explore all possibilities to render all possible help to the affected region.

State and District Level

At the state level, there are state relief commissioners who are incharge of the relief measures in wake of natural disaster in the perspective states. The chief secretary is the overall in charge of the relief operations in the state. The relief commissioner and additional relief commissioner work under his direction and control. In addition, there are number of secretaries, head of various departments who also work under the overall direction of chief secretary. At the district level, districts are headed by District Collector or district magistrate who is responsible for the overall supervision and monitoring of relief measures and preparation of disaster management plans. At the tehsil level DSO/SDM take care of the disaster management.

Despite there being a general tardiness about the manner in which we respond to disasters, there has been

ISBN: 978-93-8830-599-0

significance progress in this area and there have been many experiments and success stories worth emulating,

I. Learning from the Latur earthquake calamity, Maharastra hss launched India's first disaster management infor-network. Soon after this quake, state government launched the Maharastra Emergency Earthquake Rehabilitation programme. The programme aimed at achieving preparedness through an information-network so that unpredictable and uncontrolled disaster impacts could be offset with planned and manageable disaster mitigation efforts. This info-network links the state government machinery wilh all its tehsils and districts along with other strategically and economically important agencies based in the state. The state has been mapped for potential disasters. Statistics for potential natural calamity zones are now being complied. Record for the tide movements, potential typhoons and earthquake prone zones are being linked up with geographical information system to mitigate the disaster. The Multi-hazards Disaster Mitigation Plan will create a disaster management information at emergency operation centre at state government headquarters. Apart from forewarning of calamities like flood, earthquake, etc; post disaster relief and rehabilitation is another area of use of this network. It will help in co-ordinating among hospitals, voluntary organizations, ambulances, fire brigades and government relief measures.

II. Some State Government have got their acts together, learning from past experiences. In 1991, A.P. Government was able to implement previously planned programme to evacuate 6 lakh people from the path of an approaching cyclone with 52 hours. Fatalities numbered less than on tenth of what could have otherwise been. This was achieved through a planned approach combing both traditional and advanced channels.

III. The IMD has set up a National Seismic Telemetry Network to anticipate threats from seismic disturbances. After the Gujarat quake, 10 new seismological observation equipped with latest facilities were set up and 14 of the 45 existing observatories were upgraded with state of the art digital seismograph for better monitoring of effects of earthquake in the seismic zones.

IV. The IMD has set up cyclone warning centres along many coastlines. Information on cyclone warning is furnished to the central control room in the Ministry of Agriculture. Besides, high powered cyclone detection radars are installed at various places on the coastal belt, that can track disturbances within a range of 400 KM. Satellite imagery is another tool used when cyclone are beyond the range of the coastal radars. The ISRO has placed 250 storm warning receivers all along the Indian coast. In a time of crisis, these receivers are switched on via satellite and broadcast siren and local language warnings.

V. Measures for flood mitigation were taken from 1950 onwards, As against the total of 40 million hectares prone to floods, area of about 15 million hectares have been protected by construction of embankment. The State Government have been assisted to take up mitigation programmed like construction of raised platforms etc. Flood continues to be a menace however mainly because of the huge quantum of silt being carried by the rivers emanating from the Himalayas. This silt has raised the bed level in many rives to above the level of countryside. Embankment have also given rise to problem of drainage with heavy rainfall leading to water logging in area outside the embankment. To evolve both short-term and long term strategies for flood management / erosion control, Government of India have recently constituted Central Task Force under the chairmanship of Central Water Commission. The task for will examine causes of the problem of recurring floods and erosion in States and region prone to the flood and erosion; and suggest short term and long term measures.

VI. Due to erratic behavior of monsoons, both low and medium rainfall regions are vulnerable to periodical drought. Experience has been that almost every third year is a drought. However, in some of the States, there may be successive drought years enhancing the vulnerability of population in these areas. Local communities have devised indigenous safety mechanism and drought oriented farming methods in many parts of the country. From the experience of managing the past droughts particularly severe drought of 1987, a number of programme have been launched by the Government to mitigate the impact of drought in the long run. These programmes includes Drought Prone Area Programme (DPAP), Desert Development Programme (DDP), Integrated Water Development Projects (IWDP) etc.

VII. In order to respond effectively to floods, Ministry of Home Affairs have initiated National Disaster

Risk Management Programme in all the flood prone States. Assistance is being provided to the States to draw up disaster management plans at the State, District, Block/Taluka and village levels. Awareness generation campaigns to sensitize all the stakeholders on the need for flood preparedness and mitigation measures. Elected representative and officials are being trained in flood disaster management under the programme. Bihar, Orissa, West Bengal, Assam, and Uttar Pradesh are among the 17 multi hazard prone States where this programme is being implemented with UNDP, USID and European Commission.

VIII.   A Comprehensive programme has been taken up for earthquake risk mitigation. Although, the BIS has laid down the standard for construction in the seismic zones, these are not being followed. The building construction in urban and suburban areas is regulated by the Town and Country Planning Act and Building Regulations. In many cases, Building regulations do not incorporate the BIS codes. Even where they do, the lack of knowledge regarding seismically safe construction among the architects and engineers as well as lack of awareness regarding their vulnerability among the population led to most of the construction in urban and suburban areas being without reference to BIS standards. In the rural areas, the bulk of the housing is non-engineering construction. The mode of construction in rural areas has also changed from mud and thatch to brick and concrete construction thereby increasing the vulnerability. The increasing population has led to settlement in vulnerable areas close to the river bed which are prone to liquefaction. The Government have moved to address these issues. A National Core Group for Earthquake Risk Mitigation has been constituted consisting of experts in earthquake engineering and administrators. The core group has been assigned with the responsibility of drawing up a strategy and plan of action for mitigating the impacts of earthquakes' providing advice and guidance to the States on various aspects of earthquake mitigation; developing/organizing the preparation of handbooks/pamphlet/types designs for earthquake resistance construction.; working out systems for assisting the States in the seismically vulnerable zone to adopt/integrate appropriate BIS code in their buildings; evolving systems in the training of municipal engineers as also practicing architecs and engineers in the private sectors in the salient features of BIS codes; Evolving a system of certification of architects/engineers for testing their knowledge of quake resistance construction; evolving systems for training of masons and carry out intensive awareness generation campaigns.

IX.   Hospital preparedness is crucial to any disaster response system. Each hospital should have an emergency preparedness plan to deal with mass casualty incidents and the hospital administration/ doctor trained for the emergency. The curriculum for medical doctors does not include hospitals preparedness for emergencies. Therefore, capacity building through in service training of the current health managers and medical personnel in hospitals preparedness for emergencies or mass casualty's incidents management is essential. At the same time, the future health managers must acquire these skills systematically through the inclusion of health emergency management in undergraduate and post-graduate medical curricula. For the same, tow comities have been constituted for preparation of curriculum for introduction of emergency health management in MBBS curriculum, and preparation in service training of hospital managers and professions. Rajiv Gandhi University of health Sciences Karnataka has been identified as the lead national resources institution for the purpose.

X.   While above mitigation measures will take care of the new constructions, the problem of unsafe existing building stock would still remain. It will not be possible to address the whole existing building; therefore, the most important buildings such as hospitals, schoold, cinema halls, muti-storied apartment are being focused on. The States have been instructed and advices to have such buildings assessed and where necessary retrofitted. The ministry of Civil Aviatioin, Railways, Telecommunication, Power and Heath and Family Welfare have been instructed to take up necessary action for detailed evaluation and retrofitting of lifeline buildings located in seismically vulnerable zones so as to ensure that they comply with the BIS norms. Plan have been drawn by such ministries for detailed vulnerability analysis/evaluation and retrofitting/ strengthening of building and structures. The Finance Ministry have been requested to advice financial institutions to give loans for retrofitting on easy terms. Accordingly, the Finance Ministry had advised RBI to issue suitable instructions to tall the banks and financial institutions to see that BIS codes laws are scrupulously followed while financing/refinancing construction activities in seismically prone zones. An Earthquake Mitigation Project has been drawn up; the project has been given principle clearance by the Planning Commissions. The programme includes detailed evaluation and retrofitting of lifeline buildings such as hospitals, schoolds, water and power supply units, telecommunication buildings, airports, railway stations, bus stands and important administrative buildings in the seismic zone IV, V. The programme also includes training of masons in earthquake resistant constructions. Besides, assistance will be provided under this projects to the state Government to put in place appropriate techno legal regime.

Urban earthquake vulnerability reduction programme has been taken up in 38 cities in seismic zones III, IV, V with population a million and avove. 447 orientation programmes have been organized for senior officers and representative of the local planning and development bodies to sensitize them on earthquake preparedness and mitigation measures. The training programme for engineers and architects are being organized to impart knowledge about seismically safe construction and implementation of BIS norms. For enhanced school safety, education programmes have been organized in schools, colleges and other educational institutions. This programme will be further extended to 166 earthquake prone districts in seismic zones IV, V. Awareness generation programmes, community and neighborhood organizations have been started in these cities. These cities are also assisted to review and amend their building laws to incorporate multi hazard safety provisions. City Disaster Management Plans are being developed under these projects.

Rural housing and community assets for vulnerable sections of the population created at a fairly large scale by the Ministry of Rural Development under the Indira Awas Yojna and Sampooran Grameen Rojgar Jojna. AbOUT 250 thousand small but compact housing are constructed every year, besides community assets such as community centre, recreation centers, anganwadi centres etc. Technology support is provided by about tow hundred rural housing centers spread over the entire country. The Ministry of Home Affairs is working with the Ministry of Rural Development for changing the guidelines so that the housed constructed under IAY or school buildings/ community buildings constructed under SGRY are earthquake/cyclone/flood resistant; as also that the schemes addressing mitigation are given priority under SGRY. Ministry of Rural Development are carrying out an exercise for this purpose. This initiative is expected to go along way in popularization of seismically safe construction at village/block levels.

XI. A project for cyclone mitigation has been drawn up in consultation with the cyclone prone states. This projects envisages construction of cyclone shelters, costal shelters belt plantation in areas which are prone to storm surges, strengthening of warning system, training and education etc. This project has also been given in principle clearance by the Planning Commission and is being taken up with World Bank assistance.

XII. A national core group has been constituted under the chairmanship of secretary, Border Management and comprising of Secretary, DST; Road Transport &Highways, and the heads of GSI and NRSA for drawing up a strategy and plan of action for mitigating the impact of landslide, provide advise and guidance to the State Government on various aspects of landslide mitigation, monitor the activities relating to landslide mitigation including landslide hazard zonation and to evolve early warning system and protocol for landslide/landslide risk reduction. The Government has designated GSI as the nodal agency responsible for coordinating/undertaking geological studies, landslide hazard zonation, monitoring landslide/avalanches, studying the factor responsible and suggesting precautionary and preventive measures. The States/UTs have been requested to share the list of habitation close to landslide prone areas for the purpose of landslide hazard zonation being carried out by them. A national strategy for mitigating landslide hazard in the country is being drawn up in consultation with all the agencies concerned.

XIII. A disaster risk management programme has been taken up in 169 districts iin 17 multi -hazard prone states with the assistance of UNDP, USAID and EU. Under this project, the States are being assisted to draw up State, District, Block level disaster management plans; village disaster management plans are being developed in conjunction with the Panchayati Raj institutions and disaster management tem consisting of village volunteers are being trained in various preparedness and response functions such as search and rescue, first aid, relief coordination, shelter management etc. Equipment needs for district and state emergency operation centers have been identified by the state nodal agencies and equipment is being provided to equip them. Orientation training of masons, engineers and architects in disaster resistant technologies have been initiated in these districts and construction of model demonstration buildings have been started.

Under this programme disaster management plans have been prepared for 8643 village, 1046 Gram Panchayat, many blocks and districts. More than 29000 elected representative Panchayati Raj Institutions have already been trained, besides imparting training to member of voluntary organizations. About 18000 Government functionaries have been trained in disaster mitigation and preparedness at different level. 885 engineers and 425 architects have been trained under this programme in vulnerability assessment and retrofitting of lifeline buildings. 600 master trainers and 1200 teachers have already been trained in different districts in disaster preparedness and mitigation. Disaster management committees consisting of elected representatives, civil society members, Civil defense volunteers and Government functionary have been constituted all levels including village/urban local body/ward levels. With the creation of awareness generation on disaster mitigation, the community will be able to function as a well-knit unit in case of any emergency.

XIV. The Government has initiated a national wide awareness generation compaign as part of its overall disaster risk management strategy. In order to devise an effective and holistic campaign, a steering committee for mass media campaign has been constituted at the national level with due representation of experts from diverse stream of communication. The committee has formulated a campaign strategy aimed at changing peoples' perception of natural hazards and had consulted the agencies and experts associated with advertising and media to instill a culture of safety against natural hazards. Apart from the use of print and electronic media, it is proposed to utilize places with high public visibility namely hospitals, schools, railway stations and bus terminals, airports, and post offices, commercial complexes and municipality offices etc to make people aware of their vulnerability and promote creation of a safe living environment. A novel method being tried is the use of government stationary namely postal letters, bank stationary, railways tickets, airline boarding cards and tickets etc for disseminating the message of disaster risk reduction. Slogans and messages for this purpose have already been developed and have been communicated to concerned Ministries/agencies for printing and dissemination. The mass media campaign will help build knowledge, attitude and skills of the people in vulnerability reduction and sustainable disaster risk management measures.

XV. Disaster management as a subject in Social Sciences has been introduced in the school curriculum for class VIII &IX. The CBSE which has introduced the curriculum runs a very large number of schools throughout the country. The teachers are being trained to teach disaster management for   class X.

XVI. In order to assist the State Government in capacity building and awareness generation activities and to learn from the past experiences including sharing of best practices, the Ministry of Home Affairs has compiled/prepared a set of resource material   developed   by   various organization/institutions to be replicated and disseminated by State Government based on their vulnerability after translating it into the local languages. The voluminous material which runs in more than 10000 pages has been divided into four sections. These sections cover planning to cope with disaster; education and training; construction toolkit; and education and communication toolkit including multi-media resources on disaster mitigation and preparedness. The planning section contains material for analyzing community risk, development of preparedness. Mitigation and disaster management plans coordinating available resources and implementing measure of risk reduction.

Development of Response System

Mitigation and preparedness measures go hand in hand for vulnerability reduction and rapid response to disaster. Several inadequacies of response were noted in the aftermath of Bhuj earthquake, 2001. The govt. decided to remove the inadequacies to maintain preparedness at all times. Major response initiatives include:

(i) Preparation of Special Response Teams: The central Govt. is now in the process of training and equipping specialist and rescue teams. Each team includes doctors, paramedics, structural engineers etc. These teams will be stationed in different parts of the country.

(ii) Incident Command System: In order to professionalize the response system, it is proposed to develop incident command system. It is a very effective system in which the most experienced and knowledgeable person at a disaster site is designated as incident commander who is charged with the responsibility of inter agency coordination and management of the incident.

(iii) Standard Operating Procedure: Standard operating procedure are being laid down to ensure that all step need to be taken for disaster management are put in place. Each department/sector will have their own SOP's for each level of functionaries.

(iv) Trigger Mechanism: The high powered committee on disaster management has incorporated trigger mechanism as an emergency quick response mechanism. It has been envisaged as a preparedness plan whereby the receipt of a signal of an impending disaster would simultaneously energize and activate the mechanism for response and mitigation without loss of crucial time.

(v) Emergency Operation Centre: It has also been recommended for setting up of emergency operation centers at the national capitals, state capitals and district headquarters. EOC will function as nerve centres for integrated command and control structure. They will be convergence points for all inter agency coordination and will be equipped with the state of the art communication network.

Technological Developments

Technological innovations are vital for effective disaster management, the DST, Govt. of India is taking several measure to upgrade technological inputs. The important developments include:

(i) India Disaster Resource Network: This is a web enabled centralized data base which will ensure quick access to resources to minimize response time tune in emergencies. This database will be available at National, State and district level simultaneously. Police network is another important communication network to be used for disaster management. In emergency, mobile satellite based units which can be transported to disaster sited are being procured.

(ii) Development of GIS based National Data base for Disaster Management: The GIS is an effective tool for emergency responders to access information in terms of crucial parameters for the disaster affected areas. This includes location of public facilities, communication links, transport network etc. The GIS data is already available with government agencies, it is currently being upgraded. Comprehensive data district wise, multi layered maps based on this data are being generated.

(iii) Installation of Early Warning and Hazard Detection Equipment: Early warning system have already been installed for cyclones and floods in the country by IMD and CWC. There is a well established organizational set up for detecting, tackling and forecasting cyclones. There are six cyclone warning centres at Kolkata, Bhubneshwar, Vishakapatnam, Chennai, Mumbai and Ahemdabad. Cyclone tracking is done with the help of INSAT satellite. Cyclone detection radars are located at ten centres in different coastal areas. CWC does flood forecasting. There are nearly 700 station from where hydrological and hydro-meteorological data are collected. Now, govt. has also succeeded in acquiring and installing the Tsunami warning and detection system in the aftermath of Tsunami disaster of 2004.

What India Needs

In the view of the frequency of disaster striking India, there is a need for continued vigilance, preparedness and conscious efforts to reduce the occurrence and for mitigation of impact of natural disaster. What is requires is a planned approach to disaster management; its management is a fundamental component of sustainable development because the reduction of disaster equivalent to increased development. The following suggestions can be offered for effective disaster management system in India:

I. There should be a proper multi-tier organizational structure in a focussed and co-ordinated manner responsible for the overall management at national, state, districts and village levels.

ISBN: 978-93-8830-599-0

II.   The basic design of disaster management should consist of planned co-ordinated efforts in following important areas:
-Identification and prediction   -Early warning system –Evacuation
-Relief   -Rescue   -Rehabilitation   -Compensation   -Reconstruction   -Preparedness

III.   There is a need to share the expertise and experiences so that states can learn from each other. There is also a need for training personnel likely to face natural disaster and those who deal with the relief operations.

Conclusions

India in the recent years have made significant development in the area of disaster management. A new culture of preparedness, quick response, strategic thinking and prevention is being ushered. The administrative framework is being streamlined to deal with the various disasters. Effort are also being made to make disaster management a community movement wherein where is greater participation of the people. However, a lot more need to be done to make disaster management a mass movement in near future.

References

1.   D.E. Tallman, G.G. Wallace, Synth. Met. 90 (1997) 13.
2.   H.W. Kroto, J.E. Fischer, D.E. Cox, The Fullerenes, Pergamon, Oxford, 1993.
3.   A.G. MacDiarmid, A.J. Epstein, in W.R. Salaneck, D.T. Clark, E.J. Samuelson, (eds.), Science and Applications of Conducting Polymers, Adam Hilger, Bristol, 1991, p.117.
4.   D.I. Eaton, Porous glass support material, US Patent No. 3 904 422 (1975).
5.   http://upload.wikimedia.org/wikipedia/commons/2/2d/2004-tsunami.jpg
6.   http://proxied.changemakers.net/journal/300510/dis8.jpg
7.   http://www.lwsi.org/images/cs-disas_prepared.jpg
8.   http://1. .bp.blogspot.com/_fabeCWoNzTg/SLfTDs6JtII/AAAAAAAAAjw/Wlm2avBtNIM/s400/bihar_flood14.jpg
9.   IFRC and WPNS, Well Prepared National Society Self Assessment, 2003
10.   ISDR, ADB, AU, NEPAD, Guidelines for Mainstreaming
11.   Disaster Risk Reduction into Development, 2004,www.unisdr.org/eng/risk-reduction/sustainable-development/cca-undaf/cca-undaf.htm
12.   ISDR, Words into Action: A Guide For Implementing the Hyogo Framework for Action, United Nations, 2007, www.unisdr.org ISDR, Living with Risk, 2004, www.unisdr.org
13.   GOVT OF INDIA(2001), "High Powered Committee on Disaster Management-Report", Department of Agriculture & Cooperation, Ministry of Agriculture, New Delhi
14.   GOVT OF INDIA(2003), "Disaster Risk Reduction-The Indian Model", Ministry of Home Affairs, Govt of India, New Delhi.
15.   GOVT OF INDIA(2004), Disaster Management Status Report 2004, Ministry ofHome Affairs, Govt of India, New Delhi.
16.   SARKAR SUBHRADIPTA, SARMA ARCHANA(2006), "Disaster Management Act 2005",Economic and Political Weekly, Mumbai, pp 3760-3763, 2nd September 2006 SHARMA VINOD(2001), "Disaster Management", Indian Institute of Public Administration, New Delhi
17.   Ravindra K. Pande, Participation in practice and disaster management: experience of Uttaranchal (India),Disaster Prevention and Management, Vol.14, 3, 2006

ISBN: 978-93-8830-599-0

A Preliminary Study on the Treatment of Real-Time Canteen Wastewater and Reduction of Cr^{+6} by Microbial Fuel Cell

Gunaseelan. K[1] and S. Gajalakshmi[*]

[1&*]Sustainable Fuel Cells Technology Lab, Centre for Pollution Control & Environmental Engineering, Pondicherry University, Puducherry, India.
gunsie3@gmail.com, dr.s.gajalakshmi@gmail.com

Abstract

In this work, for the first time, an attempt of treating the real-time canteen wastewater and reduction of synthetic cr6$^+$ solution at an abiotic condition was made by microbial fuel cell technology. The feasibility and efficacy of managing both the wastewater sources were evaluated through open circuit voltage (OCV), Coulombic efficiency, COD removal, Power density and UV -spectroscopy methods for reducing the hexavalent chromium solution.

The performance of the reactor with both the wastewater sources was evaluated for 40 days under the batch mode of operation for five days per cycle. The Cr^{+6} containing potassium dichromate concentration of 50mg/L was synthetically prepared. Potassium dichromate is taken in cathode chamber and real-time canteen wastewater of COD ranging from 260-300mg/l was made in the anode chamber with anaerobic microorganisms as bio- The reactor resulted in the generation of the higher volumetric power density of 408 mW/m^3 and also with the average volumetric power density of 83.1 ± 26.4 mW/ m^3.

This work verifies the possibility of current production and simultaneous cathodic Cr^{+6} reduction in dual-chambered microbial fuel cell with low-cost CMI 700 polymer membrane as a separator rather than high-cost Nafion membrane

I. Introduction

Microbial fuel cell (MFC) is a biological electrochemical system which is capable of converting chemical energy into electrical energy due to exoelectrogenic bacteria's bio-catalytic process under t anaerobic conditions [1], [2].

MFC systems can be fabricated as per our requirements by using various materials and configurations. The anodic oxidation and cathodic reduction reactions are the key functions of the MFC bioreactors. The anode compartment of the MFC reactor will be associated with the generation of protons and electrons. Finally, the generated protons reach the cathode chamber through the proton exchange membrane (PEM) and end up with the developed potential difference.

In the early nineteenth century, it was expected that only a few microorganisms, namely exoelectrogens, would be utilized to produce the electricity. With the advancement in research, the positive potentials of several diverse microorganisms as a biocatalyst in MFC has been testified.

Early MFC theory was demonstrated for the first time by Potter in 1910 with *Escherichia coli* and *Saccharomyces sp.* Bacterial cultures for generating electricity with the costlier platinum electrodes[3]. However, it didn't get much appreciations and attention till the 1980s, but this concept of MFC drew an immense response only after the advent and use of potential electron mediators, for improving the electricity generation with multiple folds [4].

The bacterial communities are utilized to generate electricity in MFC reactors through the degrading the organic substances or substrates or wastes in the wastewater [4], [5]. MFC reactors, various categories of wastewaters have been treated successfully, which includes domestic, brewery, agro, tannery, animal, and food processing wastewaters [6]. The removal mechanism of the organic pollutants in wastewaters is along with the generation of valuable forms of energy, namely electrical power or hydrogen gas or valuable algal lipids extraction for the production of biodiesel.

Hexavalent Chromium is one of the most hazardous pollutant metal ions which got discharged into the environment by industries namely as Mining, Electroplating, Metal finishing, Leather tanning, and dye/ pigment manufacturing, etc. Cr(VI) is one of the well-known mutagen, teratogen, and carcinogen being highly water soluble and mobility.

In the recent days, even with several limitations encountered during the operation of the MFCs in the real-time, this technology has come with the assured advancements in the removal of the organic pollutants and electricity generation from the provided wastewater during the wastewater treatment process. With the definitive motto of performing the preliminary studies on the treatment of real-time canteen wastewater and

ISBN: 978-93-8830-599-0

reduction of cr^{+6} by the microbial fuel cell, this study was conceived.

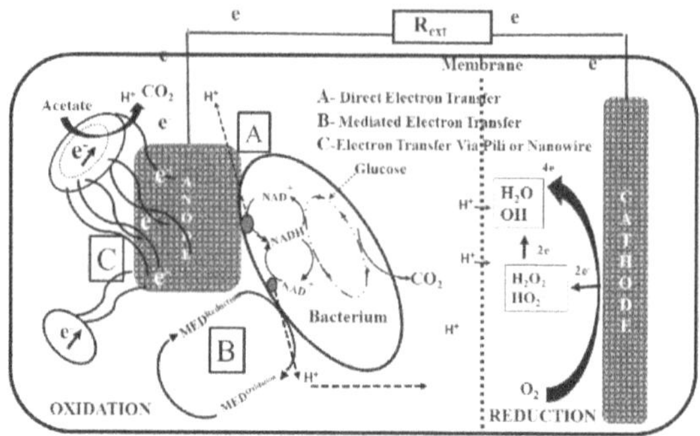

Fig. 1 Schematics of the Dual chamber MFC

II.MATERIALS AND METHODS

A. Construction of two chambered MFC and operation

This two-chambered MFC comprising of anodic and cathodic chambers with the working volume of 500ml. Both the chambers of the MFC reactor were provided with two ports at the top of the reactors as a provision for doing sampling, the addition of influent and for reference electrode wires.

The graphite plate with an effective surface area of 20 cm^2 was employed as the anode and cathode electrode. The circular CMI 700 polymer membrane of a projected diameter of 9 cm was inserted between the two chambers of the MFC reactor for ensuring the movement of H$^+$ ions from anodic to cathodic chamber's reaction interface exchange through a firmly coupled waste PET bottles without any leaks.

Both the chambers and the membrane electrode assembly (MEA) of the MFC was held together firmly with the nuts-bolts joints for ensuring leakage. The septic tank's sludge was employed as bio catalyst for breaking down the provided organic loads into free ions and electrons. The anode and cathode of this MFC coupled with a high-grade concealed copper wires sealed with silica gel to avoid the oxidation and also with a 1000 ohms resistor as an external load after the 20th days of the reaction cycle.

This MFC fabricated using the waste PET bottles was operated for 40 days under batch mode with five days per batch reaction cycle under the room temperature ranging from 33 to 37 °C. The preliminary observation study was taken for numerous reaction cycles to understand the challenges to treat the real-time canteen wastewater and reduction of the cr^{+6} solution along with the electricity generation.

B. Utilization of real-time canteen wastewater as Anodic solution

For this study, the real-time canteen wastewater was collected from the outlet pipe of the Canteen No-2, Pondicherry University. The collected real-time wastewater was strained through 500-micrometre sieve mesh to avoid the bulk solid food particles and debris etc. Further, the collected wastewater's COD was estimated.

C. Utilization of synthetic Cr^{+6} solution as a cathodic solution

For this study, synthetic wastewater of 50mg/l of Cr^{+6} was prepared from Potassium Dichromate salts. Potassium Dichromate solution is one of the potent reducing agent used in the cathodic chamber of the MFC reactors of maximum power production. The reduction of Cr^{+6} to Cr^{+3} ions can be easily estimated by the UV-spectroscopy method.

III. RESULT AND DISCUSSION

A. Voltage and Power generation

The MFC reactor with the real-time canteen wastewater as the anodic solution and the synthetic Cr6+ solution as the cathodic solution were resulted with the highest operating voltage of 452 mv across a 1000 ohms external resistance at the 20th day of operation after achieving a very stable OCV of above 924mv. Further, the reactor was operated for 40 days with total of 8 batch cycles with a duration of 5 days per cycle and resulted with the higher volumetric power density. The reactor resulted with the generation of the higher volumetric power density of 408 mW/m^3 and also with the average volumetric power density of 83.1 ± 26.4 mW/ m^3.

Table 1 Electrical parameters observed in dual chamber mfc

Avg. OCV (mv)	Max. OCV (mv)	Avg.OV (mv)	Max. OV (mv)	Avg. Volumetric P.D (mW/m^3)	Max. Volumetric P.D (mW/m^3)
442.5±256.	924	203.9± 115	452	83.1±26.4	408

OCV- open circuit voltage OV- operating voltage P.D- power density

B. Anodic Real-time wastewater Treatment efficiency

The MFC reactor made with waste PET bottles was inoculated with anaerobic inoculum collected from the septic tank has shown average COD removal efficiency of 39.125 ± 12.44 % . Moreover, the COD removal efficiency found in the MFC reactor was appreciable,

ISBN: 978-93-8830-599-0

which may be due to the potential of the selected inoculum to oxide the provided organic loads in the real-time canteen wastewater and challenges in adapting to the anaerobic condition in the anodic chamber. From this study, the COD removal reported with a maximum average of 58.5% during the 8th cycle.

Table 2 Cod Removal & Cr^{+6} Reduction Efficiency Observed in Dual Chamber Mfc

Separator	Avg.COD removal (%)	Maximum COD removal (%)	Avg. Cr^{+6} reduction (%)	Maximum Cr^{+6} reduction (%)
CMI 700	39.12±12.44	58.5	53.917±21.815	80.5

C. Cathodic Synthetic Hexavalent chromium reduction efficiency

The real-time wastewater in the anodic chamber release protons that migrate from anode to cathode through a polymer separator which by then are partially consumed by Cr^{+6} to form Cr^{+3}. The most widely used quantitative method for the determination of hexavalent chromium is diphenyl carbazide absorption method using a UV-Visible spectrophotometer at 540nm. The potassium dichromate solution of 50 mg/L was taken in reactor and the amount of hexavalent chromium reduced is analysed for four cycles at an interval of 5 days and showed higher reduction percentage of 80.5%.

IV. CONCLUSION

This preliminary assessment on the treatment efficiency of real-time canteen wastewater and reduction of cr^{+6} by microbial fuel cell shows that MFC reactor has exhibited higher power density, COD removal, Cr^{+6}reduction and significant coulombic efficiency during the highly stable operating condition. This study's findings provide the scope of using of real-time canteen wastewater and reduction of the synthetic cr^{6+} solution as a better way of treating both the types of wastewaters along with the production of electricity.

REFRENECE

1. K. Rabaey, N. Boon, M. Höfte, and W. Verstraete, "Microbial phenazine production enhances electron transfer in biofuel cells," *Environ. Sci. Technol.*, vol. 39, no. 9, pp. 3401–3408, 2005.

2. J. Heilmann and B. E. Logan, "Production of Electricity from Proteins Using a Microbial Fuel Cell," *Water Environ. Res.*, vol. 78, no. 5, pp. 531–537, 2006.

3. M. C. Potter, "Electrical Effects Accompanying the Decomposition of Organic Compounds," *Proc. R. Soc. B Biol. Sci.*, vol. 84, no. 571, pp. 260–276, 1911.

4. Y. Tekle and A. Demeke, "Review on Microbial Fuel Cell," *Basic Res. J. Microbiol.*, vol. 1, no. 2, pp. 1–32, 2015.

5. D. H. Park and J. G. Zeikus, "Electricity Generation in Microbial Fuel Cells Using Neutral Red as an Electronophore Electricity Generation in Microbial Fuel Cells Using Neutral Red as an Electronophore," vol. 66, no. 4, pp. 1292–1297, 2000.

6. S. E. Oh and B. E. Logan, "Proton exchange membrane and electrode surface areas as factors that affect power generation in microbial fuel cells," *Appl. Microbiol. Biotechnol.*, vol. 70, no. 2, pp. 162–169, 2006.

Open Street Map and GIS

Mohammed Sami Uz Zaman[1], Mohammed Baseer Uz Zaman[2] and Ballu Harish[3]
[1]Geo-Informatics and Surveying Technology, CSIT-IST
[2]Al Habeeb College of Engineering and Technology
[3]Assistant Professor, Centre for Spatial Information Technology, IST
Jawaharlal Nehru Technological University, Hyderabad
zaman1057@yahoo.com, harishballu11@gmail.com

Abstract

Availability of spatial data is to display special pieces of information or an unlimited amount of essential mapping information (layers) applied to the knowledge base of any given professional field. The study suggests the possibilities in using **OpenStreetMap (OSM)**, a collaborative project to create a free editable map of the world. The creation and growth of OSM has been motivated by restrictions on use or availability of map information across much of the world, and the advent of inexpensive portable satellite navigation devices. OSM is considered a prominent example of volunteered geographic information. The data from OSM is available for use in both traditional applications, like its usage by Facebook, Craigslist, OsmAnd, Geocaching, MapQuest Open and Foursquare, and more unusual roles like replacing the default data included with GPS receivers. It has many applications in scientific research and humanitarian aids etc.

Keywords: Volunteered geo-information, JOSM, Humanitarian Aid, GPS, OSM Xml and OSM Editors.

INTRODUCTION

The GIS enables to provide the required geographic feature and its location details. In Maps, geo-locational details and precise information content etc. are important. Access to spatial data and cartographic products has changed radically over the last decade or so. Traditionally, governmental agencies, cartographic centers, and commercial agencies were the only sources for end-users seeking spatial data. One of the most formidable barriers to more widespread access to these geodata were created by often prohibitive high fees and license charges in combination with time- and purpose-limited copyright restrictions imposed. Changes in Information and Communication Technology (ICT) brought about by the Internet and social media and the availability of inexpensive portable satellite navigation devices has seen this traditional geodata business model challenged. One of the key driving forces in this change has been the OpenStreetMap (OSM) project. OSM was launched in 2004 with the mission of creating an editable map of the whole world and released with an open content license.In general, OSM aims at building and maintaining a free editable map database of the world in a collaborative manner so that people and end-users are not forced to buy geodata in the traditional way and subsequently be subjected to restrictive copyright and license commitments. OSM started initially with a focus on mapping streets and roads. Since then it has moved far beyond these entities and it now contains a very rich variety of geographical objects (e.g., buildings, land use, Points of Interest) from all over the planet being mapped by thousands of volunteer contributors to the project.

OSM is often referred to as the Wikipedia map of the world. As it is built on many of the same ICT structures as Wikipedia it offers its project contributors the possibility of (a) almost immediate updating of the map database as well as very frequent updating of associated editing software and other tools; (b) importing geodata recorded from Global Positioning System (GPS)-enabled devices, smart phones, and other digital maps tools; (c) access to the full history of mapping activities in OSM over its lifetime; and finally (d) collaboration with other OSM users and contributors through various communication channels including mailing lists, discussion forums, and physical meetings (Mooney and Corcoran 2013a).

The OSM community is actively involved in much more than collecting geodata to build and maintain this global geodatabase. In addition, the community is involved in, for example, humanitarian work, open source software development to support OSM and the GIS community, and in building a network of support for those using and contributing to the OSM project.

Research on OSM has shown that its geodata in some parts of the world are more complete and locationally and semantically more accurate than the corresponding proprietary datasets (e.g., Zielstra and Zipf 2010; Neis and Zipf 2012; Helbich et al. 2012)

Contributors

The project has a geographically diverse user-base, due to emphasis of local knowledge and ground truth in the process of data collection. Many early contributors were cyclists who survey with and for bicyclists, charting cycleroutes and navigable trails. Others are GIS professionals who contribute data with Esri tools.

ISBN: 978-93-8830-599-0

Street-Level Image Data

In addition to several different sets of satellite image backgrounds available to OSM editors, data from several street-level image platforms are available as map data photo overlays: Bing Streetside 360° image tracks, and the open and crowdsourced Mapillary and OpenStreetCam platforms, generally smartphone and other windshield-mounted camera images. Additionally, a Mapillary traffic sign data layer can be enabled; it is the product of user-submitted images.

Government Data

Some government agencies have released official data on appropriate licences. This includes the United States, where works of the federal government are placed under public domain.

In the United States, OSM uses Landsat 7 satellite imagery, Prototype Global Shorelines from NOAA, and TIGER from the Census. In the UK, some Ordnance Survey OpenData is imported, while Natural Resources Canada's CanVec vector data and GeoBase provide landcover and streets. Out-of-copyright maps can be good sources of information about features that do not change frequently.

Copyright periods vary, but in the UK Crown copyright expires after 50 years and hence Ordnance Survey maps until the 1960s can legally be used. A complete set of UK 1 inch/mile maps from the late 1940s and early 1950s has been collected, scanned, and is available online as a resource for contributors.

Route Planning

In February 2015, OpenStreetMap added route planning functionality to the map on its official website. The routing uses external services, namely OSRM, GraphHopper and MapQuest.

There are other routing providers and applications listed in the official Routing wiki.

Map usage

Software for Viewing Maps

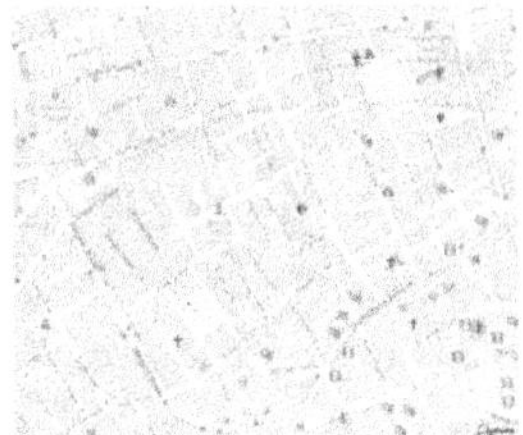

OpenStreetMap of Soho, central London, shown in "standard" OpenStreetMap layer

Same as above, shown in Mapbox Streets layer

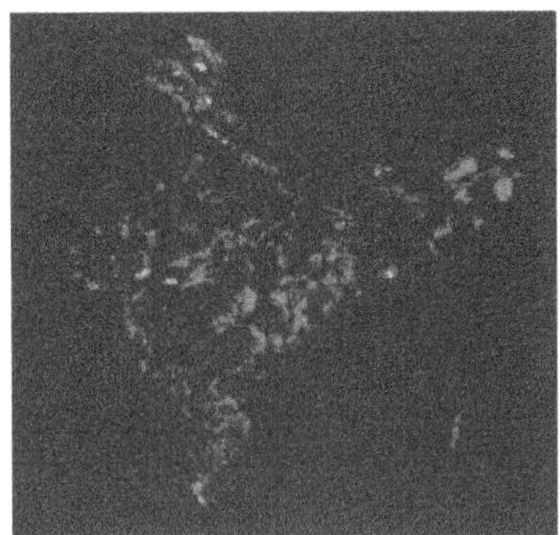

Raw OpenStreetMap data of India loading in QGIS for analysis and mapmaking

Web Browser

Data provided by the OpenStreetMap project can be viewed in a web browser with JavaScript support via Hypertext Transfer Protocol (HTTP) on its official website. The basic map views offered are: Standard, Cycle map, Transport map and Humanitarian. Finer map display and category options are available using OpenStreetBrowser.

- **OsmAnd**
  OsmAnd is free software for Android and iOS mobile devices that can use offline vector data from OSM. It also supports layering OSM vector data with prerendered raster map tiles from OpenStreetMap and other sources.

- **Maps.me**
  Maps.me is free software for Android and iOS mobile devices that provides offline maps based on OSM data.

- **GNOME Maps**
  GNOME Maps is a graphical front-end written in JavaScript and introduced in GNOME 3.10. It provides a mechanism to find the user's location with the help of GeoClue, finds directions via GraphHopper and it can deliver a list as answer to queries.

- **Marble**
  Marble is a KDE virtual globe application which received support for OpenStreetMap.

ISBN: 978-93-8830-599-0

- **FoxtrotGPS**

  FoxtrotGPS is a GTK+-based map viewer that is especially suited to touch input. It is available in the SHR or Debian repositories.

The web site OpenStreetMap.org provides a slippy map interface based on the Leaflet JavaScript library (and formerly built on OpenLayers), displaying map tiles rendered by the Mapnik rendering engine, and tiles from other sources including OpenCycleMap.org.

Custom maps can also be generated from OSM data through various software including Jawg Maps, Mapnik, Mapbox Studio, Mapzen's Tangrams.

Open Street Map maintains lists of online and offline routing engines available, such as the Open Source Routing Machine. OSM data is popular with routing researchers, and is also available to open-source projects and companies to build routing applications (or for any other purpose).

Humanitarian Aid

The 2010 Haiti earthquake has established a model for non-governmental organisations (NGOs) to collaborate with international organisations. OpenStreetMap and Crisis Commons volunteers using available satellite imagery to map the roads, buildings and refugee camps of Port-au-Prince in just two days, building "the most complete digital map of Haiti's roads". The resulting data and maps have been used by several organisations providing relief aid, such as the World Bank, the European Commission Joint Research Centre, the Office for the Coordination of Humanitarian Affairs, UNOSAT and others. NGOs, like the Humanitarian OpenStreetMap Team and others, have worked with donors like United States Agency for International Development (USAID) to map other parts of Haiti and parts of many other countries, both to create map data for places that were blank, and to engage and build capacity of local people.

After Haiti, the OpenStreetMap community continued mapping to support humanitarian organisations for various crises and disasters. After the Northern Mali conflict (January 2013), Typhoon Haiyan in the Philippines (November 2013), and the Ebola virus epidemic in West Africa (March 2014), the OpenStreetMap community has shown it can play a significant role in supporting humanitarian organisations.

The Humanitarian OpenStreetMap Team acts as an interface between the OpenStreetMap community and the humanitarian organisations.

Along with post-disaster work, the Humanitarian OpenStreetMap Team has worked to build better risk models and grow the local OpenStreetMap communities in multiple countries including Uganda, Senegal, the Democratic Republic of the Congo in partnership with the Red Cross, Médecins Sans Frontières, World Bank, and other humanitarian groups.

Data Format

OpenStreetMap uses a topological data structure, with four core elements (also known as *data primitives*):

- *Nodes* are points with a geographic position, stored as coordinates (pairs of latitude and a longitude) according to WGS 84. Outside of their usage in ways, they are used to represent map features without a size, such as points of interest or mountain peaks.
- *Ways* are ordered lists of *nodes*, representing a polyline, or possibly a polygon if they form a closed loop. They are used both for representing linear features such as streets and rivers, and areas, like forests, parks, parking areas and lakes.
- *Relations* are ordered lists of nodes, ways and relations (together called "members"), where each member can optionally have a "role" (a string). Relations are used for representing the relationship of existing nodes and ways. Examples include turn restrictions on roads, routes that span several existing ways (for instance, a long-distance motorway), and areas with holes.
- *Tags* are key-value pairs (both arbitrary strings). They are used to store metadata about the map objects (such as their type, their name and their physical properties). Tags are not free-standing, but are always attached to an object: to a node, a way or a relation. A recommended ontology of map features (the meaning of *tags*) is maintained on a wiki. New tagging schemes can always be proposed by a popular vote of a written proposal in OpenStreetMap wiki, however, there is no requirement to follow this process. There are over 89 million different kinds of tags in use as of June 2017.

Data Storage

The OSM data primitives are stored and processed in different formats.

The main copy of the OSM data is stored in OSM's main database. The main database is a PostgreSQL database with PostGIS extension, which has one table for each data primitive, with individual objects stored as rows. All edits happen in this database, and all other formats are created from it.

ISBN: 978-93-8830-599-0

For data transfer, several database dumps are created, which are available for download. The complete dump is called planet.osm. These dumps exist in two formats, one using XML and one using the Protocol Buffer Binary Format (PBF).

The LinkedGeoData data uses the GeoSPARQL and well-known text (WKT) RDF vocabularies to represent OpenStreetMap data. It is a work of the Agile Knowledge Engineering and Semantic Web (AKSW) research group at the University of Leipzig, a group mostly known for DBpedia.

Because OpenStreetMap is an open platform with an editing API, there are many other editors to choose from, some with a simplified sub-set of functionality, some operating on specific platforms such as mobile devices. The following table lists some of the options. As with all such lists on this wiki presence on the list should not be considered a recommendation, have a look at the editor usage stats to see what is currently actually in use, typically only those with a larger current user base are consistently maintained and support current good mapping practices.

Osm Editors

Name	Screenshot	Platform	Add POIs	Edit / Delete POIs	Edit arbitrary tags of existing OSM objects	Edit geometries	Support imagery offset DB	Upload to OSM
Vespucci [29]		Android	yes	yes	yes	yes	yes	yes
ShareNav [28]		Android, J2ME, Windows (including Windows 2000 and Windows XP), Linux, macOS	yes	yes	yes	Not directly, will open web browser editor for area.	?	yes
RawEditor [27]		Web	yes	yes	yes	yes	?	yes
QGIS [30]		Windows, Linux, macOS, BSD	yes	yes	yes	yes	?	yes
Pushpin OSM [26]		iOS	yes	yes	yes	no	?	yes

Contd...

ISBN: 978-93-8830-599-0

Potlatch 2 [25]		Web-based (Flash)	yes	yes	yes	yes	no	yes
Potlatch 1 [24]		Windows, Linux, macOS	yes	yes	yes	yes	no	yes
Pic4Review [23]		Web-based (JavaScript)	no	yes	yes	no	no	yes
OSMyBiz [22]		Web-based (JavaScript)	yes	yes	no	no	no	yes
OsmAnd [21]		Android, iOS	yes	yes	Yes, for nodes and closed ways with common tag combinations.	no	no	yes
OSM2Go [20]		Linux; Maemo (N800, N900)	yes	yes	yes	yes	no	yes
OpenMaps [19]		iOS	yes	yes	yes	no	?	yes
Nomino [18]		Windows, Linux, macOS, Syllable, BSD, illumos, BeOS, Blackberry OS, iOS, Solaris, QNX, etc.	no	no	yes	no	?	yes
Mumpot [17]		Linux, OpenMoko, GPE	?	yes	?	?	?	yes

Editor		Platform						
Merkaartor [16]		Windows, Linux, macOS	yes	yes	yes	yes	no	yes
MapStalt Mini [15]	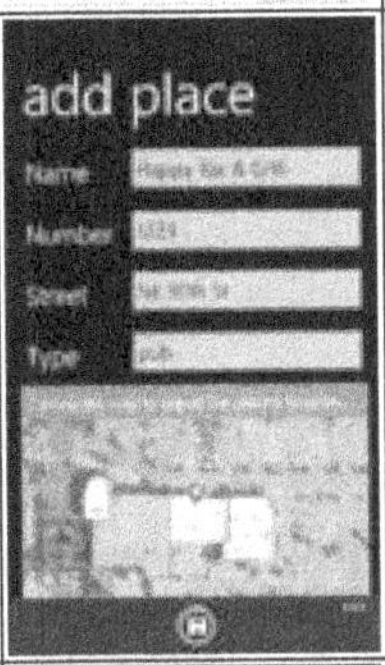	Windows Phone	yes	yes	yes	no	?	yes
MAPS.ME [14]]	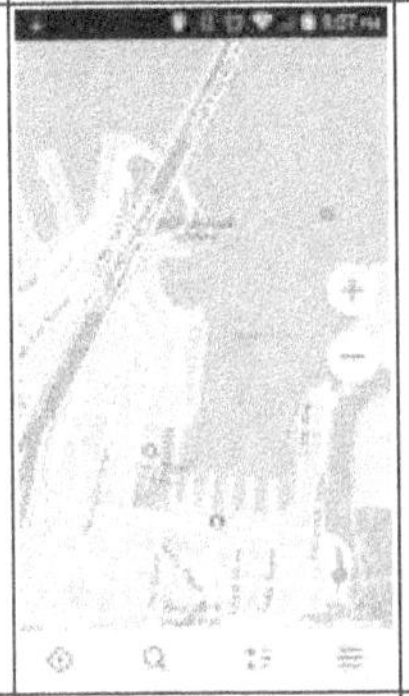	Android, iOS	yes	yes	yes	no	?	yes
MAPCAT [13]		Web	yes	yes	yes	yes	no	yes
Level0 [12]	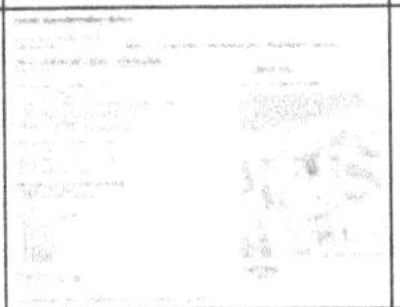	Web	yes	yes	yes	yes	?	yes
JOSM [11]		Windows, Linux, macOS	yes	yes	yes	yes	yes	yes
iD-strava [10]		Web-based (JavaScript)	yes	yes	yes	yes	no	yes
iD-indoor [9]	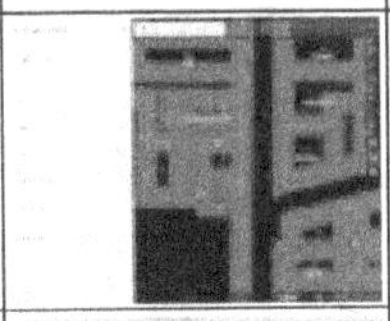	Web-based (JavaScript)	yes	yes	yes	yes	no	yes
iD [8]	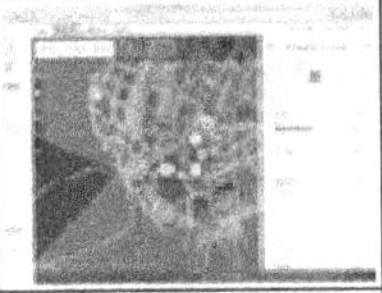	Web-based (JavaScript)	yes	yes	yes	yes	no	yes
GpsMid [7]	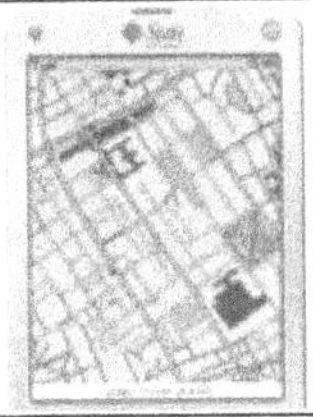	J2ME, Android	yes	yes	yes	Not directly, will open web browser editor for area.	no	yes

Contd…

GPSMapEdit [6]		Windows	yes	yes	yes	yes	?	yes
Goosm [5]		Android	yes	yes	yes	no	no	yes
Go Map!!		iOS 7+	yes	yes	yes	yes	?	yes
GNOME Maps [4]		Linux	yes	yes	no	no	no	yes
Fireyak [3]		Android	yes	?	?	?	no	yes
ArcGIS Editor for OSM [2]		Windows (Windows 2000, Windows XP, Windows Mobile 2003)	yes	yes	yes	yes	?	yes
Amenity Editor [1]		Windows, Linux, macOS, BSD, BeOS, Syllable, AmigaOS, OpenIndiana, etc.	yes	yes	yes	no	no	yes

Osm File Formats

The most important formats are:

- OSM XML – xml-format provided by the API
- PBF Format – highly compressed, optimized binary format similar to the API
- o5m – for high-speed processing, uses PBF coding, has same structure as XML format
- Overpass JSON – JSON variant of OSM XML
- Level0L - more human readable OSM XML and lowered redundancy

Osm Xml

The **.osm** file format is specific to OpenStreetMap. Many geographic data formats predate the modern internet era, and are designed for quick access and querying like one would query a database. OSM data, on the other hand is designed to be easily sent and received across the internet in a standard format. Hence, **.osm** files are coded in XML, and contain geographic data in a structured, ordered format. XML is a so called meta format to provide even human readable data interexchange formats. Various file formats use this data tree structure to embedd their datas like XHTML, SVG, ODT, etc. Human readable because of clear structure. Machine independent due to exact definitions, e.g. character sets, XML schema definitions, DTD, namespaces, etc. Ready to use parsers for general XML that can be customized for a concrete file format and has good compression ratio.

PBF Format

PBF Format ("Protocolbuffer Binary Format") is primarily intended as an alternative to the XML format.

ISBN: 978-93-8830-599-0

It is about half of the size of a gzipped planet and about 30% smaller than a bzipped planet. It is also about 5x faster to write than a gzipped planet and 6x faster to read than a gzipped planet. The format was designed to support future extensibility and flexibility. A lot of software used in the OSM project already supports PBF in addition to the original XML format, plus there are several tools to convert from PBF to OSM XML and vice versa.

o5m

The **.o5m** data format was designed to be a compromise between .osm and .pbf format. It has the same structure as .osm format, therefore input and output procedures of existing OSM data processing applications can be adapted to this format with small effort.

Overpass JSON

The Overpass API (formerly known as *OSM Server Side Scripting*, or *OSM3S* before 2011) is a read-only API that serves up custom selected parts of the OSM map data. It acts as a database over the web: the client sends a query to the API and gets back the data set that corresponds to the query.

Unlike the main API, which is optimized for editing, Overpass API is optimized for data consumers that need a few elements within a glimpse or up to roughly 10 million elements in some minutes, both selected by search criteria like e.g. location, type of objects, tag properties, proximity, or combinations of them. It acts as a database backend for various services. The OSM editing process functions because of what is known as an API, which allows editing software to communicate with the central server. For example, when you are using JOSM and you select the area you want to map, an API call is sent to the server, requesting all of the data that exists within the area that you have selected.

In fact, when you download data in JOSM, you are extracting the data from a specific area of the world. The data is then sent to you in **.osm** format, which you can then edit in JOSM. If you download data in JOSM and then save it, you will see that the file type is **.osm**.

Level0L

Level0L (or l0l in short) is a format compatible with OSM XML. It is used in Level0 editor. It is a plain text file with a stream of objects.

Conversion between Different Osm Map Data Formats

Software	OSM XML	PBF	o5m	Discussion / comments
Osmosis	yes	yes	no	
osmconvert	yes	yes	yes	own PBF implementation (doesn't use PBF library)
Osmium	yes	yes	read only	use the Osmium command line tool or see osmium_convert in examples folder
osm4j	yes	yes	no	

Data Downloading

It is possible to download map data from the OpenStreetMap dataset in a number of ways. The full dataset is available from the OpenStreetMap website download area. It is also possible to select smaller areas to download. Data normally comes in the form of XML formatted .osm files. If you just want to use a "map" (e.g. for a GPS device) then you likely do not want to download this raw data, instead see other OSM download options. Extracting Value added information from OSM database has become another emerging research topic for many academic studies. (Hagenauer and helbich 2012; Mooney and Corcoran 2012; Mooney et al. 2013; Jokar Asranjani et al. 2015).

Osm And Qgis

QGIS (formerly Quantum GIS) is a full-featured, open-source, cross-platform Geographic Information System. With QGIS you can access up-to-date OSM data whenever you want, select the tags you want to include, and easily export it into an easy-to-use SQLite database or Shapefile. This process makes it easy to get up-to-date OSM data and pull it into QGIS. Once you have layers in QGIS, it is possible to save them as shapefiles, execute filters and queries, and so forth and analyze the data.

ISBN: 978-93-8830-599-0

Imagery and OSM

Global Coverage

Name	Coverage	Age	Resolution	DOP	URLs (within editors etc)
Bing Satellite Imagery	global	Varies	Varies		Potlatch2 : Yes (Built-in) JOSM : Yes (Built-in)
DigitalGlobe / MapBox	World	ongoing	0.3m - 38m (Z19-Z12)	yes	JOSM : tms[17]:http://{switch:a,b,c}.tiles.mapbox.com/v3/openstreetmap.map-4wvf9l0l/{zoom}/{x}/{y}.png
DigitalGlobe Premium Imagery	global				JOSM : Yes (Built-in)
DigitalGlobe Standard Imagery	global				JOSM : Yes (Built-in)
Imagico.de Imagery	Various places where good imagery is not readily available				JOSM : tms:http://imagico.de/map/osmim_tiles.php?z={zoom}&x={x}&y={-y}
Esri World Imagery	global	Varies	Varies		JOSM : Yes (Built-in)

APPLICATIONS

The core of OpenStreetMap is a collection of map data which can be used for many different purposes.

Maps

Many people get to know OpenStreetMap through its world map. People can easily integrate this map on their website, with the possibility to add their own features. Wikipedia does so, and many other people too. Further, OpenStreetMap provides the raw map data to create customized maps from the scratch. Numerous third party online maps are based on OpenStreetMap data. You can also use maps without internet connection on desktop computers or mobile devices such as smart phones or even the Play Station Portable. There are tools and online services helping to create high resolution paper maps. Maps can be used in very different ways. With OpenStreetMap, people get artistic or help on humanitarian activity. Within the OpenStreetMap Community, people focus on mapping a variety of particular topics, from accessibility to Special Populations. Third party services provide maps specialised on accessibility (wheelchair, blind, ...), on sports (hiking, biking, canoeing, skiing, ...), on History and many more. List of OSM based Services tries to give a comprehensive collection.

Many projects continuously improve their maps. Since they are all limited by the available information in the OpenStreetMap data collection, contributing map data is a good way to support them. ÖPNVkarte, a free public transport map, is a good example.

3D Maps

Another innovation was the development of 3D city models from OSM (Over et al. 2010; Goetz 2013). The third dimension is a growing topic at OSM. It's already possible to add detailed buildings and a lot of minor objects. Further research has focused on attempting to extend the current OSM spatial data model by working on extensions such as: 3D (Goetz and Zipf 2012), indoor mapping (Goetz and Zipf 2012), or wheelchair routing (Neis 2014) and using the results from this in a range of applications.

Comparing Maps

- Transparent Map Comparison - Compare transparent overlays of OSM maps to other maps or satellite imagery from yahoo! and Google
- Another transparent map comparison - Compare transparent overlays of OSM maps to satellite imagery from Google, Google maps labels and other data
- Map Compare - Side by side compare of different maps provides by Geofabrik.de

Routing

Many different routing services (see also this list of routing services) are built on OpenStreetMap data, not only for cars. Based on OpenStreetMap, you can find your way in wheelchair, on bike or horse, during your next hike, using public transport. The Look and Listen Map provides an online map and routing service for blind, visually impaired and sighted persons. While still in a development status, the project already won some prizes.

ISBN: 978-93-8830-599-0

These applications provide turn-by-turn navigation and, in many cases, voice guidance:

Open Source

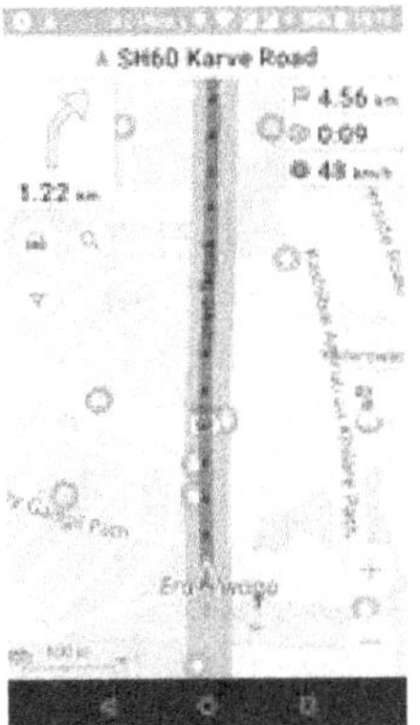

OsmAnd

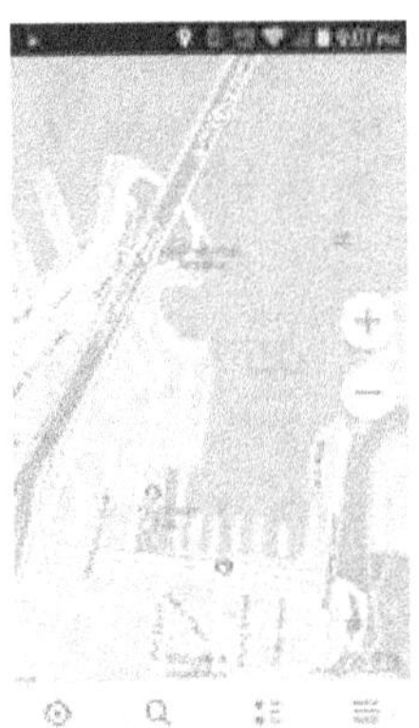

MAPS.ME

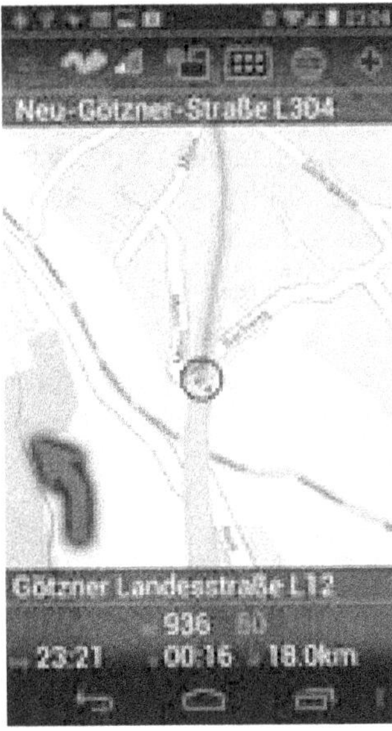

Navit

Gis Software

Many available GIS software is proprietary, closed source, and quite expensive. Often they operate with data in a proprietary data format. OpenStreetMap data, in contrast, is open. To achieve interoperability, there is a whole bunch of tools converting map data from/ to OpenStreetMap data, including support for the popular ESRI Shapefiles and PostGIS format. We try to document any special interoperability with OpenStreetMap for each GIS software. Try a search with the form on the top right of this page for your particular GIS software, or browse our GIS software category (use Category:Software as long as the previous doesn't exist). There is also open source GIS software working together with OpenStreetMap.

Education

OpenStreetMap is being used within education, in schools, universities and colleges in a wide range of disciplines. Some projects involve only the use of existing OpenStreetMap data, while others result in additional data within the OpenStreetMap data collection. The OpenStreetMap project is used in teaching geography, mathematics, ecology, community planning and technology.

Games

These games rely on OpenStreetMap data to power gameplay:

BucketMan

Pokémon Go

Ingress

ISBN: 978-93-8830-599-0

Gps Accuracy

If you are recording GPS tracks with an Android phone, it can be useful to have an application to help you troubleshoot your GPS data. A GPS testing application can help you determine which GNSS satellite networks your phone supports (Beidou, Galileo, GLONASS, GPS, QZSS, etc.). It can also help you determine the satellites it can see and is using at any given time, which frequencies your device listens on (whether it supports dual band signals, for example), and the accuracy of the data your phone reports.

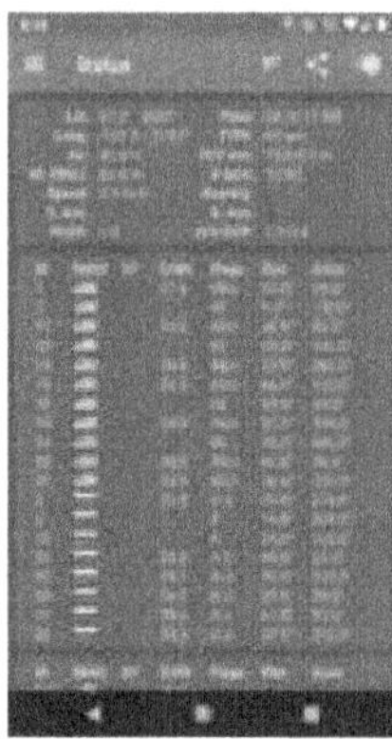

Research

OpenStreetMap data is being used for research all round the world.

LIMITATIONS

Imagery Offset

As a background layer within editors, we can use Bing, IRS, Landsat, and some others. All of these can be used to create and modify objects, but consider that each may have an offset or other kinds of distortions.

Initially, tiled imagery starts life as individual photos, taken from a satellite or airplane, tagged with a center coordinate (which itself is subject to errors, see below on GPS). The photo's corner coordinates can usually be derived from height-above-ground and lens properties. With that, special software can be used to transform it into a projection that is more commonly used with OSM map making. This transformation is not perfect, and may induce distortion and position errors.

Because there is, in any case (even if Earth was flat), perspective in the picture, objects may appear to skew to the side the higher *above the ellipsoid* they are, so this affects high-rise buildings, mountains, and roads going up/down hills. Perspective exists because the camera is only above one certain point at any time. The perspective problem can only be countered by taking more overhead pictures per distance, and combining them.

When contributing to OpenStreetMap by tracing from aerial imagery — a kind of "armchair mapping" — be aware of outdated, warped or misplaced photos. Verified data is the better data to have. The most valuable kind of contribution involves mapping an area based on visiting an area and doing ground survey to gather data. Your local knowledge can contribute, but consider that "knowledge" itself can get outdated. The exact placement of roads and other features can be assisted by recording GPS tracks on the ground. Aerial images clearly provide useful assistance in the mapping process, but it can be a bad idea to move map objects created by others just to match them to aerial images. Aerial imagery may be distorted and shifted relative to real object locations.

OSM XML

- Very huge files when decompressed.
- Might need to decompress before processing (data compression/decompression may be performed on the fly by the transport protocol such as HTTP)
- Parsing *may* take a lot of time and memory resources, but only when using the full XML capabilities; basic XML (without DTD and namespaces) however is very fast and can be performed efficiently on the fly, without computing the DOM for the whole XML document (e.g. with simple SAX parsers)

Editors such as iD or JOSM have one core function that they are very good at - making it easy for users to edit OpenStreetMap. But they are not software meant for analyzing or querying data - this function is best left to other applications. GIS software, such as the free and open source Quantum GIS (QGIS), allows users to design good-looking maps, to query and analyze data, and much more. GIS software can also be used for editing geodata, but it is much easier to edit OpenStreetMap with the dedicated OSM editors.

User Skills and Knowledge

Any person with or without formal training can make edits which could result in chaos.

CONCLUSION

OpenStreetMap is fascinating and vast in terms of scale and variety of the data offered, the technology built, and uses found for the data. But most of all, it is a fascinating and vast community. From a research perspective, OpenStreetMap is deservedly attracting a lot of attention from universities and similar institutions, and from academic scholars in 'geomatics', 'neogeography', 'volunteered geo-information' (VGI), 'digital cartography', 'participatory geoinformatics' etc. It is being used within education, in schools,

ISBN: 978-93-8830-599-0

universities and colleges in a wide range of disciplines. It significanct in disaster management has many applications in scientific research and humanitarian aids etc.

ACKNOWLEDGEMENT

The authors would like to thank **Dr. K. Manjulavani**, Professor & Head Civil Engineering Department, JNTUH, & **Mr. L.Ravi** (Research scholar), CSIT-IST, JNTU, for their timely guidance. Special thanks to **Mr. M. G. Zaman** (Former Deputy Director, NGRI-CSIR) and **Dr. V. Raghavaswamy** (Former Deputy Director, NRSC-ISRO) for their mentorship. They would also like to thank Mrs. Khaisar Sayeed (Ex-Principal, Zaheer Memorial High School) and Mrs. Kareemunnisa (Ex-Vice Principal, Zaheer Memorial High School) for laying good foundations at the beginning of life i.e during schooling.

REFERENCES

1. Goetz M (2013) Towards generating highly detailed 3D CityGML models from OpenStreet-Map. Int J Geogr Inf Sci 27:845–865

2. Goetz M, Zipf A (2012) Using crowd sourced indoor geodata for agent-based indoor evacuation simulations. ISPRS Int J Geo-Inf 1(2):186–208. doi:10.3390/ijgi1020186

3. Hagenauer, Julian & Helbich, Marco. (2012). Mining urban land use patterns from volunteered geographic information using genetic algorithms and artificial neural networks. International Journal of Geographical Information Science. 26.

963-982. 10.1080/13658816.2011.619501.

4. Helbich M, Amelunxen C, Neis P (2012) Comparative spatial analysis of positional accuracy of OpenStreetMap and proprietary geodata. In: Jekel T, Car A, Strobl J, Griesebner G (eds) Geospatial crossroads @ GI_Forum 2012, Wichmann, Heidelberg, Salzburg, pp 24–33

5. Jokar Arsanjani J, Helbich M, Bakillah M, Loos L (2015a) The emergence and evolution of OpenStreetMap: a cellular

6. Mooney P, Corcoran P (2013a) Analysis of interaction and co-editing patterns amongst OpenStreetMap contributors. Trans GIS

7. Mooney P, Rehrl K, Hochmair H (2013) Action and interaction in volunteered geographic information: a workshop review. J Locat Based Serv 7:291–311

8. Mooney P, Corcoran P (2012) Characteristics of heavily edited objects in OpenStreetMap. Future Internet 4:285–305.

9. Neis P (2014) Measuring the reliability of wheelchair user route planning based on volunteered geographic information. Trans GIS n/a–n/a, doi:10.1111/tgis.12087

10. Neis P, Zipf A (2012) Analyzing the contributor activity of a volunteered geographic information project—the case of OpenStreetMap. ISPRS Int J Geo-Inf 1:146–165.

11. Over M, Schilling A, Neubauer S, Zipf A (2010) Generating web-based 3D city models from OpenStreetMap: the current situation in Germany. Comput Environ Urban Syst 34:496–507

12. Zielstra D, Zipf A (2010) Quantitative studies on the data quality of OpenStreetMap in Germany.